Antimicrobial Resistance Factors to Findings

Vijay Soni • Ajay Suresh Akhade
Editors

Antimicrobial Resistance: Factors to Findings

Omics and Systems Biology Approaches

Springer

Editors
Vijay Soni
Division of Infectious Diseases
Weill Department of Medicine
Weill Cornell Medicine
New York, NY, USA

Ajay Suresh Akhade
Institute for Systems Biology
Seattle, WA, USA

ISBN 978-3-031-65985-0 ISBN 978-3-031-65986-7 (eBook)
https://doi.org/10.1007/978-3-031-65986-7

This Springer imprint is published by the registered company Springer Nature Switzerland AG
The registered company address is: Gewerbestrasse 11, 6330 Cham, Switzerland

Preface

The evolutionary arms race is a constant battle for life and death. The monarch butterfly, for example, warns predators of its poisonous nature via its bright orange color. This method protects the monarchs only until their natural predators develop a tolerance or resistance to the poison. With the poison no longer a deterrent, the monarchs are free game yet again. Pathogens are the same way. Bacteria, viruses, fungi, and parasitic worms all aim to find a comfortable home within you so that they can grow and proliferate to their (proverbial) heart's content. Host animals have evolved with these infectious agents to overcome their attacks. As pathogens evolve to overcome host strategies, humans use their mighty prefrontal cortices to develop alternative methods to protect ourselves from these pathogens. We wash our hands multiple times a day, apply a generous layer of Neosporin on open wounds, and receive vaccinations against life-threatening pathogens. We've developed a wide array of antibiotics through careful study. However, the incorrect use of antibiotics has been consistently reducing their efficacy as bacteria evolve to subvert their effects. Bacteria are beginning to one-up us.

This book, "Antimicrobial Resistance: Factors to Findings—Omics and Systems Biology Approaches," describes the story of antimicrobial resistance from its origins to what scientists are doing to combat it today. Each chapter highlights different aspects of this story, with meaningful analyses of the costs and benefits of certain actions and recommendations for the next step forward written by leading researchers in the field. Chapter 1 begins with an introduction to antimicrobial resistance (AMR), outlines the emergence of drug resistance, and reports the consequences of this resistance on life as we know it. It delves into the mechanisms of the evolution of AMR and the diagnostic methods used to detect AMR in clinical practice. In continuity, Chap. 2 gives an overview of genetic tools like plasmid overexpression and targeted chromosome mutagenesis used to study AMR, using *Mycobacteria* to illustrate their uses. We move on to learning more about the evolutionary arms race in Chap. 3, as well as what scientists are doing to identify novel drug targets to gain a leg-up on evolving pathogens. The authors expand on the heterogeneity in bacterial gene expression that contributes to AMR, in addition to other factors. Next, moving toward the protein side of the story, Chap. 4 provides an introduction to proteomics and its methods in the context of AMR, including methods that have revealed new classes of virulence factors in recent years. Since proteins are the last step of the central dogma and responsible for the biochemical processes, Chap. 5 introduces metabolomics with a focus on bacterial metabolism, drug metabolism, and the use of metabolomics in diagnostic applications of AMR. Further, Chap. 6 addresses the topic of the microbiome and its role in AMR, including the impact of antibiotic use on the microbiome, microbiome-based therapies for AMR, and the limitations of microbiome-based approaches to AMR. The progression of AMR within natural ecosystems and the impact of human activities, such as waste management and industrial practices, facilitate the evolution of AMR. Hence, Chap. 7 highlights the importance of genomic epidemiology in public health and the latest genomic technologies that assist in epidemiological investigations. Further, Chap. 8 discusses the integration of multiomics in AMR research and communicates the computational and statistical tools to combine different types of omics data. Subsequently, Chap. 9 emphasizes the need to look at the bigger picture by advocating a system biology based approach towards AMR. Specifically, the chapter delves into modeling heterogeneity in bacterial populations as well as modeling the evolutionary landscapes that contribute to AMR. Besides bacterial factors, host also plays a central role in the bacterial adaptation against antibiotic challenges. Thus, Chap. 10 depicts AMR from the host's perspective, describing the effects of bacterial infection and subsequent drug treatments on the host. The chapter emphasizes the importance of a host-direct approach toward AMR and treatment research. Toward advanced developments of data analytics in AMR-system biology, Chap. 11 examines the controversial role of artificial intelligence (AI) and machine learning (ML) in omics biology, and how they can be used to develop novel therapeutics against AMR and its diagnosis. It also expands on the limitations and ethical considerations in the use of AI and ML in the biological sciences. Consequently, Chap. 12 relates information about the latest innovative technologies used in drug discovery against AMR and the use of omics knowledge in identifying drugs that can be effectively repurposed for AMR. Finally, Chap. 13 concludes the book with an overview of omics-systems biology in the study of AMR, the various pipelines that can be utilized to fast-track drug discovery, and recommendations and future directions for the field overall.

Altogether, in this book, we intend to convey the story of AMR as it stands today such that readers can utilize the given information in their lives and at work. AMR is a pressing global issue and requires all the attention we can give it. The evolution of drug-resistant bacteria and their spread to geographical areas where antibiotics are already in scarce supply is a recipe for disaster. We aim to provide researchers, both in academia and industry, with a compiled and up-to-date reference that they can use to investigate AMR and develop novel therapeutics against it.

New York, NY, USA Vijay Soni
Seattle, WA, USA Ajay Suresh Akhade

Acknowledgments

This book serves as a testament to the collaborative efforts of experts in the field who have dedicated their research endeavors to unraveling the complexities of AMR. This book project was not possible alone. We express our deep gratitude to many contributors, researchers, and mentors for their guidance, support, and inspiration throughout this project.

First and foremost, we would like to thank all the contributors and authors who agreed to take on the challenge of writing chapters. Their dedication to advancing the understanding of antimicrobial resistance through Omics and Systems Biology approaches has enriched this book. Further, their consent efforts, generosity, patience, time, and enthusiasm made this project possible. Additional thanks to those exceptional authors who agreed to write at the last moment, despite time constrain, and successfully made it on time.

We would like to express our gratitude to the publisher's team at Springer Nature and Alison Ball, for her guidance, patience, expertise, and professionalism. Their commitment to excellence and their tireless efforts to ensure the success of this project has been invaluable.

Special thanks to our panel of reviewers and authors who constantly helped us to access the scientific quality of the manuscripts of every chapter.

A heartfelt thank you to our family for their unwavering support, encouragement, and patience. Their love and belief in us have been the driving force behind our work, and we could not have completed this project without them. We are also indebted to our friends, colleagues, and mentors who have been a constant source of motivation and inspiration. Their constructive feedback and insightful comments have helped us to refine the structure of the book.

We would like to express our acknowledgment to the broader system biology research community for their ongoing contributions to this dynamic field. Lastly, we would like to thank our readers for their interest in this book. It is an honor to share our ideas with you, and we hope that this book will enrich your understanding and inspire you to explore new horizons in the field of AMR system biology.

New York, NY, USA — Vijay Soni
Seattle, WA, USA — Ajay Suresh Akhade

Contents

Contributors

Upasana Das Adhikari Department of Immunology, Harvard University, Boston, MA, USA

Ragon Institute of MGH, Harvard University and Massachusetts Institute of Technology, Boston, MA, USA

Anusha Aditya Irving Medical Centre, Columbia University, New York, NY, USA

Ajay Suresh Akhade Institute for Systems Biology, Seattle, WA, USA

Nicholas Bartelo Department of Physiology and Biophysics, Weill Cornell Medicine, Institute for Computational Biomedicine, Englander Institute for Precision Medicine, New York, NY, USA

Srijani Basu Department of Medicine, Weill Cornell Medicine, New York, NY, USA

Chandrima Bhattacharya Tri-Institutional Program in Computational Biology and Medicine, New York, NY, USA

The Bin Talal Bin Abdulaziz Alsaud Institute for Computational Biomedicine, Weill Cornell Medicine, New York, NY, USA

Cristiano Valim Bizarro Centro de Pesquisas em Biologia Molecular e Funcional (CPBMF) and Instituto Nacional de Ciência e Tecnologia em Tuberculose (INCT-TB), Escola de Ciências da Saúde e da Vida, Pontifícia Universidade Católica do Rio Grande do Sul (PUCRS), Porto Alegre, Brazil

Programa de Pós-Graduação em Biologia Celular e Molecular, Pontifícia Universidade Católica do Rio Grande do Sul, Porto Alegre, Brazil

Juan Calvet-Seral Grupo de Genética de Micobacterias, Departamento de Microbiología, Facultad de Medicina, Universidad de Zaragoza IIS-Aragón, Zaragoza, Spain

CIBER Enfermedades Respiratorias, Instituto de Salud Carlos III, Madrid, Spain

Biomedical Sciences and Engineering Laboratory, Departmento de Bioingeniería, Universidad Carlos III de Madrid, Madrid, Spain

Braden Carroll Center for Global Infectious Disease Research, Seattle Children's Research Institute, Seattle, WA, USA

Department of Chemical Engineering, University of Washington, Seattle, WA, USA

Gresia Cervantes Polytechnic University of Guanajuato, Cortazar, Guanajuato, Mexico

Nagasuma Chandra Department of Biochemistry, Indian Institute of Science, Bangalore, Karnataka, India

Department of Bioengineering, Indian Institute of Science, Bangalore, Karnataka, India

Pallavi Chandra Department of Medicine, Division of Infectious Diseases, Washington University in St. Louis, St. Louis, MO, USA

Riddhi Chaudhuri Purdue Institute for Drug Discovery, West Lafayette, IN, USA

Department of Chemistry, Purdue University, West Lafayette, IN, USA

Estefanía Crespo-Yuste Grupo de Genética de Micobacterias, Departamento de Microbiología, Facultad de Medicina, Universidad de Zaragoza IIS-Aragón, Zaragoza, Spain

CIBER Enfermedades Respiratorias, Instituto de Salud Carlos III, Madrid, Spain

Amber Dahlin Channing Division of Network Medicine, Department of Medicine, Brigham and Women's Hospital and Harvard Medical School, Boston, MA, USA

Sohini Das Department of Medicine, Maimonides Medical Center, New York, NY, USA

Mehmed Taha Dinc Department of Medicine, Boston Medical Centre, Boston, MA, USA

Andréa Beltrami Doltrario Division of Infectious Diseases, Weill Department of Medicine, Weill Cornell Medicine, New York, NY, USA

Anna Eydinova Center for Global Infectious Disease Research, Seattle Children's Research Institute, Seattle, WA, USA

Chenlian Fu Tri-Institutional Program in Computational Biology and Medicine, New York, NY, USA

Memorial Sloan Kettering Cancer Center, New York, NY, USA

Jesús Gonzalo-Asensio Grupo de Genética de Micobacterias, Departamento de Microbiología, Facultad de Medicina, Universidad de Zaragoza IIS-Aragón, Zaragoza, Spain

CIBER Enfermedades Respiratorias, Instituto de Salud Carlos III, Madrid, Spain

Braden Griebel Center for Global Infectious Disease Research, Seattle Children's Research Institute, Seattle, WA, USA

Department of Chemical Engineering, University of Washington, Seattle, WA, USA

Dan Luo Shanghai Key Laboratory of Veterinary Biotechnology, School of Agriculture and Biology, Shanghai Jiao Tong University, Shanghai, China

Shanghai Collaborative Innovation Center of Agri-Seeds/School of Agriculture and Biology, Shanghai Jiao Tong University, Shanghai, China

Shuyi Ma Center for Global Infectious Disease Research, Seattle Children's Research Institute, Seattle, WA, USA

Department of Chemical Engineering, University of Washington, Seattle, WA, USA

Department of Pediatrics, University of Washington, Seattle, WA, USA

Pathobiology Graduate Program, Department of Global Health, University of Washington, Seattle, WA, USA

Christopher E. Mason The Bin Talal Bin Abdulaziz Alsaud Institute for Computational Biomedicine, Weill Cornell Medicine, New York, NY, USA

Physiology, Biophysics, and Systems Biology PhD Program, Weill Cornell Medicine, New York, NY, USA

World Quant, New York, NY, USA

Alfonso Mendoza-Losana Biomedical Sciences and Engineering Laboratory, Departmento de Bioingeniería, Universidad Carlos III de Madrid, Madrid, Spain

Adriana Canedo Miranda Centro de Pesquisas em Biologia Molecular e Funcional (CPBMF) and Instituto Nacional de Ciência e Tecnologia em Tuberculose (INCT-TB), Escola de Ciências da Saúde e da Vida, Pontifícia Universidade Católica do Rio Grande do Sul (PUCRS), Porto Alegre, Brazil

Programa de Pós-Graduação em Biologia Celular e Molecular, Pontifícia Universidade Católica do Rio Grande do Sul, Porto Alegre, Brazil

Saurabh Mishra Department of Microbiology and Immunology, Weill Cornell Medicine, New York, NY, USA

Thato Motlhalamme South African Medical Research Council Drug Discovery and Development Research Unit, Department of Chemistry, University of Cape Town, Rondebosch, South Africa

Kehilwe Nakedi Division of Infectious Diseases, Weill Department of Medicine, Weill Cornell Medicine, New York, NY, USA

Murugesh Padmanarayana Department of Biochemistry, Stanford University, Beckman Center for Molecular and Genetic Medicine, Stanford, PA, USA

Sukriti Pal Department of Biochemistry, Indian Institute of Science, Bangalore, Karnataka, India

Rupobrata Panja Center for Computational and Integrative Biology, Rutgers University-Camden, Camden, NJ, USA

Lynthia Paul Division of Medical Microbiology, Department of Pathology, University of Cape Town, Observatory, South Africa

Ayush Praveen Elucidata Data Consulting Pvt Ltd., New Delhi, India

Jake Qiu New York Genome Center, New York, NY, USA

Physiology, Biophysics, and Systems Biology PhD Program, Weill Cornell Medicine, New York, NY, USA

Shivangi Rastogi Department of Cell Biology and Molecular Genetics, University of Maryland, College Park, MD, USA

Rahul Singh Rawat Eukaryotic Gene Expression Laboratory, National Institute of Immunology, New Delhi, Delhi, India

Eric H. Rosenn Department of Biomedical Engineering, Boston University School of Engineering, Boston, MA, USA

Division of Infectious Diseases, Weill Department of Medicine, Weill Cornell Medicine, New York, NY, USA

Prabhat Ranjan Singh Department of Microbiology and Immunology, Weill Cornell Medicine, New York, NY, USA

Vinayak Singh South African Medical Research Council Drug Discovery and Development Research Unit, Department of Chemistry, University of Cape Town, Rondebosch, South Africa

Holistic Drug Discovery and Development (H3D) Centre, University of Cape Town, Rondebosch, South Africa

Institute of Infectious Disease and Molecular Medicine (IDM), University of Cape Town, Observatory, South Africa

Biplab Singha Department of Microbiology and Physiology Systems, University of Massachusetts Chan Medical School, Worcester, MA, USA

Vijay Soni Division of Infectious Diseases, Weill Department of Medicine, Weill Cornell Medicine, New York, NY, USA

Erick Tieu Center for Global Infectious Disease Research, Seattle Children's Research Institute, Seattle, WA, USA

Department of Chemical Engineering, University of Washington, Seattle, WA, USA

Dung Thuy Tran Channing Division of Network Medicine, Department of Medicine, Brigham and Women's Hospital and Harvard Medical School, Boston, MA, USA

Ramya Venkataraman Laboratory of Innate Immunity, National Institute of Immunology, New Delhi, India

Longlong Wang Shanghai Key Laboratory of Veterinary Biotechnology, School of Agriculture and Biology, Shanghai Jiao Tong University, Shanghai, China

Shanghai Collaborative Innovation Center of Agri-Seeds/School of Agriculture and Biology, Shanghai Jiao Tong University, Shanghai, China

Zhe Wang Shanghai Key Laboratory of Veterinary Biotechnology, School of Agriculture and Biology, Shanghai Jiao Tong University, Shanghai, China

Shanghai Collaborative Innovation Center of Agri-Seeds/School of Agriculture and Biology, Shanghai Jiao Tong University, Shanghai, China

Weile Xie Shanghai Key Laboratory of Veterinary Biotechnology, School of Agriculture and Biology, Shanghai Jiao Tong University, Shanghai, China

Shanghai Collaborative Innovation Center of Agri-Seeds/School of Agriculture and Biology, Shanghai Jiao Tong University, Shanghai, China

Yufan Xu Shanghai Key Laboratory of Veterinary Biotechnology, School of Agriculture and Biology, Shanghai Jiao Tong University, Shanghai, China

Shanghai Collaborative Innovation Center of Agri-Seeds/School of Agriculture and Biology, Shanghai Jiao Tong University, Shanghai, China

Chapter 1
Antimicrobial Resistance and Factors: An Introduction

Vijay Soni, Andréa Beltrami Doltrario, Eric H. Rosenn, Sohini Das, Biplab Singha, Rahul Singh Rawat, and Saurabh Mishra

Abbreviations

AMR	Anti-Microbial Resistance
MDR	Multi-Drug Resistant
GLASS	Global Antimicrobial Resistance and Use Surveillance System
NGS	next-generation sequencing
WGS	Whole Genome Sequencing
GEMs	Genome-Scale Metabolic Models
QSP	Quantitative Systems Pharmacology
FQ	Fluoroquinolone
XDR-TB	Extensively drug resistant tuberculosis
PPE	personal protective equipment
ARG	Antibiotic resistant genes
AMPs	Antimicrobial peptides

V. Soni (✉) · A. B. Doltrario
Division of Infectious Diseases, Weill Department of Medicine, Weill Cornell Medicine, New York, NY, USA
e-mail: vis2032@med.cornell.edu

E. H. Rosenn
Department of Biomedical Engineering, Boston University School of Engineering, Boston, MA, USA

S. Das
Department of Medicine, Maimonides Medical Center, New York, NY, USA

B. Singha
Department of Microbiology and Physiology Systems, University of Massachusetts Chan Medical School, Worcester, MA, USA

R. S. Rawat
Eukaryotic Gene Expression Laboratory, National Institute of Immunology, New Delhi, Delhi, India

S. Mishra
Department of Microbiology and Immunology, Weill Cornell Medicine, New York, NY, USA

V. Soni, A. S. Akhade (eds.), *Antimicrobial Resistance: Factors to Findings*,
https://doi.org/10.1007/978-3-031-65986-7_1

ASP	Antimicrobial Stewardship Programs
MIC	minimum inhibitory concentration
SDA	strand displacement amplification
TMA	transcription-mediated amplification
NASBA	nucleic acid sequence-based amplification
RCA	rolling circle amplification
RPA	recombinase polymerase amplification
LAMP	loop-mediated isothermal amplification
HDA	helicase-dependent amplification
MALDI-TOF MS	Matrix-Assisted Laser Desorption/Ionization Time-Of-Flight Mass Spectrometry
NMR	nuclear magnetic resonance spectroscopy

1.1 Introduction

> If we use antibiotics when not needed, we may not have them when they are most needed. – Tom Frieden, former director of the Centers for Disease Control and Prevention (CDC)

According to the World Health Organization (WHO), *"antimicrobial resistance is a natural phenomenon that occurs when microorganisms no longer respond to antibiotics to which they were previously susceptible and that were previously active in treating infections caused by these microorganisms."* The WHO estimates that, if unchecked, by 2050 more people will die from infections with multi-drug-resistant (MDR) bacteria than from both automobile accidents and all incidences of cancer combined [2] (Fig. 1.1). In 2019, the WHO called attention to the problem by naming anti-microbial resistance (AMR) as one of the top ten threats to global health [3]. At the turn of the century, Louis Pasteur and his contemporaries theorized that germs release substances to kill other bacteria. However, it was not till the 1930s that a systematic search for microbes producing these chemicals and their classification began; leading to the "golden age" of antibiotic discovery [4] (Fig. 1.2). The abundance of natural molecules serving as lead compounds seemed endless, but by the 1970s, new drug development had stalled, and science was confronted with the ensuing growth of microbial resistance. Realizing the diminishing potential for discoveries, pharmaceutical companies divested antimicrobial research departments, which compounded the problem [5]. Even so, antibiotics continued to be prescribed liberally, not only for treating human pathologies but also for inoculating livestock and treating processed foods. As a result of misuse in hospitals, nosocomial infections emerged [6], but a greater factor in MDR development was the release of these compounds into the environment through agricultural use and improper waste management [7]. The effects of this on shifting pressures and genetic influences on

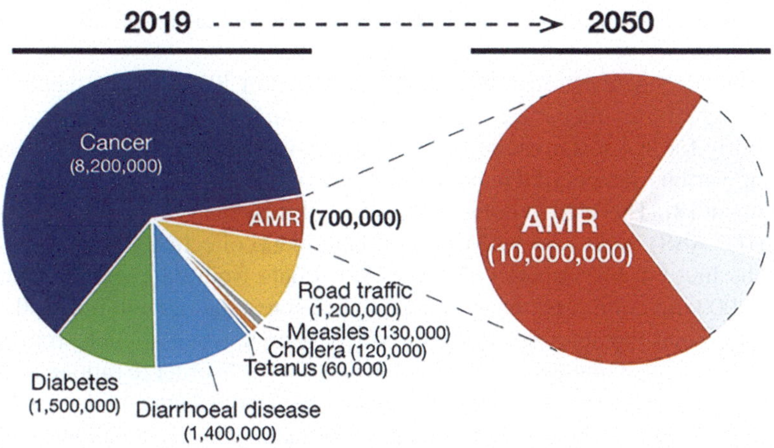

Fig. 1.1 Number of deaths and the main causes (left) in 2019 and the projection of the number of deaths due to AMR infections in 2050 (in red in the right). Gray areas represent other causes of deaths. (The figure and caption are reprinted from [1])

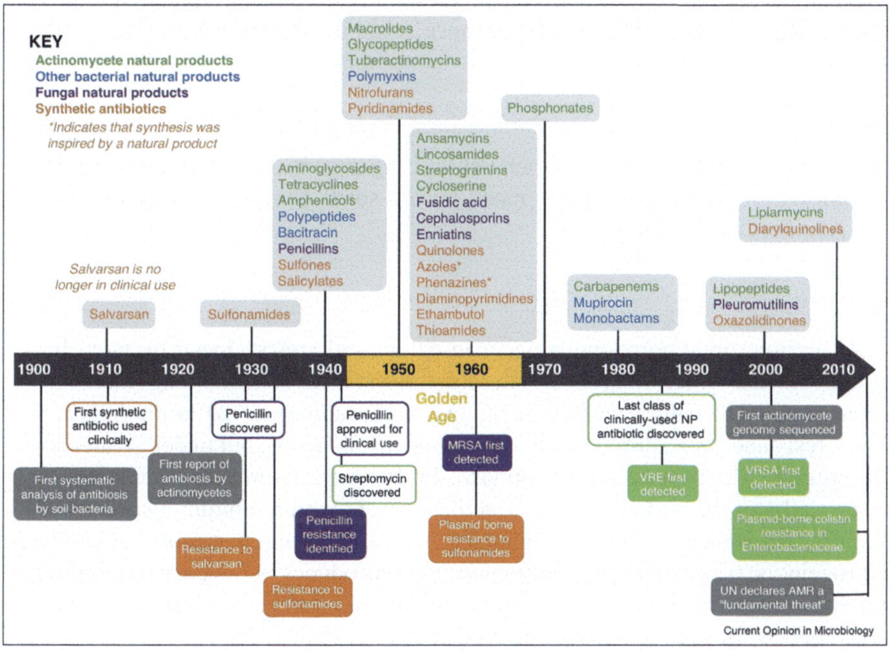

Fig. 1.2 Timeline showing the decade new classes of antibiotics reached the clinic. The antibiotics are colored per their source: green = actinomycetes, blue = other bacteria, purple = fungi, and orange = synthetic. At the bottom of the timeline are key dates relating to antibiotic discovery and antimicrobial resistance, including the first reports of drug-resistant strains such as methicillin-resistant *S. aureus* (MRSA), vancomycin-resistant enterococci (VRE), vancomycin-resistant *S. aureus* (VRSA), and plasmid-borne colistin resistance in Enterobacteriaceae. (The figure and caption are reprinted from [4])

interacting bacteria within ecological systems and biota across the world are unquantifiable.

The WHO's global action plan on AMR, launched in 2015, aims to regulate antimicrobial agent use, encourage investment in research for new agents, and ensure accessibility for low-resource areas. The plan emphasizes a "one health" approach, involving various sectors like human and veterinary medicine, agriculture, and the environment [8]. The WHO Global Antimicrobial Resistance and Use Surveillance System (GLASS) standardizes AMR surveillance data collection and sharing worldwide. The first GLASS report in 2018 included data from 22 countries, detailing over 500,000 infections and the system has since spread to 109 countries [9].

Antimicrobial therapies are still an effective avenue when applied strategically and there is renewed interest in combination drugs [10]. In order to rationally design new drugs and properly apply new therapies, we need a better understanding of bacterial biology [11]. Studying mechanisms of pathogenesis can allow us to predict if AMR is likely to develop, identifying at-risk populations, potential avenues for spread, and likely reservoirs [12]. Epidemiological surveys strive to track the "movement" of MDR strains across the globe as well as analyze pathogen populations for mutations to predict the emergence of new strains [13]. This allows for preparations such as prophylactic vaccinations or research on proper antibiotic therapy protocols to be developed in advance, establishing procedures to stifle the spread of a new strain [14].

> There is probably no chemotherapeutic drug to which, in suitable circumstances, the bacteria cannot react by in some way acquiring fastness. – Alexander Flemming

As antimicrobial compounds occur in nature and are produced by microbes to defend against other microbes, microbial resistance is something that can be acquired through natural evolution [2, 15]. When administered in pharmaceutical doses, resistance is acquired through the same mechanisms, but under greater selective pressure, the process speeds up [16]. In this way, natural resistance can occur intrinsically or be induced by the activation of certain ubiquitous genes [17]. Acquired resistance can occur through an adaptive mutation that is passed vertically to progeny or by horizontal gene transfer that introduces new genetic material [6]. Activation of dormant genes or activating mutations confers the ability to resist drug molecules by the following typical strategies [19]:

(i) The pathogen may produce enzymes or toxins that inactivate a drug by hydrolysis, inducing structural modification (e.g., aminoglycoside-modifying enzymes), or directly destroying the antibiotic (e.g., β-lactamases).
(ii) By decreasing membrane permeability or overexpressing efflux pumps, a pathogen can hinder a drug from entering or reduce its residence time within the bacteria, reducing the drug's action or/or preventing it from reaching a target site.

(iii) The pathogen can change the antibiotic target (changing the amino acid sequence at a reactive site on an enzyme), such as to make it inaccessible or non-reactive, by inducing genetic changes altering the primary transcript, or via post-translational modifications.

Transient phenotypic and fundamental genetic changes may both confer resistance, and furthermore, the occurrence of one adaptation may lead to the stable mutagenesis of the other [18]. Not only do internal genetic factors need to be considered but also inter-organismal interactions such as the formation of biofilms, another method of resistance. Any of these resistance mechanisms may allow for prolonged survival that may act as a "stepping stone" for genetic AMR development [19].

Systems biology—has exactly the same goals and purposes as general biology—with a whole new arsenal of tools.
– Eberhard Voit

Almost 20 years ago, Dr. Trey Ideker defined systems biology as *"the use of systematic genomic, proteomic, and metabolomic technologies to acquire data for developing models of complex biologic systems and pathologies"* [20]. Systems biology thus integrates many fields, amalgamating the studies of biophysical chemistry, data science, and engineering, among others, resulting in a distinct ethos. Understanding the multilevel effects involved in a biologic pathway and its interrelation with other processes allows for comprehensive modeling of the system [21]. This holistic perspective yields information about the most basic components of the system, such as cellular mechanisms, which in turn leads to more accurate inferences about how small changes, like inducing a slight pH shift, can result in large effects on a whole organism. This approach has many implications for solving specific problems, like predicting pharmacokinetics, as well as providing solutions for broader human health and environmental concerns. Integrative omics approaches have been essential in the development of the systems biology approach [22] (Fig. 1.3). It is important to understand how techniques in genomics, transcriptomics, proteomics, and metabolomics and the integration of these fields in modeling the interactome contribute to fighting MDR.

The advent of next-generation sequencing (NGS), via high-throughput, parallel sequencing of DNA fragments, has allowed pathogen genomes to be determined rapidly and at comparatively low cost [23]. Genomics looks at whole genome data and encompasses techniques most commonly used in the experimental and clinical identification of bacterial species, as well as in establishing and analyzing bacterial phylogenetic relationships [24]. Comparative phylogenetic analysis can be exploited to quantify the degree of relatedness, which, overlaid with epidemiological data, can elucidate the temporospatial dynamics of AMR and transmission. Whole Genome Sequencing (WGS) is being increasingly employed to address the public health challenge of AMR, supporting surveillance, outbreak investigation, and contributing to improved diagnostics.

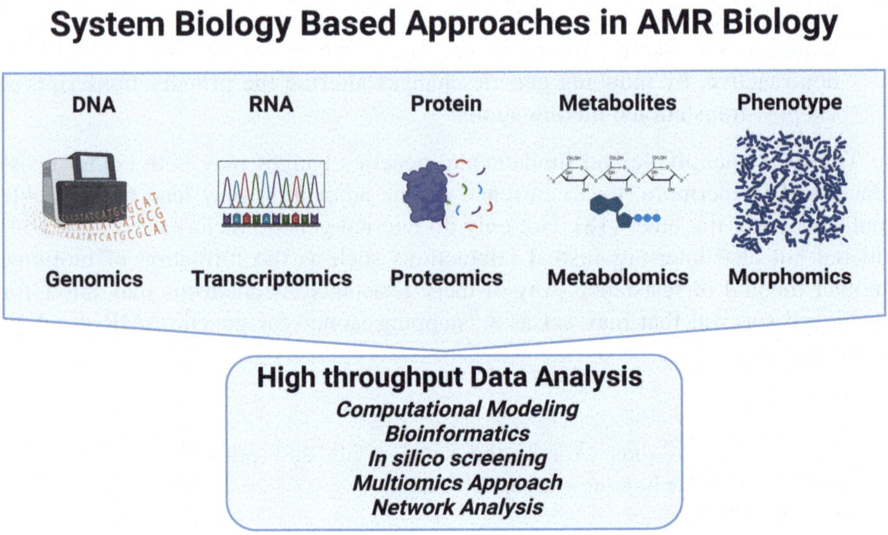

Fig. 1.3 Overview of system biology and its role to understand the microbiology of AMR (Created using Biorender.com)

Proteomics information such as proteome constitution or the relative abundance of isoforms can yield information on pathogenesis as well as aid in identifying strains [25]. Analysis of protein pathways in MDR bacteria has led to predictions of proteins in the same pathway that are likely to undergo additional mutagenesis. Furthermore, the analysis of a mutated protein can inform the probability of AMR development in homologs [26].

Metabolomics views the metabolome as the complete set of small molecules involved in cellular biochemical processes [27]. The metabolome closely corresponds with the phenotype and gives a snapshot of intracellular changes in response to antibiotics and adaptations that can confer resistance [18]. Shifts in metabolism underlie stages of pathogenesis and mechanisms of immune invasion that may confer resistance. For instance, many bacteria can switch from a high-energy state utilizing aerobic respiration to an anaerobic state when resources are scarce or depleted. This can also serve as a strategy for evasion of host defenses and confer resistance to aminoglycosides. If bacteria revert to a "dormant" state, antimicrobials that utilize active cellular processes for uptake to reach their targets are less effective. Likewise, antibiotics such as β-lactams show increased effects in conjunction with higher bacterial respiration [28].

Increases in metabolism may also confer resistance, as bacteria can prompt energetically costly changes in their own cell envelopes [22]. Regulating permeability makes antibiotic access difficult and bacteria can induce specific alterations to membrane structures targeted by antibiotics. For instance, alterations in lipid biosynthesis in Gram-negative bacteria can result in a modified Lipid A [29], the regular binding site for polymyxins [30]. Another active process is the upregulation and production of efflux pumps. In either case of increased or decreased metabolism that confers resistance, there is a fitness trade-off that occurs. Precise modeling of the new metabolic equilibrium caused by these adaptations could identify state-specific homeostatic foci as novel targets in MDR bacteria [31].

As inhibition or excitation of metabolism can both act as methods conferring MDR, a complex balance must be maintained to effectively resist drugs. Understanding this balance may be the key to understanding a method to disrupt the bacterial lifecycle, making them more susceptible to pharmaceuticals. Tracking spatiotemporal shifts in phenotype by proteomics and metabolite measurements with genome-scale metabolic models (GEMs) can help predict MDR emergence [22]. Further analysis of the fitness tradeoff involved in acquiring resistance can determine the method and rate of population stabilization, allowing for threat assessment and proactive action against dissemination [32].

There are many factors to consider that contribute to AMR—a complex system of interactions. From subcellular changes within an organism to the commensal effects of bacterial communities to ecological and whole biome considerations that affect the spatiotemporal dispersion of bacterial lineages and mutagenic influences, a vast amount of data must be considered [33]. Forming a more comprehensive model of these multilevel interactions is key to revealing how specific MDR mechanisms have developed and predicting the emergence of new AMR capabilities [34]. A system-level understanding allows us to address the problem of MDR from multiple avenues. First, with increasingly accurate modeling, a quantitative systems pharmacology (QSP) approach can be used which may potentially lead to the discovery of many new drugs or combination therapies [35]. Second, a better understanding of the environmental factors involved in MDR may highlight ways humans may be influencing the evolution of MDR strains without even realizing; and conversely, how to manipulate this evolution [36]. Lastly, understanding the development of the most resistant pathogens and their interactions with their human hosts over an evolutionary timescale may yield insights into bacterial adaptability and aid us in predicting their future behavior [37].

The objective of this chapter is to explore the general scope of AMR and the factors behind it. We have included a comprehensive discussion on bacterial evolution in the context of AMR, interpreting its connections with diverse clinical factors. Furthermore, we highlighted the implications of basic knowledge to understand the global consequences of AMR, encompassing both the disease aspect and policy considerations. The chapter further extends to different diagnostic methods for AMR, followed by an assessment of its impact on healthcare. We have also reviewed the pivotal role of systems biology in the battle against AMR. We envision this chapter as a simple guide for all readers of this book, serving as a resource to develop a deeper understanding of the fundamental biology of AMR and the imminent challenges that demand timely attention.

1.2 Evolution of AMR: Environmental and Biological Factors

A band of bacterial brothers, swigging ATP with some others. In a jocular fit, they laughed til' they split, now they're all microbial mothers. – Richard Cowen

In the introduction section, we indicated some pivotal points in the history of AMR development. In recent years, the rapid spread of AMR bacteria has given us a clear view of its devastating trajectory. We have also mentioned the implications for progressive therapeutic innovation necessary to combat AMR, stressing the need for a deeper understanding of AMR mechanisms as well as continued research into new antimicrobial treatments. When we consider the misuse of antibiotics and other environmental impacts of humans that have caused the more rapid emergence of AMR, we mainly focus on the last 100 or so years. This is quite a short period on an evolutionary timeline. While the direct human impact on the spread of AMR is evident, we must also realize that millennia of evolution have taken place that primed these bacteria to be able to mutate so rapidly in response to stress.

From the beginning of mankind, there has been a co-evolution of pathogenic bacteria [38]. In fact, it is posited that many modern diseases arose after agricultural revolution, as large and concentrated human populations were required to maintain the pathogens' minimal viable populations [39]. Subsequentially, the immune system of a modern human is quite distinct from that of our ancestors, even a few thousand years ago. Likewise, new bacterial strains emerged in response to these new defense mechanisms [40]. The intertwined nature of human and bacterial coevolution has resulted in myriad complicated relationships; this can be seen in modern bacterial species that range from symbiotic gut microbes to pathogens so hostile that mere contact is a death sentence.

Yersinia pestis, for instance, which has caused three major pandemics throughout history, including the bubonic plague, has been isolated from corpses dated 3800 years old and remains endemic to some regions today [41]. It was not until the 1930s through the analysis of epidemiologic data that it was understood that rats not only could carry the disease but were also acting as reservoirs, allowing it to persist and cause sporadic outbreaks. Such reservoirs present another important aspect influencing the evolution of AMR. Revelations like these have underlined the importance of historical microbiology and research into past bioevents [42].

We can see over shorter time spans how pathogens exclusive to animal hosts can mutate and gain the ability to be transmitted to humans, as was determined to be the case regarding the COVID-19 pandemic. Additionally, the phylogenic relationship between the animal host and humans has been shown to be a possible factor affecting the transmissibility and even infectivity of a pathogen [43]. By comparing differences in immune responses between other host species and humans, the immune evasive mechanisms of the pathogen can be better understood, in addition to its mode of transmission. Practically, this could reveal a novel inducible immune factor that might aid in resistance to AMR bacteria, as well as expose new strategies for disrupting transmission [44].

The integration of environmental and genetic influences on and between humans and bacteria reveals the emergent properties of a system that can explain processes like the response of bacteria to selective pressures or give insight into the evolution of natural resistance mechanisms such as gene transfer between bacterial species. Moreover, there is importance in identifying the locus of evolutionarily conserved genetic elements and other genetic features such as points of rapid mutation that seem to incorporate many active genes without being overly detrimental to the organism [45]. Having identified these possibly significant genes, a network-based approach can be applied using omics techniques to elucidate genetic interaction pathways [46]. If any of these pathways are related to drug-resistance genes, then the other pathways related to the locus can be considered high-profile leads for further understanding the molecular basis of AMR. Changes in bacteria through the process of gaining AMR present a number of fitness trade-offs that can be exploited in therapeutic development [47]. For instance, the gain of AMR function is usually accompanied by morphologic and metabolic changes that may present new modalities for drug action [48].

It is also important to note the intrinsic resistome of many bacteria, which is an independent general defense mechanism that has evolved through natural evolution and also functions in AMR. *P. aeruginosa,* for instance, has developed such adaptability that it is nearly ubiquitous in every environment and species [49]. Moreover, populations cultured under various conditions do not represent distinct strains, but rather fluid physiological and metabolic changes driven by highly adaptable expression systems [50]. The mutational resistome also displays intrinsic versatility in that a large number of loci represent functional phenotypic shifts that contribute to acquired immunity. Under stress from any particular antimicrobial, a vast number of single mutations are possible that can confer resistance [51]. Many other pathogens display similar adaptability in that established intrinsic defenses are highly effective in interfering with antibiotics, despite this being a novel functional role. It is posited that these intrinsic mechanisms are simply far too adaptable for any one antimicrobial formulation to be an effective therapeutic solution. Subsequently, current research is examining the metabolic changes that occur in response to regimens of multiple antibiotics and how these changes might build off each other to expose a novel target or "weak point." [52] In addition to this evolutionary approach, by studying historic changes in metabolism under environmental pressure, we might predict how changing nutrient availability in disease states might affect susceptibility to antimicrobials and other pharmaceuticals [53].

1.2.1 Scope of Influences Driving AMR

> Working Upward multicellular organisms such as you and I are social groups of Eukaryotic cells. Working Downward bacteria are social groups of genes. —David Sloan Wilson

Under the theory of multi-level selection, natural selection may occur not only between individuals in a population but at various levels of organization [54]. In a simple example of "group selection," a population with greater genetic diversity is more "fit" than a homogenous population and is more likely to have surviving members when confronted with a disease. While many scientists have critiqued the theory, we will use its definition in a limited scope here to discuss the phenomenon of emergent traits of levels of organization within bacteria and consider distinct "selective" pressures at these levels that have influenced the development of antibiotic resistance genes.

As we have described previously, bacteria are highly adaptable due to plasticity at many levels. At the genetic level, we see a high mutation rate and a fluid genome with transmissible elements [55]. At higher levels of organization, we see promiscuous proteins, asymmetric phenotypic expression, tradeoffs between variation and privileged environmental access, regional strain variants displaying unique characteristics, and emergent strains with significant functional distinctions. Each of these represents an emergent and variable trait of a hierarchical organization, with each level experiencing its own selective pressures. An important facet of multilevel selection is that we consider selective pressures at the level of organization where they are most useful in a particular context. Take our population example above; a more infectious but homogenous strain populating an organ may contain more fit, rapidly reproducing, individuals but can be more easily killed with an antibacterial than a more temperate strain that grows slower while generating greater variation.

There are so many processes producing adaptive variability that seem redundant throughout the hierarchical structure of bacterial species; however, there is an order to the madness. In examining the number of naturally occurring resistance genes as well as those that can be recruited through other transfer mechanisms, there is a question posed about why an even higher number of AMR traits are not observed [56]. In what is referred to in some works as a bottleneck, there is high competition between AMR elements, which are selected by bacterial genetic recruitment mechanisms based on stringent criteria. One of the major factors is the ability of the new genetic material to be easily integrated into bacterial epigenetic mechanisms, which is the basis for many higher-level functions.

Bacteria multiply and spread rapidly through various environments, resulting in population growth under many different influences [57]. There will thus be divergent evolution between populations, resulting in greater genetic variation within the species or strain. A larger population will be more stable as it will encompass greater variation as well as be more adaptable as there is increased competition within the group. However, an opposing situation may occur when smaller subpopulations are formed, which may result in more radical adaptations. Consider the case of small amounts of bacteria being passed from one host to another or dispersed throughout a host body. In sufficiently small, isolated subpopulations, genetic drift may occur, resulting in the incubation of a particular phenotype [58]. A weak AMR gene with a neutral fitness effect may become highly resistant, and subsequent exposure may introduce this novel resistance to the greater bacterial population.

Another way in which AMR can develop in a non-Darwinian fashion is through phenotypic mutations. Intrapopulation phenotypic heterogeneity occurring from noise in expression mechanisms can result in greater adaptive variation as well as novel AMR resistance traits. If these are concurrent with some antimicrobial pressure, the mutation may be passed on via epigenetic inheritance. Furthermore, through gene amplification, small genotype variances can have a disproportionate but reversible phenotypic response and result in heteroresistance to moderate antibiotic pressure or a high probability of adaptive mutation under strong antibiotic stress. This epigenetic-based phase variation underlies *Neisseria meningitidis* susceptibility to ceftazidime and ciprofloxacin [59].

At the various levels of organizational hierarchy, we see cases where adaptive variation has facilitated exaptations that have ultimately evolved into AMR mechanisms. We can tell transposases' primary function is in environmental adaptation as only a small percent carry AMR element; however, it does so highly efficiently, resulting in a large effect on AMR [60]. Promiscuous proteins in ancient bacteria were also able to bind and disrupt natural antimicrobials, an initial exaptation forming a portion of intrinsic resistance, and later gained broader functions. In fact, the ability of most portions of the intrinsic resistome to function in clinical antibiotic resistance represents many exaltations.

Looking at how many ways AMR elements can result from various mechanisms of adaptive variation, it seems bacteria might be able to shift any survival tool at their disposal toward the development of AMR under sufficient pressure. Examining AMR from an evolutionary perspective has given us great insight into a highly complex organism. In going forward in developing new therapies to circumvent AMR, we should be extremely considerate of these adaptive mechanisms as well as sources of random mutation that could present dangerous obstacles.

1.2.2 Evolutionary Development-Informed Strategies for AMR Treatment and Prevention

> Imagining a world without vaccines, anesthetics, contraception, and anti-infectives reveals how medicines revolutionized humanity. How to manipulate conditions that favor such discoveries is worth consideration. —John Warren

Understanding the basis of AMR development is important for the rational and effective development of all therapeutics to treat bacterial infections. As discussed, understanding the past evolution of bacteria and AMR informs us how certain stressors might affect a pathogen, thus allowing us to incorporate artificial selection into the rational design of a therapeutic mechanism [61]. In addition to manipulating a pathogen in this manner, we may also take advantage of fitness tradeoffs and important genomic expression systems we have uncovered.

Cross-sensitivity is a fitness trade-off effect observed when a new adaptive trait causes the weakening or reversal of a previously acquired adaptive trait. Collateral sensitivity (previously induced sensitivity) specifically describes increased sensitivity to one antibiotic when a bacterium gains resistance against another antibiotic [62]. In 1952, Szybalski and Bryson first observed that *E. coli* cultures could not develop increased resistance to both polymyxin B and chloromycetin. They later established that increased resistance to one agent caused an obligate increased sensitivity to the other [62].

The use of collateral sensitivity has since been shown to be a viable strategy for eliminating AMR strains. In particular, when a disproportionate collateral sensitivity relationship is observed, a treatment involving altering exposure between the antimicrobials involved can result in hypersensitivity to one of the agents that effectively clears the pathogen. In one experiment, for instance, MDR *P. aeruginosa* strains were treated with an alternating tobramycin/ceftazidime regimen [47]. Tobramycin administration first kills any bacteria in the population that have heterogeneous ceftazidime resistance. Ceftazidime treatment then causes hypersensitivity to tobramycin and a subsequent tobramycin administration then kills off the bacteria entirely.

Research is being conducted to identify conserved evolutionary networks involving the collateral resistance phenotype and inform the rational design of drug pairs with a predictable range of target strains [63]. However, the information generated has significant gaps due to processes like epistasis, polymorphisms, and allelic expression that confound predictions based on genetic similarity. Another problem is the potential heterogeneous emergence of collateral sensitivity, leading to intrapopulation variation in the robustness of induced sensitivity [64]. Convergent collateral sensitivity is a strategy employing a regimen comprising "steps" of controlled exposure to additional antibiotic compounds [65]. This takes advantage of the process of phenotypic convergence, in which a diverse population, or populations, subject to the same stress will adopt a similar phenotype. Applying a sequence of antibiotic stressors causes a significant shift toward a similar phenotype, making collateral sensitivity a more feasible approach for the treatment of diverse clinical strains as well as heterogeneous systematic infections.

Another evolutionary strategy examining metabolic changes in response to antimicrobial challenges might reveal combinations of drugs causing metabolic shifts that can be exploited [66, 67]. For instance, even an antimicrobial to which a particular bacterium has developed complete resistance might still induce some metabolic change that makes the bacteria more susceptible to another antibiotic. An eco-evolutionary angle is also being applied by considering antimicrobial resistance elements' original function, such as how drug efflux pumps may have functioned in bacterial metabolism before [68]. Understanding this state can help identify adjuvants, such as nutrient compounds that affect the element's function, that could be co-administered with an antibiotic and result in a greater effect. In addition to antibiotics and adjuvants, another way in which artificial selection for more susceptible bacteria might be accomplished is through alteration of the environment itself by affecting the host microbiome.

1.3 Clinical Pressures in the Development of Drug Resistance Pathogens

Clinical pressures are a crucial issue behind the development of drug-resistant pathogens. Factors such as antibiotic misuse, inadequate dosage, and improper prescribing practices contribute to the selective advantage of AMR. Hospital environments, global travel, and the lack of new antibiotic development further amplify this challenge. These pressures underline the pressing need for wide-ranging approaches to alleviate antibiotic resistance, emphasizing judicious use, enhanced surveillance, and the development of novel treatment methods to ensure the sustained effectiveness of antimicrobial therapies.

1.3.1 Key Factors

Antibiotic Misuse and Wrong Clinical Practices

The use of antibiotics results in selective microbial pressure that promotes the growth of antibiotic-resistant organisms [69]. Negative consequences outweigh the benefits when antibiotics are used despite no clinical indication. Antibiotics are frequently used for the treatment of respiratory infections, even if they are not indicated. Shafrin et al. found that ciprofloxacin was prescribed to 15% of women with UTIs in the United States [70]. Apart from AMR, inappropriate antibiotics lead to poor patient outcomes and an increase in healthcare costs [70, 71]. Inappropriate antibiotic use includes the wrong antibiotic, dose, and duration.

The key question is, "If inappropriate antibiotic therapy has adverse effects and no benefits, why are antibiotics prescribed in these scenarios?" It is found that time pressure, diagnostic uncertainty, patient demand/satisfaction, concern for secondary bacterial infections, and high patient volumes lead to inappropriate antibiotic prescriptions [72, 73]. Physician-related factors associated with a higher likelihood of antibiotics include a long duration of practice (more than 10 years), medical liability, and the perception of severe illness. Often, taking a few minutes to talk to the patient is enough to address their concerns about the need for antibiotics [74]. Patients with a history of corticosteroid use, fever, and multiple comorbidities are likely to be treated with antibiotics. More than 85% of antibiotics are prescribed in clinics, of which 41% are prescribed in primary care [75–79]. Self-medication is common in countries where antibiotics are available over-the-counter. Measures to curb inappropriate use of antibiotics include the sale of antibiotics as prescription only, enabling prescribers to provide the exact dose and duration, patient education about appropriate antibiotic use by providers and pharmacists, antibiotic use and resistance surveillance systems, and public education campaigns [80].

Food Animals

Antibiotic use in food animals (poultry, swine, cattle, and sheep) is a significant contributor to AMR. Antibiotics are used for metaphylaxis, prophylaxis, treatment, and as growth promoters. Metaphylaxis is the mass medication of healthy animals when an infection is present in one or more individuals in the herd. Prophylaxis is antimicrobial administration in healthy animals to avert disease. Growth promoters are antibiotics (often at subtherapeutic levels) used to increase the feed-to-muscle conversion rate [81, 82]. Selective pressure leads to the predominance of resistant bacteria, which are transmitted to humans through food or the environment.

Fluoroquinolone (FQ) resistance rates in *E. coli* are less than 5% in the United States, where its use is prohibited in food animals. However, more than 40% of *E. coli* are FQ-resistant in Brazil, Europe, and China, which use FQ in food animals [83].

Overcrowding and Poor Sanitation
Overcrowding, poor sanitation, a lack of handwashing, and the drainage of contaminated water into rivers and lakes increase the spread of AMR. This is especially relevant in urban slums in LMICs and refugee camps. Overcrowding is associated with increased respiratory and MRSA skin and soft tissue infections (SSTI) [84, 85]. Emphasis on good hygiene practices is an important measure to prevent the spread of AMR.

1.3.2 Impact of Clinical Pressures on the Emergence of AMR

The rapid rate of emergence of AMR has outpaced the development of new antibiotics. As we have discussed in Sect. 1.2, this creates the possibility of a frightening situation like a hundred years ago, prior to the availability of antibiotics, when common infections like pneumonia, urinary tract infections, and meningitis had no effective treatment. The World Health Organization (WHO) declared AMR a top-ten global public health crisis in 2019 [86]. Overuse and misuse of antibiotics, poor infection prevention and control measures, poor sanitation and handwashing practices, limited access to vaccines, and effective antimicrobials hasten the development and spread of AMR [87].

During the initial phase of the COVID-19 pandemic, healthcare facilities were overwhelmed by critically ill patients requiring mechanical ventilation, central venous lines, and urinary catheters. Rates of carbapenem-resistant Enterobacterales, extended-spectrum β-lactamases (ESBLs) Enterobacterales, and MDR *P. aeruginosa* increased by 35%, 32%, and 32%, respectively. It is alarming that 47% of ESBL Enterobacterales infections were community-acquired. Vancomycin-resistant Enterococcus (VRE) and methicillin-resistant *S. aureus* (MRSA) increased by 14% and 13%, respectively.

This was attributed to [88]:

- Broad-spectrum use of antibiotics for sick patients before the availability of evidence-based COVID-19 pneumonia therapies
- Repurposing of AMR resources to tracking and testing for COVID-19
- Limited appointments in clinics, which delayed treatment for common infections

Tuberculosis poses another challenge in clinics. MDR tuberculosis (MDR-TB) is defined as resistance to both rifampicin and isoniazid. Extensively drug-resistant tuberculosis (XDR-TB) is defined as resistance to rifampicin, isoniazid, fluoroquinolones, and one second-line injectable drug. Compared to drug-susceptible tuberculosis, MDR-TB and XDR-TB are difficult to treat, with longer treatment courses, severe adverse effects of second-line antituberculosis drugs, and higher morbidity and mortality. Healthcare workers and family members who take care of/live with these patients are at risk of acquiring the disease. After years of decline, there was an increase in MDR-TB cases from 2020 to 2021, from 437,000 to 451,000. This was attributed to the COVID-19-associated disruption of tuberculosis detection and treatment services [89]. In a South African study that included 273 patients with XDR TB, 74% had poor treatment outcomes that comprised treatment failure, default, relapse, or death on treatment. More than half (63%) were discharged into the community from the hospital. The median time to death after home discharge was 10 months. 20% of discharged patients had positive sputum smear results at 3 weeks, indicating that they were a potential source of the spread of infection [90].

1.3.3 Strategies to Address Clinical Pressures Contributing to Drug Resistance

Appropriate Use of Antibiotics Appropriate antibiotic prescribing is the cornerstone of slowing down AMR. These include:

- The initial antibiotic should be chosen keeping in mind the common etiologic organism(s).
- Use local antibiograms to select empiric antibiotics before culture results are available. An antibiogram is an aggregate of sensitivity reports collected from individual isolate data. Antibiograms can be created for each hospital as well as at a regional, national, and global level. They provide valuable information about antimicrobial susceptibility in a specific region and trends in AMR at a national and international level [91].
- Obtain a microbiological diagnosis with cultures. After culture reports are available, de-escalate antibiotics from a broad spectrum to a narrow spectrum.
- Use the appropriate dose and duration. Higher than recommended doses increase the risk of adverse effects. Suboptimal doses run the risk of therapeutic failure and can induce resistant pathogens. Treatment duration varies by site of infection [92, 93].
- Always aim for source control. Antibiotics without source control expose patients to the adverse effects of antibiotics, with the risk of persistent infection.
- Keep in mind antibiotic penetration in specific tissues while treating infections.
- Assess the need for central venous catheters, arterial lines, urinary catheters, and other devices daily. Remove catheters and lines when no longer required.

Antibiotic Stewardship The CDC defines antimicrobial stewardship as measures to ensure the appropriate use of antibiotics for providing the best possible care to the patient as well as to keep in check the threat of antimicrobial resistance [94]. Antibiotic stewardship applies to healthcare settings including hospitals, outpatient clinics, dental and urgent care, dialysis centers, long-term care facilities, and veterinary medicine in food, farm, and companion animals.

Infection Prevention and Control Measures Standard precautions must be followed in all patient-care settings. These are measures to prevent exposure to blood, body secretions, mucosa, and non-intact skin. Two essential components of standard precautions are hand hygiene and the use of personal protective equipment (PPE). Other components of standard precautions include cough etiquette principles, safe disposal of sharps, handling textiles and laundry carefully, cleaning patient care equipment, and safe injection practices [95]. Transmission-based precautions have three categories: airborne, contact, and droplet precautions. These are additional precautions taken over and above standard precautions; a risk assessment at hospital admission must be done to assess if a patient requires transmission-based precautions [96].

Water, Sanitation, and Hygiene Measures (WaSH) As compared to developed countries, low- and middle-income countries (LMICs) treat less than 10% of their wastewater [97]. Suboptimal waste disposal leads to contamination of surface and groundwater with human and animal excreta and exposes the population to antibiotics and AMR bacteria. Factors that promote the spread of AMR include open defecation, septic tanks, pit latrines, a lack of isolation of animal excreta, and antibiotics that may get incorporated into plant produce when wastewater reaches crops [98]. Developing basic sanitation facilities like toilets, drainage systems, and proper waste disposal systems will help reduce the spread of AMR. Local community engagement, toilet ownership by households, and investment in infrastructure helped promote better WaSH behaviors [99]. Further resources dedicated to inspecting food preparation, food products, restaurants, and eating places will improve food hygiene [100].

1.4 Global Impact and Consequences of AMR in Disease Control and Treatment

During the COVID-19 pandemic, resources allocated to AMR control were redirected to the pandemic. Significant AMR testing backlogs occurred, and many samples remained untested [101]. The impact of AMR on the global health system is profound and multifaceted. AMR presents a serious threat to the efficiency of antibiotics and other antimicrobial agents, threatening our ability to control bacterial, viral, parasitic, and fungal infections. The increasing incidence of drug-resistant strains undermines standard treatment protocols, leading to prolonged illnesses, increased mortality rates, and heightened healthcare costs. Moreover, AMR sabotages the success of disease prevention measures, as regular health practices such as surgeries, chemotherapy, and organ transplants become riskier due to the potential for untreatable infections. The complicated relationship between AMR and global health demands collaborative efforts to implement rigorous antibiotic stewardship, improve surveillance, and foster the development of novel antimicrobial therapies, emphasizing the critical need for a unified, international approach to address this intensifying public health crisis.

1.4.1 General Impacts

Increased Morbidity and Mortality

Pneumonia, bloodstream infections, and intra-abdominal infections together contributed to 80% of the AMR burden. Lower respiratory chest infections alone led to 400,000 deaths due to AMR. Six pathogens (*E. coli, S. aureus, K. pneumoniae, Streptococcus pneumoniae, Acinetobacter baumanii,* and *P. aeruginosa*) were responsible for more than 75% of deaths attributable to AMR [102]. The absolute number of deaths due to AMR in 2019 was more than that due to tuberculosis, malaria, or AIDS. It is crucial to obtain surveillance data to devise region-specific strategies to fight AMR [102]. As we develop new therapies and technologies, the number of transplants and cancer chemotherapy patients will increase. These immunocompromised patients rely on antimicrobials to fight infections. Antimicrobials are used as prophylaxis for invasive procedures and surgeries. AMR jeopardizes the safety of immunocompromised patients and invasive procedures and surgeries.

Impacts on Public Health

AMR has several adverse effects on global public health [103–106]. AMR causes infections like respiratory and urinary tract infections and tuberculosis, which are difficult to treat. This results in a longer duration of exposure for patients to the toxic effects of antimicrobials. Similarly, patients with influenza and COVID-19 are at risk of secondary bacterial infection. With AMR, patients with these illnesses are at increased risk of death. Further, surgical procedures have increased the chance of infectious complications as the effectiveness of antibiotics recedes. Immunosuppressed patients—those on cancer chemotherapy and transplant recipients—are at risk of dying from AMR infections.

1.4.2 Global Efforts and Action Plans to Address AMR

One Health Approach The foundation of One Health is communication, collaboration, and coordination among human, animal, and environmental health professionals and their partners [107]. Antibiotic-resistant genes (ARG) circulate among human, animal, and environmental microbiomes, which form the different sectors of the One Health concept. Human and animal activities shape the environmental resistome, which is the reservoir of ARGs [108]. The One Health Model for Dairy

Farming aims to raise food animals, taking measures to protect the health of the animals, workers, consumers, local communities, and ecosystem [109]. One of the overlooked consequences of a robust One Health system is poverty alleviation. A successful One Health collaboration will reduce poverty, prevent infectious diseases, and reduce direct and indirect expenses [110]. The COVID-19 pandemic highlighted the need for the sharing of information across health systems and countries. One Digital Health Framework was initiated in 2021 to facilitate easy accessibility of data and information across various healthcare settings [111].

Research and Development of New Antimicrobials The pipeline for new antimicrobials is dry. Most novel antibiotics are in Phase 1 or Phase 2 clinical trials, far away from being released on the market. Factors that impede the research and development of new antimicrobials include:

- Antimicrobial drugs are available for most infections except for MDR infections.
- The time for the development of new antibiotics is usually more than a decade and costs more than one billion [112].
- Antibiotic stewardship measures promote the use of new antimicrobial agents as a last resort. This results in a lower volume of new antimicrobials consumed, even if a new drug is released on the market.
- Limited support and funding exist for drug development companies developing new antimicrobials.
- Antimicrobials are used for a limited duration to treat infections. This is unlike other medications for chronic diseases, which are used for years/decades and generate more revenue [113, 114].

In the United States, the PASTEUR (Pioneering Antimicrobial Subscriptions to End Upsurging Resistance Act) was introduced recently. This Act provides federal funding to drug development companies upfront for research and development of new antimicrobial agents. This act will be based on a subscription-style model in exchange for the availability and judicious use of novel antimicrobial agents [115]. The National Health Service in the United Kingdom is planning to initiate a similar scheme based on subscription-style payments after a successful pilot project [116].

Novel Antimicrobial Therapies With increasing capacity to overcome antibiotic challenges in bacterial pathogens, it is important to find alternative therapeutics to control AMR (Fig. 1.4). Antimicrobial peptides (AMPs) are found in many organisms as part of the innate immune system. Chemically diverse, AMPs work by various mechanisms such as immune modulation, nucleic acid inhibition, cell wall synthesis inhibition, membrane disruption, and inhibition of biofilm formation. AMPs can delay resistance to antibiotics. Some AMPs have in vitro synergistic activity with antibiotics. Limitations include high production costs and renal toxicity [117, 118]. In the same direction, several bacteriophage-based therapies are in clinical trials to treat periprosthetic joint infections, cardiac device-related infections, urinary tract infections, and respiratory tract infections [119, 120].

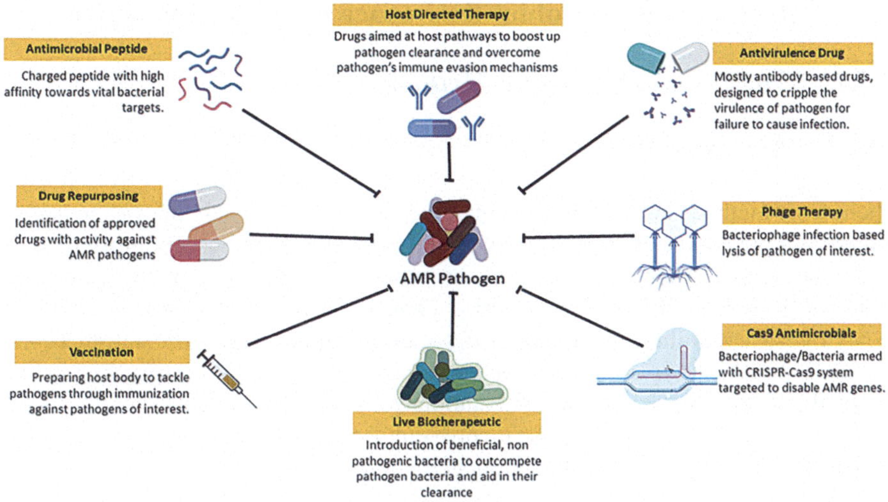

Fig. 1.4 Schematic representations of newer approaches to tackle AMR pathogen. These are the set of approaches that are currently under research for efficient treatment of AMR pathogen-based infection and its spread. (The figure and caption are reprinted from [124])

Finally, monoclonal antibodies work by neutralizing virulence factors, lipopolysaccharides, capsular polysaccharides, and protein targets and also present a hope to control the AMR infections. For example, bezlotoxumab has been approved by the FDA for the treatment of recurrent *Clostridium difficile* infections. It binds to toxin B, which plays a role in virulence and recurrent infection [121]. Similarly, MEDI4893 is another human monoclonal antibody that targets the alpha toxin of *S. aureus*. It is being studied for its role in prophylaxis against systemic *S. aureus* disease [122, 123].

Antimicrobial Stewardship Programs (ASP) As discussed in the previous section, antimicrobial stewardship is the responsible use of antimicrobials, balancing the individual's need for treatment of infection with societal needs for continual access to effective therapy [125]. Core elements of ASPs as defined by the CDC include [94] financial support, training clinicians, surveillance, public awareness, and the implementation of healthcare standards in patient care units and pharmacies. Research shows that ASPs reduce hospital costs, including decreased antibiotic spending and decreased indirect costs, and this can offset the costs of funding the program [126]. Further, ASPs have been shown to reduce in-hospital length of stay (LOS), costs, and antibiotic consumption [127, 128]. There is evidence that ASPs are beneficial for adults and children as well as in university and community hospitals. Even in cancer patients, premature infants, transplant recipients, patients in the ICU, and long-term care facilities, ASPs are safe and effective and should be implemented [129–135].

1.5 Detecting Antimicrobial Resistance in Clinical Practice

1.5.1 Traditional Diagnostic Methods

1.5.1.1 Phenotypic Methods

Phenotypic methods are widely used in clinical microbiology laboratories to assess the phenotypic characteristics of bacteria and determine their susceptibility to antimicrobial agents. These methods rely on observing bacterial growth, measuring zones of growth inhibition, or analyzing changes in bacterial growth patterns to verify the susceptibility or resistance of bacterial isolates to specific antimicrobial agents [136]. Phenotypic methods include broth dilution tests, antimicrobial gradient methods, disk diffusion tests, and automated instrument systems such as Vitek2 [136]. Manual phenotypic methods involve culture steps, increasing the turnaround time to 18–36 h compared to the automated platforms, which can return a result in 2–24 h [137].

1.5.1.2 Disc Diffusion

This technique involves testing isolated bacteria derived from patient samples by applying the bacterial inoculum directly to the surface of an agar plate. After placing paper discs containing specific antibiotics on the agar, the plates are incubated, allowing the antibiotics to diffuse into the surrounding area. The diameter of the growth inhibition zones around each disc is measured, which indicates the susceptibility of the bacteria to the tested antibiotic [136]. Diameter measurement cutoffs classify bacterial isolates into three categories (susceptible, intermediate, or resistant), rather than specifying the minimum inhibitory concentration (MIC). The interpretation of these zones is guided by the CLSI (Clinical and Laboratory Standards Institute) or EUCAST criteria [136, 138–140]. The disk diffusion method is simple and well-standardized. It provides flexibility in selecting disks for testing, and the categorical result can guide clinicians in choosing an adequate antibiotic regimen. It is also cost-effective and one of the least costly among the susceptibility tests. The test does not require specialized equipment [136, 140]. However, the lack of mechanization or automation can be considered a drawback of the method. Other limitations of the test include a lower efficiency in detecting certain types of resistance, nor is it suitable for testing all bacteria [136].

1.5.1.3 Broth Dilution

The broth dilution method is another type of phenotypic test and one of the earliest antimicrobial susceptibility testing methods. The test is performed by incubating two-fold dilutions of antibiotics in liquid growth media inoculated with standardized bacterial suspensions containing a known concentration of the isolated bacteria

[136, 140]. After overnight incubation at 35 °C, the test tube is examined for visible growth by turbidity, and the lowest concentration of antibiotic that presents visible bacterial growth represents the MIC. Unlike the disc diffusion method, this test provides a quantitative result that indicates the susceptibility of the bacteria to the tested antibiotic [136]. This method, alongside the disk diffusion test, is well-standardized and cost-effective. It can be used for multiple antibiotics testing in a single tray, providing reproducibility and convenience, with prepared panels also providing the MIC [136, 140]. However, this traditional test involves the manual preparation of the antibiotic solutions and requires many reagents and physical space. All this considered, precision problems and errors are always possible during solution preparation.

1.5.1.4 E-Test

The E-test is an example of a commercial antimicrobial gradient method. A central piece of the method is a thin plastic strip that contains an impregnated side with a dried antibiotic concentration gradient and a concentration scale on the upper surface. An agar plate after inoculation with a standardized bacterial suspension will usually receive more than one strip (up to six strips in a 150 mm Petri dish) placed radially for an overnight incubation [136, 140]. The MIC number is determined by the intersection of the lower part of the ellipse-shaped growth inhibition area around the test strip [136, 140]. This test is easy to perform and can still be performed for bacteria that require special enriched growth media or incubation setups or if they are slow growers. It is also ideal if only a few antibiotics need to be tested. However, the strips can be more expensive than the disks for diffusion tests, and the turnaround of up to 1 day can be a limiting disadvantage [136, 140].

1.5.1.5 Automated Instruments

One of the purposes of the development of automated instruments is to standardize the reading of endpoint results while also providing faster susceptibility test results than manual readings. Several automated platforms are approved for use in the United States by the FDA (Food and Drug Administration) [136, 140]. Each platform contains its particularities and uses a diverse range of susceptibility testing and detection. Generally, they are self-contained incubators and reader devices that can verify bacterial growth in the presence of diverse antibiotic panels. Bacterial growth is detected in microdilution trays through photometry or fluorometry (MicroScan WalkAway). Sensititre ARIS 2X uses standard microdilution plates, and the bacterial growth is determined by fluorescence measurement. The BD Phoenix Automated Microbiology System uses turbidimetric and colorimetric growth detection methods and offers susceptibility testing for various panels containing antibiotic dilutions for different bacterial groups. The Vitek2 System uses compact plastic reagent cards containing antibiotics and test media and employs turbidimetric monitoring for growth detection [136, 140]. All these automated instruments have in common the need for computer software for interpreting susceptibility results and energy connections, which can limit their use in low-resource settings. However, the rapid turnaround of results can lead to timely changes in antimicrobial therapy and cost savings [136, 140].

1.5.2 Genotypic Methods

Genotypic tests are molecular-based methods that detect known antimicrobial genes associated with resistance. These tests are gaining space in routine diagnostic laboratories because of the precise characterization and detection of resistant associated genes with a multiplex targeting methodology and the possibility of doing that in non-purified human samples [137, 140]. This advanced technology also has limitations, mainly because genotypic tests cannot define a MIC for the antibiotics that present a mutation associated with resistance. Importantly, the tests can miss newly evolved mutations associated with resistance. The cost of these tests is also considerably higher than that of traditional phenotypic assays. Methods based on DNA amplification can generate a result as low as 0.5–4 h [137].

1.5.2.1 PCR

The polymerase chain reaction, known as PCR, is a widely used molecular biology technique for amplifying specific DNA sequences through temperature cycles. It involves using the DNA polymerase enzyme, forward and reverse primers, and a DNA template containing the target sequence [141]. In the context of antimicrobial resistance detection, various types of assays (real-time PCR, qPCR, multiplex PCR, and high-throughput quantitative PCR (HT-qPCR)) have been developed to detect gene regions with mutations associated with antibiotic resistance [137]. Real-time PCR carries the advantage of simultaneous monitoring of the amplification using fluorescent dye [141]. Multiplex platforms can simultaneously detect multiple genes. Different primer combinations can be used to, for example, test resistance to multiple antibiotics used against Neisseria gonorrhea or to detect beta-lactamase genes in Gram-negative bacteria [137, 141]. Xpert® MTB/RIF (Cepheid) detects *M. tuberculosis* rifampicin-resistance mutations directly from sputum samples from patients suspected of pulmonary tuberculosis. Other test versions also detect mutations associated with resistance to other antibiotics for treating this disease (aminoglycosides, fluoroquinolones, pyrazinamide, and isoniazid) [142].

PCR-based methods can provide results in just a few hours (4–5 h) [141] and can be performed in non-purified samples, and the assays can be quickly modified to detect new resistance factors. However, these tests cannot provide a specific MIC and miss mutations not precisely searched for in the assay. The increased experimental design complexity translates into a higher cost associated with the development of the assays, especially when involving multiplex targeting, elevating the cost of performing these tests routinely [137].

1.5.2.2 Isothermal Amplification

The isothermal amplification methods also work by amplifying specific targeted DNA regions. However, this newer development eliminates the thermocycling required in traditional PCR tests, using a constant temperature throughout the process [137, 141]. Several types of assays for DNA amplification can be performed: strand displacement amplification (SDA), transcription-mediated amplification (TMA), nucleic acid sequence-based amplification (NASBA), rolling circle amplification (RCA), recombinase polymerase amplification (RPA), loop-mediated isothermal amplification (LAMP), and helicase-dependent amplification (HDA) [137, 141]. LAMP assays can be used to detect ESBLS and carbapenemases and can be performed on pure isolates and non-purified sampled material [141]. Isothermal amplification assays eliminate the need for thermocycling, reducing power consumption and resulting in shorter analysis times. It is fast and sensitive and requires simplified reaction conditions, with the potential to be used in limited resource settings as a point-of-care test. Nevertheless, some methods involve complex reactions, and implementing multiplex approaches for isothermal amplification methods is more challenging than PCR tests [137].

1.5.2.3 Hybridization Probe

The various methods that detect antimicrobial resistance through hybridization have a slide with labeled DNA probes of genomic regions of interest that represent wild-type genes or genes with specific mutations that confer resistance to chosen antibiotics [140, 141]. The labeled DNA probes represent restrained complementary oligonucleotides on the platform surface [140]. The binding or not of the tested DNA to specific DNA probes of the test translates into the presence or absence of each gene, which together can profile the tested strain as resistant or susceptible to specific antibiotics [141]. Line Probe Assays (LPAs) for detecting resistance to first- and second-line antibiotics to treat TB and drug-resistant TB are approved for clinical use by the WHO [143]. It can be performed on purified cultures or sputum samples that are known to be positive for *M. tuberculosis* by another test. It contains the most common mutations, and the classification of resistance or susceptibility depends on the pattern of bands of hybridized DNA [143]. INNO-LiPA® Rif.TB (Innogenetics) and GenoType® MTBDR (plus) (Hain LifeScience GmbH) are examples of commercially available LPA for TB use [142].

Another technique that uses the hybridization concept is DNA microarrays. DNA microarrays contain DNA probes of a reference strain, and the tested DNA is identified by comparing the hybridization pattern with the reference genome [137]. This technique has evolved a lot with the implementation of simultaneous testing of antimicrobial-resistant genes labeled with fluorescence. Hybridization assays can be fast, bypassing the need for purified cultures and simultaneously detecting multiple antimicrobial resistance-related genes. However, trained personnel are needed to perform the test, and, as for the other genotypic tests, the presence of mutations sometimes correlates differently with phenotypic resistance. Additionally, mutations that are not explicitly searched cannot be detected, and therefore, a resistant strain could be erroneously classified as susceptible [137, 140, 143].

1.5.3 Advanced Diagnostic Methods

1.5.3.1 Matrix-Assisted Laser Desorption/Ionization Time-of-Flight Mass Spectrometry (MALDI-TOF MS)

The MALDI-TOF MS technique has made significant progress in bacterial characterization and detection of antimicrobial resistance. This technique involves the analysis of the proteome of bacterial cells to profile microorganisms, and the turnaround to a result is between 2 and 4 hours [137]. After purification and cultivation, human samples can be mixed with a matrix (an energy-absorbent solution) to crystallize upon drying. The matrix containing the sample is then hit by a laser beam to produce ions charged with protons. The ions are accelerated and separated based on their mass-to-charge ratio (m/z), determined by the time each ion moves along the specific tube to obtain the TOF. Each microorganism is identified by comparing the ion profile generated (peptide mass fingerprint (PMF) or molecular signature) to a database that contains known, previously characterized profiles [137, 141]. The most widespread use of this technique is for pathogen identification in just a few hours [141]. However, it can also characterize antimicrobial resistance when the reference databases contain specific spectra of peptides from resistant strains and susceptible strains to compare [137]. Indirectly, the activity of an antibiotic can be determined by searching for products of degradation resulting from the activity of the drug [141]. Though MALDI TOF MS is a promising technique involving sophisticated, reliable, and fast results, it also presents severe limitations to implementing the technology in limited resource settings [137, 140].

1.5.3.2 DNA Sequencing and Whole Genome Sequencing (WGS)

Some of the most common methods to detect antimicrobial resistance are Sanger sequencing, NGS, third-generation sequencing, and WGS. The complexity of the methods increased from low throughput single-base resolution with one sample analysis at a time to multiplexing, simultaneous analysis of samples without prior isolation of bacteria, and genome characterization of complex microbial communities [137]. Frederick Sanger developed the Sanger method, which remains widely utilized for clinical sample sequencing. Next-generation sequencing has the advantage of increased throughput, analyzing multiple samples simultaneously compared to the Sanger method [137]. Sequencing entire bacteria genomes (WGS) allows the creation of databases of reference strains that can be compared to each newly sequenced strain to establish a resistance genotypic profile [140]. This advanced technology contributes to surveillance systems and the detection of resistance from Gram-positive or Gram-negative bacteria, and recent studies show a high correlation between genotypic profiles and phenotypic resistance [137, 140]. Another advantage is that WGS allows adding new targeted genomic regions to the analysis more quickly than microarray technology [141]. DNA sequencing and WGS, along with genomic database availability, correspond to significant improvements in resistance detection and characterization, and there are an increasing number of bioinformatic tools available for the alignment and detection of AMR genes [141]. However, the routine application of these technologies in clinical laboratories still needs to be improved due to the high cost of this powerful technology [137, 140].

1.5.4 Challenges and Limitations of Diagnostic Methods for AMR

Antimicrobial resistance diagnostic tests face several challenges and limitations, many of which have already been explored in this chapter while describing each available tool. The goal of this section is to describe a few other implementations and clinical limitations related to antimicrobial resistance. One significant challenge is the disproportional availability of tests in hospital and community settings. Additionally, there are a few diagnostic platforms that can be called portable and offer automated procedures [137]. However, both settings have the high cost of more advanced diagnostic technologies as a barrier to widespread implementation [137]. The clinical relevance of the isolated pathogen and the AMR result also depend, among other details, on the appropriateness of sample collection [144]. Before performing any test, a clinically significant indication for obtaining biospecimens for a test should be pursued. The quality of the collected sample and the distinction between actual infection and contamination or asymptomatic carriage should be carefully evaluated to avoid misinterpretation of the results and unnecessary antibiotic use [145].

Another challenge is that for many tests to accurately detect resistance, at least one kind of sample pre-treatment is needed [137] to obtain pure isolates or a high bacterial count. This condition still needs to be met in many laboratories, especially in limited-resource settings. Additionally, long periods of incubation and detection time [137] limit access to timely results for therapy adjustment. Molecular tests demand caution and interpretation since detecting genes or mutations associated with resistance does not always represent phenotypic resistance. Current molecular tests also cannot determine with 100% accuracy which organism carries a resistance gene, which demands clinical inference. On the other hand, phenotypic tests are usually more operator-dependent, and the subjective interpretation of the test can limit the application of these tests in point-of-care scenarios [145].

1.5.5 Future Directions for AMR Diagnostics

Promising platforms such as MALDI TOF, flow cytometry, and isothermal microcalorimetry have shown considerable potential in AMR surveillance and identification [140]. Current text technologies are limited by their inability to provide both phenotypic and genotypic results simultaneously. Addressing this limitation, a potential future advancement lies in the development of integrated platforms capable of offering comprehensive information on both the antimicrobial susceptibility of pathogens and the underlying genetic mechanisms of resistance [145]. Such all-in-one platforms would empower healthcare professionals to make faster and more informed treatment decisions, leading to better patient outcomes and more effective management of antimicrobial use. Furthermore, advancements in diagnostic test panels that can combine multiple technologies and identify a range of pathogens hold tremendous promise [145]. These versatile panels could be tailored for syndrome evaluations, such as respiratory or gastrointestinal infections, offering a comprehensive assessment of potential pathogens and their resistance profiles.

Another crucial aspect of future AMR diagnostics is the development of point-of-care disposable tests that can be conducted during a single clinical visit [146]. These rapid and on-site tests would eliminate the need for lengthy laboratory processing and waiting times, allowing for timely and appropriate treatment decisions. Moreover, these tests would ideally be performed directly on clinical samples without the requirement for extensive purification or culture isolation steps [146]. Simplified testing processes would not only expedite diagnosis but also facilitate AMR surveillance in resource-limited settings. The use of PCR is already used for wastewater surveillance, for example, against *vanA* (encoding vancomycin resistance), *mecA* (encoding methicillin resistance), and *ampC* (encoding ampicillin resistance) [141]. The successful realization of these ambitious advancements requires a collaborative effort between the public and private sectors [146]. Ensuring accessibility and affordability, especially in low-resource settings, remains a priority to combat AMR effectively on a global scale.

1.6 Role of Omics and System Biology in the Understanding of AMR

Understanding the mechanisms involved in antimicrobial resistance acquired by microbes at the system level is critical for obtaining answers and exploring solutions to counteract this resistance. The utilization of robust whole-genome sequencing methodologies, as well as other omics techniques such as transcriptomics, proteomics, and metabolomics, is critical for providing basic insights into antibiotic resistance mechanisms. Since the amount of data generated by various omics techniques can be enormous and thus difficult to interpret, genomic models have been developed that can collect the massive amount of experimentally generated information into mathematical models that can be explained [147–149]. Recent attempts have demonstrated that the use of omics technology and systems biology as quantitative and rigorous hypothesis-generating frameworks can help us better understand antibiotic resistance (Table 1.1).

Current advances in NGS performance, along with lower costs, have enabled WGS of a diverse range of microbes. WGS studies allow for the identification of new ARGs [150] as well as the investigation of isolated phylogeny and epidemiological features, which is crucial in understanding AMR [151].

Transcriptomic studies emphasize general changes in gene expression in response to environmental events, allowing for a snapshot of the genes that function under a specific scenario. In the case of antibiotic resistance, this picture is the result of variation in gene expression patterns, which provides insight into the mechanism of action of the specific antimicrobial under investigation as well as the general physiological responses of bacteria to stress caused by the presence of such molecules. Further understanding of the mechanism of action of existing antibiotics might help reveal new techniques to control the propagation of resistance genes [152].

Although transcriptomics is the most commonly utilized approach for studying bacterial responses to antibiotic stress, proteomics has been widely used to advance our understanding of the steady-state proteome of antibiotic-resistant clinical isolates [22]. In recent years, numerous researchers have concentrated their efforts on discovering and developing protein biomarkers associated with susceptibility or resistance. These advancements are made feasible by enhanced proteomics methods, such as MALDI-TOF MS. Proteomics provides a more comprehensive molecular perspective of AMR than conventional techniques; nevertheless, in many situations, the utilization of these methodologies is related to other forms of omics, and the relationship with WGS has been shown in multiple publications.

As we have discussed before, bacterial metabolism plays an important role in AMR and may contribute to the development and changes in AMR due to exposure to these antimicrobials [18]. For example, high metabolic activity is essential for promoting resistance events, as it requires activation of a wide range of molecular mechanisms such as cell wall modification, mutation stabilization, transport, energy production, and efflux pump overexpression. In this sense, understanding the metabolic processes involved in this mechanism may help to strategically modify bacterial metabolism and re-sensitize.

Through approaches such as metabolomics, it is possible to obtain a complete picture of all the molecules involved in metabolism, which are closely related to the phenotype of an organism. Knowing this snapshot of the molecules involved in metabolic processes reveals the early response to antibiotic stress and the adaptations necessary to maintain the AMR mechanism. Like other omics technologies, metabolomics can be used in a variety of ways and approaches. In general, mass

Table 1.1 Role of multi-omics techniques in understanding AMR in disease-causing microorganisms

Multi-omics technique	Microorganism	Cause of AMR	Resistance against	Reference
Genomics	*Neisseria gonorrhoeae*	Mutation of the *penA* and *mtrR* alleles in mosaic	Cefixime and azithromycin	[153, 154]
	Helicobacter pylori	Mutation in the 23S rRNA gene	Clarithromycin	[155]
	Proteus cibarius strain G11	Horizontal transfer of the florfenicol resistance (*floR*) gene	Florfenicol	[156]
	Staphylococcus lentus strain H29	11 genes conferring resistance, including mobile genetic element (MGE)-related *floR* genes *cfr* (IS256-cfr), and *fexA* (*radC-tnpABC*-hp-*fexA*)	Multidrug-resistant (MDR)	[157]
	Arthrobacter nicotianae OTC-16	Eight genes related to antibiotic resistance	Tetracycline	[158]
	N. gonorrhoeae isolates	Presence of *gyrA* S91F and *parC* S87R mutations, mitations in *rpsJ V57M* and *tetM*, the *penA*-34.001 mosaic mutation, and the presence of *mtrR*, *porB1b*, and G101D genes; and *pbp1* L421Pla mutations	Ciprofloxacin, tetracycline, benzylpenicillin, and β-lactamases	[159]
	Campylobacter jejuni isolates	Presence of *cmeABC* (related to efflux pumps), tet(O) (tetracycline resistance gene), *aadE* (streptomycin resistance gene), and a point mutation, *gyrA* T861	Tetracycline, streptomycin, and quinolone	[160]
	Klebsiella pneumoniae	Mutation in *mgrB*, *pmrB*, *phoP*, *phoQ*, and *K3X*	Colistin	[161]
	Staphylococcus capitis	G2576T mutation in 23S rRNA	Aminoglycosides and methicillin	[162]

Transcriptomics	*Pseudomonas aeruginosa*	G183D mutation in the β-lactamase *ampC* and DNA metylation	Ceftazidime/avibactam (CZA), ceftolozane/tazobactam (C/T), and piperacillin/tazobactam (P/T)	[163]
	Salmonella typhimurium	Increased expression of *phoP* and *phoQ*	Polymyxin	[164]
		Synergistic effect of CpxAR and efflux pumps	Colistin	[158]
	Salmonella enterica serovar enteritidis	*purBCDFHKLMNT* were the core genes	Fluoroquinolone	[165]
	Staphylococcus aureus	*agr* and other virulence factors mediate quorum-sensing (qs) systems	Vancomycin and other related drugs	[166]
	A. baumannii	Efflux pumps (*craA*)	Chloramphenicol	[167]
Proteomics	*Heliobacter pylori*	Role of *rpoB* and FBPAII	Clarithromycin and Metronidazole	[168]
	E. coli	Downregulation of *lamB*	Ciprofloxacin	[169]
Metabolomics	*Pseudomonas aeruginosa*	A role of trehalose-6-phosphate levels	Polymyxin	[170]

spectrometry (MS) and nuclear magnetic resonance spectroscopy (NMR) are the most widely used approaches for metabolite detection, and a wide variety of combinations of analytical tools are possible that can be combined with mass spectrometry [25]. Considering the limitations, each approach used in studies related to antimicrobial resistance requires experimental planning to decide which method is most suitable to obtain the desired answers about the involvement of AMR in the microbial population.

1.7 Consequences of Antimicrobial Resistance

Antimicrobial resistance AMR is one of the most challenging health problems facing countries worldwide [171]. Unless action is taken now, the burden of antibiotic resistance-related deaths may rise to ten million annually by 2050 [172]. Even though the causes of AMR are multifaceted, there is no doubt that antibiotic overuse has played a significant role. Between 2000 and 2015, antibiotic use increased by 65% globally, primarily driven by a substantial increase across low-income countries [173]. As a result of AMR, public health and global healthcare systems are severely affected. Patients with infections due to antimicrobial-resistant organisms have higher costs than patients with infections caused by antimicrobial-susceptible organisms; when comparing patients with antimicrobial-resistant organisms with patients without infections, the cost difference is even more significant. There is a need to develop strategies to prevent the emergence and spread of antimicrobial-resistant organisms in nosocomial settings [174].

1.7.1 Elevated Morbidity and Mortality

AMR disproportionately affects health outcomes, particularly respiratory and thoracic infections, across 47 countries, 23 bacterial pathogens, and 88 pathogen-drug combinations. The highest mortality rates were observed in the Central African Republic, Lesotho, and Eritrea. It also highlighted the critical challenges in surveillance and laboratory capacity in the region, advocating for improved data collection and tailored healthcare policies to mitigate the impact of AMR [102].

1.7.2 Increased Healthcare Costs

AMR significantly increases healthcare costs [175]. The CDC reports that treating bacterial infections in antibiotic-resistant patients can add approximately $1400 to their hospital bills in the United States [176]. Annually, these costs could surge to over $2 billion. Globally, estimates suggest that 2050 AMR could cost between

$300 billion and $1 trillion annually [177]. The direct financial impact of AMR on healthcare includes the high costs of intensive treatments and increased resource use [177, 178]. In addition, outbreaks of resistant pathogens can lead to the closure of hospital sections and the cancellation of elective surgeries, further increasing costs [106]. A comprehensive strategy is required to mitigate the increased healthcare costs associated with AMR.

1.7.3 Higher Spread of Infections

AMR leads to the increased spread of infections by portraying common antimicrobial treatments as less effective. This resistance allows infections to persist and spread more easily, creating challenges in treating diseases that were once easily managed with standard antibiotics. AMR exacerbates the severity and duration of infections, increasing the risk of disease transmission [178]. A report has suggested that approximately 30% of travelers return home with antimicrobial-resistant bacteria due to international travel [179]. Among the key risk factors are travel destinations, antibiotic use, and traveler's diarrhea. In light of this, efforts must be made to educate travelers, promote responsible antibiotic use, and enhance surveillance to monitor and mitigate the spread of AMR. To prevent the spread of AMR organisms internationally, it is crucial to understand and address these factors.

1.8 Conclusion

Humans have co-evolved with bacteria, navigating a prolonged route of adaptations, selections, and diseases. Our interaction with pathogenic bacteria led to the discovery of antibiotics, a significant cure that primarily contained pandemics. However, the intrinsic abilities of bacterial systems and evolutionary forces eventually drove resistance against these treatments, portraying a tough challenge. With the advent of novel technologies in molecular and systems biology, we find ourselves at the early stages of understanding and projecting our next course of action. Omics sciences, in their emerging phase, provide the best alternative tools to discover bacterial survival and resistance mechanisms for antimicrobial treatments. AMR is a multifactorial issue, involving policies, healthcare systems, disease surveillance, government regulations, and sanitation quality. In this chapter, we reviewed the evolutionary factors driving AMR, emphasizing their impact and schemes for remedy and avoidance. Subsequently, we explained the clinical pressures aiding AMR development, highlighting key factors, impacts, and strategies for intervention. In the same direction, we have also addressed the global impact and consequences of AMR and emphasized the requirements of combined global endeavors and action plans. Turning to diagnostics, both traditional and advanced methods for detecting AMR were reviewed, along with general methodologies, associated challenges, and future directions. Lastly, we explored the paramount role of omics and systems biology in solving AMR mechanisms. The consequences of AMR, evident in elevated mortality and morbidity, increased healthcare costs, and amplified infection spread, underscore the urgency for collaborative and proactive measures. As this chapter concludes, it sets a platform for advanced exploration into AMR biology, highlighting the requirement for constant innovations, research, and collective processes to moderate its impact on public health and avoid any future pandemics.

References

1. Rosini R, et al. Vaccines Against Antimicrobial Resistance. Front Immunol. 2020;11:1048.
2. Urban-Chmiel R, et al. Antibiotic resistance in bacteria: a review. Antibiotics. 2022;11(8):1079.
3. Eclinicalmedicine. Antimicrobial resistance: a top ten global public health threat. eClinicalMedicine. 2021;41:101221.
4. Hutchings MI, Truman AW, Wilkinson B. Antibiotics: past, present and future. Curr Opin Microbiol. 2019;51:72–80.
5. Butler MS, et al. Analysis of the clinical pipeline of treatments for drug-resistant bacterial infections: despite progress, more action is needed. Antimicrob Agents Chemother. 2022;66(3):e01991–21.
6. Mancuso G, et al. Bacterial antibiotic resistance: the most critical pathogens. Pathogens. 2021;10:1310.
7. Irfan M, Almotiri A, AlZeyadi ZA. Antimicrobial resistance and its drivers: a review. Antibiotics. 2022;11(10):1362.
8. Cars O, et al. Resetting the agenda for antibiotic resistance through a health systems perspective. Lancet Glob Health. 2021;9(7):e1022–7.
9. Charani E, et al. An analysis of existing national action plans for antimicrobial resistance-gaps and opportunities in strategies optimising antibiotic use in human populations. Lancet Glob Health. 2023;11(3):e466–74.
10. Breijyeh Z, Karaman R. Design and synthesis of novel antimicrobial agents. Antibiotics (Basel). 2023;12(3):628.
11. Duval RE, Grare M, Demoré B. Fight against antimicrobial resistance: we always need new antibacterials but for right bacteria. Molecules. 2019;24(17):3152.
12. Wani AK, et al. Microbial adaptation to different environmental conditions: molecular perspective of evolved genetic and cellular systems. Arch Microbiol. 2022;204(2):144.
13. Núñez-Samudio V, et al. Molecular epidemiology of Escherichia coli clinical isolates from Central Panama. Antibiotics. 2021;10(8):899.
14. He M, et al. Genetic diversity and population structure of vibrio parahaemolyticus isolated from clinical and food sources. Front Microbiol. 2021;12:708795.
15. Smith WPJ, et al. Bacterial defences: mechanisms, evolution and antimicrobial resistance. Nat Rev Microbiol. 2023;21:519.
16. Tedijanto C, et al. Estimating the proportion of bystander selection for antibiotic resistance among potentially pathogenic bacterial flora. Proc Natl Acad Sci USA. 2018;115(51):E11988–e11995.
17. Trubenová B, et al. Population genetics, biofilm recalcitrance, and antibiotic resistance evolution. Trends Microbiol. 2022;30(9):841–52.
18. Kok M, et al. Unraveling antimicrobial resistance using metabolomics. Drug Discov Today. 2022;27(6):1774–83.
19. Baran A, Kwiatkowska A, Potocki L. Antibiotics and bacterial resistance: a short story of an endless arms race. Int J Mol Sci. 2023;24(6):5777.

20. Xie W, et al. Mycobacterium smegmatis, a promising vaccine vector for preventing TB and other diseases: Vaccinomics insights and applications. Vaccine. 2023;11(8):1302.
21. Tavassoly I, Goldfarb J, Iyengar R. Systems biology primer: the basic methods and approaches. Essays Biochem. 2018;62(4):487–500.
22. Mack SG, Turner RL, Dwyer DJ. Achieving a predictive understanding of antimicrobial stress physiology through systems biology. Trends Microbiol. 2018;26(4):296–312.
23. Janiszewska D, et al. "Omic" approaches to bacteria and antibiotic resistance identification. Int J Mol Sci. 2022;23(17):9601.
24. Waddington C, et al. Exploiting genomics to mitigate the public health impact of antimicrobial resistance. Genome Med. 2022;14(1):15.
25. Francine P. Systems biology: new insight into antibiotic resistance. Microorganisms. 2022;10(12):2362.
26. Foudraine DE, et al. Exploring antimicrobial resistance to beta-lactams, aminoglycosides and fluoroquinolones in E. coli and K. pneumoniae using proteogenomics. Sci Rep. 2021;11(1):12472.
27. Ruppé E, d'Humières C, Armand-Lefèvre L. Inferring antibiotic susceptibility from metagenomic data: dream or reality? Clin Microbiol Infect. 2022;28(9):1225–9.
28. Lopatkin AJ, et al. Bacterial metabolic state more accurately predicts antibiotic lethality than growth rate. Nat Microbiol. 2019;4(12):2109–17.
29. Aye SM, et al. Lipid A profiling and metabolomics analysis of paired polymyxin-susceptible and -resistant MDR Klebsiella pneumoniae clinical isolates from the same patients before and after colistin treatment. J Antimicrob Chemother. 2020;75(10):2852–63.
30. Ayoub Moubareck C. Polymyxins and bacterial membranes: a review of antibacterial activity and mechanisms of resistance. Membranes (Basel). 2020;10(8):181.
31. Périchon B, Courvalin P. Antibiotic resistance. In: Schaechter M, editor. Encyclopedia of microbiology. 3rd ed. Oxford: Academic Press; 2009. p. 193–204.
32. Fortuin S, Soares NC. The integration of proteomics and metabolomics data paving the way for a better understanding of the mechanisms underlying microbial acquired drug resistance. Front Med (Lausanne). 2022;9:849838.
33. Li L. Accessing hidden microbial biosynthetic potential from underexplored sources for novel drug discovery. Biotechnol Adv. 2023;66:108176.
34. Garcia E, et al. Moving from point-based analysis to systems-based modeling: integration of knowledge to address antimicrobial resistance against MDR bacteria. Clin Pharmacol Ther. 2021;110(5):1196–206.
35. Chernov VM, et al. Omics of antimicrobials and antimicrobial resistance. Expert Opin Drug Discov. 2019;14(5):455–68.
36. De Wit G, et al. Microbial interspecies interactions and their impact on the emergence and spread of antimicrobial resistance. Ann Rev Microbiol. 2022;76:179–92.
37. Grote A, Earl AM. Within-host evolution of bacterial pathogens during persistent infection of humans. Curr Opin Microbiol. 2022;70:102197.
38. Woolhouse MEJ, et al. Biological and biomedical implications of the co-evolution of pathogens and their hosts. Nat Genet. 2002;32(4):569–77.
39. Wolfe ND, Dunavan CP, Diamond J. Origins of major human infectious diseases. Nature. 2007;447(7142):279–83.
40. Schrader SM, Botella H, Vaubourgeix J. Reframing antimicrobial resistance as a continuous spectrum of manifestations. Curr Opin Microbiol. 2023;72:102259.
41. Glatter KA, Finkelman P. History of the Plague: an ancient pandemic for the age of COVID-19. Am J Med. 2021;134(2):176–81.
42. Kambouris ME, et al. Historical microbiology: researching past bioevents by integrating scholarship (re)sources with paleomicrobiology assets. Future Microbiol. 2023;18:681–93.
43. Valenzuela Sánchez A, et al. Why disease ecology needs life-history theory: a host perspective. Ecol Lett. 2021;24:876.

44. Bean AGD, et al. Studying immunity to zoonotic diseases in the natural host — keeping it real. Nat Rev Immunol. 2013;13(12):851–61.
45. Denisov NS, Kamenskikh EM, Fedorova OS. Trends in population-based studies: molecular and digital epidemiology (review). Sovrem Tekhnologii Med. 2022;14(4):60–70.
46. Diaz-Beltran L, et al. Systems biology as a comparative approach to understand complex gene expression in neurological diseases. Behav Sci (Basel). 2013;3(2):253–72.
47. Hernando-Amado S, Sanz-García F, Martínez JL. Rapid and robust evolution of collateral sensitivity in Pseudomonas aeruginosa antibiotic-resistant mutants. Sci Adv. 2020;6(32):eaba5493.
48. Genova R, et al. Collateral sensitivity to Fosfomycin of Tobramycin-resistant mutants of Pseudomonas aeruginosa is contingent on bacterial genomic background. Int J Mol Sci. 2023;24(8):6892.
49. Sanz-García F, et al. Coming from the wild: multidrug resistant opportunistic pathogens presenting a primary, not human-linked, environmental habitat. Int J Mol Sci. 2021;22(15):8080.
50. Jones DC, et al. One health and global health view of antimicrobial susceptibility through the "eye" of Aeromonas: systematic review and meta-analysis. Int J Antimicrob Agents. 2023;62(2):106848.
51. López-Causapé C, et al. Evolution of the Pseudomonas aeruginosa Aminoglycoside mutational resistome in vitro and in the cystic fibrosis setting. Antimicrob Agents Chemother. 2018;62(4):e02583.
52. Jahn LJ, et al. Compatibility of evolutionary responses to constituent antibiotics drive resistance evolution to drug pairs. Mol Biol Evol. 2021;38(5):2057–69.
53. Laborda P, Martínez JL, Hernando-Amado S. Evolution of habitat-dependent antibiotic resistance in Pseudomonas aeruginosa. Microbiol Spectr. 2022;10(4):e0024722.
54. Clejan I, Congleton CD, Lerch BA. The emergence of group fitness. Evolution. 2022;76(8):1689–705.
55. Fodor A, et al. Multidrug Resistance (MDR) and collateral sensitivity in bacteria, with special attention to genetic and evolutionary aspects and to the perspectives of antimicrobial peptides-a review. Pathogens. 2020;9(7):522.
56. Baquero F, Tedim AP, Coque TM. Antibiotic resistance shaping multi-level population biology of bacteria. Front Microbiol. 2013;4:15.
57. Custer GF, Bresciani L, Dini-Andreote F. Ecological and evolutionary implications of microbial dispersal. Front Microbiol. 2022;13:855859.
58. Waters JM, et al. Dispersal reduction: causes, genomic mechanisms, and evolutionary consequences. Trends Ecol Evol. 2020;35(6):512–22.
59. Wang X, Yu D, Chen L. Antimicrobial resistance and mechanisms of epigenetic regulation. Front Cell Infect Microbiol. 2023;13:1199646.
60. Ramamurthy T, et al. Deciphering the genetic network and programmed regulation of antimicrobial resistance in bacterial pathogens. Front Cell Infect Microbiol. 2022;12:952491.
61. Lässig M, Mustonen V, Nourmohammad A. Steering and controlling evolution - from bioengineering to fighting pathogens. Nat Rev Genet. 2023;24(12):851–67.
62. Szybalski W, Bryson V. Genetic studies on microbial cross resistance to toxic agents. I. Cross resistance of Escherichia coli to fifteen antibiotics. J Bacteriol. 1952;64(4):489–99.
63. Ardell SM, Kryazhimskiy S. The population genetics of collateral resistance and sensitivity. elife. 2021;10:e73250.
64. Chevereau G, et al. Quantifying the determinants of evolutionary dynamics leading to drug Resistance. PLoS Biol. 2015;13(11):e1002299.
65. Wee GN, et al. Phenotypic convergence of bacterial adaption to sub-lethal antibiotic treatment. Front Cell Infect Microbiol. 2022;12:913415.
66. Mehta HH, Prater AG, Shamoo Y. Using experimental evolution to identify druggable targets that could inhibit the evolution of antimicrobial resistance. J Antibiot (Tokyo). 2018;71(2):279–86.
67. Singha B, et al. Metabolic rewiring of mycobacterium tuberculosis upon drug treatment and antibiotics resistance. Meta. 2024;14(1):63.

68. Sanz-García F, et al. Translating eco-evolutionary biology into therapy to tackle antibiotic resistance. Nat Rev Microbiol. 2023;21(10):671–85.
69. Antimicrobial Resistance Questions and Answers. https://www.cdc.gov/antibiotic-use/antibiotic-resistance.html]. Available from: https://www.cdc.gov/antibiotic-use/antibiotic-resistance.html.
70. Shafrin J, et al. Impact of suboptimal or inappropriate treatment on healthcare resource use and cost among patients with uncomplicated urinary tract infection: an analysis of integrated delivery network electronic health records. Antimicrob Resist Infect Control. 2022;11:133.
71. Greenhouse I, et al. Long-term outcomes of inappropriate antibiotic therapy for upper urinary tract infections caused by extended-spectrum beta-lactamase-producing Enterobacteriaceae: a retrospective cohort study. Diagn Microbiol Infect Dis. 2017;89(3):222–9.
72. Richards AR, Linder JA. Behavioral economics and ambulatory antibiotic Stewardship: a narrative review. Clin Ther. 2021;43(10):1654–67.
73. Tillekeratne LG, et al. Antibiotic overuse for acute respiratory tract infections in Sri Lanka: a qualitative study of outpatients and their physicians. BMC Fam Pract. 2017;18(1):37.
74. Biezen R, et al. Dissonant views - GPs' and parents' perspectives on antibiotic prescribing for young children with respiratory tract infections. BMC Fam Pract. 2019;20:46.
75. Antimicrobial consumption dashboard (ESAC-Net). 2010 2010/06/23/T11:33:54+0200; Available from: https://www.ecdc.europa.eu/en/antimicrobial-consumption/surveillance-and-disease-data/databasefiles/47/database.html.
76. NethMap. SWAB; Available from: https://swab.nl/nl/nethmapfiles/49/nethmap.html.
77. Suda KJ, et al. Antibiotic prescriptions in the community by type of provider in the United States, 2005-2010. J Am Pharm Assoc. 2016;56(6):621–626.e1.
78. Duffy E, et al. Antibacterials dispensed in the community comprise 85%-95% of total human antibacterial consumption. J Clin Pharm Ther. 2018;43(1):59–64.
79. Chua K-P, Linder JA. Prevalence of inappropriate antibiotic prescribing by antibiotic among privately and publicly insured non-elderly US patients, 2018. J Gen Intern Med. 2021;36(9):2861–4.
80. Alhomoud F, et al. Self-medication and self-prescription with antibiotics in the Middle East-do they really happen? A systematic review of the prevalence, possible reasons, and outcomes. Int J Infect Dis. 2017;57:3–12.
81. Patel SJ, et al. Antibiotic stewardship in food-producing animals: challenges, Progress, and opportunities. Clin Ther. 2020;42(9):1649–58.
82. Toldra F, Reig M. Growth promotor - an overview. ScienceDirect Topics. 2016:266–9.
83. Roth N, et al. The application of antibiotics in broiler production and the resulting antibiotic resistance in Escherichia coli: a global overview. Poult Sci. 2019;98(4):1791–804.
84. Mavrouli M, et al. Respiratory infections following earthquake-induced Tsunamis: transmission risk factors and lessons learned for disaster risk management. Int J Environ Res Public Health. 2021;18(9):4952.
85. Beltrán MA, et al. Relationship between crowding and community acquired skin and soft tissues infections. Medicina. 2017;77(6):465–8.
86. WHO. Ten threats to global health in 2019. Available from: https://www.who.int/news-room/spotlight/ten-threats-to-global-health-in-2019.
87. Antibiotic resistance. Available from: https://www.who.int/news-room/fact-sheets/detail/antibiotic-resistancefiles/201/antibiotic-resistance.html.
88. COVID-19: U.S. Impact on antimicrobial resistance, special report 2022. 2022, National Center for Emerging and Zoonotic Infectious Diseases.
89. Drug-resistant TB. Available from: https://www.who.int/teams/global-tuberculosis-programme/tb-reports/global-tuberculosis-report-2022/tb-disease-burden/2-3-drug-resistant-tbfiles/211/2-3-drug-resistant-tb.html.
90. Dheda K, et al. The epidemiology, pathogenesis, transmission, diagnosis, and management of multidrug-resistant, extensively drug-resistant, and incurable tuberculosis. Lancet Respir Med. 2017:S2213-2600(17)30079-6.

91. Simner PJ, et al. What's new in antibiograms? Updating CLSI M39 guidance with current trends. J Clin Microbiol. 2022;60(10):e0221021.
92. Metlay JP, et al. Diagnosis and treatment of adults with community-acquired pneumonia. An official clinical practice guideline of the American Thoracic Society and Infectious Diseases Society of America. Am J Respir Crit Care Med. 2019;200(7):e45–67.
93. Berbari EF, et al. 2015 Infectious Diseases Society of America (IDSA) clinical practice guidelines for the diagnosis and treatment of native vertebral osteomyelitis in Adultsa. Clin Infect Dis. 2015;61(6):e26–46.
94. Core Elements of Hospital Antibiotic Stewardship Programs | Antibiotic Use | CDC. 2023 2023/09/07/T07:46:23Z; Available from: https://www.cdc.gov/antibiotic-use/core-elements/hospital.htmlfiles/606/hospital.html.
95. Standard Precautions for All Patient Care | Basics | Infection Control | CDC, 2019.
96. Transmission-Based Precautions | Basics | Infection Control | CDC 2020.
97. Luby SP, et al. Broad approaches to cholera control in Asia: water, sanitation and handwashing. Vaccine. 2020;38(Suppl 1):A110–7.
98. Singer AC, et al. Review of Antimicrobial Resistance in the environment and its relevance to environmental regulators. Front Microbiol. 2016;7:1728.
99. Tseole NP, et al. Barriers and facilitators to Water, Sanitation and Hygiene (WaSH) practices in Southern Africa: a scoping review. PLoS One. 2022;17(8):e0271726.
100. Musoke D, et al. The role of environmental health in preventing antimicrobial resistance in low- and middle-income countries. Environ Health Prev Med. 2021;26(1):100.
101. Centers for Disease, C. and Prevention. Antibiotic resistance threats in the United States, 2019. Centers for Disease Control and Prevention (U.S.); 2019.
102. Murray CJL, et al. Global burden of bacterial antimicrobial resistance in 2019: a systematic analysis. Lancet. 2022;399(10325):629–55.
103. Environment, U.N., Antimicrobial resistance: a global threat.. UNEP - UN Environment Programme, 2020.
104. Canada, P.H.A.o. Antibiotic resistance and risks to human health. [education and awareness] 2014 2014/09/30/; Available from: https://www.canada.ca/en/public-health/services/antibiotic-antimicrobial-resistance/impacts-antibiotic-resistance.htmlfiles/619/impacts-antibiotic-resistance.html.
105. Five reasons to care about antimicrobial resistance (AMR). 2023 2023/09/20/; Available from: https://www.consilium.europa.eu/en/infographics/antimicrobial-resistance/files/621/antimicrobial-resistance.html.
106. Friedman ND, Temkin E, Carmeli Y. The negative impact of antibiotic resistance. Clin Microbiol Infect. 2016;22(5):416–22.
107. One Health Basics | One Health | CDC. 2023 2023/05/10/T02:52:21Z; Available from: https://www.cdc.gov/onehealth/basics/index.htmlfiles/451/index.html.
108. Kim D-W, Cha C-J. Antibiotic resistome from the One-Health perspective: understanding and controlling antimicrobial resistance transmission. Exp Mol Med. 2021;53(3):301–9.
109. Douphrate DI. Animal agriculture and the one health approach. J Agromedicine. 2021;26(1):85–7.
110. Kakkar M, et al. Opportunities for One Health policies to reduce poverty. Revue Scientifique Et Technique (International Office of Epizootics). 2019;38(1):135–44.
111. Scott P, et al. One health in a digital world: technology, data, information and knowledge. In: Yearbook of medical informatics; 2023.
112. Iskandar K, et al. Antibiotic discovery and resistance: the chase and the race. Antibiotics (Basel, Switzerland). 2022;11(2):182.
113. Clancy CJ, Nguyen MH. Buying time: the AMR action fund and the state of antibiotic development in the United States 2020. Open Forum Infect Dis. 2020;7(11):ofaa464.
114. Dheman N, et al. An analysis of antibacterial drug development trends in the United States, 1980–2019. Clin Infect Dis. 2021;73(11):e4444–50.
115. Ahead of HELP Committee Hearing on Superbugs, Bennet, Young Call on Congress to Pass Their PASTEUR Act. Michael Bennet; Available from: https://www.bennet.senate.gov/

public/index.cfm/2023/7/ahead-of-help-committee-hearing-on-superbugs-bennet-young-call-on-congress-to-pass-their-pasteur-actfiles/533/ahead-of-help-committee-hearing-on-superbugs-bennet-young-call-on-congress-to-pass-their-pasteu.html.
116. NHS England » NHS steps up battle against life-threatening infections following successful world-first pilot. Available from: https://www.england.nhs.uk/2023/07/nhs-steps-up-battle-against-life-threatening-infections-following-successful-world-first-pilot/files/537/nhs-steps-up-battle-against-life-threatening-infections-following-successful-world-first-pilot.html.
117. Browne K, et al. A new era of antibiotics: the clinical potential of antimicrobial peptides. Int J Mol Sci. 2020;21(19):7047.
118. Baëtz B, et al. Alternatives to fight vancomycin-resistant Staphylococci and Enterococci. Antibiotics (Basel, Switzerland). 2021;10(9):1116.
119. Luong T, Salabarria A-C, Roach DR. Phage therapy in the resistance era: where do we stand and where are we going? Clin Ther. 2020;42(9):1659–80.
120. Suh GA, et al. Considerations for the use of phage therapy in clinical practice. Antimicrob Agents Chemother. 2022;66(3):e0207121.
121. Lee Y, et al. Bezlotoxumab (Zinplava) for Clostridium Difficile Infection. Pharm Therap. 2017;42(12):735–8.
122. Ruzin A, et al. Characterisation of anti-alpha toxin antibody levels and colonisation status after administration of an investigational human monoclonal antibody, MEDI4893, against Staphylococcus aureus alpha toxin. Clin Transl Immunol. 2018;7(1):e1009.
123. Tkaczyk C, et al. Alanine scanning mutagenesis of the MEDI4893 (Suvratoxumab) epitope reduces alpha toxin lytic activity in vitro and Staphylococcus aureus fitness in infection models. Antimicrob Agents Chemother. 2018;62(11):e01033–18.
124. Bhandari V, Suresh A. Next-generation approaches needed to tackle antimicrobial resistance for the development of novel therapies against the deadly pathogens. Front Pharmacol. 2022;13:838092.
125. Dyar OJ, et al. What is antimicrobial stewardship? Clin Microbiol Infect. 2017;23(11):793–8.
126. Standiford HC, et al. Antimicrobial stewardship at a large tertiary care academic medical center: cost analysis before, during, and after a 7-year program. Infect Control Hosp Epidemiol. 2012;33(4):338–45.
127. Baur D, et al. Effect of antibiotic stewardship on the incidence of infection and colonisation with antibiotic-resistant bacteria and Clostridium difficile infection: a systematic review and meta-analysis. Lancet Infect Dis. 2017;17(9):990–1001.
128. Nathwani D, et al. Value of hospital antimicrobial stewardship programs [ASPs]: a systematic review. Antimicrob Resist Infect Control. 2019;8:35.
129. Muratore E, et al. Antimicrobial Stewardship interventions in pediatric oncology: a systematic review. J Clin Med. 2022;11(15):4545.
130. Mardani M, Abolghasemi S, Shabani S. Impact of an antimicrobial stewardship program in the antimicrobial-resistant and prevalence of clostridioides difficile infection and amount of antimicrobial consumed in cancer patients. BMC Res Notes. 2020;13(1):246.
131. Itoh N, et al. Effects of infectious disease consultation and antimicrobial stewardship program at a Japanese cancer center: an interrupted time-series analysis. PLoS One. 2022;17(1):e0263095.
132. Rajar P, et al. Antibiotic Stewardship in premature infants: a systematic review. Neonatology. 2020;117(6):673–86.
133. Lindsay PJ, et al. Antimicrobial Stewardship and intensive care unit mortality: a systematic review. Clin Infect Dis. 2019;68(5):748–56.
134. Feldstein D, Sloane PD, Feltner C. Antibiotic Stewardship programs in nursing homes: a systematic review. J Am Med Dir Assoc. 2018;19(2):110–6.
135. Crespo-Rivas JC, et al. Are antimicrobial stewardship interventions effective and safe in long-term care facilities? A systematic review and meta-analysis. Clin Microbiol Infect. 2021;27(10):1431–8.

136. Jorgensen JH, Ferraro MJ. Antimicrobial susceptibility testing: a review of general principles and contemporary practices. Clin Infect Dis. 2009;49(11):1749–55.
137. Kaprou GD, et al. Rapid methods for antimicrobial resistance diagnostics. Antibiotics (Basel). 2021;10(2):209.
138. TECAST, The European Committee on Antimicrobial Susceptibility Testing, Breakpoint tables for interpretation of MICs and zone diameters. Version 13.1, 2023. http://www.eucast.org.
139. CLSI. Performance standards for antimicrobial susceptibility testing. CLSI supplement M100 Vol. 33 ed. USA: Clinical and Laboratory Standards Institute; 2023.
140. Gajic I, et al. Antimicrobial susceptibility testing: a comprehensive review of currently used methods. Antibiotics (Basel). 2022;11(4):427.
141. Anjum MF, Zankari E, Hasman H. Molecular methods for detection of antimicrobial resistance. Microbiol Spectr. 2017;5(6):1–17.
142. Nandlal L, Perumal R, Naidoo K. Rapid molecular assays for the diagnosis of drug-resistant tuberculosis. Infect Drug Resist. 2022;15:4971–84.
143. WHO. In: Programme GT, editor. Line probe assays for detection of drug-resistant tuberculosis: interpretation and reporting manual for laboratory staff and clinicians. Geneva: World Health Organization; 2022.
144. Septimus EJ. Antimicrobial Resistance: an antimicrobial/diagnostic Stewardship and infection prevention approach. Med Clin North Am. 2018;102(5):819–29.
145. Ann Van den Bruel PT. Forecast diagnostics for antimicrobial resistance (AMR). NIHR Diagnostic Evidence Cooperative Oxford, University of Oxford; 2016.
146. Banerjee R, Patel R. Molecular diagnostics for genotypic detection of antibiotic resistance: current landscape and future directions. JAC Antimicrob Resist. 2023;5(1):dlad018.
147. Hendriksen RS, et al. Using genomics to track global antimicrobial resistance. Front Public Health. 2019;7:242.
148. Heinemann M, et al. In silico genome-scale reconstruction and validation of the Staphylococcus aureus metabolic network. Biotechnol Bioeng. 2005;92(7):850–64.
149. Gil-Gil T, et al. Antibiotic resistance: time of synthesis in a post-genomic age. Comput Struct Biotechnol J. 2021;19:3110–24.
150. Hwang SM, et al. Whole-genome sequencing for investigating a health care-associated outbreak of Carbapenem-resistant Acinetobacter baumannii. Diagnostics (Basel). 2021;11(2):201.
151. Sukhum KV, Diorio-Toth L, Dantas G. Genomic and metagenomic approaches for predictive surveillance of emerging pathogens and antibiotic resistance. Clin Pharmacol Ther. 2019;106(3):512–24.
152. Brazas MD, Hancock RE. Using microarray gene signatures to elucidate mechanisms of antibiotic action and resistance. Drug Discov Today. 2005;10(18):1245–52.
153. Grad YH, et al. Genomic epidemiology of Neisseria gonorrhoeae with reduced susceptibility to cefixime in the USA: a retrospective observational study. Lancet Infect Dis. 2014;14(3):220–6.
154. Grad YH, et al. Genomic epidemiology of Gonococcal resistance to extended-spectrum Cephalosporins, macrolides, and fluoroquinolones in the United States, 2000-2013. J Infect Dis. 2016;214(10):1579–87.
155. Marques AT, et al. Trends in Helicobacter pylori resistance to clarithromycin: from phenotypic to genomic approaches. Microb Genom. 2020;6(3):e000344.
156. Zhu T, et al. Genomic and functional characterization of fecal sample strains of Proteus cibarius carrying two floR antibiotic resistance genes and a multiresistance plasmid-encoded cfr gene. Comp Immunol Microbiol Infect Dis. 2020;69:101427.
157. Wu C, et al. Characterization of florfenicol resistance genes in the coagulase-negative Staphylococcus (CoNS) isolates and genomic features of a multidrug-resistant Staphylococcus lentus strain H29. Antimicrob Resist Infect Control. 2021;10(1):9.
158. Zhang X, et al. Genomic insights into the antibiotic resistance pattern of the tetracycline-degrading bacterium, Arthrobacter nicotianae OTC-16. Sci Rep. 2021;11(1):15638.

159. Boiko I, et al. Genomic epidemiology and antimicrobial resistance determinants of Neisseria gonorrhoeae isolates from Ukraine, 2013-2018. APMIS. 2020;128(7):465–75.
160. Rokney A, et al. WGS-based prediction and analysis of antimicrobial resistance in campylobacter jejuni isolates from Israel. Front Cell Infect Microbiol. 2020;10:365.
161. D'Onofrio V, et al. Epidemiology of colistin-resistant, carbapenemase-producing Enterobacteriaceae and Acinetobacter baumannii in Croatia. Infect Genet Evol. 2020;81:104263.
162. Lee T, et al. A three-year whole genome sequencing perspective of Enterococcus faecium sepsis in Australia. PLoS One. 2020;15(2):e0228781.
163. Xu Z, et al. Comparative proteomic profiling of Methicillin-susceptible and resistant Staphylococcus aureus. Proteomics. 2020;20(2):e1900221.
164. Li L, et al. Mechanisms of polymyxin resistance induced by Salmonella typhimurium in vitro. Vet Microbiol. 2021;257:109063.
165. Gu Y, et al. The evolution of fluoroquinolone resistance in Salmonella under exposure to sub-inhibitory concentration of Enrofloxacin. Int J Mol Sci. 2021;22(22):12218.
166. Subramanian D, Natarajan J. RNA-seq analysis reveals resistome genes and signalling pathway associated with vancomycin-intermediate Staphylococcus aureus. Indian J Med Microbiol. 2019;37(2):173–85.
167. Kroger C, et al. The primary transcriptome, small RNAs and regulation of antimicrobial resistance in Acinetobacter baumannii ATCC 17978. Nucleic Acids Res. 2018;46(18):9684–98.
168. Kittisenachai S, et al. FBPAII and rpoBC, the two novel secreted proteins identified by the proteomic approach from a comparative study between antibiotic-sensitive and antibiotic-resistant helicobacter pylori-associated gastritis strains. Infect Immun. 2021;89(6):e00053.
169. Li W, et al. An integrated quantitative proteomic and metabolomics approach to reveal the negative regulation mechanism of LamB in antibiotics resistance. J Proteome. 2019;194:148–59.
170. Han ML, et al. Comparative metabolomics and Transcriptomics reveal multiple pathways associated with Polymyxin killing in Pseudomonas aeruginosa. mSystems. 2019;4(1):e00149.
171. Wagenlehner FME, Dittmar F. Re: global burden of bacterial antimicrobial resistance in 2019: a systematic analysis. Eur Urol. 2022;82(6):658.
172. Antimicrobial Resistance C. Global burden of bacterial antimicrobial resistance in 2019: a systematic analysis. Lancet. 2022;399(10325):629–55.
173. Klein EY, et al. Global increase and geographic convergence in antibiotic consumption between 2000 and 2015. Proc Natl Acad Sci USA. 2018;115(15):E3463–70.
174. Cosgrove SE. The relationship between antimicrobial resistance and patient outcomes: mortality, length of hospital stay, and health care costs. Clin Infect Dis. 2006;42:82–9.
175. Shrestha P, et al. Enumerating the economic cost of antimicrobial resistance per antibiotic consumed to inform the evaluation of interventions affecting their use. Antimicrob Resist Infect Control. 2018;7:98.
176. Thorpe KE, Joski P, Johnston KJ. Antibiotic-resistant infection treatment costs have doubled since 2002, now exceeding $2 billion annually. Health Aff (Millwood). 2018;37(4):662–9.
177. Chokshi A, et al. Global contributors to antibiotic resistance. J Glob Infect Dis. 2019;11(1):36–42.
178. Prestinaci F, Pezzotti P, Pantosti A. Antimicrobial resistance: a global multifaceted phenomenon. Pathog Glob Health. 2015;109(7):309–18.
179. Sridhar S, et al. Antimicrobial-resistant bacteria in international travelers. Curr Opin Infect Dis. 2021;34(5):423–31.

Chapter 2
A Historical Perspective to Decrypt AMR in *Mycobacterium*: From Classic Discoveries to Genomics Approaches

Juan Calvet-Seral, Estefanía Crespo-Yuste, Alfonso Mendoza-Losana, and Jesús Gonzalo-Asensio

Abbreviations

AMR	antimicrobial resistance
WHO	World Health Organization
MTBC	*Mycobacterium tuberculosis* Complex
TB	tuberculosis
Mtb	*M. tuberculosis*
NTM	Nontuberculous mycobacteria
STR	streptomycin (STR)
PAS	para-aminosalicylic acid
INH	isoniazid
PZA	pyrazinamide
EMB	ethambutol
RIF	rifampicin
MOX	moxifloxacin

J. Calvet-Seral
Grupo de Genética de Micobacterias, Departamento de Microbiología, Facultad de Medicina, Universidad de Zaragoza IIS-Aragón, Zaragoza, Spain

CIBER Enfermedades Respiratorias, Instituto de Salud Carlos III, Madrid, Spain

Biomedical Sciences and Engineering Laboratory, Departmento de Bioingeniería, Universidad Carlos III de Madrid, Madrid, Spain

E. Crespo-Yuste · J. Gonzalo-Asensio (✉)
Grupo de Genética de Micobacterias, Departamento de Microbiología, Facultad de Medicina, Universidad de Zaragoza IIS-Aragón, Zaragoza, Spain

CIBER Enfermedades Respiratorias, Instituto de Salud Carlos III, Madrid, Spain
e-mail: jagonzal@unizar.es

A. Mendoza-Losana
Biomedical Sciences and Engineering Laboratory, Departmento de Bioingeniería, Universidad Carlos III de Madrid, Madrid, Spain

V. Soni, A. S. Akhade (eds.), *Antimicrobial Resistance: Factors to Findings*,
https://doi.org/10.1007/978-3-031-65986-7_2

LVX	levofloxacin
BDQ	bedaquiline
LZD	linezolid
MDR-TB	Multidrug-resistant TB
XDR-TB	Extensively drug-resistant TB
DST	Drug susceptibility testing
NGS	next-generation sequencing
WGS	whole genome sequencing
GWAS	genome-wide association study

2.1 Introduction: AMR Spread in *Mycobacterium tuberculosis* and Nontuberculous Mycobacteria

2.1.1 Tuberculosis Infections and the Alarming Emergence of Drug Resistance

Tuberculosis (TB) has plagued humankind since antiquity, and it historically remains the leading cause of mortality among bacterial, viral, and parasitic infections worldwide [93]. Although TB is curable and preventable, it still causes 1.3 million deaths globally, surpassing AIDS and malaria [147].

As per the last 2022 World Health Organization (WHO) report, approximately 10.6 million individuals developed TB in 2021, resulting in 1.6 million deaths (4000 daily TB deaths) worldwide [143].

TB is an airborne disease caused by bacteria from the *Mycobacterium tuberculosis* Complex (MTBC), a group of phylogenetically clonal bacteria capable of infecting humans and animals [99]. Although *M. tuberculosis* (*Mtb*) "sensu stricto" is the primary etiologic agent in humans, other members of the MTBC are also clinically relevant.

Mtb is an intracellular pathogen that infects humans as its natural host [48]. Infection occurs when the bacilli reach the lungs after inhalation of aerosols expelled by individuals with active pulmonary TB. *Mtb* bacilli are then engulfed by alveolar macrophages, which ultimately result in the activation and recruitment of other immune components, provoking a granuloma formation which usually contains TB progression. The immune suppression or other yet unknown factors lead to *Mtb* replication outside the granulomatous lesions, resulting in active TB disease [25]. Pulmonary TB is the most prevalent form of the disease, but *Mtb* can spread and cause extrapulmonary TB in virtually any part of the body [90].

The first drugs targeting TB were discovered in the mid-twentieth century, including streptomycin (STR), para-aminosalicylic acid (PAS), isoniazid (INH), pyrazinamide (PZA), ethambutol (EMB), and rifampicin (RIF). However, as Alexander Fleming anticipated in 1945, the inappropriate use of antibiotics would lead to the immediate emergence of drug-resistant strains. Accordingly, the initial optimism for effective TB treatment was overshadowed by the rise of AMR [90].

Treatment of resistant TB involves the use of second-line compounds, which show higher toxicity and reduced efficacy, representing a major challenge for patients and healthcare systems [144]. Drug-resistant TB is classified into five groups by the WHO, based on the resistance profile of the causative *Mtb* strains [90, 140]:

1. INH-resistant TB: *Mtb* strains resistant to INH
2. RIF-resistant TB (RR-TB): *Mtb* strains resistant to RIF
3. Multidrug-resistant TB (MDR-TB): *Mtb* strains resistant to both INH and RIF, the cornerstone drugs for TB treatment
4. Pre-extensively drug-resistant TB (pre-XDR-TB): *Mtb* strains that fulfill the definition of MDR/RR-TB and are also resistant to second-line anti-TB fluoroquinolone drugs such as moxifloxacin (MOX) or levofloxacin (LVX)
5. Extensively drug-resistant TB (XDR-TB): *Mtb* strains that fulfill the definition of MDR/RR-TB and are also resistant to any fluoroquinolone (MOX or LVX), plus at least one of the recently approved drugs bedaquiline (BDQ) or linezolid (LZD)

Globally, the estimated number of people who developed MDR-TB or RR-TB (MDR/RR-TB) each year remained relatively stable between 2015 and 2020. However, there were an estimated 450,000 incident cases in 2021—an increase of 3.1% relative to 2020—possibly linked to the impact of the COVID-19 pandemic on TB detection and treatment [143].

Treatments for patients with MDR/RR-TB are far from satisfactory. Compared to drug-susceptible TB treatments, these regimens require longer treatment duration, higher pill burden, and the use of medicines with a higher toxicity profile. As a result, patients suffer significant adverse events, and almost 15% of patients with MDR/RR-TB still die from the disease [143].

In 2018, the WHO updated the treatment guidelines for MDR/RR-TB. These guidelines recommended two options for fully oral regimens: either shorter, standardized MDR regimens lasting 9–12 months, or longer, individualized treatment regimens spanning 18–20 months [144]. Concerning XDR-TB, the need for more effective regimens has driven multiple assays with new, current, or repurposed medicines. Two randomized controlled trials assayed a 6-month BDQ, pretomanid, and LZD regimen, allowing the WHO to outline updated recommendations for the treatment of XDR-TB [145].

2.1.2 *Increasing Relevance of Nonttuberculous Mycobacteria in Clinics and Drug Resistance*

The global significance of nontuberculous mycobacteria (NTM) as a source of opportunistic infections is increasing [1]. NTM refers to mycobacterial species other than the MTBC and *Mycobacterium leprae* [88]. Over 200 species of NTM have been discovered, and certain species present notable risks of infection [115].

Different NTM species exhibit distinct phenotypes, including varying growth rates. Some species grow rapidly (e.g., *Mycobacterium abscessus*, *M. chelonae*, and *M. fortuitum*), while others are slow growers (e.g., *M. avium* complex and *M. kansasii*). This diversity of NTM species implies that effective diagnostics measures are crucial [126] for accurately determining appropriate antibiotic susceptibility and drug regimens.

Drug susceptibility testing (DST) is a decisive tool for identifying effective drug regimens against NTM infections [11]. However, discrepancies between in vitro DST results and actual drug efficacy in vivo exist, except for macrolides and amikacin [132]. These discrepancies can be attributed to the lack of standardized laboratory methods, control strategies, and monitoring [67].

The recommended drug regimen for most slow growers typically includes EMB, RIF, and a macrolide, with the addition of amikacin or STR in severe cases. For fast growers, drug therapy is primarily based on in vitro DST results for macrolides, aminoglycosides, fluoroquinolones, amikacin, imipenem, tetracyclines, LZD, and trimethoprim-sulfamethoxazole [126].

Resistance mechanisms have been identified for NTM drug regimens, including macrolides, quinolones, clofazimine, LZD, BDQ, aminoglycosides, sulfonamides, and tetracyclines [126]. Therefore, there is an urgent need to develop novel, safe, and more effective drug regimens for addressing NTM-resistant infections.

2.1.3 Importance of AMR Diagnostics to Prescribe the Correct Treatment for Mycobacterial Diseases

The WHO updated its guidelines for TB diagnosis in 2021, providing a comprehensive summary of the current evidence and recommendations for TB identification [142]. Diagnosis is a major challenge within the cascade of care. Detection of pulmonary active drug-resistant TB typically involves routine clinical assessment and targeted sampling, such as the collection of sputum or tracheobronchial lavage samples for respiratory TB in adult patients [28]. However, it is important to note that these assays do not detect the specific AMR, being necessary to perform DSTs—which are based on phenotypic assays—to determine the bacterial ability to grow in the presence of specific antimicrobials. However, the main limitation of these phenotypic DSTs is the slow growth rate of mycobacteria, resulting in a significant delay between the initial diagnosis and the identification of appropriate treatment.

Since 2015, the WHO has recommended the use of rapid molecular tests for the diagnosis of MDR-TB [148]. New molecular platforms can simultaneously detect both MTBC DNA and RIF resistance in a single test. One of the WHO-endorsed tests is Xpert MTB/RIF, which has revolutionized the screening and confirmation of RIF-resistant TB. Subsequent advances in molecular testing include lowering the detection limit (Xpert MTB/RIF Ultra) thereby improving TB diagnosis and AMR, especially in children [53]. Drug-resistant strains can be further analyzed using the

Xpert MTB/XDR, which screens for AMR mutations against INH, fluoroquinolones, ethionamide, and injectable aminoglycosides [49].

Gene arrays and next-generation sequencing (NGS)-based techniques are gradually replacing probe assays for the detection of TB drug resistance in developed countries [42]. However, the emergence of drug resistance to new and repurposed drugs poses a threat to recent advances in MDR-TB treatment, as molecular tests to detect resistance against these drugs are often unavailable [41]. Additionally, the insufficient resources to detect TB drug resistance in developing countries is even more alarming.

The WHO highlights the threat of AMR in the context of tuberculosis and emphasizes the importance of research as part of the collective effort to address this challenge [146]. With the decreasing cost of whole genome sequencing (WGS), new mutations in drug-resistant isolates are being discovered daily. It is crucial to characterize these new mutations to understand their impact on bacterial biology and, more specifically, to verify their involvement in AMR. By confirming the genotype–phenotype relationship of candidate mutations and ruling out other factors such as drug tolerance—which is based on the ability to develop drug resistance regardless of genetic variations—we could gain insights into new AMR mutations. This knowledge is essential for developing effective diagnostic and treatment strategies [149], which ideally should be implemented in formats suitable for use in both developed and developing countries.

In this chapter, we will provide an overview of the main events in the discovery of AMR determinants in *Mycobacterium*. First, we revise the seminal discoveries of the mechanism of action of the main antituberculous drugs. Then, we summarize the main applications of this knowledge to design molecular tests to detect AMR. Finally, we provide a genomic perspective of the impact of NGS on the identification of AMR determinants in a worldwide level.

2.2 Methods Used to Identify Genetic Determinants Responsible for AMR in Mycobacteria

2.2.1 Historic Events in the Identification of Drug Targets for Antituberculous Drugs

Today, worldwide implementation of NGS and lowering costs in DNA sequencing have revolutionized the discovery of AMR determinants. Genomics has been undoubtedly useful in understanding how resistance determinants emerge, evolve, and spread in the population. Thus, genomics is gaining momentum as a preferred source of data to diagnose and treat mycobacterial infections. Nevertheless, we should not forget those seminal discoveries that contributed to identifying the mechanism of action of antituberculous drugs, and which paved the way for current genomic knowledge.

AMR in mycobacteria began in the mid-1950s, early after the introduction of the first antibiotics to treat mycobacterial infections [46]. Accordingly, the molecular mechanisms of drug resistance against RIF, STR, INH, EMB, PZA, and fluoroquinolones were inferred even before the advent of DNA sequencing in 1977. Here we provide a historical overview of the key discoveries that led to the identification of the molecular targets of the first anti-TB drugs (Fig. 2.1).

2.2.1.1 Rifampicin

In the early 1970s, using partially purified extracts of RNA polymerase complexes, White et al. were able to demonstrate that the RNA-polymerase activity of *Mycobacterium smegmatis* was inhibited in the presence of RIF [139]. Using a similar strategy, Konno et al. also observed that RIF inhibited the activity of RNA polymerases from *M. bovis* BCG [69].

The assignment of a specific gene responsible for RIF resistance was elucidated based on previously identified *rpoB* mutations in *Escherichia coli* resistant to this antibiotic. Using genetic probes derived from the *E. coli rpoB* sequence (and other bacterial species such as *Bacillus subtilis* and various *Streptomyces* spp.), the presence of the *rpoB* ortholog in a cosmid library from the *M. leprae* genome was also identified [36, 59]. In addition, sequencing of the *rpoB* gene from five drug-susceptible and nine drug-resistant *M. leprae* isolates revealed that all RIF-resistant strains had mutations in the *rpoB* gene [60, 127].

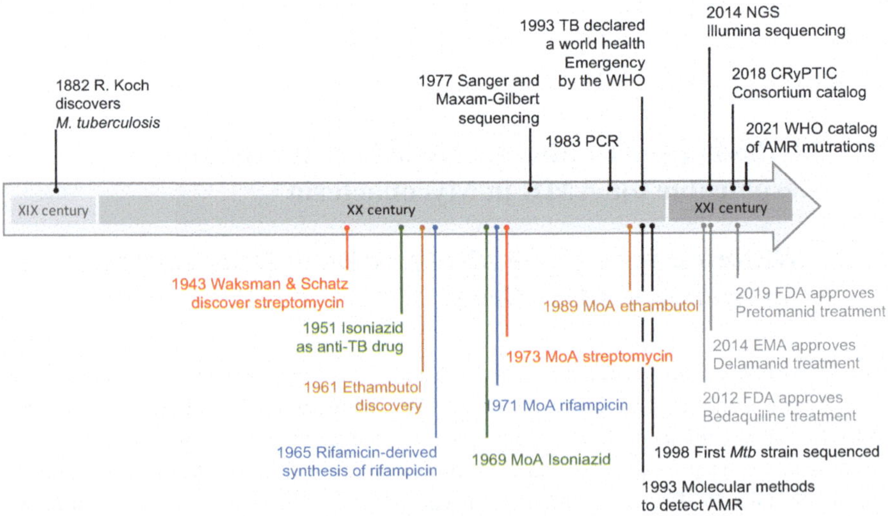

Fig. 2.1 Timeline showing the main events in antimicrobial treatment of diseases caused by *Mycobacterium*. The image shows dates of the main scientific discoveries regarding the main antituberculous drugs, their mechanism of action (MoA), or scientific hallmarks in the treatment of TB

2.2.1.2 Streptomycin

Similar to RIF, the observation that STR was able to inhibit protein synthesis in *Mtb* and that spontaneous-resistant mutants were able to retain protein synthesis activity in the presence of STR led to the suggestion that the mechanism of resistance was related to ribosomal alterations [112]. This was confirmed by testing purified ribosomal subunits from susceptible and resistant mutants of *M. smegmatis* [151].

Based on the observation that *E. coli* STR-resistant mutants carried mutations in the ribosomal S12 protein, encoded by the *rpsL* gene, the ortholog gene was located in mycobacteria. Using conserved coding sequences of the *rpsL* gene in *E. coli* and *Micrococcus luteus*, a PCR strategy was designed to amplify the coding sequence from *Mtb* DNA, leading to a partial amplification and sequencing of the mycobacterial *rpsL* gene. To locate the complete *rpsL* gene, a genomic DNA library from *Mycobacterium intracellulare* was screened with probes based on the previously described partial *rpsL* sequence, which later resulted in the discovery of the *Mtb rpsL* gene. Sequencing the *rpsL* gene from STR-susceptible and STR-resistant *Mtb* strains confirmed that a missense mutation in the *rpsL* gene was responsible for STR resistance only in some *Mtb*-resistant genetic backgrounds, and therefore, other mutations may be contributing to STR resistance in *Mtb* [80].

Additional studies with STR-susceptible and STR-resistant *Mtb* strains led to the identification of new missense mutations not only in the *rpsL* gene, but also in the *rrs* gene, which encodes the 16S rRNA subunit. These polymorphisms responsible for STR resistance in *Mtb* were homologous to resistance mutations previously observed in *E. coli* [43].

2.2.1.3 Isoniazid

Another illustrative example was the assignment of the target for INH. This molecule was described to inhibit mycolic acid biosynthesis, a specific component of the mycobacterial cell wall [152], and therefore, no previous knowledge could be gained from model microorganisms such as *E. coli*.

On one hand, empirical observations that some *Mtb* clinical isolates resistant to INH lost their catalase activity led to the transformation with a shuttle cosmid library from an INH susceptible *Mtb* strain into an *M. smegmatis* resistant strain. Hypersusceptible *M. smegmatis* transformants were confirmed by Southern blot using an oligonucleotide probe based on the amino acid sequence of the hydroperoxidase I enzyme from *E. coli*, encoded by the *katG* gene. When new oligonucleotide probes were designed based on the *Mtb katG* sequence, they failed to hybridize in some *Mtb*-resistant isolates, indicating loss of *katG*, which also explained the loss of catalase activity. However, since this enzyme is not involved in lipid biosynthesis, KatG was not proposed as the target of INH, but as a required enzyme to activate the INH prodrug into an active form [155].

On the other hand, in a complementary approach, the transformation of cosmid libraries from INH-resistant strains of *M. smegmatis* and *M. bovis* into their susceptible parent strains led to the isolation of INH-resistant transformants. Molecular characterization of the cosmids responsible for the resistance phenotype led to the identification of a highly conserved gene involved in lipid biosynthesis, named *inhA*. Further studies in *Mtb* confirmed the enoyl-ACP reductase (*inhA*) as the target of the active form of INH, and missense mutations, mutations in regulatory sequences, or gene duplications, resulted in InhA overexpression and low-level INH resistance [7].

2.2.1.4 Ethambutol

EMB is another specific drug used to treat TB and opportunistic infections of some NTM such as *M. avium*. It has been described that EMB inhibits the incorporation of D-arabinose into the arabinogalactan of the mycobacterial cell wall, preventing the proper attachment of the mycolic acids [124]. The transformation of a genomic library from a resistant strain of *M. avium* into a susceptible strain of *M. smegmatis* led to the identification of a genetic region specifically involved in EMB resistance. This region, containing *embA* and *embB* homologous genes proposed to encode for arabinosyl transferases, was conserved among different mycobacterial strains, including *Mtb* and *M. leprae*, and increased expression levels of *embAB* genes using a multi-copy plasmid correlated with higher EMB resistance compared to physiological expression using a single-copy integrative plasmid [9].

Another study used a genomic library from an *M. smegmatis* EMB-resistant mutant which was transformed into the *wild-type* (WT) *M. smegmatis* strain. The genes identified corresponded to the homologous *embAB* operon from *M. avium*, but *M. smegmatis* contained an additional gene forming the *embCAB* operon, which was also shown to be present in *Mtb* and *M. leprae*. Sequencing of the *emb* operon from resistant strains of *M. smegmatis* and *Mtb* revealed that cumulative mutations in the three genes were responsible for increased levels of resistance and that *M. leprae*, which is intrinsically resistant to EMB, has single nucleotide polymorphisms (SNPs) at the *embB* locus compatible with resistance [129]. Collectively these studies contributed to identifying *embCAB* as the target for EMB in mycobacteria.

2.2.1.5 Pyrazinamide

PZA is a particular case. Despite its lack of in vitro bactericidal activity against *Mtb*, the evidence of its antituberculous activity in mice led to its inclusion in antitubercular regimens [119]. The observation that protein extracts from PZA-resistant mycobacterial strains lacked pyrazinamidase activity led to the conclusion that pyrazinoic acid, the product of the pyrazinamidase reaction, was the active antitubercular compound [68]. To identify the genetic mechanisms of PZA resistance in mycobacteria, the presence of the pyrazinamidase enzyme was analyzed using a transposon library in *E. coli*, which is known to possess this enzymatic activity. This led to the identification of the *E. coli pncA* gene [47]. Subsequently, the *E. coli pncA* gene was used to identify the *Mtb pncA* gene. Paradoxically, this gene is also present in *M. bovis*, which is intrinsically resistant to PZA. Sequencing of the *pncA* gene from *Mtb* and *M. bovis* strains revealed a missense mutation exclusively present in *M. bovis*, explaining PZA resistance in the latter. Further supporting the activation of the PZA prodrug by PncA, some resistant *Mtb* strains also contained mutations in this gene. Introduction of the *pncA* gene from a susceptible strain into *M. bovis* BCG, or *Mtb pncA*-resistant mutants, restored their susceptibility to the drug [110].

2.2.1.6 Other Mechanisms of Resistance

Later on, other mechanisms of AMR were described in *Mtb* and the fast-growing *M. smegmatis* [85]. They involve direct modification and inactivation of the drug, such as (i) constitutive expression of the β-lactamase BlaC, which hydrolyzes β-lactams [44] and renders *Mtb* intrinsically resistant to this important family of antibiotics; (ii) overexpression of the acetyltransferase Eis in clinical isolates, which acetylates and inactivates kanamycin [153]; or (iii) the mutations in a transcription factor, Rv2887, a transcriptional repressor of the MarR family, which upregulates the expression of Rv0560c, a putative methyltransferase that N-methylates a novel DprE1 inhibitor abrogating its bactericidal activity [138]. Target modification also plays a role in macrolide resistance through the methylation of 23S rRNA by the Erm family of methyltransferases [12]. Active efflux of the antibiotic from the bacterial cytosol to the extra bacterial environment represents another mechanism of resistance which will be later explained.

2.2.2 Application of Genetic Knowledge to the Diagnostic of Drug resistance in the Pre-genomic Era

At the same time that scientists were gaining knowledge about drug targets and resistance mutations, they were also applying this genetic information to develop molecular assays capable of rapidly identifying resistance mutations in clinical isolates. This was made possible by the discovery of DNA sequencing and other molecular biology techniques, such as PCR. The development of molecular DST for screening the drug resistance genotype avoiding the need for phenotypic DST, mainly based on the long-lasting measurement of bacterial growth, proved to be of paramount importance for the proper clinical management of the patient after the initial diagnosis. Thus, molecular strategies have been developed to rapidly identify different loci associated with drug resistance in *Mtb* (Fig. 2.1).

Different methods have been used to develop tests for antimicrobial resistance.

2.2.2.1 PCR-Based Methods

Telenti et al. [128] reported some *rpoB* mutations in RIF-resistant *Mtb* isolates, and they also described a method for rapid screening of these mutations based on PCR single-strand conformation polymorphisms (PCR-SSCP). These single nucleotide polymorphisms could be differentiated by the different electrophoretic mobilities of the single DNA strands, allowing susceptible and resistant genotypes to be distinguished by a molecular test. This method could also be adapted for automated analysis.

The previously described PCR-SSCP method was adapted by several groups mainly to identify *rpoB* mutations responsible for RIF resistance [56, 89, 108, 122]; for the identification of mutations in *pncA,* responsible for PZA resistance [109]; in *gyrA* responsible for fluoroquinolone resistance [121, 125]; in *rpsL* responsible of STR resistance [150], or in *katG*, *inhA*, and *ahpC* genes, responsible for INH resistance [66].

2.2.2.2 Restriction Enzyme Digestion-Based Techniques

Nair et al. [80] developed a different method to detect the mutation in the 43rd codon of *rpsL* responsible for STR resistance in *Mtb*. After amplification of the *rpsL* gene by PCR, the product was digested with the restriction enzyme *Mbo*II, which recognizes and cleaves the susceptible WT sequence but not the resistant allele, allowing genotypic differentiation by agarose gel electrophoresis. Other assays based on PCR amplification coupled with restriction enzyme digestion were established to detect mutations in *katG* from clinical isolates resistant to INH [66].

2.2.2.3 Hybridization-Based Tests

One of the first standardized methods introduced as a commercial kit for the detection of resistance mutations in *Mtb* was the line probe assay (LiPA). This assay is based on complementary hybridization, in which different oligonucleotides targeting different regions of the *Mtb* genome sequence under investigation are immobilized at known positions on a nitrocellulose strip. LiPA strips contain complementary oligonucleotides against WT or resistant alleles allowing discrimination between them. Denatured *Mtb* DNA incubated with the nitrocellulose strips results in complementary hybridization and subsequent visualization of specific band patterns representing the different genomic polymorphisms in the *Mtb* samples. Prior amplification of the *Mtb* target DNA by PCR using biotinylated primers has been shown to improve detection [123].

This methodology was adapted to detect mutations in the *rpoB* gene from patient sputum samples, being able to predict RIF resistance in the clinic with over 97% accuracy, later confirmed by phenotypic DST testing [27]. This assay was commercialized under the name of the "INNO-LiPA Rif TB Assay." No discrepancies were found between LiPA and sequencing results, although LiPA was not always able to confirm the precise mutation present in the sample [102]. Another commercial rapid molecular detection kit is based on the specific activity of the enzyme RNase A, which cleaves unpaired bases of RNA duplexes, making it particularly useful for detecting mutations in ribosomal RNA. Amplification of the target sequence by PCR using primers containing the T7 or SP6 RNA polymerases promoter region at the 5′-end allowed the subsequent in vitro transcription of the RNA sequence to be analyzed. Hybridization with a WT reference complementary RNA sequence and subsequent RNase A digestion allowed the detection of single mutations after electrophoretic visualization. Nash and Inderlied [82] adapted this assay to detect macrolide resistance mutations in the 23S rRNA gene of *M. avium* strains and to detect RIF resistance mutations in the *rpoB* gene of *Mtb* [81].

2.3 Genetic Tools to Deconvolute Genomic Data: Targeted Identification of AMR Determinants in *Mycobacterium*

The availability of the complete genome sequence of *Mtb* since 1998 [21] has exponentially increased scientific knowledge and improved our understanding of drug resistance mechanisms in *Mtb*. In the post-genomic era, sequencing of resistant isolates revealed a large number of potential candidate mutations that may confer drug resistance. At the same time, the development of specific genetic engineering technologies allowed scientists to precisely confirm the phenotype–genotype relationship of these mutations.

2.3.1 Need to Establish Genotype–Phenotype Associations in AMR

The relevance of WGS in the study of AMR has been discussed in several reviews [19, 33, 57, 77, 96]. The consensus of these reviews is consistent with the purpose of this chapter: all the identified mutations predicted to be relevant in silico should be experimentally verified to fully understand their relevance in drug resistance. This applies not only to anti-TB drugs already in clinical use but also to drug candidates under development. These latter candidates should be thoroughly characterized at the molecular level to predict future resistance mutations before their broad clinical application. In this way, we will be prepared to anticipate resistance using molecular DST.

Essentially, an adapted molecular version of Koch's postulates should be followed to confirm the potential resistance phenotypes of the mutations identified by WGS. Accordingly, (1) the candidate mutation responsible for drug resistance should be present in resistant bacteria but absent in susceptible bacteria, and (2) the introduction of the mutation into susceptible bacteria should lead to the development of drug resistance. Translating these postulates into *Mycobacterium* is challenging. While sequencing technologies have greatly improved our knowledge to address the first postulate, for the second one, it is required to introduce foreign genetic modifications into a susceptible bacterium, which is particularly challenging in mycobacteria. It is also important to remember that the availability of genetic tools for *Mtb* is restricted, especially when compared to commonly used genetic model microorganisms such as *E. coli*. Genetic manipulation of NTM is even more limited, although there has been an improvement in recent years [15].

A traditional and direct approach often employed to validate the relationship between a mutant-resistant allele and its phenotype is through the introduction of the mutant allele into a susceptible strain using a plasmid. However, this method creates a merodiploid strain carrying both the mutant and susceptible alleles, potentially confounding phenotypic observations. Moreover, the expression of a gene product at nonphysiological levels may result in aberrant phenotypes that do not accurately represent the clinical context. Therefore, a preferred approach to validate associations between the genetic content and a phenotype in AMR, or any other biological trait under investigation, involves direct genetic modification of the bacterial chromosome. These and other strategies are discussed in the following sections.

Drug resistance mutations in pathogenic mycobacteria have been usually tested by performing genetic modifications directly in drug-susceptible reference strains of the same species (e.g., *Mtb* H37Rv, *Mtb* CDC1551, *M. abscessus* ATCC19977) or by using safer surrogates of the same (e.g., *Mtb* H37Ra) or different (*M. smegmatis*) species. Therefore, in this section, we will review a variety of studies that have used genetic assays to study drug-resistance mutations identified by WGS in different mycobacteria, as well as the pros and cons of these methodologies.

2.3.2 Classic Genetic Vectors to Study AMR

To study the specific contribution of a single mutation to a bacterial phenotype, it is necessary to introduce that specific mutation into a controlled genetic background, usually using laboratory-adapted strains. The pioneering classical genetic tools used to identify AMR determinants in mycobacteria have been already mentioned in previous sections. These methods served as the cornerstone for the development of additional genetic tools to study mycobacterial biology which are summarized below.

The introduction of foreign DNA into *Mycobacterium* is challenging due to the distinctive thick and hydrophobic cell wall of this genus. Thus, the first strategies to introduce DNA into mycobacteria relied on the use of mycobacteriophages, which are capable of injecting the packaged DNA directly into mycobacteria. Cosmids, carrying the *cos* ends of mycobacteriophages, can be packaged into phage particles that are later used to infect the desired mycobacteria. Even though the introduction of cosmid DNA into *Mycobacterium* is biologically efficient, the whole process of constructing and delivering cosmid DNA is cumbersome and has been mostly limited to nonpathogenic mycobacteria. Thus, one of the major advances in the genetic manipulation of mycobacteria was the implementation of electroporation, which allowed the direct transformation of DNA vectors, even into the pathogenic *Mycobacterium*, without the need for mycobacteriophages [116].

Phasmids were one of the first extrachromosomal vectors used in mycobacteria. They consist of genetic components of mycobacteriophages cloned into a cosmid vector capable of replicating as a plasmid in *E. coli*. Phasmids can be used to clone large DNA sequences within the mycobacteria in a stable form [64, 65]. They were successfully used to identify the first INH resistance mutations in mycobacteria. Delivery of genomic libraries from susceptible *Mtb* H37Rv DNA into an *M. smegmatis-resistant* strain [155], or genomic libraries from INH-resistant *M. smegmatis*, and *M. bovis* BCG, into their susceptible parent strains [7], led to the identification of candidate mutations in *katG* and *inhA*, respectively. Phasmids have also been used to identify mutations in *embB* responsible for EMB resistance [9, 129]. However, phasmids are large DNA molecules that present inherent difficulties for genetic manipulation or transformation, prompting the development of smaller vectors.

The discovery of a plasmid replicon in *M. fortuitum*, the pAL5000 origin [72], allowed for the construction of a series of different pAL5000-based shuttle plasmids capable of extrachromosomal replication in mycobacteria and *E. coli* [55, 95]. These vectors allowed for genetic manipulation of the DNA of interest in *E. coli*, which could later be transferred into mycobacteria. Development of integrative plasmids carrying site-specific integration systems from mycobacteriophages, such as L5 [79]; or carrying a homologous sequence to the chromosome, allows integration by homologous recombination, in which a single copy of the foreign DNA is stably introduced into the chromosome.

Plasmids, both extrachromosomal and integrative, allowed specific cloning of drug target genes into mycobacteria for study. Expression of the *pncA* gene using a pAL5000-derived plasmid in *M. bovis* BCG restored PZA susceptibility and confirmed the genotype–phenotype relationship of the WT allele of *pncA* from *Mtb* strains and its role in the susceptibility to PZA [110].

Plasmids have also helped to understand and characterize the natural resistance mechanisms of mycobacteria, such as efflux pumps. Cloning the *MMR* gene from the *Mtb* into *M. smegmatis* increased its resistance to several cationic dyes and erythromycin. This gene encodes an efflux pump that actively extrudes compounds toxic to the bacteria, as the accumulation of a radiolabeled molecule inside *M. smegmatis* cells is dramatically reduced when it contains multiple copies of the *MMR* gene [29]. Similar effects were observed when other predicted efflux pump genes were cloned and overexpressed in mycobacteria [103]. Cloning the *tap* genes from *M. fortuitum* and *M. tuberculosis* into *M. smegmatis* resulted in a slight increase in resistance to tetracycline, but only the *tap* allele from *M. fortuitum* increased resistance levels to STR and gentamycin when expressed in *M. smegmatis* [3]. The *Mtb* *Rv2333c/spt* gene expressed in *M. bovis* BCG using a multicopy plasmid caused an increase in resistance levels to spectinomycin and tetracycline associated with the overexpression of this gene [94].

The use of plasmids to (over)express mutant alleles in wild-type susceptible mycobacteria leads to a phenomenon of heteroploidy, in which both susceptible and resistant alleles are expressed in the same bacterium, and depending on the dominance relationship between the two alleles, the observed phenotype may be resistant or susceptible. For example, Dupont et al. [35] decided to validate that some mutations in the *atpE* gene of *M. abscessus* confer BDQ resistance. When the two desired alleles were cloned using a multicopy plasmid, the resulting recombinants did not show the expected resistant phenotype due to the dominance of the WT susceptible allele. In another example, Hartkoorn et al. [54] identified a mutation in the *Rv0678* gene from spontaneous *Mtb* H37Rv mutants resistant to clofazimine. This gene encodes for a transcriptional repressor that regulates the expression of the neighboring genes *mmpS5* and *mmpL5*, which encode transporter proteins that can act as efflux pumps. *Rv0678* mutants showed an increased expression of *mmpS5* and *mmpL5* and increased clofazimine and BDQ resistance. However, when the mutant allele of *Rv0678* was transformed into a susceptible *Mtb* H37Rv using a multicopy plasmid, an intermediate resistance phenotype was observed as it carried both WT and mutant genotypes.

2.3.3 Gene Inactivation Methods to Study AMR

A commonly used strategy to avoid these heteroploid genotypes is to genetically inactivate, or *knockout*, the WT allele. Inactivation of the target gene can also be used to confirm the role of a gene in AMR resistance. Complementation of the resulting *knockout* with the WT or mutant alleles allows the comparison of the WT-restored phenotype with the variations produced by the genetic mutation(s) under study.

The use of suicide vectors has been critical to understanding bacterial biology and drug resistance phenotypes. These vectors lack a mycobacterial origin of replication and consequently are therefore unable to replicate in mycobacteria. Consequently, suicide vectors rely on the homologous recombination pathway to specifically integrate into the chromosome allowing them to *knock out* the genes into which they have recombined [92, 97]. Another genetic tool developed specifically to manipulate the chromosomes of slow-growing mycobacteria is specialized transduction, which consists of the construction of mycobacteriophages carrying a genomic region containing the desired mutation. Construction of the desired DNA is performed in *E. coli* using shuttle phasmids, and functional mycobacteriophages are produced in *M. smegmatis* after electroporation and selection of the phasmid. Engineered mycobacteriophages introduce the desired mutation into slow-growing mycobacteria by their infectious process, and once inside the bacteria, the mutation is incorporated into the chromosome by allelic exchange via homologous recombination [8].

After 2007, manipulation of the mycobacterial chromosome was facilitated by the laboratory of Graham Hatfull [133] adapting to mycobacteria the recombineering method, originally described for *E. coli* [26]. Recombineering involves the use of proteins from mycobacteriophages that specifically promote the recombination of allelic exchange substrates (AES) into the chromosome. The Che9c mycobacteriophage proteins, GP60 and GP61, are conditionally expressed from replicative plasmids and allow the specific incorporation of AES into the chromosome. The gene *knockout* AES is designed as a linear double-stranded DNA molecule (dsAES) with two long (500–1500 bp) homology arms to the chromosomal *locus* to be edited, flanking an antibiotic cassette, which will be used as a selectable marker of recombinant mycobacteria. Recombineering allows targeted chromosomal recombination to be performed in a single step in both, fast- and slow-growing mycobacteria, yielding desired recombinants faster than with the previous methods. The efficiency of recombineering depends mainly on the transformation efficiency of the electroporated DNA. One of the main technical difficulties of recombineering is related to the construction of the AES. In this context, a complementary method called BAC recombineering has been developed using a Bacterial Artificial Chromosome (BAC) library from *Mtb* H37Rv transformed into *E. coli* [51]. The specific *E. coli* clone carrying the BAC containing the *Mtb* region of interest is then transformed using the *E. coli* recombineering system [26] to specifically replace the *Mtb* H37Rv gene(s) of interest with an antibiotic cassette. This modified BAC is used in a PCR reaction to amplify the AES which can be used for recombineering in mycobacteria [2]. Today, the decreasing costs of DNA synthesis have made it possible to chemically synthesize AES without the need for genetic manipulation in the laboratory.

The development of these chromosome-targeting tools has been crucial in studying the role of specific alleles or mutations in avoiding heteroploidy. Inactivation of the efflux pump gene *spt* using suicide vectors in *M. bovis* BCG led to increased susceptibility of BCG against spectinomycin and tetracycline [94]. Similarly, inactivation of the *lfrA* efflux pump gene in *M. smegmatis* results in increased susceptibility to ciprofloxacin, whereas complementation of KO-*lfrA* strains with a multicopy plasmid carrying the WT genotype restored susceptibility and even increased resistance due to higher gene expression [106].

More recently, Hicks et al. [58] used genome-wide association studies to evaluate polymorphisms involved in drug resistance and tolerance from a set of over 500 resistant clinical isolates from China. They identified the *prpR* gene, which encodes a transcription factor that regulates propionate metabolism, as a potential target involved in multidrug resistance. To test the contribution of some of the different identified mutations, the first step was to generate the *prpR knockout* in *Mtb*. Subsequently, a panel of isogenic mutants was obtained by transformation with integrative plasmids carrying the WT or the different polymorphisms/mutations associated with resistance. Phenotypic evaluation in different carbon sources and infection models revealed that mutations in the *prpR* gene are associated with multidrug resistance/tolerance.

There are multiple examples that follow the "*knock out* plus complementation" strategy, but it is important to remember that this strategy is not always feasible because drug resistance mutations are often present in essential genes that cannot be *knocked out* to obtain viable bacteria. Some examples of genes that could not be studied with this strategy are *dnaA, gyrA, gyrB, rpoB, rpsL, rrs, rrl, inhA, embB, atpE,* and *dprE1* [30]. It is not surprising to find in this list the target genes of anti-tuberculous drugs already either in use or in development, since the chemical inactivation of essential bacterial functions is the basis of antibiotic therapy. In these cases, (over)expression of the mutant allele in a plasmid could allow the establishment of a genotype–phenotype relationship, but this strategy has technical limitations, especially when WT-susceptible alleles are dominant. In these cases, specific replacement of the susceptible alleles in the chromosome is mandatory to link candidate mutations to AMR.

2.3.4 Targeted Introduction of Chromosomal Mutations Into Essential Genes to Study AMR

To study AMR determinants in essential genes, different approaches have been developed to introduce the specific mutation(s) directly into the mycobacterial chromosome. Vilchèze et al. [135] performed the first validation of transferring a point mutation conferring INH resistance using specialized transduction. They used a shuttle phasmid to introduce an AES carrying an *inhA* mutant allele flanked by a hygromycin resistance marker. When *Mtb* was transduced with this cosmid and selected for hygromycin resistance, most recombinants showed INH resistance due to the replacement of the WT allele with the mutant allele contained in the cosmid.

A similar approach could be performed using the recombineering technique, where dsAES carrying the AMR candidate mutation is flanked by an antibiotic cassette. In addition, recombineering also allows the introduction of foreign DNA into the chromosome using single-stranded DNA (ssDNA) molecules as AES (ssAES). These ssAES are designed to anneal as Okazaki fragments in the lagging strand of the replication fork and can contain single mutations, or small insertions or deletions. In Hatfull's laboratory [134], they were able to validate the versatility of this technique by introducing single point mutations in the *gyrA*, *rpoB*, *rpsL,* and *inhA* genes in *M. smegmatis* conferring resistance to ofloxacin, RIF, STR, and INH, respectively; and in *Mtb* by introducing mutations in the *rpoB* and *rpsL* genes.

This strategy has also been applied to new drug candidates against *Mtb*. Ioerger et al. [63] selected eight drug candidates in a high-throughput whole-cell inhibition screen. Spontaneous resistant mutants to the drugs were then isolated to characterize their mode of action. WGS of different resistant colonies to each drug revealed different candidate genes responsible for resistance. Recombineering in susceptible *Mtb*, using ssAES carrying several of the identified mutations, served to validate the results obtained by WGS.

Similarly, after a high-throughput screen, Singh et al. [114] discovered a new chemical class active against *Mtb*. When spontaneous resistant mutants were sequenced, all mutations were found in the *wag31* locus. Different mutations were associated with different levels of resistance and the mutation that conferred the strongest phenotype was subjected to further characterization. Interestingly, the authors demonstrated that when WT and mutant alleles were in heteroploidy, the resulting bacteria exhibited susceptible phenotypes, indicating that the WT *wag31* allele was dominant. Consequently, this is another example of the need for direct chromosomal targeting to establish genotype–phenotype associations. The authors then used ssDNA recombineering, where the ssAES also contained a secondary, additional synonymous mutation, close to the mutated codon. This secondary mutation was useful to ensure that the clones recovered after recombineering did not arise from a spontaneous mutation, since the sequencing of the obtained clones should identify not only the candidate mutation but also the secondary mutation used as a reporter.

Another example of the use of ssDNA recombineering was performed in *M. abscessus*. As a proof of concept, several resistance mutations were introduced into the *atpE* gene of *M. abscessus* conferring bedaquiline resistance. In addition, the authors demonstrated the utility of incorporating multiple synonymous mutations around the mutation of interest acting as a specific genetic barcode. These barcoded secondary mutations were easily recognizable by PCR technologies, allowing confirmation of the *M. abscessus* recombinants in a timely manner, and without the need for sequencing [13].

At this point, it is important to note that the introduction of secondary synonymous mutations linked to the candidate AMR mutation in recombineering-based assays could only be applied to coding regions. Regarding noncoding regions, such as ribosomal RNA subunits, which are the targets of some antimicrobials, this methodology could hardly be used unless specific modifications are introduced.

2.3.5 Other Genetic Tools to Study Essential Genes as Potential Drug Targets

In addition to characterizing putative AMR mutations, genetic tools can also be used to identify potential new targets for drug development. As mentioned above, essential genes are required for *Mtb* viability. Thus, the random inactivation of different genes throughout the genome can be used to predict the essentiality of the different genes from MTBC or NTM bacteria. One of the most widely used techniques in this context is transposon mutagenesis. The construction of transposon libraries and the subsequent sequencing of the resulting population allows the identification of loci where transposons are never inserted, indicating possible essential regions for bacterial survival [30]. Although these genes are virtually impossible to inactivate by constructing classical *knockouts*, their essentiality can be further confirmed by constructing conditional *knockouts* or *knockdowns*. In this context, a relevant technique has been the use of CRISPR interference, which directs the Cas9 complex to block the transcription of the desired target gene(s) [16, 101]. Another complementary technique is to replace the natural promoter of the essential gene with an inducible or repressible promoter, such as the TetON or TetOFF strategy, in which the addition of anhydrotetracycline can induce or repress the expression of the essential gene, respectively [10, 39].

2.4 Genomics to Track the Emergence of AMR in *Mycobacterium*

In recent years, we have witnessed the progressive development of NGS technologies. These, bolstered by advancements in technical and computational power, have enabled the application of WGS to unravel multiple questions about bacterial infections and drug resistance [6, 71, 83, 86] (Fig. 2.1).

WGS is being applied not only to identify TB drug resistance determinants of drugs currently in use or under development but also to study other aspects of AMR that could not be studied before. As illustrative examples, WGS is helping to understand whether the infecting mycobacteria consist of a uniform or heterogeneous population. WGS has also allowed us to understand how resistance evolves within patients and how it spreads to new hosts, differentiating between transmission cases and de novo mutations. During this clinical evolution of AMR, WGS allows us to study the chronological order of acquisition of resistance and compensatory mutations, and how they become fixed in the mycobacterial population [19].

In one of the largest retrospective studies, Walker et al. analyzed 3651 genomes from drug-sensitive and drug-resistant *Mtb* clinical isolates. They successfully associated 232 resistance mutations in 23 genes with phenotypic resistance against first and second-line drugs. This study concluded that drug resistance could be accurately predicted based on interrogation of genomic data, and suggested that most resistance mechanisms are explained by previously known mutations [137]. This study was soon followed by similar, and more ambitious projects, such as the CRyPTIC Consortium and the 100,000 Genomes Project, which focused on resistance against first-line drugs in worldwide distributed *Mtb* clinical strains [4] (Fig. 2.1). The main genomic studies contributing to the understanding of drug-resistant TB were nicely documented by Cohen et al. [19] and are summarized below (Table 2.1).

These and other WGS studies have served as cornerstones to elaborate datasheets about drug resistance mechanisms. The WHO has recently published a comprehensive catalogue of MTBC mutations and their correlation with phenotypic drug resistance [141]. This catalogue serves as a reference standard to comprehend mutations linked to resistance toward first-line and several second-line drugs. The report summarizes the analysis of more than 38,000 isolates from over 40 countries, combining WGS and phenotypic DST data for 13 anti-MTBC drugs. It includes information on more than 17,000 mutations, their frequency, association with a specific resistance, and details of the methods used for their identification [141] (Fig. 2.1). Today, NGS diagnostics, coupled with bioinformatic pipelines for genetic data analysis, offer an accurate prediction of TB drug resistance. This approach will probably become the new standard for resistance detection and therapeutic recommendations, providing more accurate and tailored treatment strategies [52].

As the WHO catalogue of AMR mutations refers to the MTBC, it would be advisable to also extend the data to NTMs due to the increasing emergence of AMR in the latter. In this context, recent studies are paving the way to gain knowledge about the drug susceptibility of NTM based on WGS data [118].

The WHO catalogue has otherwise allowed for simplifying the process of identification of AMR candidate mutations by using targeted NGS of specific chromosomal regions. This technique, also known as amplicon sequencing, allows the direct interrogation of a gene set putatively involved in susceptibility or resistance against first and second-line drugs [23]. Since targeted NGS simplifies the process compared with WGS, it could be more extensively implemented.

Current WGS pipelines in the MTBC exclude up to 10% of the genome due to the repetitive nature of certain genomic regions (i.e., PE and PEE gene families, or insertion sequences). Despite there is not an obvious association between these repetitive regions and the development of AMR, some biological observations support their implications in drug resistance mechanisms. A striking example comes from the clinical observation of a rare outbreak responsible for XDR-TB in the early 1990s in Spain. The strain responsible for the outbreak was not related to a *Mtb* isolate but rather represented a zoonotic transmission of *Mycobacterium bovis*, the causative agent of TB in cattle [61, 100, 105]. Since *M. bovis* is rarely transmitted between humans, the clinical isolate responsible for the outbreak was molecularly characterized, and unexpectedly, it was found to contain 2 copies of the insertion sequence *IS*6110, while *M. bovis* usually contains a unique *IS*6110 element. The second additional copy of this insertion sequence lay upstream of the *phoPR* virulence system [120] and was proposed to act as a mobile promoter enhancing the transcription of these virulence genes [50]. Despite there is not a direct link between the *IS*6110-mediated increased transcription of *phoPR* and the emergence of drug resistance in the MTBC, this clinical observation serves as an argument to include repetitive regions of the genome in future WGS studies.

The *Mtb* H37Rv strain represented the first WGS of an MTBC strain and one of the first bacterial genomes to be sequenced in 1998 [21]. This pioneering discovery fueled the use of H37Rv as a reference for biological and genomic studies. Despite the MTBC being highly clonal, using *Mtb* H37Rv as a reference genome might be inappropriate since this strain does not represent the whole variability of the MTBC, and it contains exclusive polymorphisms, including some mutations in AMR-related genes, as the *gyrA* S95T mutation. In this context, two possible alternatives have been proposed as reference genomes. On one hand, it is possible to assemble a "pangenome" covering the genomic diversity of the known MTBC lineages. This approach could be facilitated by the clonality observed between the MTBC and the absence of horizontal gene transfer between strains. On the other hand, it would be possible to deduce the ancestral genome from which all known MTBC strains

Table 2.1 Summary of the main genomic studies about AMR in *M. tuberculosis*

Description	References
Identifying M. tuberculosis drug-resistance determinants	
Large-scale WGS project: sequencing of 116 genomes from around the globe	[40]
Large-scale WGS project: sequencing of 161 genomes from China	[154]
Analysis of 23 candidate resistance genes from 3651 clinical isolates	[137]
Use of a combination of the correlated evolution test and a GWAS framework to identify drug-resistance-associated mutations in 498 genomes from China and South Africa	[31]
GWAS study of 6465 *M. tuberculosis* clinical isolates from more than 30 countries	[22]
Prediction of first-line-drug susceptibility in a dataset of 10,209 clinical isolates from 16 countries	[4]
Within-patient evolution of resistance	
WGS of nine serial isolates cultured from a single patient over a 42-month period	[38]
Very deep WGS of serial sputum specimens from patients receiving treatment for TB	[131]
Transmission versus de novo evolution of resistance	
Literature review including a meta-analysis of 12 studies published between 2005 and 2014	[87]
WGS of 14 phenotypically diverse strains within the Beijing lineage in South Africa	[62]
Sequencing of more than 400 strains from South Africa	[111]
WGS of a set of 5310 isolates, with the diverse geographical origins, genetic backgrounds, and drug-resistance profiles	[75]
Geographic spread of multidrug-resistance	
Further analysis of geographic trends in MDR strains within the set of 5310 strains from Manson et al. [75]	[18]
Sequencing of 344 patients with XDR-TB, combined with global positioning system coordinates	[84]
Order of acquisition of resistance mutations	
WGS and drug-susceptibility testing on 337 clinical isolates collected in Kwazulu-Natal, South Africa	[17]
WGS of all 252 available clinical isolates from an outbreak in Argentina	[37]
WGS of 5310 isolates with diverse geographical origin, genetic background, and drug-resistance profiles	[75]
Evolution of compensatory and stepping-stone mutations	
Review paper that discusses the evolution of compensatory mutations that can ease fitness effects caused by resistance	[45]
Comparison of the genome sequences of ten clinical rifampicin-resistant isolates to those of the corresponding rifampicin-susceptible isolates from the same individual at an earlier time point	[24]
Large-scale analysis of 1000 strains from Russia	[14]
WGS and drug-susceptibility testing of 337 clinical isolates collected in Kwazulu-Natal, South Africa	[17]
Sequencing of highly resistant TB strains from Central Asia	[78]
GWAS study of 6465 *M. tuberculosis* clinical isolates from more than 30 countries	[22]
Genetically and biochemically characterized strains selected in vitro for ethambutol resistance	[104]

(continued)

Table 2.1 (continued)

Description	References
Understanding mixed infections and spatial heterogeneity within a patient	
WGS for rapid drug-susceptibility testing of a patient with XDR-TB	[70]
Deep WGS of serial sputum isolates within a patient	[74]
Sequencing of samples from post-mortem biopsies from different body sites	[73]
Sequencing of samples biopsied from seven different body sites, as well as pre-treatment and serial sputum samples	[32]
Assessed methods for detecting mixed infections using WGS data from in vitro and in silico artificially mixed *M. tuberculosis* samples	[117]
Bench to bedside with WGS	
Prospective study evaluating the use of WGS for diagnosis	[91]
Sequencing-based detection without culturing	[34]
Evaluation of Oxford Nanopore sequencing for diagnostic or surveillance purposes	[136]

Abbreviations: *GWAS* genome-wide association study, *MDR* multidrug-resistant, *MIC* minimum inhibitory concentration, *PAS* para-aminosalicylic acid, *SNP* single nucleotide polymorphism, *TB* tuberculosis, *XDR* extensively drug-resistant

emerged [77]. Additionally, we can also envisage a future in which decreasing costs in sequencing-by-synthesis technologies allow de novo sequencing of clinical isolates without the need for reference genomes.

Regardless of these limitations, WGS nowadays is crucial in the detection of AMR and has been endorsed by the WHO for both epidemiological and drug resistance studies. Nearly two decades after the introduction of the current NGS platforms, WGS of TB-causing bacteria is increasingly moving from research to the clinic, and consequently, it is essential to establish standardized international practices to ensure consistency, comparability, and validation across laboratories.

2.5 Challenges and Future Recommendations

Understanding the genetic and molecular mechanisms of mycobacterial biology is essential for the development of new therapeutic and diagnostic strategies. It has been known since the 1950s that most drugs are less effective against *Mtb* during infection than in vitro [76]. This may be due to the robust immune response aiming at encapsulating *Mtb*-infected cells in multicellular aggregates called granulomas, and/or to the diverse strategies employed by *Mtb* to generate phenotypically distinct bacterial subpopulations in vivo, each of them with different levels of drug tolerance [20]. Taking together, this represents a technical challenge to identify potential drug targets in vitro whose activity could be readily translated into in vivo efficacy. In addition, the requirements of animal facilities within biosafety laboratories to test drug candidates in vivo make the search for new drugs even more challenging. In this context, cellular models that mimic in vivo situations, such as in vitro 2D and 3D granulomas, have been used to evaluate antimicrobial activity against *Mtb* or to study *Mtb* tolerance to immunomodulatory drugs and could represent an attractive replacement surrogate model for in vivo infections [5, 113]. Organ-on-a-chip models have also been proposed as strategies to replace animal models, with the successful recreation of a lung-on-a-chip to study the early stages of *Mtb* infection [130].

Unlike other priority AMR pathogens, *Mtb* is not known to exchange AMR determinants by horizontal gene transfer, and in addition, genomes of MTBC isolates are considered highly clonal with 99.95% identity. These observations may seem paradoxical in the context of the development of AMR. However, two different *Mtb* strains selected at random would differ by approximately 4000 nucleotides, which explains the opportunity to acquire mutations in antimicrobial target genes. This global heterogeneity among *Mtb* strains could play an important role in the accumulation and spread of AMR, antimicrobial tolerance, treatment failure, and the evolution of resistance [98]. The development of WGS has led to a dramatic increase in the number of available *Mtb* genomes, which today represents a double-edged sword, as knowledge gained from the plethora of genomic data is defiant. In this context, it would be very helpful if WGS data were accompanied by detailed epidemiologic information about drug regimens, adherence, or even physiologic data of the patient. In this way, navigation in WGS data could be used to answer specific biological questions and would facilitate the bioinformatics analyses.

The development of genetic tools must lead to an improvement in our understanding as scientists of the different mechanisms of drug resistance, but more importantly, it must have an impact on TB patients. Understanding the genotype–phenotype relationships should lead to an improvement in the molecular DST and early diagnostic by expanding the current tests, whether they are based on PCR (Xpert and Truenat test), LiPA (Nipro and GenoType test), or targeted NGS (AmpliSeq, Illumina, and Genoscreen test) [33].

In contrast to *Mtb*, there is a need for deeper knowledge about NTM infections. The urgency to discover and validate new therapies against NTM is driven by the increasing number of infections reported worldwide and the ineffectiveness of current treatments. NTM exhibit inherent resistance mechanisms such as hydrophobic cell walls, drug-neutralizing enzymes, efflux pumps, and adaptive mechanisms, making the development of effective antimycobacterial drugs challenging [107]. To accelerate drug discovery, various techniques and models can be used at different stages of the drug development process. These include WGS to identify potential targets, selection of drug candidates, testing on clinical isolates, and adaptation of animal models to understand the mechanism of action of drugs and to select effective treatment strategies. In addition, the implementation of genetic tools in NTM would help not only to gain insight into the biology of these pathogens, but also to establish genotype–phenotype associations between drugs and targets, or between potential AMR mutations and their associated phenotypes.

2.6 Conclusions

The fight against AMR in *Mtb* and NTM should be a priority, not only due to the emergence of acquired resistance to the limited available anti-tubercular drugs, but also because of the challenges of accurately diagnosing, and adhering to treatment, cost, and accessibility of appropriate drugs. The coexistence of comorbidities further compounds these issues. We should not forget that although mycobacterial infections are globally distributed, they mainly affect low and medium-income countries, where the AMR is even more worrying than in developed countries. Consequently, addressing these challenges requires a global approach in all countries to improve diagnostic systems, develop new drugs, promote treatment adherence, and ensure equitable access to medicines.

Mycobacterial infections either caused by tuberculous, or nontuberculous mycobacteria, require long-term treatments with multiple medicines to minimize the emergence of AMR. Most of these medicines were discovered in the mid-twentieth century, and their inadequate use has led to a worrying accumulation of drug-resistant strains. Paradoxically, this growing number of drug-resistant strains is overshadowed by the lack of new antimicrobial discoveries. Even more worrying, the recent introduction of BDQ and nitroimidazoles (delamanid and pretomanid) in the last decade has already resulted in the emergence of the first resistant strains.

The advent of NGS methods, coupled with reducing the cost of DNA sequencing have paved the way for genomics approaches to identify AMR determinants. Different genomic methods have served to either verify previously known resistance polymorphisms or identify yet unknown genome variations conferring resistance. It is tempting to speculate that genomics approaches could be applied to predict the putative resistance determinants of drugs that are under development. This way, the introduction of new antimicrobials in the clinic would benefit from anticipating its resistance mutations by genomics.

Cumulative knowledge gained through almost a century of using antimicrobials against TB has provided researchers with comprehensive catalogs of AMR determinants. Today, molecular tests and genomics allow the rapid identification of drug-resistance mutations without the necessity of laborious and sluggish bacterial cultures. Undoubtedly, the growing utilization of WGS to detect AMR will surely replace other methodologies, but it is important to acknowledge the historic seminal discoveries that lay the foundations of drug resistance in *Mycobacterium*.

References

1. Abubakar I, Gupta RK, Rangaka MX, Lipman M. Update in tuberculosis and nontuberculous mycobacteria 2017. Am J Respir Crit Care Med. 2018;197(10):1248–53. https://doi.org/10.1164/rccm.201801-0106UP.
2. Aguilo N, Gonzalo-Asensio J, Alvarez-Arguedas S, Marinova D, Gomez AB, Uranga S, Spallek R, Singh M, Audran R, Spertini F, Martin C. Reactogenicity to major tuberculosis

antigens absent in BCG is linked to improved protection against Mycobacterium tuberculosis. Nat Commun. 2017;8(May):16085. https://doi.org/10.1038/ncomms16085.

3. Aínsa JA, Blokpoel MCJ, Otal I, Young DB, De Smet KAL, Martín C. Molecular cloning and characterization of Tap, a putative multidrug efflux pump present in Mycobacterium fortuitum and Mycobacterium tuberculosis. J Bacteriol. 1998;180(22):5836–43. https://doi.org/10.1128/jb.180.22.5836-5843.1998.
4. Allix-Béguec C, Arandjelovic I, Bi L, Beckert P, Bonnet M, Bradley P, Cabibbe AM, Cancino-Muñoz I, Caulfield MJ, Chaiprasert A, Cirillo DM, Clifton DA, Comas I, Crook DW, De Filippo MR, de Neeling H, Diel R, Drobniewski FA, Faksri K, et al. Prediction of susceptibility to first-line tuberculosis drugs by DNA sequencing. N Engl J Med. 2018;379(15):1403–15. https://doi.org/10.1056/NEJMoa1800474.
5. Arbués A, Brees D, Chibout SD, Fox T, Kammüller M, Portevin D. TNF-α antagonists differentially induce TGF-β1-dependent resuscitation of dormant-like Mycobacterium tuberculosis. PLoS Pathog. 2020;16(2) https://doi.org/10.1371/JOURNAL.PPAT.1008312.
6. Babiker A, Mustapha MM, Pacey MP, Shutt KA, Ezeonwuka CD, Ohm SL, Cooper VS, Marsh JW, Doi Y, Harrison LH. Use of online tools for antimicrobial resistance prediction by whole-genome sequencing in methicillin-resistant Staphylococcus aureus (MRSA) and vancomycin-resistant enterococci (VRE). J Global Antimicrobi Resist. 2019;19(31):136–43. https://doi.org/10.1016/j.jgar.2019.04.006.
7. Banerjee A, Dubnau E, Quemard A, Balasubramanian V, Um KS, Wilson T, Collins D, De Lisle G, Jacobs WR. inhA, a gene encoding a target for isoniazid and ethionamide in Mycobacterium tuberculosis. Science. 1994;263(5144):227–30. https://doi.org/10.1126/science.8284673.
8. Bardarov S, Bardarov S, Pavelka MS, Sambandamurthy V, Larsen M, Tufariello JA, Chan J, Hatfull G, Jacobs WR. Specialized transduction: an efficient method for generating marked and unmarked targeted gene disruptions in Mycobacterium tuberculosis, M. bovis BCG and M. smegmatis. Microbiology. 2002;148(10):3007–17. https://doi.org/10.1099/00221287-148-10-3007.
9. Belanger AE, Besra GS, Ford ME, Mikusová K, Belisle JT, Brennan PJ, Inamine JM. The embAB genes of Mycobacterium avium encode an arabinosyl transferase involved in cell wall arabinan biosynthesis that is the target for the antimycobacterial drug ethambutol. Proc Natl Acad Sci USA. 1996;93(21):11919–24. https://doi.org/10.1073/pnas.93.21.11919.
10. Boldrin F, Casonato S, Dainese E, Sala C, Dhar N, Palù G, Riccardi G, Cole ST, Manganelli R. Development of a repressible mycobacterial promoter system based on two transcriptional repressors. Nucleic Acids Res. 2010;38(12) https://doi.org/10.1093/nar/gkq235.
11. Brill FHH, Lenz J, Lach C, Radischat N, Paßvogel L, Goroncy-Bermes P, Gabriel H, Steinmann J, Steinhauer K. Improved method for tuberculocidal and mycobactericidal activity testing of disinfectants based on the European Standard EN 14348. J Hosp Infect. 2021;111:176–9. https://doi.org/10.1016/j.jhin.2021.02.008.
12. Buriánková K, Doucet-Populaire F, Dorson O, Gondran A, Ghnassia JC, Weiser J, Pernodet JL. Molecular basis of intrinsic macrolide resistance in the Mycobacterium tuberculosis complex. Antimicrob Agents Chemother. 2004;48(1):143. https://doi.org/10.1128/AAC.48.1.143-150.2004.
13. Calvet-Seral J, Crespo-Yuste E, Mathys V, Rodriguez-Villalobos H, Ceyssens P-J, Martin A, Gonzalo-Asensio J. Targeted chromosomal barcoding establishes direct genotype-phenotype associations for antibiotic resistance in Mycobacterium abscessus. Microbiol Spectr. 2023;11(3):e0534422. https://doi.org/10.1128/spectrum.05344-22.
14. Casali N, Nikolayevskyy V, Balabanova Y, Harris SR, Ignatyeva O, Kontsevaya I, Corander J, Bryant J, Parkhill J, Nejentsev S, Horstmann RD, Brown T, Drobniewski F. Evolution and transmission of drug-resistant tuberculosis in a Russian population. Nat Genet. 2014;46(3):279–86. https://doi.org/10.1038/ng.2878.
15. Chimukuche NM, Williams MJ. Genetic manipulation of non-tuberculosis Mycobacteria. Front Microbiol. 2021;12(February):1–12. https://doi.org/10.3389/FMICB.2021.633510/BIBTEX.

16. Choudhary E, Thakur P, Pareek M, Agarwal N. Gene silencing by CRISPR interference in mycobacteria. Nat Commun. 2015;6(1):6267. https://doi.org/10.1038/ncomms7267.
17. Cohen KA, Abeel T, Manson McGuire A, Desjardins CA, Munsamy V, Shea TP, Walker BJ, Bantubani N, Almeida DV, Alvarado L, Chapman SB, Mvelase NR, Duffy EY, Fitzgerald MG, Govender P, Gujja S, Hamilton S, Howarth C, Larimer JD, et al. Evolution of extensively drug-resistant tuberculosis over four decades: whole genome sequencing and dating analysis of Mycobacterium tuberculosis isolates from KwaZulu-Natal. PLoS Med. 2015;12(9):e1001880. https://doi.org/10.1371/journal.pmed.1001880.
18. Cohen KA, Manson AL, Abeel T, Desjardins CA, Chapman SB, Hoffner S, Birren BW, Earl AM. Extensive global movement of multidrug-resistant M. tuberculosis strains revealed by whole-genome analysis. Thorax. 2019;74(9):882–9. https://doi.org/10.1136/thoraxjnl-2018-211616.
19. Cohen KA, Manson AL, Desjardins CA, Abeel T, Earl AM. Deciphering drug resistance in Mycobacterium tuberculosis using whole-genome sequencing: progress, promise, and challenges. Genome Med. 2019;11(1):45. https://doi.org/10.1186/s13073-019-0660-8.
20. Colangeli R, Arcus VL, Cursons RT, Ruthe A, Karalus N, Coley K, Manning SD, Kim S, Marchiano E, Alland D. Whole genome sequencing of Mycobacterium tuberculosis reveals slow growth and low mutation rates during latent infections in humans. PLoS One. 2014;9(3) https://doi.org/10.1371/JOURNAL.PONE.0091024.
21. Cole ST, Brosch R, Parkhill J, Garnier T, Churcher C, Harris D, Gordon SV, Eiglmeier K, Gas S, Barry CE, Tekaia F, Badcock K, Basham D, Brown D, Chillingworth T, Connor R, Davies R, Devlin K, Feltwell T, et al. Deciphering the biology of Mycobacterium tuberculosis from the complete genome sequence. Nature. 1998;396(6707):190. https://doi.org/10.1038/24206.
22. Coll F, Phelan J, Hill-Cawthorne GA, Nair MB, Mallard K, Ali S, Abdallah AM, Alghamdi S, Alsomali M, Ahmed AO, Portelli S, Oppong Y, Alves A, Bessa TB, Campino S, Caws M, Chatterjee A, Crampin AC, Dheda K, et al. Genome-wide analysis of multi- and extensively drug-resistant Mycobacterium tuberculosis. Nat Genet. 2018;50(2):307–16. https://doi.org/10.1038/s41588-017-0029-0.
23. Colman RE, Anderson J, Lemmer D, Lehmkuhl E, Georghiou SB, Heaton H, Wiggins K, Gillece JD, Schupp JM, Catanzaro DG, Crudu V, Cohen T, Rodwell TC, Engelthaler DM. Rapid drug susceptibility testing of drug-resistant Mycobacterium tuberculosis isolates directly from clinical samples by use of amplicon sequencing: a proof-of-concept study. J Clin Microbiol. 2016;54(8):2058–67. https://doi.org/10.1128/JCM.00535-16.
24. Comas I, Borrell S, Roetzer A, Rose G, Malla B, Kato-Maeda M, Galagan J, Niemann S, Gagneux S. Whole-genome sequencing of rifampicin-resistant Mycobacterium tuberculosis strains identifies compensatory mutations in RNA polymerase genes. Nat Genet. 2012;44(1):106–10. https://doi.org/10.1038/ng.1038.
25. Cooper AM. Cell-mediated immune responses in tuberculosis. Annu Rev Immunol. 2009;27:393–422. https://doi.org/10.1146/annurev.immunol.021908.132703.
26. Datsenko KA, Wanner BL. One-step inactivation of chromosomal genes in Escherichia coli K-12 using PCR products. Proc Natl Acad Sci USA. 2000;97(12):6640–5. https://doi.org/10.1073/pnas.120163297.
27. De Beenhouwer H, Lhiang Z, Jannes G, Mijs W, Machtelinckx L, Rossau R, Traore H, Portaels F. Rapid detection of rifampicin resistance in sputum and biopsy specimens from tuberculosis patients by PCR and line probe assay. Tuber Lung Dis. 1995;76(5):425–30. https://doi.org/10.1016/0962-8479(95)90009-8.
28. De Haas P, Yenew B, Mengesha E, Slyzkyi A, Gashu Z, Lounnas M, Tesfaye E, Bedru A, Tiemersma E, Kremer K, Amare M, Diriba G, Zerihun B, Gudina T, Tegegn B, Bonnet M, Negeri C, Klinkenberg E. The Simple One-Step (SOS) stool processing method for use with the Xpert MTB/RIF assay for a child-friendly diagnosis of tuberculosis closer to the point of care. J Clin Microbiol. 2021;59(8) https://doi.org/10.1128/JCM.00406-21.

29. De Rossi E, Branzoni M, Cantoni R, Milano A, Riccardi G, Ciferri O. mmr, a Mycobacterium tuberculosis gene conferring resistance to small cationic dyes and inhibitors. J Bacteriol. 1998;180(22):6068–71. https://doi.org/10.1128/jb.180.22.6068-6071.1998.
30. Dejesus MA, Gerrick ER, Xu W, Park SW, Long JE, Boutte CC, Rubin EJ, Schnappinger D, Ehrt S, Fortune SM, Sassetti CM, Ioerger TR. Comprehensive essentiality analysis of the Mycobacterium tuberculosis genome via saturating transposon mutagenesis. MBio. 2017;8(1) https://doi.org/10.1128/MBIO.02133-16.
31. Desjardins CA, Cohen KA, Munsamy V, Abeel T, Maharaj K, Walker BJ, Shea TP, Almeida DV, Manson AL, Salazar A, Padayatchi N, O'Donnell MR, Mlisana KP, Wortman J, Birren BW, Grosset J, Earl AM, Pym AS. Genomic and functional analyses of Mycobacterium tuberculosis strains implicate ald in D-cycloserine resistance. Nat Genet. 2016;48(5):544–51. https://doi.org/10.1038/ng.3548.
32. Dheda K, Lenders L, Magombedze G, Srivastava S, Raj P, Arning E, Ashcraft P, Bottiglieri T, Wainwright H, Pennel T, Linegar A, Moodley L, Pooran A, Pasipanodya JG, Sirgel FA, van Helden PD, Wakeland E, Warren RM, Gumbo T. Drug-penetration gradients associated with acquired drug resistance in patients with tuberculosis. Am J Respir Crit Care Med. 2018;198(9):1208–19. https://doi.org/10.1164/rccm.201711-2333OC.
33. Domínguez J, Boeree MJ, Cambau E, Chesov D, Conradie F, Cox V, Dheda K, Dudnyk A, Farhat MR, Gagneux S, Grobusch MP, Gröschel MI, Guglielmetti L, Kontsevaya I, Lange B, van Leth F, Lienhardt C, Mandalakas AM, Maurer FP, et al. Clinical implications of molecular drug resistance testing for Mycobacterium tuberculosis: a 2023 TBnet/RESIST-TB consensus statement. Lancet Infect Dis. 2023; https://doi.org/10.1016/s1473-3099(22)00875-1.
34. Doughty EL, Sergeant MJ, Adetifa I, Antonio M, Pallen MJ. Culture-independent detection and characterisation of Mycobacterium tuberculosis and M. africanum in sputum samples using shotgun metagenomics on a benchtop sequencer. PeerJ. 2014;2:e585. https://doi.org/10.7717/peerj.585.
35. Dupont C, Viljoen A, Thomas S, Roquet-Banères F, Herrmann JL, Pethe K, Kremer L. Bedaquiline inhibits the ATP synthase in mycobacterium abscessus and is effective in infected zebrafish. Antimicrob Agents Chemother. 2017;61(11) https://doi.org/10.1128/AAC.01225-17.
36. Eiglmeier K, Honoré N, Woods SA, Caudron B, Cole ST. Use of an ordered cosmid library to deduce the genomic organization of Mycobacterium leprae. Mol Microbiol. 1993;7(2):197–206. https://doi.org/10.1111/j.1365-2958.1993.tb01111.x.
37. Eldholm V, Monteserin J, Rieux A, Lopez B, Sobkowiak B, Ritacco V, Balloux F. Four decades of transmission of a multidrug-resistant Mycobacterium tuberculosis outbreak strain. Nat Commun. 2015;6(1):7119. https://doi.org/10.1038/ncomms8119.
38. Eldholm V, Norheim G, von der Lippe B, Kinander W, Dahle UR, Caugant DA, Mannsåker T, Mengshoel AT, Dyrhol-Riise AM, Balloux F. Evolution of extensively drug-resistant Mycobacterium tuberculosisfrom a susceptible ancestor in a single patient. Genome Biol. 2014;15(11):490. https://doi.org/10.1186/s13059-014-0490-3.
39. Evans JC, Mizrahi V. The application of tetracyclineregulated gene expression systems in the validation of novel drug targets in Mycobacterium tuberculosis. Front Microbiol. 2015;6(AUG) https://doi.org/10.3389/FMICB.2015.00812.
40. Farhat MR, Shapiro BJ, Kieser KJ, Sultana R, Jacobson KR, Victor TC, Warren RM, Streicher EM, Calver A, Sloutsky A, Kaur D, Posey JE, Plikaytis B, Oggioni MR, Gardy JL, Johnston JC, Rodrigues M, Tang PKC, Kato-Maeda M, et al. Genomic analysis identifies targets of convergent positive selection in drug-resistant Mycobacterium tuberculosis. Nat Genet. 2013;45(10):1183–9. https://doi.org/10.1038/ng.2747.
41. Farooq HZ, Cirillo DM, Hillemann D, Wyllie D, Van der Werf MJ, Ködmön C, Nikolayevskyy V. Limited capability for testing Mycobacterium tuberculosis for susceptibility to new drugs. Emerg Infect Dis. 2021;27(3):985–7. https://doi.org/10.3201/EID2703.204418.
42. Feuerriegel S, Kohl TA, Utpatel C, Andres S, Maurer FP, Heyckendorf J, Jouet A, Badalato N, Foray L, Kamara RF, Conteh OS, Supply P, Niemann S. Rapid genomic first- and second-line

drug resistance prediction from clinical Mycobacterium tuberculosis specimens using Deeplex-MycTB. Eur Respir J. 2021;57(1) https://doi.org/10.1183/13993003.01796-2020.
43. Finken M, Kirschner P, Meier A, Wrede A, Böttger EC. Molecular basis of streptomycin resistance in Mycobacterium tuberculosis: alterations of the ribosomal protein S12 gene and point mutations within a functional 16S ribosomal RNA pseudoknot. Mol Microbiol. 1993;9(6):1239–46. https://doi.org/10.1111/j.1365-2958.1993.tb01253.x.
44. Flores AR, Parsons LM, Pavelka MS. Genetic analysis of the β-lactamases of Mycobacterium tuberculosis and Mycobacterium smegmatis and susceptibility to β-lactam antibiotics. Microbiology. 2005;151(2):521–32. https://doi.org/10.1099/mic.0.27629-0.
45. Fonseca JD, Knight GM, McHugh TD. The complex evolution of antibiotic resistance in Mycobacterium tuberculosis. Int J Infect Dis. 2015;32:94–100. https://doi.org/10.1016/j.ijid.2015.01.014.
46. Fox W, Ellard GA, Mitchison DA. Studies on the treatment of tuberculosis undertaken by the British Medical Research Council tuberculosis units, 1946-1986, with relevant subsequent publications. Int J Tuberc Lung Dis. 1999;3(10 Suppl 2):S231–79. http://www.ncbi.nlm.nih.gov/pubmed/10529902
47. Frothingha R, Meeker-O'Connell WA, Talbot EAS, George JW, Kreuzer KN. Identification, cloning, and expression of the Escherichia coli pyrazinamidase and nicotinamidase gene, pncA. Antimicrob Agents Chemother. 1996;40(6):1426–31. https://doi.org/10.1128/aac.40.6.1426.
48. Gagneux S. Host–pathogen coevolution in human tuberculosis. Philos Trans R Soc B Biol Sci. 2012;367(1590):850–9. https://doi.org/10.1098/rstb.2011.0316.
49. Georghiou SB, Penn-Nicholson A, de Vos M, Macé A, Syrmis MW, Jacob K, Mape A, Parmar H, Cao Y, Coulter C, Ruhwald M, Pandey SK, Schumacher SG, Denkinger CM. Analytical performance of the Xpert MTB/XDR® assay for tuberculosis and expanded resistance detection. Diagn Microbiol Infect Dis. 2021;101(1) https://doi.org/10.1016/J.DIAGMICROBIO.2021.115397.
50. Gonzalo-Asensio J, Malaga W, Pawlik A, Astarie-Dequeker C, Passemar C, Moreau F, Laval F, Daffé M, Martin C, Brosch R, Guilhot C. Evolutionary history of tuberculosis shaped by conserved mutations in the PhoPR virulence regulator. Proc Natl Acad Sci USA. 2014;111(31):11491–6. https://doi.org/10.1073/pnas.1406693111.
51. Gordon SV, Brosch R, Billault A, Garnier T, Eiglmeier K, Cole ST. Identification of variable regions in the genomes of tubercle bacilli using bacterial artificial chromosome arrays. Mol Microbiol. 1999;32(3):643–55. https://doi.org/10.1046/J.1365-2958.1999.01383.X.
52. Grobbel HP, Merker M, Köhler N, Andres S, Hoffmann H, Heyckendorf J, Reimann M, Barilar I, Dreyer V, Hillemann D, Kalsdorf B, Kohl TA, Sanchez Carballo P, Schaub D, Todt K, Utpatel C, Maurer FP, Lange C, Niemann S. Design of Multidrug-Resistant Tuberculosis Treatment Regimens Based on DNA sequencing. Clin Infect Dis. 2021;73(7):1194–202. https://doi.org/10.1093/CID/CIAB359.
53. Günther G, Ruswa N, Keller PM. Drug-resistant tuberculosis: advances in diagnosis and management. Curr Opin Pulm Med. 2022;28(3):211–7. https://doi.org/10.1097/MCP.0000000000000866.
54. Hartkoorn RC, Uplekar S, Cole ST. Cross-resistance between Clofazimine and Bedaquiline through upregulation of MmpL5 in Mycobacterium tuberculosis. Antimicrob Agents Chemother. 2014;58(5):2979–81. https://doi.org/10.1128/AAC.00037-14.
55. Hatfull GF. Genetic transformation of mycobacteria. Trends Microbiol. 1993;1(8):310–4. https://doi.org/10.1016/0966-842X(93)90008-F.
56. He X, Zhuang Y, Li G. Application of PCR-SSCP technique in detection of rpoB gene mutation in rifampin-resistant Mycobacterium tuberculosis. Zhonghua Jie He He Hu Xi Za Zhi = Zhonghua Jiehe He Huxi Zazhi = Chinese J Tuberc Respir Dis. 1996;19(6):338–41. https://pubmed.ncbi.nlm.nih.gov/9596814/

57. Hendriksen RS, Bortolaia V, Tate H, Tyson GH, Aarestrup FM, McDermott PF. Using genomics to track global antimicrobial resistance. Front Public Health. 2019;7:242. https://doi.org/10.3389/FPUBH.2019.00242.
58. Hicks ND, Yang J, Zhang X, Zhao B, Grad YH, Liu L, Ou X, Chang Z, Xia H, Zhou Y, Wang S, Dong J, Sun L, Zhu Y, Zhao Y, Jin Q, Fortune SM. Clinically prevalent mutations in Mycobacterium tuberculosis alter propionate metabolism and mediate multidrug tolerance. Nat Microbiol. 2018;3(9):1032–42. https://doi.org/10.1038/s41564-018-0218-3.
59. Honoré N, Bergh S, Chanteau S, Doucet-Populaire F, Eiglmeir K, Garnier T, Georges C, Launois P, Limpaiboon T, Newton S, Niang K, Portillo P, Ramesh GR, Reddi P, Ridel PR, Sittisombut N, Wu-Hunte S, Cole ST. Nucleotide sequence of the first cosmid from the Mycobacterium leprae genome project: structure and function of the Rif-Str regions. Mol Microbiol. 1993;7(2):207–14. https://doi.org/10.1111/j.1365-2958.1993.tb01112.x.
60. Honore N, Cole ST. Molecular basis of rifampin resistance in Mycobacterium leprae. Antimicrob Agents Chemother. 1993;37(3):414–8. https://doi.org/10.1128/AAC.37.3.414.
61. Iglesias MJ, Ibarz D, Cebollada A, Comín J, Jiménez MS, Vázquez MC, Samper S, Cabezas T, Reyes A, Ruiz I, García P, López MD, Cardeñoso L, de la Calle IJ, Ruiz P, Alados JC, Román J, Villa-Real R, Saavedra J, et al. The value of the continuous genotyping of multidrug resistant tuberculosis over 20 years in Spain. Sci Rep. 2020;10(1):20433. https://doi.org/10.1038/s41598-020-77249-x.
62. Ioerger TR, Feng Y, Chen X, Dobos KM, Victor TC, Streicher EM, Warren RM, Gey van Pittius NC, Van Helden PD, Sacchettini JC. The non-clonality of drug resistance in Beijing-genotype isolates of Mycobacterium tuberculosis from the Western Cape of South Africa. BMC Genomics. 2010;11(1):670. https://doi.org/10.1186/1471-2164-11-670.
63. Ioerger TR, O'Malley T, Liao R, Guinn KM, Hickey MJ, Mohaideen N, Murphy KC, Boshoff HIM, Mizrahi V, Rubin EJ, Sassetti CM, Barry CE, Sherman DR, Parish T, Sacchettini JC. Identification of new drug targets and resistance mechanisms in Mycobacterium tuberculosis. PLoS One. 2013;8(9):1–13. https://doi.org/10.1371/journal.pone.0075245.
64. Jacobs WR, Snapper SB, Tuckinan M, Bloom BR. Mycobacteriophage vector systems. Rev Infect Dis. 1989;11(April):S404–10. https://doi.org/10.1093/clinids/11.Supplement_2.S404.
65. Jacobs WR, Tuckman M, Bloom BR. Introduction of foreign DNA into mycobacteria using a shuttle phasmid. Nature. 1987;327(6122):532–5. https://doi.org/10.1038/327532a0.
66. Kiepiela P, Bishop KS, Smith AN, Roux L, York DF. Genomic mutations in the katG, inhA and aphC genes are useful far the prediction of isoniazid resistance in Mycobacterium tuberculosis isolates from Kwazulu Natal, South Africa. Tuber Lung Dis. 2000;80(1):47–56. https://doi.org/10.1054/tuld.1999.0231.
67. Koh W-J. Nontuberculous Mycobacteria—overview. Microbiol Spectr. 2017;5(1):634. https://doi.org/10.1128/microbiolspec.TNMI7-0024-2016.
68. Konno K, Feldmann FM, McDermott W. Pyrazinamide susceptibility and amidase activity of tubercle bacilli. Am Rev Respir Dis. 1967;95(3):461–9. https://doi.org/10.1164/arrd.1967.95.3.461.
69. Konno K, Oizumi K, Oka S. Mode of action of rifampin on mycobacteria. II. Biosynthetic studies on the inhibition of ribonucleic acid polymerase of Mycobacterium bovis BCG by rifampin and uptake of rifampin- 14 C by Mycobacterium phlei. Am Rev Respir Dis. 1973;107(6):1006–12. https://doi.org/10.1164/arrd.1973.107.6.1006.
70. Köser CU, Bryant JM, Becq J, Török ME, Ellington MJ, Marti-Renom MA, Carmichael AJ, Parkhill J, Smith GP, Peacock SJ. Whole-genome sequencing for rapid susceptibility testing of M. tuberculosis. N Engl J Med. 2013;369(3):290–2. https://doi.org/10.1056/NEJMc1215305.
71. Köser CU, Ellington MJ, Peacock SJ. Whole-genome sequencing to control antimicrobial resistance. Trends Genet. 2014;30(9):401–7. https://doi.org/10.1016/j.tig.2014.07.003.
72. Labidi A, David HL, Roulland-Dussoix D. Resitriction endonuclease mapping and cloning of Mycobacterium fortuitum var. fortuitum plasmid pAL5000. Annales de l'Institut Pasteur / Microbiologie. 1985;136(2):209–15. https://doi.org/10.1016/S0769-2609(85)80045-4.

73. Lieberman TD, Wilson D, Misra R, Xiong LL, Moodley P, Cohen T, Kishony R. Genomic diversity in autopsy samples reveals within-host dissemination of HIV-associated Mycobacterium tuberculosis. Nat Med. 2016;22(12):1470–4. https://doi.org/10.1038/nm.4205.
74. Liu Q, Via LE, Luo T, Liang L, Liu X, Wu S, Shen Q, Wei W, Ruan X, Yuan X, Zhang G, Barry CE, Gao Q. Within patient microevolution of Mycobacterium tuberculosis correlates with heterogeneous responses to treatment. Sci Rep. 2015;5(1):17507. https://doi.org/10.1038/srep17507.
75. Manson AL, Cohen KA, Abeel T, Desjardins CA, Armstrong DT, Barry CE, Brand J, Chapman SB, Cho S-NN, Gabrielian A, Gomez J, Jodals AM, Joloba M, Jureen P, Lee JS, Malinga L, Maiga M, Nordenberg D, Noroc E, et al. Genomic analysis of globally diverse Mycobacterium tuberculosis strains provides insights into the emergence and spread of multidrug resistance. Nat Genet. 2017;49(3):395–402. https://doi.org/10.1038/ng.3767.
76. McCune RM, Tompsett R. Fate of mycobacterium tuberculosis in mouse tissues as determined by the microbial enumeration technique. J Exp Med. 1956;104(5):737–62. https://doi.org/10.1084/jem.104.5.737.
77. Meehan CJ, Goig GA, Kohl TA, Verboven L, Dippenaar A, Ezewudo M, Farhat MR, Guthrie JL, Laukens K, Miotto P, Ofori-Anyinam B, Dreyer V, Supply P, Suresh A, Utpatel C, van Soolingen D, Zhou Y, Ashton PM, Brites D, et al. Whole genome sequencing of Mycobacterium tuberculosis: current standards and open issues. Nat Rev Microbiol. 2019;17(9):533–45. https://doi.org/10.1038/s41579-019-0214-5.
78. Merker M, Barbier M, Cox H, Rasigade J-P, Feuerriegel S, Kohl TA, Diel R, Borrell S, Gagneux S, Nikolayevskyy V, Andres S, Nübel U, Supply P, Wirth T, Niemann S. Compensatory evolution drives multidrug-resistant tuberculosis in Central Asia. elife. 2018;7 https://doi.org/10.7554/eLife.38200.
79. Lee MH, Pascopella L, Jacobs WR, Hatfull GF, Lee MH, Pascopella L, Jacobs WR, Hatfull GF. Site-specific integration of mycobacteriophage L5: integration-proficient vectors for Mycobacterium smegmatis, Mycobacterium tuberculosis, and bacille Calmette-Guerin. Proc Natl Acad Sci USA. 1991;88(8):3111–5. https://doi.org/10.1073/pnas.88.8.3111.
80. Nair J, Rouse DA, Bai G, Morris SL. The rpsL gene and streptomycin resistance in single and multiple drug-resistant strains of Mycobacterium tuberculosis. Mol Microbiol. 1993;10(3):521–7. https://doi.org/10.1111/j.1365-2958.1993.tb00924.x.
81. Nash KA, Gaytan A, Inderlied CB. Detection of rifampin resistance in Mycobacterium tuberculosis by use of a rapid, simple, and specific RNA/RNA mismatch assay. J Infect Dis. 1997;176(2):533–6. https://doi.org/10.1086/517283.
82. Nash KA, Inderlied CB. Rapid detection of mutations associated with macrolide resistance in Mycobacterium avium complex. Antimicrob Agents Chemother. 1996;40(7):1748–50. https://doi.org/10.1128/aac.40.7.1748.
83. Ndagi U, Falaki AA, Abdullahi M, Lawal MM, Soliman ME. Antibiotic resistance: bioinformatics-based understanding as a functional strategy for drug design. RSC Adv. 2020;10(31):18451–68. https://doi.org/10.1039/D0RA01484B.
84. Nelson KN, Shah NS, Mathema B, Ismail N, Brust JCM, Brown TS, Auld SC, Valley Omar S, Morris N, Campbell A, Allana S, Moodley P, Mlisana K, Gandhi NR. Spatial patterns of extensively drug-resistant tuberculosis transmission in KwaZulu-Natal, South Africa. J Infect Dis. 2018; https://doi.org/10.1093/infdis/jiy394.
85. Nguyen L, Thompson CJ. Foundations of antibiotic resistance in bacterial physiology: the mycobacterial paradigm. Trends Microbiol. 2006;14(7):304–12. https://doi.org/10.1016/J.TIM.2006.05.005.
86. NIHR Global Health Research Unit on Genomic Surveillance of AMR. Whole-genome sequencing as part of national and international surveillance programmes for antimicrobial resistance: a roadmap. BMJ Glob Health. 2020;5(11):e002244. https://doi.org/10.1136/bmjgh-2019-002244.

87. Nikolayevskyy V, Kranzer K, Niemann S, Drobniewski F. Whole genome sequencing of Mycobacterium tuberculosis for detection of recent transmission and tracing outbreaks: a systematic review. Tuberculosis. 2016;98:77–85. https://doi.org/10.1016/j.tube.2016.02.009.
88. Nour-Neamatollahie A, Ebrahimzadeh N, Siadat SD, Vaziri F, Eslami M, Akhavan Sepahi A, Khanipour S, Masoumi M, Sakhaee F, Ghazanfari Jajin M, Bahrmand A, Fateh A. Distribution of non-tuberculosis mycobacteria strains from suspected tuberculosis patients by heat shock protein 65 PCR–RFLP. Saudi J Biol Sci. 2017;24(6):1380–6. https://doi.org/10.1016/j.sjbs.2016.02.001.
89. Ohno H, Koga H, Kohno S, Higashiyama Y, Miyazaki Y, Ogawa K, Yanagihara K, Yamamoto Y, Noda T, Miyamoto J, Hashimoto A, Tomono K, Kaku M, Hara K. Evaluation for rapid detection of rifampicin-resistant mycobacterium tuberculosis by polymerase chain reaction-single strand conformation polymorphism. Kekkaku : [Tuberculosis]. 1994;69(12):773–8. https://pubmed.ncbi.nlm.nih.gov/7844934/
90. Pai M, Behr MA, Dowdy D, Dheda K, Divangahi M, Boehme CC, Ginsberg A, Swaminathan S, Spigelman M, Getahun H, Menzies D, Raviglione M. Tuberculosis. Nat Rev Dis Prim. 2016;2(1):16076. https://doi.org/10.1038/nrdp.2016.76.
91. Pankhurst LJ, del Ojo Elias C, Votintseva AA, Walker TM, Cole K, Davies J, Fermont JM, Gascoyne-Binzi DM, Kohl TA, Kong C, Lemaitre N, Niemann S, Paul J, Rogers TR, Roycroft E, Smith EG, Supply P, Tang P, Wilcox MH, et al. Rapid, comprehensive, and affordable mycobacterial diagnosis with whole-genome sequencing: a prospective study. Lancet Respir Med. 2016;4(1):49–58. https://doi.org/10.1016/S2213-2600(15)00466-X.
92. Parish T, Stoker NG. Use of flexible cassette method to generate a double unmarked Mycobacterium tuberculosis tlyA plcABC mutant by gene replacement. Microbiology. 2000;146(8):1969–75. https://doi.org/10.1099/00221287-146-8-1969.
93. Paulson T. A mortal foe. Nature. 2013;502(7470):S2. https://doi.org/10.1038/502S2a.
94. Ramón-García S, Martín C, De Rossi E, Aínsa JA. Contribution of the Rv2333c efflux pump (the Stp protein) from Mycobacterium tuberculosis to intrinsic antibiotic resistance in Mycobacterium bovis BCG. J Antimicrob Chemother. 2007;59(3):544–7. https://doi.org/10.1093/jac/dkl510.
95. Ranes MG, Rauzier J, Lagranderie M, Gheorghiu M, Gicquel B. Functional analysis of pAL5000, a plasmid from Mycobacterium fortuitum: construction of a 'mini' mycobacterium-Escherichia coli shuttle vector. J Bacteriol. 1990;172(5):2793–7. https://doi.org/10.1128/jb.172.5.2793-2797.1990.
96. Raphenya AR, Robertson J, Jamin C, de Oliveira Martins L, Maguire F, McArthur AG, Hays JP. Datasets for benchmarking antimicrobial resistance genes in bacterial metagenomic and whole genome sequencing. Sci Data. 2022;9(1):341. https://doi.org/10.1038/s41597-022-01463-7.
97. Reyrat JM, Berthet FX, Gicquel B. The urease locus of Mycobacterium tuberculosis and its utilization for the demonstration of allelic exchange in Mycobacterium bovis bacillus Calmette- Guerin. Proc Natl Acad Sci USA. 1995;92(19):8768–72. https://doi.org/10.1073/pnas.92.19.8768.
98. Richards JP, Cai W, Zill NA, Zhang W, Ojha AK. Adaptation of Mycobacterium tuberculosis to biofilm growth is genetically linked to drug tolerance. Antimicrob Agents Chemother. 2019;63(11) https://doi.org/10.1128/AAC.01213-19.
99. Riojas MA, McGough KJ, Rider-Riojas CJ, Rastogi N, Hazbón MH. Phylogenomic analysis of the species of the Mycobacterium tuberculosis complex demonstrates that Mycobacterium africanum, Mycobacterium bovis, Mycobacterium caprae, Mycobacterium microti and Mycobacterium pinnipedii are later heterotypic synonyms of Mycob. Int J Syst Evol Microbiol. 2018;68(1):324–32. https://doi.org/10.1099/ijsem.0.002507.
100. Rivero A, Marquez M, Santos J, Pinedo A, Sanchez MA, Esteve A, Samper S, Martin C. High rate of tuberculosis reinfection during a nosocomial outbreak of multidrug-resistant tuberculosis caused by Mycobacterium bovis Strain B. Clin Infect Dis. 2001;32(1):159–61. https://doi.org/10.1086/317547.

101. Rock JM, Hopkins FF, Chavez A, Diallo M, Chase MR, Gerrick ER, Pritchard JR, Church GM, Rubin EJ, Sassetti CM, Schnappinger D, Fortune SM. Programmable transcriptional repression in mycobacteria using an orthogonal CRISPR interference platform. Nat Microbiol. 2017;2(February):1–9. https://doi.org/10.1038/nmicrobiol.2016.274.
102. Rossau R, Traore H, De Beenhouwer H, Mijs W, Jannes G, De Rijk P, Portaels F. Evaluation of the INNO-LiPA Rif. TB assay, a reverse hybridization assay for the simultaneous detection of Mycobacterium tuberculosis complex and its resistance to rifampin. Antimicrob Agents Chemother. 1997;41(10):2093–8. https://doi.org/10.1128/aac.41.10.2093.
103. De Rossi E, Aínsa JA, Riccardi G. Role of mycobacterial efflux transporters in drug resistance: an unresolved question. FEMS Microbiol Rev. 2006;30(1):36–52. https://doi.org/10.1111/j.1574-6976.2005.00002.x.
104. Safi H, Lingaraju S, Amin A, Kim S, Jones M, Holmes M, McNeil M, Peterson SN, Chatterjee D, Fleischmann R, Alland D. Evolution of high-level ethambutol-resistant tuberculosis through interacting mutations in decaprenylphosphoryl-β-D-arabinose biosynthetic and utilization pathway genes. Nat Genet. 2013;45(10):1190–7. https://doi.org/10.1038/ng.2743.
105. Sagasti S, Millán-Lou MI, Soledad Jiménez M, Martín C, Samper S. In-depth analysis of the genome sequence of a clinical, extensively drug-resistant Mycobacterium bovis strain. Tuberculosis. 2016;100(1):46–52. https://doi.org/10.1016/j.tube.2016.06.005.
106. Sander P, De Rossi E, Böddinghaus B, Cantoni R, Branzoni M, Böttger EC, Takiff H, Rodriquez R, Lopez G, Riccardi G. Contribution of the multidrug efflux pump LfrA to innate mycobacterial drug resistance. FEMS Microbiol Lett. 2000;193(1):19–23. https://doi.org/10.1016/S0378-1097(00)00456-0.
107. Saxena S, Spaink HP, Forn-Cuní G. Drug resistance in nontuberculous Mycobacteria: mechanisms and models. Biology. 2021;10(2):96. https://doi.org/10.3390/biology10020096.
108. Scarpellini P, Braglia S, Brambilla AM, Dalessandro M, Cichero P, Gori A, Lazzarin A. Detection of rifampin resistance by single-strand conformation polymorphism analysis of cerebrospinal fluid of patients with tuberculosis of the central nervous system. J Clin Microbiol. 1997;35(11):2802–6. https://doi.org/10.1128/JCM.35.11.2802-2806.1997.
109. Scorpio A, Lindholm-Levy P, Heifets L, Gilman R, Siddiqi S, Cynamon M, Zhang Y. Characterization of pncA mutations in pyrazinamide-resistant Mycobacterium tuberculosis. Antimicrob Agents Chemother. 1997;41(3):540–3. https://doi.org/10.1128/AAC.41.3.540.
110. Scorpio A, Zhang Y. Mutations in pncA, a gene encoding pyrazinamidase/nicotinamidase, cause resistance to the antituberculous drug pyrazinamide in tubercle bacillus. Nat Med. 1996;2(6):662–7. https://doi.org/10.1038/nm0696-662.
111. Shah NS, Auld SC, Brust JCM, Mathema B, Ismail N, Moodley P, Mlisana K, Allana S, Campbell A, Mthiyane T, Morris N, Mpangase P, van der Meulen H, Omar SV, Brown TS, Narechania A, Shaskina E, Kapwata T, Kreiswirth B, Gandhi NR. Transmission of extensively drug-resistant tuberculosis in South Africa. N Engl J Med. 2017;376(3):243–53. https://doi.org/10.1056/NEJMoa1604544.
112. Shaila MS, Gopinathan KP, Ramakrishnan T. Protein synthesis in Mycobacterium tuberculosis H37Rv and the effect of streptomycin in streptomycin-susceptible and -resistant strains. Antimicrob Agents Chemother. 1973;4(3):205–13. https://doi.org/10.1128/AAC.4.3.205.
113. Silva-Mira M, Ekaza E, Breiman A, Asehnoune K, Barros-Aguirre D, Pethe K, Ewann F, Brodin P, Ballell-Pages L, Altare F. High-content screening technology combined with a human granuloma model as a new approach to evaluate the activities of drugs against Mycobacterium tuberculosis. Antimicrob Agents Chemother. 2015;59(1):693–7. https://doi.org/10.1128/AAC.03705-14.
114. Singh V, Dhar N, Pató J, Kolly GS, Korduláková J, Forbak M, Evans JC, Székely R, Rybniker J, Palčeková Z, Zemanová J, Santi I, Signorino-Gelo F, Rodrigues L, Vocat A, Covarrubias AS, Rengifo MG, Johnsson K, Mowbray S, et al. Identification of aminopyrimidine-sulfonamides as potent modulators of Wag31-mediated cell elongation in mycobacteria. Mol Microbiol. 2017;103(1):13–25. https://doi.org/10.1111/mmi.13535.

115. Smirnova T, Ustinova V, Andreevskaya S, Larionova E, Kiseleva E, Chernousova L, Varlamov D, Sochivko D, Ergeshov A. Evaluation of a new assay for nontuberculous mycobacteria species identification in diagnostic material and cultures. Tuberculosis. 2021;130(September):102124. https://doi.org/10.1016/j.tube.2021.102124.
116. Snapper SB, Lugosi L, Jekkel A, Melton RE, Kieser T, Bloom BR, Jacobs WR. Lysogeny and transformation in mycobacteria: stable expression of foreign genes. Proc Natl Acad Sci USA. 1988;85(18):6987–91. https://doi.org/10.1073/pnas.85.18.6987.
117. Sobkowiak B, Glynn JR, Houben RMGJ, Mallard K, Phelan JE, Guerra-Assunção JA, Banda L, Mzembe T, Viveiros M, McNerney R, Parkhill J, Crampin AC, Clark TG. Identifying mixed Mycobacterium tuberculosis infections from whole genome sequence data. BMC Genomics. 2018;19(1):613. https://doi.org/10.1186/s12864-018-4988-z.
118. Solanki P, Lipman M, McHugh TD, Satta G. Whole genome sequencing and prediction of antimicrobial susceptibilities in non-tuberculous mycobacteria. Front Microbiol. 2022;13 https://doi.org/10.3389/fmicb.2022.1044515.
119. Solotorovsky M, Gregory FJ, Ironson EJ, Bugie EJ, O'neill RC, Pfister K. Pyrazinoic Acid Amide-An Agent Active Against Experimental Murine Tuberculosis. Proc Soc Exp Biol Med. 1952;79(4):563–5. https://doi.org/10.3181/00379727-79-19447.
120. Soto CY, Menéndez MC, Pérez E, Samper S, Gómez AB, García MJ, Martín C. IS6110 mediates increased transcription of the phoP virulence gene in a multidrug-resistant clinical isolate responsible for tuberculosis outbreaks. J Clin Microbiol. 2004;42(1):212–9. https://doi.org/10.1128/JCM.42.1.212-219.2004.
121. Sougakoff W, Lemaître N, Cambau E, Szpytma M, Revel V, Jarlier V. Nonradioactive single-strand conformation polymorphism analysis for detection of fluoroquinolone resistance in mycobacteria. Eur J Clin Microbiol Infect Dis. 1997;16(5):395–8. https://doi.org/10.1007/BF01726372.
122. Srivastava LM, Khanna M, Goel A, Jain RC. Genotypic detection of mutations in rifampicin resistant clinical isolates ofMycobacterium tuberculosis. Indian J Clin Biochem. 1997;12(Suppl 1):83–5. https://doi.org/10.1007/BF02873069.
123. Stuyver L, Rossau R, Wyseur A, Duhamel M, Vanderborght B, Van Heuverswyn HV, Maertens G. Typing of hepatitis C virus isolates and characterization of new subtypes using a line probe assay. J Gen Virol. 1993;74(6):1093–102. https://doi.org/10.1099/0022-1317-74-6-1093.
124. Takayama K, Kilburn JO. Inhibition of synthesis of arabinogalactan by ethambutol in Mycobacterium smegmatis. Antimicrob Agents Chemother. 1989;33(9):1493–9. https://doi.org/10.1128/AAC.33.9.1493.
125. Takiff HE, Salazar L, Guerrero C, Philipp W, Huang WM, Kreiswirth B, Cole ST, Jacobs WR, Telenti A. Cloning and nucleotide sequence of Mycobacterium tuberculosis gyrA and gyrB genes and detection of quinolone resistance mutations. Antimicrob Agents Chemother. 1994;38(4):773–80. https://doi.org/10.1128/AAC.38.4.773.
126. Tarashi S, Siadat SD, Fateh A. Nontuberculous mycobacterial resistance to antibiotics and disinfectants: challenges still ahead. In: BioMed Research International, vol. 2022. Hindawi Limited; 2022. https://doi.org/10.1155/2022/8168750.
127. Telenti A, Imboden P, Marchesi F, Lowrie D, Cole ST, Colston MJ, Matter L, Schopfer K, Bodmer T. Detection of rifampicin-resistance mutations in Mycobacterium tuberculosis. Lancet. 1993;341(8846):647–51. https://doi.org/10.1016/0140-6736(93)90417-F.
128. Telenti A, Imboden P, Marchesi F, Schmidheini T, Bodmer T. Direct, automated detection of rifampin-resistant Mycobacterium tuberculosis by polymerase chain reaction and single-strand conformation polymorphism analysis. Antimicrob Agents Chemother. 1993;37(10):2054–8. https://doi.org/10.1128/AAC.37.10.2054.
129. Telenti A, Philipp WJ, Sreevatsan S, Bernasconi C, Stockbauer KE, Wieles B, Musser JM, Jacobs WR. The emb operon, a gene cluster of Mycobacterium tuberculosis involved in resistance to ethambutol. Nat Med. 1997;3(5):567–70. https://doi.org/10.1038/nm0597-567.
130. Thacker VV, Dhar N, Sharma K, Barrile R, Karalis K, McKinney JD. A lung-on-chip model of early Mycobacterium tuberculosis infection reveals an essential role for alveolar

epithelial cells in controlling bacterial growth. elife. 2020;9:1–73. https://doi.org/10.7554/ELIFE.59961.

131. Trauner A, Liu Q, Via LE, Liu X, Ruan X, Liang L, Shi H, Chen Y, Wang Z, Liang R, Zhang W, Wei W, Gao J, Sun G, Brites D, England K, Zhang G, Gagneux S, Barry CE, Gao Q. The within-host population dynamics of Mycobacterium tuberculosis vary with treatment efficacy. Genome Biol. 2017;18(1) https://doi.org/10.1186/s13059-017-1196-0.
132. Van Ingen J, Kuijper EJ. Drug susceptibility testing of nontuberculous mycobacteria. Future Microbiol. 2014;9(9):1095–110. https://doi.org/10.2217/fmb.14.60.
133. van Kessel JC, Hatfull GF. Recombineering in Mycobacterium tuberculosis. Nat Methods. 2007;4(2):147–52. https://doi.org/10.1038/nmeth996.
134. Van Kessel JC, Hatfull GF. Efficient point mutagenesis in mycobacteria using single-stranded DNA recombineering: characterization of antimycobacterial drug targets. Mol Microbiol. 2008;67(5):1094–107. https://doi.org/10.1111/j.1365-2958.2008.06109.x.
135. Vilchèze C, Wang F, Arai M, Hazbón MH, Colangeli R, Kremer L, Weisbrod TR, Alland D, Sacchettini JC, Jacobs WR. Transfer of a point mutation in Mycobacterium tuberculosis inhA resolves the target of isoniazid. Nat Med. 2006;12(9):1027–9. https://doi.org/10.1038/nm1466.
136. Votintseva AA, Bradley P, Pankhurst L, del Ojo Elias C, Loose M, Nilgiriwala K, Chatterjee A, Smith EG, Sanderson N, Walker TM, Morgan MR, Wyllie DH, Walker AS, Peto TEA, Crook DW, Iqbal Z. Same-day diagnostic and surveillance data for tuberculosis via whole-genome sequencing of direct respiratory samples. J Clin Microbiol. 2017;55(5):1285–98. https://doi.org/10.1128/JCM.02483-16.
137. Walker TM, Kohl TA, Omar SV, Hedge J, Del Ojo Elias C, Bradley P, Iqbal Z, Feuerriegel S, Niehaus KE, Wilson DJ, Clifton DA, Kapatai G, Ip CLC, Bowden R, Drobniewski FA, Allix-Béguec C, Gaudin C, Parkhill J, Diel R, et al. Whole-genome sequencing for prediction of Mycobacterium tuberculosis drug susceptibility and resistance: a retrospective cohort study. Lancet Infect Dis. 2015;15(10):1193–202. https://doi.org/10.1016/S1473-3099(15)00062-6.
138. Warrier T, Kapilashrami K, Argyrou A, Ioerger TR, Little D, Murphy KC, Nandakumar M, Park S, Gold B, Mi J, Zhang T, Meiler E, Rees M, Somersan-Karakaya S, Porras-De Francisco E, Martinez-Hoyos M, Burns-Huang K, Roberts J, Ling Y, et al. N -methylation of a bactericidal compound as a resistance mechanism in Mycobacterium tuberculosis. Proc Natl Acad Sci. 2016;113(31):E4523–30. https://doi.org/10.1073/pnas.1606590113.
139. White RJ, Lancini GC, Silvestri LG. Mechanism of action of rifampin on Mycobacterium smegmatis. J Bacteriol. 1971;108(2):737–41. https://doi.org/10.1128/jb.108.2.737-741.1971.
140. WHO. Meeting report of the WHO expert consultation on the definition of extensively drug-resistant tuberculosis. In: Wold Health Organization (Issue October); 2020. https://www.who.int/publications/i/item/meeting-report-of-the-who-expert-consultation-on-the-definition-of-extensively-drug-resistant-tuberculosis.
141. WHO. Catalogue of mutations in Mycobacterium tuberculosis complex and their association with drug resistance; 2021. https://www.who.int/publications/i/item/9789240028173.
142. WHO. WHO operational handbook on tuberculosis. Module 3: diagnosis - rapid diagnostics for tuberculosis detection 2021 update. In: Module 3: Diagnosis Rapid diagnotics for tuberculosis diagnosis; 2021. https://apps.who.int/iris/bitstream/handle/10665/340256/9789240022614-eng.pdf.
143. WHO. Global tuberculosis report 2022. In: Wold Health Organization, vol. 4, Issue 1; 2022. https://www.who.int/teams/global-tuberculosis-programme/tb-reports/global-tuberculosis-report-2022.
144. WHO. Rapid communication: key changes to the treatment of drug-resistant tuberculosis. In: World Health Organization (Issue WHO/UCN/TB/2022.2.); 2022. https://www.who.int/about/licensing.
145. WHO. WHO consolidated guidelines on tuberculosis. Module 4: treatment - drug-resistant tuberculosis treatment, 2022 update. In: WHO Press; 2022.
146. WHO. Global research agenda for antimicrobial resistance in human health; 2021.

147. WHO. Global Tuberculosis Report 2023; 2023.
148. WHO. Rapid diagnostic test and shorter, cheaper treatment signal new hope for multidrug-resistant tuberculosis patients. Geneva: World Health Organization; 2016. https://www.who.int/news/item/12-05-2016-rapid-diagnostic-test-and-shorter-cheaper-treatment-signal-new-hope-for-multidrug-resistant-tuberculosis-patients.
149. Windels EM, Van Den Bergh B, Michiels J. Bacteria under antibiotic attack: different strategies for evolutionary adaptation. PLoS Pathog. 2020;16(5):e1008431. https://doi.org/10.1371/journal.ppat.1008431.
150. Wu X, Zhuang Y, He X. Detection of M. tuberculosis streptomycin-resistant gene. Zhonghua Jie He He Hu Xi Za Zhi = Zhonghua Jiehe He Huxi Zazhi = Chinese J Tubercu Respir Dis. 1996;19(6):342–5. http://www.ncbi.nlm.nih.gov/pubmed/9596815
151. Yamada T, Masuda K, Shoji K, Hori M. Pleiotropic antibiotic resistance mutations associated with ribosomes and ribosomal subunits in Mycobacterium smegmatis. Antimicrob Agents Chemother. 1974;6(1):46–53. https://doi.org/10.1128/AAC.6.1.46.
152. Youatt J. A review of the action of isoniazid. Am Rev Respir Dis. 1969;99(5):729–49. https://doi.org/10.1164/arrd.1969.99.5.729.
153. Zaunbrecher MA, Sikes RD, Metchock B, Shinnick TM, Posey JE. Overexpression of the chromosomally encoded aminoglycoside acetyltransferase eis confers kanamycin resistance in Mycobacterium tuberculosis. Proc Natl Acad Sci USA. 2009;106(47):20004–9. https://doi.org/10.1073/PNAS.0907925106.
154. Zhang H, Li D, Zhao L, Fleming J, Lin N, Wang T, Liu Z, Li C, Galwey N, Deng J, Zhou Y, Zhu Y, Gao Y, Wang T, Wang S, Huang Y, Wang M, Zhong Q, Zhou L, et al. Genome sequencing of 161 Mycobacterium tuberculosis isolates from China identifies genes and intergenic regions associated with drug resistance. Nat Genet. 2013;45(10):1255–60. https://doi.org/10.1038/ng.2735.
155. Zhang Y, Heym B, Allen B, Young D, Cole ST. The catalase-peroxidase gene and isoniazid resistance of Mycobacterium tuberculosis. Nature. 1992;358(6387):591–3. https://doi.org/10.1038/358591A0.

Chapter 3
Deciphering Microbial Regulation of Heterogeneity and Drug Adaptation with Meta-transcriptomics and Other System-Level Tools

Erick Tieu, Braden Griebel, Braden Carroll, Anna Eydinova, and Shuyi Ma

Abbreviations

MIC	minimum inhibitory concentration
ROS	reactive oxygen species
RNS	reactive nitrogen species
DNA	deoxyribonucleic acid
RNA	ribonucleic acid
mRNA	messenger RNA
tRNA	transfer RNA
rRNA	ribosomal RNA
ATP	adenosine triphosphate
TA	toxin–antitoxin
VBNC	viable but nonculturable

E. Tieu · B. Griebel · B. Carroll
Center for Global Infectious Disease Research, Seattle Children's Research Institute, Seattle, WA, USA

Department of Chemical Engineering, University of Washington, Seattle, WA, USA

A. Eydinova
Center for Global Infectious Disease Research, Seattle Children's Research Institute, Seattle, WA, USA

S. Ma (✉)
Center for Global Infectious Disease Research, Seattle Children's Research Institute, Seattle, WA, USA

Department of Chemical Engineering, University of Washington, Seattle, WA, USA

Department of Pediatrics, University of Washington, Seattle, WA, USA

Pathobiology Graduate Program, Department of Global Health, University of Washington, Seattle, WA, USA
e-mail: shuyi.ma@seattlechildrens.org

V. Soni, A. S. Akhade (eds.), *Antimicrobial Resistance: Factors to Findings*,
https://doi.org/10.1007/978-3-031-65986-7_3

DD	differentially detectable
MLS_B	macrolides, lincosamides, and streptogramins B
Erm	erythromycin ribosomal methylation
TCA cycle	tricarboxylic acid cycle
NADH	nicotinamide adenine dinucleotide
SCV	small colony variant
Tn-seq	transposon sequencing
CRISPRi	CRISPR interference
sgRNAs	single guide RNAs
TRIP screen	transcriptional regulator-induced phenotype screen
TF	transcription factor
Abl	Abelson tyrosine kinase
MRSA	methicillin-resistant *Staphylococcus aureus*

Bacterial pathogens encounter diverse and dynamic stresses as they navigate interactions with their hosts, their surrounding environments, and with other microbes during their life cycle. The adaptations that bacteria have evolved to survive under these environmental stresses also contribute to their ability to adapt to antibiotic therapy. Defense mechanisms include enzymatic inactivation of a compound, modification of compound accumulation, and modulation of growth or metabolism to reduce dependency on targeted processes [1]. Ensuring effective treatment of drug-resistant bacteria remains a critical challenge in modern medicine, and understanding the multifaceted mechanisms of drug adaptation could inform the development of more effective therapeutic intervention strategies.

An important contributor to microbial drug adaptation strategies is heterogeneity [2]. We define heterogeneity here as a phenomenon in which the molecular or phenotypic behavior of a bacterial population is variable, resulting in the presence of different bacterial subpopulations. Bacteria have evolved a range of mechanisms that allow them to control the extent of heterogeneity within a population and the adaptive consequences of this heterogeneity. The formation of subpopulations increases the diversity of molecular responses to stress and in turn improves the likelihood that at least a subset of these response strategies would be able to withstand the encountered adversity (i.e., bet-hedging) [3]. A common illustration of bet hedging is the formation of persister cells, in which some members of an isogenic population stochastically enter a dormant phase and become transiently less susceptible to antimicrobial exposure. The greater levels of variation in cellular metabolism and growth rates that persistence induces amongst the cells in the bacterial population increase population-level robustness in survival under diverse conditions because this heterogeneity increases the chances that at least a subset of cells is responding to the environment in ways conducive to survival.

Heterogeneity is commonly thought of as arising from genetic or environmental differences. However, these are not the only ways for bacteria to display phenotypic heterogeneity. Ackermann's review on phenotypic heterogeneity in microorganisms categorized mechanisms of nongenetic and nonenvironmental heterogeneity into four different modalities: stochastic state switching, cell age, oscillations, and cell-to-cell interactions, as described in Fig. 3.1 [4]. The different sources of heterogeneity can cause distinct phenotypes to arise within a population at a greater rate than expected from baseline rates of spontaneous mutation acquisition [5, 6].

Bacterial heterogeneity is an important driver of drug resistance and antibiotic treatment failure. Given the complexity of the molecular mechanisms of

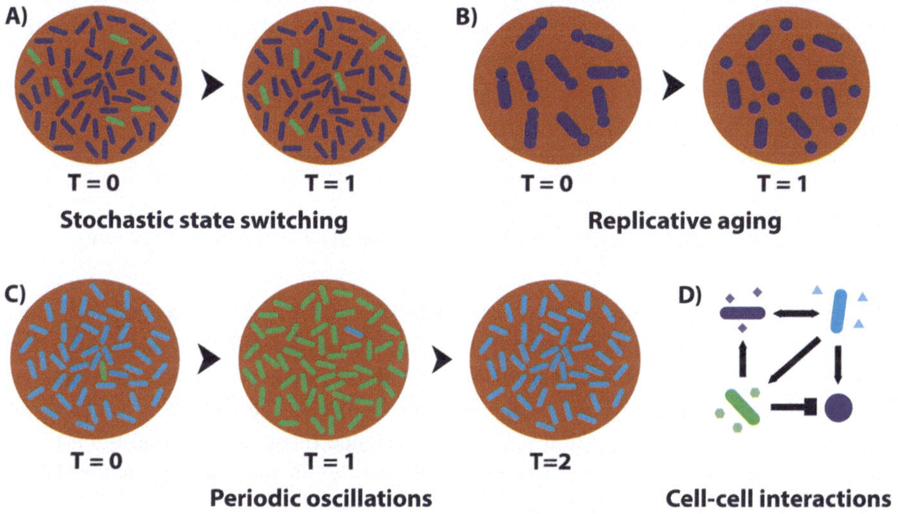

Fig. 3.1 Four types of nongenetic, nonenvironmental modalities of heterogeneity, as described in Ackermann's review [4]. (**a**) Stochastic state switching is the variation in cell state, or molecular makeup, due to the intrinsic variability associated with intracellular chemistry [7]. Here two cell states are represented in purple and green. Cells can switch between states rapidly and independently of environmental conditions, thereby eliminating the time necessary to sense and respond to external stimuli. This trait is especially useful in situations where environmental conditions are rapidly changing. This modality includes the stochastic induction of persister subpopulations, allowing bacterial populations to endure external stresses. (**b**) Replicative aging is an intrinsic cellular property of many bacterial species, caused by asymmetric cell division and subsequent differential accumulation of proteins or deleterious factors within the daughter cells [8]. (**c**) Periodic oscillations describe extrinsic changes that occur at some frequency (e.g., transition from blue state to green state). These oscillations can be induced via environmental forces, such as periodic nutrient availability or circadian regulation of host processes, and can also be regulated internally through cell-cycle-related processes. (**d**) Finally, cell-to-cell interactions between pathogen and host and between members of a polymicrobial community are a source of extrinsic variability that can induce differential subpopulation responses specific to each interaction. In this diagram, an example of a polymicrobial interaction network is presented. Arrows represent metabolites secreted by one bacterium that can be metabolized by another. Flat, "T-headed" arrows represent inhibition (e.g., via microbially derived antibiotic). The different sources of heterogeneity described here can cause distinct phenotypes to arise within a population at a greater rate than expected from baseline rates of spontaneous mutation acquisition [5, 6]

heterogeneity, it is important to elucidate the genetic regulators, the downstream effector processes, and the phenotypic consequences of heterogeneity on drug response. System biology tools, including meta-transcriptomics, single-cell transcriptomics, functional genomic screening, and genome-scale computational modeling can help us to clarify these mechanisms, which will help to inform the next generation of therapies and tools to improve treatment efficacy. In this chapter, we will discuss how specific modalities of heterogeneity can regulate the adaptation of drug response to promote delayed drug killing (i.e., tolerance), subpopulations of tolerant bacteria (i.e., persistence), and acquisition of drug resistance via heritable genetic changes that alter susceptibility to the antibiotic mode of action (Fig. 3.2) [9]. We will also highlight tools that have been developed to study the molecular drivers of these drug-adaptive responses, as well as intervention strategies that have the potential to improve treatment success by countering these drug-adaptive responses.

3.1 Heterogeneity and Drug Response in Complex Stresses

The complex environments and their associated stresses that bacteria encounter during their life cycle help to drive bacterial heterogeneity, and the adaptations that bacteria have evolved to support survival under these diverse conditions also help to promote their adaptive survival under drug treatment. In the following section, we discuss the extrinsic stresses that bacteria experience through host–pathogen interactions and polymicrobial interactions and their roles in modulating bacterial heterogeneity and drug-adaptive responses.

3.1.1 Host–Pathogen Interactions

The human body is a host to a tremendous number of bacteria. In response, we and other animals have evolved a wide array of barriers, composed of chemical and cellular sources, that deter microbes from entering or spreading throughout the body in detrimental ways. Each of these barriers induces changes in bacterial physiology, which often leads to altered population variability, as well as collateral tolerance and persistence to antibiotic exposure.

In some of these infection conditions, bacteria compete with host cells as well as with one another for nutrients. Host cells further elicit sequestration responses to limit access to key nutrients from pathogens as a modality of infection control [11]. Nutrient limitation can be due to macronutrient deficiencies, such as reduced access to carbohydrates for carbon sources, but can result from micronutrient deficiencies, such as trace metals (e.g., zinc, iron, manganese, and magnesium), which are essential for all domains of life [11, 12]. Nutrient limitation is a well-established trigger

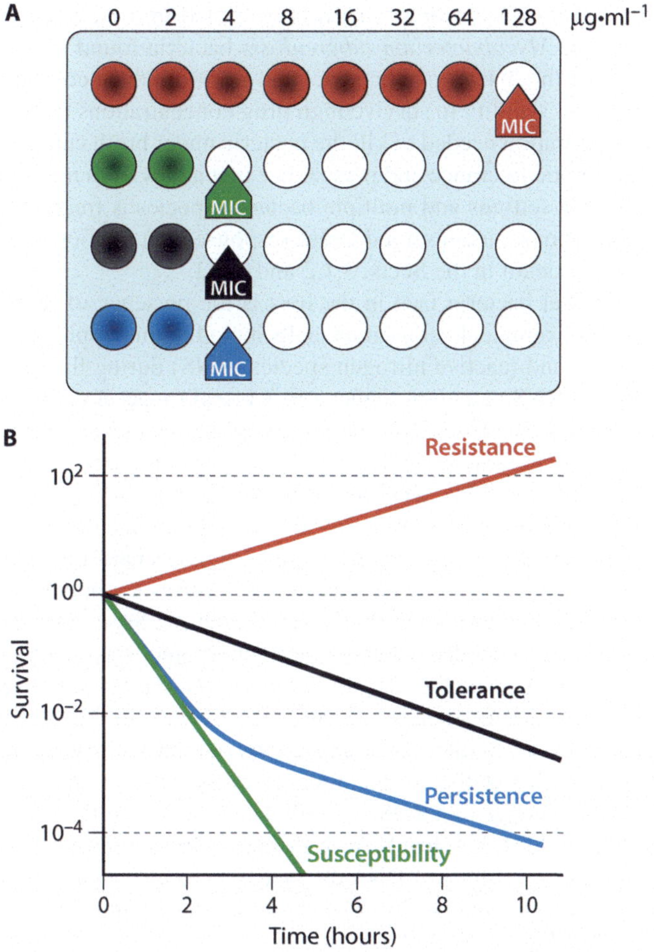

Fig. 3.2 Phenotypic differences in bacteria with resistant (red), susceptible (green), tolerant (black), and persistent (blue) behavior. (**a**) A minimum inhibitory concentration (MIC) assay plate showing bacterial growth (colored) in response to increasing antibiotic concentrations (left to right). Drug-resistant bacteria are characterized by an increased concentration of drugs required to inhibit the growth of the bacteria. Both tolerant and persistent bacterial populations show similar drug inhibitory concentrations as susceptible populations. (**b**) When cultured in the presence of a drug, resistant bacteria can grow uninhibited. Tolerant bacteria experience a reduced rate of lethality following drug exposure. Persistent bacteria are characterized by a subpopulation that is tolerant to drug exposure. When tolerant and persistent bacteria are removed from antibiotic exposure, the population returns to being drug susceptible. (Figure reproduced from [10])

for antibiotic tolerance and persistence as it is typically associated with reduced growth and bacteriostatic dormancy [13–15]. An example of a host-generated nutrient-deficient niche is the caseum found in the lungs of patients infected with tuberculosis. Caseum is a dense lipid-rich substance that results from the breakdown of foamy macrophages, which is restricted in iron, oxygen, and other key catabolites. The *Mycobacterium tuberculosis* bacteria found living in caseum are thought to exist in a bacteriostatic state, and this dormancy contributes significantly to the bacteria's ability to survive high drug concentrations (tens or hundreds of times the concentration needed to kill the bacteria under broth culture conditions) [14, 16]. The antibiotic tolerance and persistence that arises from nutrient limitation in multiple infection settings and multiple bacterial species is frequently regulated by the stringent response, a conserved stress response to starvation conditions that will be discussed in detail in the Sects. 3.1.2 and 3.2.3.

Another stress that bacteria face in the host is the presence of reactive oxygen species (ROS). For example, host immune cells including macrophages and neutrophils produce ROS and reactive nitrogen species (RNS) during their antimicrobial response. ROS and RNS can elicit damage to a broad range of cellular machinery and biomolecules, ranging from DNA damage to protein damage. In response, bacteria have evolved enzymes, including catalases, peroxiredoxins, and superoxide dismutases to directly modulate ROS and RNS levels, not only to control genomic fidelity but also to maintain the function of other biomolecules [17–19]. These enzymes scavenge O_2^- or H_2O_2 and convert them into less harmful molecules [20]. In *M. tuberculosis,* the superoxide dismutase SodA forms a complex with a membrane protein, DoxX, and predicted thiol-oxidoreductase, SseA, to regulate oxidative stress by oxygen radical detoxification and thiol homeostasis using thiol pools, a major target of ROS damage [18, 21]. In *Staphylococcus aureus*, transcription and activity of its superoxide dismutase enzymes, SodA and SodM, are dynamically increased upon exposure to internal or exogenous oxidative stressors, respectively, and the enzymes work together to maintain viability under oxidative stress [22]. Despite the cellular damage that ROS can inflict and the role of ROS in contributing to antibiotic efficacy [23], low levels of ROS may help to improve bacterial stress response by presenting a source of DNA damage which can lead to mutagenesis and evolved resistance formation [24–26]. The processes that regulate this response will be discussed in detail in the Sect. 3.2.1.

Bacterial stress experienced during host infection and immune activation can directly confer tolerance to drug exposure, and this phenotypic tolerance correlates with measurable transcriptional changes in bacterial stress response genes, including those that regulate adaptation to nutrient limitation, oxidative stress, and nitrosative stress [27]. Interestingly, studies have also found that the bacterial stress-responsive gene expression changes elicited by drug exposure during macrophage infection strongly resembled expression changes elicited during infection in macrophages that were activated with pro-inflammatory stimuli [27]. Given this interplay between bacterial response to immune pressure and to drug-imposed stress, and given the capacity for antibiotics to modulate host transcriptional response and immune activation [28], the ability to study concurrently the transcriptional changes of both pathogen and host in response to different drug treatment and infection conditions using tools like dual RNA sequencing [29–32] will help to elucidate the mechanisms of this bacterial drug adaptation in vivo.

3.1.2 Biofilms and Polymicrobial Communities

Bacterial pathogens, both obligate and opportunistic, often exist within polymicrobial communities for at least parts of their life cycle, including in host microbiomes (e.g., humans, animals) and environmental ecosystems (e.g., wastewater, soil). Biofilms are important loci of polymicrobial communities, comprising groups of microbial cells, attached to surfaces that are surrounded by an extracellular matrix [33]. Bacteria that exist in biofilms are significantly more phenotypically resistant and tolerant to drugs than those in planktonic culture [33–35], in part because of the structural properties of biofilms that restrict antibiotic dissemination. Biofilm communities can induce bacterial cell wall remodeling, as in the case of *Streptococcus anginosus,* which upregulates its biosynthesis of cell wall components and induces cell wall thickening in biofilms. This leads to decreased susceptibility of *S. anginosus* to vancomycin in biofilms [36]. Because they can restrict the diffusion of many kinds of molecules, biofilms can also promote microenvironments of nutrient limitation and dormancy, so they can harbor a high proportion of persister bacteria [37], which represents a greater risk toward the development of antimicrobial resistance.

The evolution of antibiotic-resistant bacteria is a naturally occurring result of microbially derived antibiotics being produced in the environment. Over time, bacteria have evolved defenses against microbially derived antibiotics as a means of survival [1]. While antibiotic-resistance genes are valuable for survival in microbial communities replete with microbially derived antibiotics, the heavy use of antibiotics and nonantibiotic pharmaceuticals in human medicine and agriculture has caused many polymicrobial communities to become reservoirs for antibiotic-resistance genes [38–40]. These reservoirs of antibiotic resistance genes, collectively called the "resistome," can serve as a source that mediates the direct exchange of genetic resistance strategies between microbes via horizontal gene transfer. Horizontal gene transfer is a phenomenon whereby bacteria uptake and integrate DNA from external sources into their genomes, and it is one of the main contributors to resistance in clinically relevant bacteria [41]. There are three canonical ways for DNA to enter a new bacterial cell: (1) transformation, in which environmental DNA is taken up; (2) transduction, in which a bacteriophage transfers the DNA; and (3) conjugation, in which DNA transfer is facilitated by direct contact between bacteria. In addition to these canonical mechanisms, other mechanisms have also been discovered, including transport via membrane vesicles, nanotubes, or phage-like transfer agents [42]. Polymicrobial interactions within complex ecosystems can influence each of the modalities of horizontal gene transfer and thus modulate the dynamics of drug resistance.

Within polymicrobial biofilms, horizontal gene transfer is an important mechanism leading to increased antibiotic resistance [35]. Living in biofilms has been shown to increase the rates of conjugation and transformation, and the biofilm environment also increases the stability of plasmids [35]. The bacterial populations within biofilms are not uniformly exchanging DNA; rather, specific subpopulations seem to be responsible for much of the exchange [43, 44]. For example, the exchange of DNA through integron-integrase activity can be triggered under biofilm cohabitation conditions. *Escherichia coli* will control integron-integrase expression when in biofilms via the stringent response, a stress response triggered by nutrient limitation [45]. The stringent response is regulated by the RelA/SpoT superfamily enzymes and the alarmone (p)ppGpp and ultimately regulates a transition to slow growth that can lead to antibiotic tolerance, controlled in part via toxin–antitoxin systems [46] (discussed further in the Sect. 3.2.3). The stringent response regulates integron-integrase expression via two mechanisms, which results in higher mutation rates and more rapid acquisition of antibiotic resistance mutations [47]. First, the stringent response can induce integron-integrase expression by pausing transcription elongation, which leads to DNA damage. This then induces the SOS response (a DNA damage response that will be discussed in greater detail in the Sect. 3.2.1), which ultimately leads to expression repression of the integrase *intI1*. Second, the stringent response can induce integrase expression by repressing the activity of exopolyphosphatase, which leads to the accumulation of polyphosphate. This polyphosphate accumulation forms a complex with the Lon protease, regulating its activity, and ultimately promotes the upregulated expression of *intI1* [45]. The stringent response-induced increase in integrase activity promotes the transfer of antibiotic-resistance genes in biofilm cultures [45]. The importance of the stringent response in mediating dormancy-associated drug tolerance, horizontal gene transfer, and DNA mutability means that inhibition of the stringent response can lead to increased antibiotic effectiveness. For example, the bacterial inhibition efficacy of kanamycin was significantly enhanced by using molecularly imprinted polymer nanoparticles to stop the (p)ppGpp signaling and thus the stringent response [48].

In addition to direct horizontal gene transfer of antibiotic resistance genes from the resistome, the acquisition of resistance to naturally derived antimicrobial agents can lead concomitantly to decreased susceptibility to clinically developed antibiotics. This process occurs through the phenomena of collateral tolerance and collateral resistance, in which the resistance mutation confers protection against multiple antibiotics [49, 50]. This phenomenon can influence bacterial adaptation to drugs of the same class when the resistance mutation interferes with a shared mechanism of action. Collateral resistance and tolerance can also reduce susceptibility to drugs of unrelated classes under circumstances when the resistance mutation impacts drug uptake or other secondary, off-target physiological impacts that modulate overall drug efficacy [49, 50]. A study investigating the impact of resistance to β-lactam antibiotics on responses to antibiotics from eight drug families in *E. coli* found that most of the antibiotic resistance plasmids conferred resistance both to β-lactam antibiotics and to antibiotics from unrelated families [51]. Collateral resistance is pervasive in laboratory evolution experiments with multiple bacterial species [52, 53], demonstrating significant risk in increasing rates of antibiotic resistance since exposure to environmental antimicrobials or antimicrobial resistance genes through horizontal gene transfer could lead to resistance to clinically developed antibiotics [49]. Collateral resistance and tolerance may partially contribute to the observation that antibiotic resistance genes have been found in avian gut microbiomes from extreme environments, including Antarctic penguins that live in areas that would normally be assumed to be less influenced by anthropogenic antimicrobial agents [54].

An important recent tool to study polymicrobial interactions and their effect on drug response is meta-transcriptomics, which quantifies gene expression across an entire community of bacteria [55–57]. Using meta-transcriptomics and metagenomics (quantifying the genomes in a polymicrobial community), the entire resistome of an environment of interest, such as a polymicrobial biofilm, can be profiled to study the impact of antibiotic treatment on both community structure and gene expression. Examining metagenomic data from murine gut microbiomes showed that antibiotic treatment leads to changes in the diversity of the bacterial population (calculated using the Shannon diversity index), and significant changes in antibiotic resistance gene abundance [58]. Interestingly, the molecular changes were not targeted to the administered antibiotic. For example, treatment with the cell wall-targeting antibiotic amoxicillin induced increases in antibiotic resistance genes linked to protein synthesis inhibition, as well as decreases in fosmidomycin and trimethoprim-associated antibiotic resistance genes (linked to metabolic inhibition) [58]. The observed metagenomic changes in antibiotic resistance gene composition within the community possibly reflected overall changes in microbiome composition. In contrast, meta-transcriptomic analyses found that observed changes in transcriptional response in the population aligned more closely to the administered antibiotics' direct mechanisms of action, with amoxicillin inducing increases in beta-lactam resistance gene expression, and doxycycline increasing tetracycline resistance gene abundance [58]. This highlights how meta-transcriptomics can reveal functional changes in drug adaptation behavior within microbial populations, and underscores the importance of using multiple tools to examine population-level behavior.

Subpopulation-specific transcriptome profiling approaches akin to meta-transcriptomics have empowered direct molecular characterization of persister bacterial subpopulations. In particular, recent advances in bacterial single-cell RNA sequencing have started to enable transcriptional interrogation of the distinct subpopulations that emerge in response to treatment with different antibiotics [59, 60]. Among the transcriptional changes that strongly define differences between subgroups are genes comprising mobile genetic elements that disseminate with horizontal gene transfer, as well as genes that contribute to DNA damage response and metabolic adaptation, which we will discuss further in the coming sections. Defining which genes exhibit expression behavior that correlates with specific persister subpopulations can not only help inform the understanding of how these subpopulations emerge, but they can also serve as biomarkers to facilitate direct labeling and monitoring of the subpopulations of interest (e.g., via promotor-specific induction of fluorescent reporter genes) [61]. Together, these tools empower detailed, potentially dynamic characterizations of molecular and phenotypic changes that subpopulations elicit to adapt to drug treatment, uncovering insights into the regulation of these responses.

3.2 Bacterial Adaptations That Tune Heterogeneity and Subvert Antibiotic Activity

Bacteria have evolved a diverse range of molecular mechanisms that enable them to adapt to stressful conditions incurred by both infection and drug exposure by modulating their physiology at different levels. These molecular adaptations can impart either transient or permanent, heritable changes within the population. Here, we describe several examples of bacterial physiological adaptations that influence their heterogeneity and drug adaptation (Fig. 3.3).

3.2.1 Genetic Modulation

Genomic fidelity is an important mechanism by which bacteria can control population heterogeneity. Perfect genomic fidelity would render a species less adaptable to environmental changes that favor altered gene functionality. On the other hand,

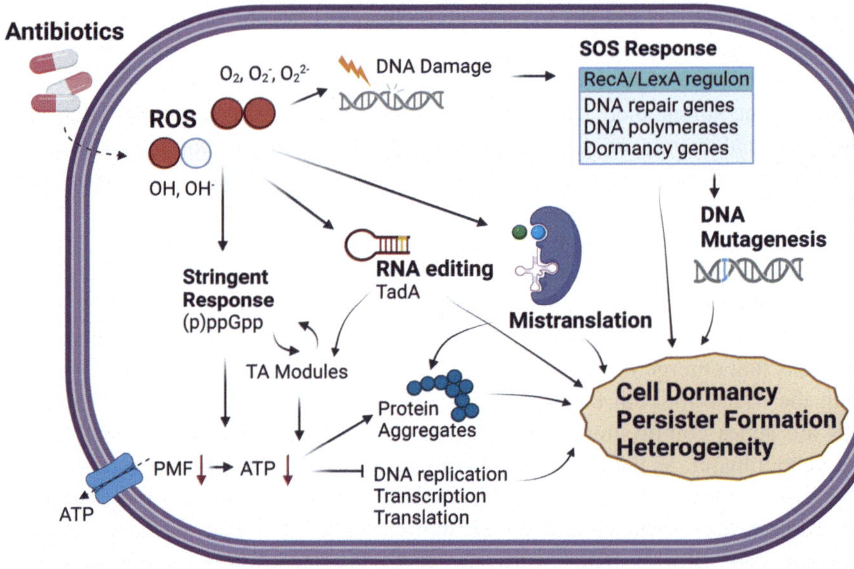

Fig. 3.3 Biological pathways that can lead to cellular heterogeneity or metabolic dormancy following the treatment with antibiotics. The introduction of antibiotics can lead to the formation of reactive oxygen species (ROS). ROS can damage DNA, and the SOS response is an adaptive mechanism initiated to repair the damage. This response can trigger the upregulation of error-prone DNA repair genes and genes associated with bacterial dormancy [62]. In addition, ROS may also trigger RNA editing, mistranslation, decreased metabolism, and cellular activity. Mistranslation and reduced intracellular ATP levels can promote protein aggregation in the cell, which can trigger the formation of persisters or bacterial dormancy. (Figure inspired by [63] and created with BioRender.com)

excessive mutations are also deleterious, as mutations have the potential to impair the functions of essential genes. As DNA repair is necessary to maintain genomic fidelity, bacteria have evolved mechanisms to regulate the accuracy of these repair mechanisms in a condition-dependent manner. Under nonstressed conditions, high-fidelity DNA repair mechanisms allow microbes to maintain genetic integrity and minimize the accumulation of deleterious mutations, while under stressed conditions, alternative, more error-prone DNA repair mechanisms are triggered to "hedge bets" and create a more genetically diverse population. Bacteria are thus often able to take advantage of errors in DNA replication to increase their rate of mutation in response to environmental factors.

DNA repair is a complex process that involves detecting the presence and type of DNA damage, as well as recruiting and activating proteins used to repair the damage. Bacteria utilize a suite of DNA repair machineries to respond to different types of DNA damage. These repair mechanisms include but are not limited to base excision repair, nucleotide excision repair, mismatched repair, nonhomologous end joining, and homologous recombination [64]. These repair mechanisms are not all active at a single time; rather, their condition-specific cellular utilization is controlled through the gene regulatory networks that mediate DNA damage responses. The most studied DNA damage response in bacteria is called the SOS response, a phylogenetically conserved stress response that is found in both Gram-negative and Gram-positive bacteria. The SOS response displays switch-like behavior—under standard conditions, genes regulated by the SOS response are generally not significantly expressed, but SOS-regulated genes can be induced upon exposure to environmental conditions that can induce double-stranded DNA breaks, including exposure ROS from host immune cells and exposure to drugs including quinolones, antivirals, and anticancer drugs such as 5-fluorouracil [65, 66].

When induced, the SOS response triggers DNA repair. The SOS response does not result in a homogenous phenotype across a population, even under conditions of stress. Rather, the SOS response can induce error-prone DNA polymerases which can increase the rate of mutation [67], generating highly mutable "gambler" subpopulations [68]. In many bacteria, this response is controlled by two proteins: RecA, an inducer, and LexA, a repressor. LexA forms a dimer that binds to the SOS box (a ~20 bp consensus sequence) proximal to SOS-regulated genes, which limits the expression of these genes. Once the bacterial cell is exposed to DNA-damaging units, single-stranded DNA begins to accumulate due to stalling at DNA replication forks. RecA then binds to the single-stranded DNA, forming an active complex that can stimulate the self-cleavage of the LexA repressor. Downstream SOS-regulated genes are now freely expressible, leading to the activation of genes involved in DNA repair, including the Uvr proteins that are involved in nucleotide excision repair. If the Uvr genes are not sufficient to resolve the levels of DNA damage, then due to the differential binding affinity of LexA, a series of other genes may be expressed. Homologous recombination repair genes such as *ruvAB* and *recN* are expressed next followed by DNA polymerase genes *polB* and *dinB*. If DNA damages are still not fully repaired, then cell division may be inhibited with SulB/FtsZ, and finally, to maintain survival, the error-prone DNA polymerase V is induced. Once the damage is repaired, then RecA no longer binds to single-stranded DNA and LexA repression of the SOS box resumes. The SOS response provides a mechanism where following DNA damage, DNA repair is regulated and provides the potential for mutations to occur through errors in DNA repair [69].

SOS response plays an important role in the mutability and development of drug resistance, but it is one of several mechanisms that regulate mutagenesis. The DNA translocase protein, Mfd (mutation frequency decline), is another crucial component of the DNA damage response and mediator of mutability. Compared to the SOS response, Mfd regulates a complementary DNA damage response that is triggered by damage-associated RNA polymerase stalling. Disrupting Mfd function leads to a reduction in bacterial mutation rate and a decrease in antibiotic resistance development across multiple drugs. Conversely, Mfd activity was found to promote hypermutation [70]. Mfd is thought to play a role in nucleotide excision repair, a process that repairs bulky lesions that are commonly the result of UV damage. In *E. coli*, recognition of damage is initiated by a UvrA–UvrB complex that recognizes distortion in the DNA strand [71]. When damage is found, UvrC replaces the UvrA subunit and forms a UvrBC complex. UvrB then cleaves the phosphate backbone four base pairs downstream of the damage and UvrC cleaves eight base pairs upstream of the damage. DNA helicase II, also known as UvrD, removes the cleaved portion by breaking the hydrogen bonds. The gap is then repaired with DNA polymerase I and DNA ligase. In transcriptionally active genes, the process of nucleotide excision repair occurs differently. Here, DNA damage is recognized when RNA polymerase stalls on the DNA strand. Mfd moves along the DNA strand and removes the RNA polymerase when it stalls. Mfd then recruits UvrA and continues the remaining steps in nucleotide excision repair [70]. Interestingly, Mfd and the SOS response both utilize the nucleotide excision repair pathway for DNA repair but differ in their detection of DNA damage and the scale of the mechanisms used to repair the damage. Notably, Mfd activity preferentially influences RNA polymerase interaction to stochastically disrupt transcription at genomic regions with highly structured regulatory RNAs, which can subsequently also alter transcriptional expression levels [72]. Interestingly, Mfd can confer increased mutagenesis to toxin genes that can control metabolic dormancy via toxin–antitoxin (TA) modules [72].

In addition to the SOS response and Mfd, some species also harbor species-specific DNA repair systems. For example, an SOS complementary system of DNA damage repair exists in some mycobacteria species in the form of the transcriptional activator, PafBC [73]. PafBC appears to activate a larger number and partially overlapping set of genes compared to the SOS response in response to DNA damage. Both pathways are necessary for bacterial survival and mutagenesis following DNA damage, while the regulation of some mutasome components, including DnaE2 and ImuB-ImuA, is regulated specifically by PafBC [73]. Interestingly, PafBC proteins are constitutively expressed so their activation mechanism must be distinct from SOS. Evidence suggests that SOS and PafBC may have distinct regulatory roles depending on the type of DNA damage. The SOS response plays a larger role in responding to UV damage, while PafBC plays a larger role following DNA gyrase inhibition.

These DNA repair mechanisms are representative examples of processes that dynamically regulate mutagenic capability in bacteria. It is through processes like these that enable bacteria to generate hypermutable subpopulations to promote faster evolution of drug resistance de novo under mutagenic stress conditions. These mutable "gamblers" are also phenotypically heterogeneous [68], which can increase transient bacterial survival via persistence during antibiotic exposure even without acquiring antibiotic resistance-conferring mutations.

3.2.2 Nongenetic Molecular Modulation

Modifications of the genome are far from the only ways that bacteria can vary phenotypic behavior. Following transcription, bacteria can utilize multiple mechanisms to modify protein activity, which can contribute to phenotypic heterogeneity and drug adaptation. Some notable mechanisms include RNA editing, mistranslation, and methylation.

RNA editing is a recently characterized source of nongenetic molecular heterogeneity in bacteria, elicited by chemical modification of RNA transcripts at specific residues prior to translation. In *E. coli*, RNA editing of both mRNA and tRNA is mediated by the tRNA-specific adenosine deaminase (*tadA*) enzyme, which catalyzes a deamination of adenosine (A) into inosine (I) [74]. Inosines are recognized by the ribosome as guanosine (G), so they can potentially alter the sequence of translated proteins. In *E. coli*, the TadA-mediated A-to-I editing events modify mRNA at loci that result in translational recoding of amino acid sequences from tyrosine (TAC) to cysteine (TGC), which can occur in up to 25% of the bacterial population under oxidative stress conditions [75]. Four of the protein-coding RNA editing loci targeted by TadA occur within the coding regions of the hok family of host-killing toxins, which belong to toxin–antitoxin modules that can confer antibiotic tolerance by eliciting bacterial membrane depolarization and subsequently suppress metabolic respiration [76]. The translational recoding of the hok toxins elicited by RNA editing appears to potentiate the toxins' biochemical activity [74]. Since RNA editing is not uniform across the population, and since the proportion of cells that are edited by TadA increases with cell density, RNA editing provides cells with a mechanism to modulate metabolic heterogeneity within the population as a function of extrinsic cues [74, 77].

Besides alterations at the mRNA level, bacteria can also introduce molecular heterogeneity by inserting modifications at the protein level. Analogous to mutagenesis arising from error-prone DNA repair, errors in translating mRNA to proteins can lead to molecular variation of the proteome. These mistranslation error rates can be as high as 10% under specific conditions such as oxidative stress (several orders of magnitude higher than DNA mutation rates from error-prone DNA repair) and contribute to the "statistical proteome" [78]. Bacteria mistranslation events of relevant proteins can yield phenotypic resistance to drug exposure. For example, bacteria from the *Streptomyces* and *Kitasatospora* genera harbor a family of "dual identity" tRNA that can translate proline amino acids into codons that usually encode asparagine or threonine [79]. Mistranslation resulting from these dual identity tRNA conferred tolerance to the cell wall-targeting antibiotic carbenicillin [79].

M. tuberculosis can actively regulate mistranslation rates at glutamine and asparagine residues under environmental stress conditions including nutrient starvation during stationary phase and low pH. The regulation of mistranslation rates can be mediated by controlling expression levels of the heterotrimeric glutamine amidotransferase enzyme, GatCAB, which contributes to the aminoacylation of glutamine and asparagine tRNAs [80, 81]. Mistranslation in *M. tuberculosis* is also controlled by the methyltransferase GidB. GidB methylates the ribosome at position 518 of the 16S rRNA, and disruption of *gidB* can improve translational fidelity, though the exact mechanism of this regulation is unknown [82]. Notably, increasing rates of mistranslation yielded rifampicin tolerance due to protein-level modulations of the drug-binding regions of the bacterial RNA polymerase [83], and GidB-mediated methylation of the ribosome also reduces the binding affinity of the ribosome-targeting aminoglycoside streptomycin, suggesting a dual role for GidB in mediating drug tolerance [82]. Conversely, chemical inhibition of *M. tuberculosis* mistranslation by administering the aminoglycoside kasugamycin potentiated killing by the antibiotic rifampicin during in vivo murine infection. Kasugamycin administration during *M. tuberculosis* mouse infection notably also caused significant *M. tuberculosis* growth restriction, even when the doses treated had yielded serum concentrations that were below in vitro minimum inhibitory concentrations. This suggests that reducing heterogeneity via mistranslation could potentially pose as a candidate anti-virulence intervention [84].

Besides influencing drug tolerance via mistranslation, methylation of the ribosome can also play a key role in bacterial adaptation to treatment with macrolides, lincosamides, and streptogramins B (MLS_B), which are used against Gram-positive bacteria and target protein synthesis by binding to the peptidyl transferase region of the 50S subunit of the bacterial ribosome. MLS_B-resistant phenotypes arise most commonly as a result of ribosome methylation by the erythromycin ribosomal methylation (Erm) proteins, with antibiotic efflux and enzymatic inactivation also playing a role. There are over 40 Erm proteins, which catalyze ribosome methylation that protect bacteria against MLS_B activity. This ribosome methylation may also cause reduced bacterial fitness due to inefficient translation. To minimize these effects, some bacteria, including *S. aureus* and *M. abscessus*, have evolved mechanisms such that the antibiotic acts as an inducer of the *erm* gene. Here, macrolide presence results in ribosome stalling, and this induces a conformation change that exposes the *ermC* ribosomal binding site, permitting its translation [85]. Once the *erm* gene is translated, these methyltransferases mono- or di-methylate the adenine residue (position A2058, *E. coli* numbering) of the 23rRNA of the 50S ribosomal subunit [85]. This methylation decreases the affinity of the macrolide to the ribosome, thereby effectively interfering with drug efficacy. Di-methylation leads to cross-resistance to lincosamides and streptogramins, since these drugs share the 23S binding site [86]. The resulting MLS_B resistance displays diverse and heterogeneous behavior, in part due to the number of distinct Erm proteins with differing methylation and regulatory behavior. Erm-associated resistance can also be transferable through horizontal gene transfer and is subject to polymicrobial interactions [87].

Nongenetic molecular modifications are regulatable processes that enable bacteria to foster transient molecular heterogeneity without eliciting heritable genomic modifications. These post-transcriptional events can decrease phenotypic susceptibility to drug efficacy by directly decreasing drug binding affinity or by eliciting changes to microbial metabolism that protect the cells from death.

3.2.3 Metabolic Modulation

As noted in previous sections, the metabolic state of the cell can play significant roles in the heterogeneity and drug response of a bacterial population. Stokes and colleagues identified three postulates that relate antibiotic efficacy and bacterial metabolism, namely "(1) antibiotics alter the metabolic state of bacteria, which contributes to the resulting death or stasis; (2) the metabolic state of bacteria influences their susceptibility to antibiotics; and (3) antibiotic efficacy can be enhanced by altering the metabolic state of bacteria [88]."

Many antibiotics can modulate bacterial metabolism through a variety of mechanisms that can contribute to bacterial death or stasis. For example, β-lactams, aminoglycosides, and quinolones all induce downstream metabolic changes in addition to their primary mechanisms of action which ultimately leads to the buildup of toxic metabolic byproducts, including ROS [23]. In the case of treatment with the quinolone antibiotic norfloxacin, the production of the lethal ROS involved the tricarboxylic acid (TCA) cycle and a transient depletion of NADH [24]. The ROS can then contribute to the lethality of the drug or sublethal levels and contribute to DNA damage and mutagenesis, as previously discussed. Production of ROS is common to antibiotics with multiple modes of action, including drugs whose primary target is not directly linked to metabolism or replication, such as protein synthesis inhibitors or cell wall inhibitors [89].

Just as secondary metabolic modulations induced by antibiotics can potentiate on-target antibiotic efficacy [90], the pre-existing metabolic state of bacteria can significantly impact the effectiveness of antibiotics' modes of action. For example, metabolic slowdown can occur with nutrient limitation via the stringent response, such as when bacteria transition to the stationary phase of growth [88, 91]. Examples linking metabolic dormancy with decreased susceptibility to antibiotic treatment relative to log-phase growth have been observed in clinical isolates from a broad range of bacteria (including *S. aureus*, *E. coli*, *Salmonella* Typhimurium, and *M. tuberculosis*) [45, 47, 48, 91–96].

Bacteria can also exhibit metabolic dormancy in conditions outside of the stationary phase. Stresses can induce another modality of metabolic dormancy described as differentially detectable (DD) or Viable But Not Culturable (VBNC) bacteria [93, 97–99]. There is a range of external influences or human interventions that can lead to the induction of VBNC/DD, including starvation, low temperatures, chemical stresses, and nonthermal plasma treatment [98]. The intracellular triggers for entering VBNC/DD states are varied, including toxin–antitoxin (TA) modules [100, 101], regulation of cell division [102], ROS accumulation [97], ATP depletion, and protein aggregation [103]. These bacteria are viable but cannot be quantified by standard culturing methods without resuscitation, which makes them difficult to detect [99]. Notably, VBNC/DD bacteria often exhibit tolerance to a broad range of antibiotics [97, 104, 105].

In *E. coli,* the frequency of persister cells correlated with the level of protein aggregates accumulated in stationary phase cultures [103]. Persister bacteria carry protein aggregates in early development, and VBNC/DD bacteria carry more mature aggregates, suggesting a partially shared physiology between persisters and VBNC/DD bacteria [104]. Progressive protein aggregation is additionally closely tied to another mechanism of entry into VBNC/DD: ATP depletion [104]. ATP depletion can cause bacteria to enter a dormant persistent state, where they become more tolerant to bactericidal antibiotics [91, 106]. Protein aggregates are observed at elevated levels when ATP is depleted, and these aggregates must be cleared for bacteria to leave the dormant VBNC/DD state [107]. Besides playing a role in regulating the transition into a VBNC/DD state, protein aggregates also correlate with how long it takes after antibiotic removal for the bacteria to exit the VBNC/DD state [107]. For example, the breakdown of protein aggregates by the DnaK-ClpB protein degradation protease machinery is required for leaving the VBNC/DD state. Since degradation of protein aggregates is required to exit the VBNC/DD state, variability in the amount of protein aggregates accrued between different individual cells changes the lag time of transition, resulting in a heterogeneity of "dormancy depths" for different subpopulations [107]. This type of lag time was found to be important with *M. tuberculosis* exiting the VBNC/DD state, in which ROS was found to contribute to VBNC/DD initiation [92]. Given this, Saito and colleagues hypothesized that allowing time to repair oxidative damage permitted these cells to recover since replication prior to the needed repairs would be "suicidal" [92]. A common theme thus emerges for the exit from the VBNC/DD state: the need to clear the initial cellular causes of entry into the VBNC/DD state, such as protein aggregation or oxidative damage, leading to population heterogeneity. In addition to the intrinsic heterogeneity caused by VBNC/DD state transitions, the multiple mechanisms of entry into VBNC/DD states suggest that there may be distinct VBNC/DD states depending on their mechanism of entry and that these states contribute to overall population heterogeneity in dormancy.

Small colony variants (SCV), a common cause of chronic and reincurring infections in multiple pathogens such as *S. aureus* and *M. tuberculosis*, are another manifestation of metabolic heterogeneity that incur clinically relevant treatment disparities. These SCVs are subpopulations of genetic mutants that are characterized by the formation of smaller colonies that take longer to grow on agar, as well as increased persistence to antibiotics, which necessitates multidrug therapy to treat [108, 109]. SCVs can be selected for a range of different stress conditions, including nutrient limitation, low oxygen, exposure to subinhibitory concentrations of antibiotics, or exposure to exotoxins via polymicrobial interactions [108, 110, 111]. SCVs exhibit genomic phase variation, reverting to their normal "large colony" genotype and physiology at a rate faster than classical mutagenesis would allow when no longer exposed to the inducing stressor [112, 113]. Evidence suggests that some SCVs may arise from insertion/deletion frameshift events in homo-polymeric tracts of the genome or from chromosomal inversions that lead to broader transcriptional changes. Because SCVs exhibit phase variation and because SCV can arise from mutations in a diverse range of genes, the full range of genetic changes that trigger SCV is difficult to pinpoint [112]. SCVs are frequently associated with metabolic changes, emerging with mutations in metabolic genes including but not limited to *hemA/H*, *menB*, or *thyA* (which confer auxotrophy in heme, menadione, and thymidine respectively) [110, 114]. This auxotrophy leads to reduced TCA and electron transport chain activity, which also leads to reduced ATP production [108, 114]. As in the case of VBNC/DD, SCV-associated mutations that reduce ATP production also lead to reduced growth and metabolic dormancy, which confers tolerance to antibiotics that depend on bacterial growth for efficacy [114]. SCV-associated mutations can also reduce drug activity via multiple other mechanisms. For example, the reduction in electron transport chain activity leads to reduced transmembrane potential, which can decrease the penetration of cationic peptides and aminoglycosides and reduce their antimicrobial activity [108, 114]. In addition to reduced metabolic activity, the specific mutation in *thyA* can cause bacterial tolerance to sulfonamide antibiotics by reducing the conversion of dUMP to dTMP, which in turn reduces bacterial dependency on the folate pathway that the sulfonamides target [114]. Furthermore, the reduction of bacterial respiration via TCA and electron transport chain activity decreases the formation of ROS [114], which as discussed earlier is an important effector of multiple bactericidal drugs. It is likely that SCVs represent a polymorphic phenotype with many pathways of induction that yield similar phenotypic outcomes, and it underscores the diversity of bacterial mechanisms that instigate population heterogeneity and drug adaptation.

Bacteria can dynamically regulate their growth dynamics through signal transduction. An example of this can be seen in *M. tuberculosis* with PknG, a serine-threonine protein kinase known to regulate central carbon and nitrogen metabolism in the bacteria. PknG phosphorylates GarA, a well-studied regulator of the TCA cycle and glutamate synthesis [115]. As discussed above, modulation of bacterial respiration via TCA cycle and intracellular NADH levels could enable PknG to alter drug susceptibility. Besides regulating respiration, PknG phosphorylation of GarA also alters the hydrophobicity of the cell wall [116]. Strains lacking PknG had significantly reduced cell wall hydrophobicity, which can also alter drug susceptibility by influencing drug uptake. This is consistent with studies showing that mycobacteria without PknG showed increased susceptibility to erythromycin, vancomycin, imipenem, ethambutol, and rifampicin [116].

In contrast to the effects of reduced metabolism, enhanced metabolism can either potentiate or reduce antibiotic activity, depending on the circumstances. For example, *E. coli* with higher than basal respiration rates are more susceptible to bactericidal antibiotics over the unmodified *E. coli,* potentially due to the creation of a futile cycle leading to ROS production due to increased aerobic respiration [117]. On the other hand, recent studies have found that fast-growing bacteria can also reduce antimicrobial efficacy by reducing antimicrobial accumulation and increasing drug efflux [118, 119].

The effect of varying growth rates has largely been studied on the clonal population scale, but intrinsic transcriptional variability at the single-cell level is where phenotypic heterogeneity often manifests. Since there are different molecular processes triggered with different growth phases, certain subpopulations in specific growth phases may be differentially tolerant to antimicrobial agents. For example, iron accumulates during the *S.* Typhimurium lag phase, and this accumulation was associated with increased sensitivity to oxidative stress [120]. This vulnerability could represent a therapeutic opportunity to be exploited by treatment targeting lag-phase bacteria.

Since bacterial metabolism plays such a critical role in both increasing and reducing antibiotic efficacy, there is strong interest in identifying host-protective intervention strategies that potentiate antibiotic efficacy via shifting bacterial metabolism into more drug-susceptible states. For example, co-administering metabolites found in upper glycolysis (e.g., glucose, mannitol, and fructose) with aminoglycoside treatment enhances bacterial proton-motive force, which was found to facilitate aminoglycoside uptake [121]. This potentiation enabled the killing of previously persistent bacteria, reducing the survival percentage from nearly 100% after a 2-hour treatment with the aminoglycoside gentamicin alone, to below 0.1% with glucose, mannitol, or fructose co-administration with gentamicin treatment [121]. Similarly, fumarate or succinate supplementation was able to activate cellular respiration, generating a proton motive force via the TCA cycle stimulation to enhance the efficacy of the aminoglycoside tobramycin [122]. Following the addition of fumarate or succinate, the survival percentage after 4 hours of tobramycin treatment was reduced from nearly 100% with tobramycin monotherapy down to below 0.1% with metabolite co-administration. The addition of exogenous alanine or glucose also enhanced the killing of multidrug resistant *Edwardsiella tarda* through a similar mechanism [123]. Beyond potentiation via increased drug uptake, the addition of metabolites can also induce a resuscitation of growth, which can rescue the efficacy of antibiotics that target growth-related processes. As discussed previously, stationary-phase bacteria are normally resistant to treatment with multiple drug classes [124], including quinolones, since the detrimental impact of the topoisomerase ligase inhibitors is reduced by the lack of replication. However, supplementation of a stationary-phase bacterial culture with glucose and a suitable terminal electron acceptor (such as fumarate or oxygen) stimulated bacterial respiration and was able to restore the quinolone effectiveness for multiple bacterial species, with oxygen and 1% glucose leading to a 0.1% survival rate compared to nearly 100% survival rate with quinolone treatment alone [125]. Given the importance of bacterial metabolism in antibiotic response, metabolic targeting may play a key role in countering drug resistance [126].

Some therapies exploit nutritional metal limitation by conjugating the bacterial inhibitor to a metal chelator, thus presenting the drug as a sort of "Trojan horse." As bacteria seek these trace metals, they may also uptake the conjugated antibiotic. For example, pyoverdine, a siderophore (iron chelator) produced by *Pseudomonas aeruginosa* [127], has previously been conjugated to ampicillin and demonstrated to exhibit antibacterial activity against ampicillin-resistant strains [128]. In another example, cefiderocol, a siderophore cephalosporin, is currently in two phase III clinical trials for the treatment of carbapenem-resistant infections and hospital-acquired bacterial pneumonia [129]. Metallic conjugation is a key illustration of how drug-resistant treatment can be improved by considering the unique stressors imposed by the host environment.

The metabolic state of a bacterium can significantly impact the response to drug treatment. Perturbing metabolism is just one of many ways in which bacterial tolerance, persistence, and resistance can be mitigated. By understanding the underlying molecular mechanisms of drug adaptation and heterogeneity, we can develop targeted therapies that counter or exploit these mechanisms to improve therapeutic outcomes.

3.3 Interventions to Mitigate Drug Adaptations

As antimicrobial resistance continues increasing in prevalence and severity, significant efforts are being made to develop novel treatment regimens. However, even as the rates of mortality due to drug-resistant infections are increasing, the rate of antibiotic development is slowing [130]. New interventions that can reduce the adaptive mechanisms that lead to drug tolerance, persistence, and resistance by means such as modulating heterogeneity have the potential for stronger clinical impact.

3.3.1 Tool Development: Identifying Drug Targets

Identifying genetic regulators and effectors that mediate heterogeneity and drug adaptation is still an active area of research. Systems biology approaches—including the meta-transcriptomics described in the Sect. 3.1.2, as well as functional genomic screening and genome-scale in silico modeling—are powerful tools that can implicate potential genes and pathways of interest that are associated with drug adaptation-promoting processes. These molecular candidates can then be prioritized for more detailed mechanistic study and evaluation of feasibility as intervention targets.

Functional genomic screening approaches, including Tn-seq and CRISPRi screening, depend on a parallelized assessment of individual gene disruptions and have identified genes that are essential for drug response processes and heterogeneity. These screening strategies take advantage of genetic perturbations that are traceable by sequencing: either the transposon genetic element insertion that mediates

Tn-seq-based disruption [131, 132] or the single guide RNAs (sgRNAs) that mediate CRISPR interference (CRISPRi)-based gene repression [133, 134]. Libraries of transposon disruption or CRISPRi mutants with genome-scale coverage can be pooled into a combined population and exposed as a pool to growth under selective stress (e.g., drug treatment) or subpopulation fractionation methods (e.g., fluorescence-activated cell sorting) [132]. Mutants that confer differential fitness or promote the formation of phenotypically distinct subpopulations would be differentially enriched under the aforementioned selective or fractionation conditions and would thus be detectable via differential relative abundance in sequencing-based quantification. Such systematic functional assessments of gene involvement with survival under drug stress or modulating capacity for phenotypic heterogeneity are particularly impactful for identifying novel candidate genes that are not broadly homologous across bacterial phylogeny or that operate via previously understudied mechanisms. Transposon-based and CRISPRi-based screening approaches have successfully identified genes involved in drug tolerance, drug resistance, and heterogeneity in multiple bacterial species [135–142].

Many adaptive processes, including those that mediate drug tolerance, are dynamically co-ordinated by networks of co-regulated genes. Phenotypic screens that specifically focus on the consequences of regulatory perturbations are sensitive to these network effects and provide a complementary functional profiling approach to Tn-seq and CRISPRi screens. An example of this is the Transcriptional Regulator Induced Phenotype (TRIP) screen [143], which uses a library of recombinant *M. tuberculosis* strains, each of which conditionally induces the expression of a different *M. tuberculosis* transcription factor (TF). Following growth, TRIP uses sequencing to determine the change in abundance of different TF-expressing recombinant strains. TRIP has the potential to uncover emergent responses when information about phenotype-inducing TFs is integrated with network information about their regulatory targets. Using this approach, the phenotypic effects of different TFs on *M. tuberculosis* survival during exposure to the frontline antitubercular antibiotic isoniazid were systematically characterized, which revealed regulators and effector target genes that modulated drug tolerance and hyper-susceptibility.

Under circumstances when functional genomic screening is not experimentally tractable, computational modeling approaches have the potential to assist in contextualizing experimental data and in generating hypotheses that can help to prioritize specific genes are pathways for additional experimental characterization. Given the importance of metabolism in influencing drug adaptation and phenotypic heterogeneity, metabolic modeling has enormous potential utility for identifying new candidate effector genes that would be therapeutically beneficial in modulating drug response [144–148]. Genome-scale metabolic modeling approaches, including flux balance analysis, have been successfully applied to contextualize bacterial transcriptomic and metabolomic data and to simulate computationally the metabolic responses of bacterial cells under different environmental conditions [149, 150]. These metabolic modeling simulations can also be used to predict the consequences of genetic perturbations under diverse environmental conditions, based on the bacteria's predicted capacity to execute key metabolic functions, such as the ability to generate ATP or to produce the metabolites necessary to sustain biomass formation [149]. These metabolic simulations can also be linked with models simulating transcriptional regulation to yield predictions relevant to TF perturbations [148, 151, 152]. Metabolic modeling approaches have successfully informed the identification of important genes and pathways that influence drug adaptation and tolerance [144–148].

3.3.2 Implementation Strategy: Combination Therapy

By combining multiple mechanisms of inhibition that are potentially responsive to multiple distinct bacterial subpopulations, multidrug regimens represent a powerful approach to tackle drug adaptation and phenotypic heterogeneity, whilst also reducing the likelihood of drug resistance evolution. In this section, we will describe some of the strategies involved in developing combinatorial interventions featuring antibiotics, as well as alternative therapeutic agents that work by host-direct intervention or by harnessing bacteriophages (Table 3.1).

Some of the most well-studied combination therapies include co-administration of a drug that counters the known bacterial resistance mechanism of the primary antibiotic inhibitory agent. For example, resistance to β-lactams can be caused by bacterial expression β-lactamases that degrade the antibiotic [156]. As such, β-lactam antibiotics are often co-administered with β-lactamase inhibitors, such as in the cases of amoxicillin-clavulanate, ampicillin-sulbactam, and piperacillin-tazobactam [157, 158]. By understanding the mechanism of resistance development, we can develop treatments that target specific subpopulations that are more likely to be resistant to antibiotics.

Table 3.1 Comparison of intervention modalities that could be incorporated into combination therapies

Treatment	Advantages	Challenges	References
Antibiotic therapy Uses small molecules to directly target the pathogen	Cheaper to manufacture, distribute Directly target bacteria	Antimicrobial resistance acquisition Side effects	[153]
Host-directed therapy Modifies host response to yield improved infection outcome	Enhances or resuscitates host immune response Less likely for a pathogen to develop resistance to therapy	Efficacy is influenced by host characteristics that can vary patient-to-patient Side effects	[154]
Phage therapy Utilizes predator–prey dynamics to control bacteria	Highly specific with little risk of off-target effects Auto-dosing due to predator–prey dynamic	Too specific—phages may be strain-specific Requires comprehensive phage characterization Phage instability may cause toxicity	[155]

When developing resistance to a drug, bacteria sometimes develop increased sensitivity to a different drug in a phenomenon called collateral sensitivity [51]. Collateral sensitivity relies on oppositional mechanisms of drug adaptation for each drug within a collaterally sensitive pair. Acquisition of collateral sensitivity can occur through mutations in an antibiotic target or mutations secondary effector mechanisms, and it can also be mediated through horizontal transfer of antibiotic-resistant plasmids [51].

Collateral sensitivity presents an important opportunity for the development of antibiotic regimes and for informing the strategic stewardship of existing antibiotic use to delay antibiotic obsolescence [159]. For example, by leveraging collateral sensitivity and drug cycling, gentamicin and cefuroxime were shown to treat and prevent resistance in pathogenic *E. coli* [160]. Hundreds of potential drug sets amenable to drug cycling have been identified [160]. Drug regimens taking advantage of collateral sensitivity have also been found to suppress the rate of resistance formation [53]. Despite these successes, the use of collateral sensitivity to combat bacterial infection is complicated by its heterogenous nature. Even among different resistance mutations to the same drug, the collateral sensitivity profile to other drugs can significantly differ [52]. This is likely at least in part due to the multiple molecular mechanisms that can lead to resistance, so different resistance mutations may lead to different collateral sensitivity and cross-resistance profiles. Molecular trends regarding collateral sensitivity in liquid culture have been found, however (e.g., susceptibility to rifampin is increased when cultures were selected using cell wall inhibitors), which suggests additional study could aid in efforts to improve rational design of collaterally sensitive drug interactions [52].

Besides modulating drug resistance behaviors, drug combinations can directly interact to impact efficacy. A drug interaction is considered synergistic when combined drug concentration is as effective as the individual drugs at a lower concentration [161]. In other words, drugs can interact synergistically with each other to yield better outcomes with lower concentrations. Such synergistic interactions can enable repurposing of existing drugs for new indications that would otherwise not be potent enough in activity when treated alone [162, 163].

Alternatively, antagonistic interactions describe decreased killing upon combinatorial treatment compared summed effects of each drug. That is, one drug hinders the activity of another [164]. Hyper-antagonist combinations are those in which the cotreatment results in weaker killing than treatment with one drug alone. Though antagonistic drug combinations seem unsuitable for clinical applications, these interactions have demonstrated utility in treating drug-resistant bacteria. For example, using hyper-antagonistic drug combinations resulted in the selective killing of tetracycline-resistant bacteria in an experiment comparing drug response in a tetracycline-resistant versus susceptible strain of *E. coli* [165]. In this study, combining the tetracycline drug doxycycline with ciprofloxacin treatment resulted in selection against the tetracycline-resistant allele. If antagonism can be exploited to treat antimicrobial-resistant infections, this affords a greater range of potential treatment options to address patient-specific needs.

3.3.3 Multidrug Regimens Incorporating Host-Directed Intervention

Drugs do not necessarily need to target the pathogen directly to effect a cure. Host-directed therapies target immune responses and other host processes that are independent of bacterial regulatory mechanisms, leading to the treatment of bacteria that have evolved resistance to antibiotics that target pathogen-centric processes. Host-directed therapies are also thought to be less prone to the development of resistance because acquiring resistance would require modifying host substrates or evolving altered pathogen utilization of host factors that the bacteria is not adapted to exploit at baseline, providing a significant barrier to resistance development [154]. As many host-directed therapies modulate the immune response, these agents can also improve disease prognosis by lessening the pathology of affected tissues that stem from overactive inflammation. Host-directed therapeutics have been successfully used to treat conditions such as *Propionibacterium acnes*-associated acne vulgaris [166], and they represent a promising intervention strategy for many bacterial infections, particularly when used in combination with antimicrobials.

Because the immune response is a major force mediating host control of infection and host–pathogen interactions, interventions designed to modify the immune response in host-protective ways have been a key targeting strategy for host-directed therapeutics. For example, under standard infection conditions, one of the virulence mechanisms that *M. tuberculosis* can adopt to subvert macrophage-mediated immune control is by secreting the bacterial phosphatase PtpA, which binds a subunit of the macrophage V-ATPase complex, a complex necessary for phagosome acidification. PtpA secretion can therefore effectively inhibit macrophage phagolysosome maturation, making it easier for *M. tuberculosis* to survive within the macrophage [167]. A host-directed therapeutic counter to this mechanism is the drug, imatinib, which is an inhibitor of host Abelson tyrosine kinase (Abl) and is primarily a licensed host-directed therapy for cancers including leukemia [168, 169]. Upon Abl inhibition, the transcription and expression of the V-H+ATP complex are increased, resulting in decreased lysosomal pH [170]. Mice treated with imatinib had decreased *M. tuberculosis* load in infected organs, and imatinib treatment was effective against rifampicin-resistant infections [171].

Given the shared similarity of some biological processes between mammals and bacteria, some therapeutic interventions that were originally developed to treat host-centric conditions have the potential to also exert host-protective effects on pathogens. A prime case study of this is with statins, a group of molecules that can interfere with lipid metabolism that have been used to lower cholesterol in humans. The bacterial pathogen *Listeria monocytogenes* requires cholesterol to enable the activity of its secreted protein listeriolysin O, a protein that creates pores in macrophage phagosomes that help the bacteria to escape the phagosome before its maturation, a process critical to establishing an infection [172]. Since cholesterol is required for *L. monocytogenes* phagosomal escape and establishment of infection, treatment of cholesterol-lowering statins should inhibit this virulence process, which has been shown to improve infection outcomes in mice. Additionally, patients on statins were less likely to develop sepsis, die from sepsis, or be placed in an intensive care unit during pneumonia treatment [173]. Lipids also form important components of bacterial cells, including functional membrane microdomains. In methicillin-resistant *S. aureus* (MRSA), these functional membrane microdomains accumulate multimeric protein complexes including PBP2a, a protein that confers penicillin resistance to the bacteria when oligomerized in the context of these functional membrane microdomains [174]. Statins inhibit the mevalonate metabolic pathway, which reduces the production of staphyloxanthin-derived lipids that are required for the formation of the functional membrane microdomains. By inhibiting the formation of these bacterial functional membrane microdomains, statins reduce the capacity for MRSA to resist penicillin activity by restricting PBP2a oligomerization [174], so statin co-administration can sensitize MRSA to conventional antibiotic therapy. As showcased by these examples, statins are compounds originally developed for host-directed intervention that may be repurposed to also impair bacterial functionality directly by targeting an evolutionarily conserved biological process. These and other potentially dual-acting interventions that concurrently modulate host and bacterial processes therapeutically for host protection during infection are attractive because their multimodal mechanisms of action impede bacterial drug resistance evolution.

3.3.4 Multidrug Regimens Incorporating Bacteriophage Intervention

In addition to host-directed therapies, bacteriophage therapy is also being explored as a treatment option, both as a primary as well as complementary treatment system [175]. Phage therapy has reported success in treating infection of multiple bacterial species in mice, and success at treating *P. aeruginosa* infections in humans has also been reported [155, 176]. Perhaps the most prominent advantages of phage therapy include mitigation of side effects due to off-target host inhibition, minimal disturbance of nonpathogenic microbiome due to their highly specific targeting (often strain-specific or species-specific), and applicability to antibiotic-resistant bacteria [177]. The vast natural diversity of phages can enable the potential to select and cultivate optimal phage properties for tailored interventions specific to an individual infection [178–180]. Furthermore, phage therapy has the potential to require only a single dose to treat acute infections due to the ability of phages to self-replicate in the presence of their target bacteria, leading to more dose-efficient elimination of targeted pathogens.

The mechanisms employed by phages to target bacteria are dissimilar enough from those of antibiotics that most bacterial mechanisms to confer antibiotic resistance cannot offer cross-resistance to phages. Although the rate of resistance formation may be slowed by co-evolution between bacteria and phage, bacteria can over time develop mechanisms to resist phages [181]. Strategic selection of phage targets can impart host-protective bacterial fitness costs upon resistance evolution, including eliciting collateral sensitivity to antibiotic treatment upon the evolution of resistance to phage targeting [182]. To reduce the likelihood of the bacteria evolving resistance, the therapeutic application of bacteriophages is primarily focused on combination therapy with antibiotics and with multiple phages meant to target the same bacteria with different mechanisms via polyphage cocktails [183]. Optimizing the complementary strengths of phages, host-directed agents, and antibiotics in combination therapies may offer treatment efficacy in the hardest-to-treat infection settings.

3.3.5 Rational Design of Combination Therapy

Systematically identifying efficacious therapeutic combinations remains a challenge since the number of possible combinations to evaluate scales exponentially with the number of agents considered, particularly when identifying the correct combinatorial concentrations [184], and when accounting for differences in efficacy between patients [185]. Even when considering antibiotics alone, the immense space of possible combinations makes exhaustive experimental characterization intractable; however, computational tools such as machine learning can help explore the space by prioritizing specific interactions with high potential efficacy for experimental characterization. Recent work has shown that drug response information collected from monotherapy or pairwise drug treatment can be used to inform predictions of three-way and other higher-order interaction outcomes between drugs [186]. These efforts have found that sometimes, nonintuitive drug combinations can lead to more effective therapies by targeting specific subpopulations in the bacterial ecosystem [187].

Besides simulating the direct consequences of gene perturbation, metabolic modeling has also been used in combination with machine learning techniques to evaluate the consequences of combinatorial drug intervention. For example, the modeling approach CARAMeL combines flux balance analysis-based metabolic modeling with random forest classification to predict drug synergy and antagonism [188]. This model uses transcriptome data measuring bacterial response to monotherapy as input and was able to predict both pairwise and higher-order multidrug bacterial inhibition levels under varying environmental conditions with high accuracy. Modeling approaches like CARAMeL that investigate synergy and antagonism by interrogating drug response transcriptional profiles can not only directly assist in designing multidrug regimens with favorable interaction profiles but they can also help to uncover genetic regulators of interaction outcomes, which opens the possibility for intervention gene targets that can induce broad potentiation of combinatorial drug efficacy [189].

Other methods have sought to develop models based on monotherapy and pairwise drug combination treatment outcome data under different environmental conditions to predict higher-order (three or four-drug) effectiveness, such as [187]. The resulting condition-specific drug response data-informed random forest classifiers that could successfully predict whether drug combinations improved on the standard of care treatment in antimicrobial efficacy as measured from murine infections and clinical trial data [187]. Interestingly, this study found that combining treatment response information from multiple environmental conditions into the computational model resulted in improved model predictive performance and in the identification of more efficacious drug combinations. For example, this study found that drug potency in butyrate-containing media and pairwise interaction outcome in cholesterol-containing media were predictive of drug combinations that were more efficacious in in vivo and clinical trial therapeutic outcomes. This implies that the most efficacious multidrug regimens might comprise subsets of agents that are efficacious at inhibiting distinct subpopulations of bacteria in different metabolic or physiological states. Designing multidrug regimens to be responsive to distinct bacterial subpopulations presents a potentially powerful approach to exert therapeutic control on bacterial heterogeneity and associated drug adaptation responses.

Combinatorial intervention targets that can take advantage of collateral sensitivity [190] in drug-resistant bacteria are particularly attractive candidate genes for development, and metabolic modeling methods have been successful at identifying these gene sets [188, 191]. Synthetic lethality, the phenomenon describing the interaction between two genes that only cause cell death when perturbed concurrently [192], can be used as a search strategy for collaterally sensitive interacting gene pairs. The underlying premise of synthetic lethality can be leveraged for the collateral sensitivity search leveraging the fact that drug resistance-conferring mutations can often partially impair the function of targeted proteins. These resulting hypomorphs can potentially reveal condition-specific collateral sensitivities in genes that would otherwise not be detrimental to pathogen fitness. So far, this concept has mainly been applied to treating cancerous cells [191], with some recent work on potential use in viral infections [193, 194]. For example, the Fast-SL method can predict synthetic lethal pairs from genome-scale metabolic models [195], and other computational methods have been developed to predict sets of genes whose collective knockout is fatal [191]. Combined with omics data from antibiotic-treated bacteria in physiologically relevant stress conditions, these methods could generate predictions of conditionally essential genes in drug-resistant strains, making these genes potentially effective drug targets for these resistant strains.

Combination therapy represents a powerful therapeutic tool to counter bacterial mechanisms of drug adaptation including heterogeneity. Development and application of rational regimen design tools such as those described here could help to optimize combinatorial interventions featuring antibiotics, host-directed therapeutics, and bacteriophages more effectively and tailor regimens to account for bacterial heterogeneity and to address individual patient needs.

3.4 Summary and Conclusions

The clinical consequences of unmitigated antimicrobial treatment failure are dire. Antibiotic resistance is currently estimated to cause over a million deaths annually worldwide [196], and mortality rates could increase to ten million people per year by 2050 based on projections [197]. The problem of antimicrobial drug adaptation and resistance is exacerbated by the restricted rates of new antibiotic development since the 1980s, due in part to the time- and resource-intensive nature of the drug development process [198]. In order to bridge the gap between the expanding set of bacterial resistance mechanisms and the dwindling complement of antimicrobial treatment options, it will be critical to identify new intervention modalities that are harder for bacteria to adapt to, whilst also managing drug stewardship to delay the progression of antibiotic obsolescence [159]. Understanding the mechanisms underlying how bacteria subvert drug action is critical to informing improved rational design of both novel therapeutic targeting strategies that can reduce bacterial capacity adaptation, as well as stewardship practices for reducing antibiotic resistance rates, by employing strategies such as combination treatment based on collateral sensitivity and drug cycling.

As discussed in this chapter, genetic, molecular, and phenotypic heterogeneity in bacteria play key roles in shaping bacterial bet-hedging and adaptation to environmental and drug-associated stresses (Fig. 3.4). A key opportunity for developing the next generation of antimicrobial interventions therefore lies in countering heterogeneity by reducing the capacity for bacteria to expand population variability or exploiting bacterial population variability, either by being effective against a diverse range of bacterial physiologies or by targeting against collateral sensitivities. Intervening specifically on the molecular regulators of bacterial drug adaptation and heterogeneity that we have already identified has yielded multiple examples of promising therapeutic outcomes in a range of preclinical and clinical contexts,

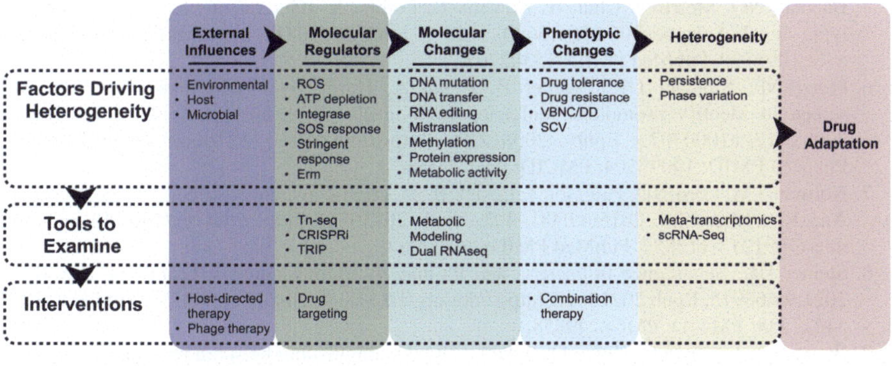

Fig. 3.4 Representation of the factors that contribute to heterogeneity and the resulting drug adaptation response. For each factor, research tools exist to examine the nature of the factor, and clinical intervention strategies exist that can be utilized to target the underlying mechanism of drug adaptation

several case studies of which were featured in this chapter [48, 84, 129, 162, 172, 182, 199]. These tantalizing advances suggest that additional intervention efforts in this area, supported by further knowledge of the molecular and cellular processes that govern heterogeneity and drug adaptation, could potentially be transformative in yielding more durably effective therapies.

Many unknowns remain regarding what molecular and cellular processes enable the diverse range of phenotypic variability achievable by bacteria and how bacteria can dynamically sense and regulate these processes under different changing conditions. By leveraging experimental and computational tools that offer a more holistic characterization of heterogeneity—including functional genetic screening, metatranscriptomics, single-cell profiling, machine learning, and metabolic network modeling—we can improve our understanding of these drug-adaptive processes and uncover new opportunities to intervene therapeutically.

References

1. Perez J, Contreras-Moreno FJ, Marcos-Torres FJ, Moraleda-Munoz A, Munoz-Dorado J. The antibiotic crisis: how bacterial predators can help. Comput Struct Biotechnol J. 2020;18:2547–55. Epub 20200915. https://doi.org/10.1016/j.csbj.2020.09.010. PubMed PMID: 33033577; PMCID: PMC7522538.
2. Weigel WA, Dersch P. Phenotypic heterogeneity: a bacterial virulence strategy. Microbes Infect. 2018;20(9–10):570–7. Epub 20180201. https://doi.org/10.1016/j.micinf.2018.01.008. PubMed PMID: 29409898.
3. Jolly MK, Kulkarni P, Weninger K, Orban J, Levine H. Phenotypic plasticity, bet-hedging, and androgen independence in prostate cancer: role of non-genetic heterogeneity. Front Oncol. 2018;8:50. Epub 20180306. https://doi.org/10.3389/fonc.2018.00050. PubMed PMID: 29560343; PMCID: PMC5845637.
4. Ackermann M. A functional perspective on phenotypic heterogeneity in microorganisms. Nat Rev Microbiol. 2015;13(8):497–508. Epub 20150706. https://doi.org/10.1038/nrmicro3491. PubMed PMID: 26145732.
5. Balaban NQ, Merrin J, Chait R, Kowalik L, Leibler S. Bacterial persistence as a phenotypic switch. Science. 2004;305(5690):1622–5. Epub 20040812. https://doi.org/10.1126/science.1099390. PubMed PMID: 15308767
6. Freed NE, Silander OK, Stecher B, Bohm A, Hardt WD, Ackermann M. A simple screen to identify promoters conferring high levels of phenotypic noise. PLoS Genet. 2008;4(12):e1000307. Epub 20081219. https://doi.org/10.1371/journal.pgen.1000307. PubMed PMID: 19096504; PMCID: PMC2588653
7. Norman TM, Lord ND, Paulsson J, Losick R. Stochastic switching of cell fate in microbes. Ann Rev Microbiol. 2015;69:381–403. Epub 20150821. https://doi.org/10.1146/annurev-micro-091213-112852. PubMed PMID: 26332088.
8. Steiner UK. Senescence in bacteria and its underlying mechanisms. Front Cell Dev Biol. 2021;9:668915. Epub 20210618. https://doi.org/10.3389/fcell.2021.668915. PubMed PMID: 34222238; PMCID: PMC8249858.
9. Brauner A, Fridman O, Gefen O, Balaban NQ. Distinguishing between resistance, tolerance and persistence to antibiotic treatment. Nat Rev Microbiol. 2016;14(5):320–30. https://doi.org/10.1038/nrmicro.2016.34. PubMed PMID: 27080241.

10. Van Melderen L. iTDtest: an easy-to-handle and visual assay to detect tolerant and Persister cells in an antibiotic combination regimen. MBio. 2022;13(3):e0036222. Epub 20220613. https://doi.org/10.1128/mbio.00362-22. PubMed PMID: 35695462; PMCID: PMC9239093.
11. Murdoch CC, Skaar EP. Nutritional immunity: the battle for nutrient metals at the host-pathogen interface. Nat Rev Microbiol. 2022;20(11):657–70. Epub 20220531. https://doi.org/10.1038/s41579-022-00745-6. PubMed PMID: 35641670; PMCID: PMC9153222.
12. Hibberd MC, Wu M, Rodionov DA, Li X, Cheng J, Griffin NW, Barratt MJ, Giannone RJ, Hettich RL, Osterman AL, Gordon JI. The effects of micronutrient deficiencies on bacterial species from the human gut microbiota. Sci Transl Med. 2017;9(390) https://doi.org/10.1126/scitranslmed.aal4069. PubMed PMID: 28515336; PMCID: PMC55241380
13. Moreno-Gamez S, Dal Co A, van Vliet S, Ackermann M. Microfluidics for single-cell study of antibiotic tolerance and persistence induced by nutrient limitation. Methods Mol Biol. 2021;2357:107–24. https://doi.org/10.1007/978-1-0716-1621-5_8. PubMed PMID: 34590255.
14. Sarathy JP, Via LE, Weiner D, Blanc L, Boshoff H, Eugenin EA, Barry CE 3rd, Dartois VA. Extreme drug tolerance of Mycobacterium tuberculosis in Caseum. Antimicrob Agents Chemother. 2018;62(2):Epub 20180125. https://doi.org/10.1128/AAC.02266-17. PubMed PMID: 29203492; PMCID: PMC5786764.
15. Walter ND, Ernest JP, Dide-Agossou C, Bauman AA, Ramey ME, Rossmassler K, Massoudi LM, Pauly S, Al Mubarak R, Voskuil MI, Kaya F, Sarathy JP, Zimmerman MD, Dartois V, Podell BK, Savic RM, Robertson GT. Lung microenvironments harbor Mycobacterium tuberculosis phenotypes with distinct treatment responses. Antimicrob Agents Chemother. 2023;67(9):e0028423. Epub 20230811. https://doi.org/10.1128/aac.00284-23. PubMed PMID: 37565762; PMCID: PMC10508168.
16. Sarathy JP, Dartois V. Caseum: a Niche for Mycobacterium tuberculosis drug-tolerant Persisters. Clin Microbiol Rev. 2020;33(3):e00159–19. Epub 20200401. https://doi.org/10.1128/CMR.00159-19. PubMed PMID: 32238365; PMCID: PMC7117546.
17. Ballal A, Manna AC. Regulation of superoxide dismutase (sod) genes by SarA in Staphylococcus aureus. J Bacteriol. 2009;191(10):3301–10. Epub 20090313. https://doi.org/10.1128/JB.01496-08. PubMed PMID: 19286803; PMCID: PMC2687179.
18. Nambi S, Long JE, Mishra BB, Baker R, Murphy KC, Olive AJ, Nguyen HP, Shaffer SA, Sassetti CM. The oxidative stress network of Mycobacterium tuberculosis reveals coordination between radical detoxification systems. Cell Host Microbe. 2015;17(6):829–37. https://doi.org/10.1016/j.chom.2015.05.008. PubMed PMID: 26067605; PMCID: PMC4465913.
19. Bakshi CS, Malik M, Regan K, Melendez JA, Metzger DW, Pavlov VM, Sellati TJ. Superoxide dismutase B gene (sodB)-deficient mutants of Francisella tularensis demonstrate hypersensitivity to oxidative stress and attenuated virulence. J Bacteriol. 2006;188(17):6443–8. https://doi.org/10.1128/JB.00266-06. PubMed PMID: 16923916; PMCID: PMC1595384.
20. Fang FC. Antimicrobial actions of reactive oxygen species. MBio. 2011;2(5):Epub 20110906. https://doi.org/10.1128/mBio.00141-11. PubMed PMID: 21896680; PMCID: PMC3171981.
21. Baba SP, Bhatnagar A. Role of Thiols in oxidative stress. Curr Opin Toxicol. 2018;7:133–9. Epub 20180321. https://doi.org/10.1016/j.cotox.2018.03.005. PubMed PMID: 30338308; PMCID: PMC6188637.
22. Gaupp R, Ledala N, Somerville GA. Staphylococcal response to oxidative stress. Front Cell Infect Microbiol. 2012;2:33. Epub 20120316. https://doi.org/10.3389/fcimb.2012.00033. PubMed PMID: 22919625; PMCID: PMC3417528.
23. Belenky P, Ye JD, Porter CB, Cohen NR, Lobritz MA, Ferrante T, Jain S, Korry BJ, Schwarz EG, Walker GC, Collins JJ. Bactericidal antibiotics induce toxic metabolic perturbations that lead to cellular damage. Cell Rep. 2015;13(5):968–80. Epub 20151022. https://doi.org/10.1016/j.celrep.2015.09.059. PubMed PMID: 26565910; PMCID: PMC4648786.
24. Dwyer DJ, Kohanski MA, Collins JJ. Role of reactive oxygen species in antibiotic action and resistance. Curr Opin Microbiol. 2009;12(5):482–9. Epub 20090731. https://doi.org/10.1016/j.mib.2009.06.018. PubMed PMID: 19647477; PMCID: PMC2761529.

25. Li H, Zhou X, Huang Y, Liao B, Cheng L, Ren B. Reactive oxygen species in pathogen clearance: the killing mechanisms, the adaption response, and the side effects. Front Microbiol. 2020;11:622534. Epub 20210204. https://doi.org/10.3389/fmicb.2020.622534. PubMed PMID: 33613470; PMCID: PMC7889972.
26. Sakai A, Nakanishi M, Yoshiyama K, Maki H. Impact of reactive oxygen species on spontaneous mutagenesis in Escherichia coli. Genes Cells. 2006;11(7):767–78. https://doi.org/10.1111/j.1365-2443.2006.00982.x. PubMed PMID: 16824196.
27. Liu Y, Tan S, Huang L, Abramovitch RB, Rohde KH, Zimmerman MD, Chen C, Dartois V, VanderVen BC, Russell DG. Immune activation of the host cell induces drug tolerance in Mycobacterium tuberculosis both in vitro and in vivo. J Exp Med. 2016;213(5):809–25. Epub 20160425. https://doi.org/10.1084/jem.20151248. PubMed PMID: 27114608; PMCID: PMC4854729.
28. Yang JH, Bhargava P, McCloskey D, Mao N, Palsson BO, Collins JJ. Antibiotic-induced changes to the host metabolic environment inhibit drug efficacy and Alter immune function. Cell Host Microbe. 2017;22(6):757–65 e3. Epub 20171130. https://doi.org/10.1016/j.chom.2017.10.020. PubMed PMID: 29199098; PMCID: PMC5730482.
29. Westermann AJ, Forstner KU, Amman F, Barquist L, Chao Y, Schulte LN, Muller L, Reinhardt R, Stadler PF, Vogel J. Dual RNA-seq unveils noncoding RNA functions in host-pathogen interactions. Nature. 2016;529(7587):496–501. Epub 20160120. https://doi.org/10.1038/nature16547. PubMed PMID: 26789254.
30. Westermann AJ, Barquist L, Vogel J. Resolving host-pathogen interactions by dual RNA-seq. PLoS Pathog. 2017;13(2):e1006033. Epub 20170216. https://doi.org/10.1371/journal.ppat.1006033. PubMed PMID: 28207848; PMCID: PMC5313147.
31. Nuss AM, Beckstette M, Pimenova M, Schmuhl C, Opitz W, Pisano F, Heroven AK, Dersch P. Tissue dual RNA-seq allows fast discovery of infection-specific functions and riboregulators shaping host-pathogen transcriptomes. Proc Natl Acad Sci USA. 2017;114(5):E791–E800. Epub 20170117. https://doi.org/10.1073/pnas.1613405114. PubMed PMID: 28096329; PMCID: PMC5293080.
32. Pisu D, Huang L, Grenier JK, Russell DG. Dual RNA-Seq of Mtb-infected macrophages in vivo reveals ontologically distinct host-pathogen interactions. Cell Rep. 2020;30(2):335–50 e4. https://doi.org/10.1016/j.celrep.2019.12.033. PubMed PMID: 31940480; PMCID: PMC7032562.
33. Hall CW, Mah TF. Molecular mechanisms of biofilm-based antibiotic resistance and tolerance in pathogenic bacteria. FEMS Microbiol Rev. 2017;41(3):276–301. https://doi.org/10.1093/femsre/fux010. PubMed PMID: 28369412.
34. Chakraborty P, Kumar A. The extracellular matrix of mycobacterial biofilms: could we shorten the treatment of mycobacterial infections? Microb Cell. 2019;6(2):105–22. Epub 20190118. https://doi.org/10.15698/mic2019.02.667. PubMed PMID: 30740456; PMCID: PMC6364259.
35. Orazi G, O'Toole GA. "It takes a village": mechanisms underlying antimicrobial recalcitrance of Polymicrobial biofilms. J Bacteriol. 2019;202(1):Epub 20191206. https://doi.org/10.1128/JB.00530-19. PubMed PMID: 31548277; PMCID: PMC6932244.
36. Tavernier S, Sass A, De Bruyne M, Baeke F, De Rycke R, Crabbe A, Vandecandelaere I, Van Nieuwerburgh F, Coenye T. Decreased susceptibility of Streptococcus anginosus to vancomycin in a multispecies biofilm is due to increased thickness of the cell wall. J Antimicrob Chemoth. 2018;73(9):2323–30. https://doi.org/10.1093/jac/dky216. PubMed PMID: 29901811.
37. Eoh H, Liu R, Lim J, Lee JJ, Sell P. Central carbon metabolism remodeling as a mechanism to develop drug tolerance and drug resistance in Mycobacterium tuberculosis. Front Cell Infect Microbiol. 2022;12:958240. Epub 20220822. https://doi.org/10.3389/fcimb.2022.958240. PubMed PMID: 36072228; PMCID: PMC9441700.
38. Wang Y, Hu Y, Liu F, Cao J, Lv N, Zhu B, Zhang G, Gao GF. Integrated metagenomic and metatranscriptomic profiling reveals differentially expressed resistomes in human, chicken,

and pig gut microbiomes. Environ Int. 2020;138:105649. Epub 20200319. https://doi.org/10.1016/j.envint.2020.105649. PubMed PMID: 32200314.
39. Wang Y, Lu J, Engelstadter J, Zhang S, Ding P, Mao L, Yuan Z, Bond PL, Guo J. Non-antibiotic pharmaceuticals enhance the transmission of exogenous antibiotic resistance genes through bacterial transformation. ISME J. 2020;14(8):2179–96. Epub 20200518. https://doi.org/10.1038/s41396-020-0679-2. PubMed PMID: 32424247; PMCID: PMC7367833.
40. Wang X, Zhang H, Long X, Xu X, Ren H, Mao D, Alvarez PJJ, Luo Y. Global increase of antibiotic resistance genes in conjugative plasmids. Microbiol Spectr. 2023;11(2):e0447822. Epub 20230322. https://doi.org/10.1128/spectrum.04478-22. PubMed PMID: 36946731; PMCID: PMC10100709.
41. Lerminiaux NA, Cameron ADS. Horizontal transfer of antibiotic resistance genes in clinical environments. Can J Microbiol. 2019;65(1):34–44. Epub 20180924. https://doi.org/10.1139/cjm-2018-0275. PubMed PMID: 30248271.
42. Arnold BJ, Huang IT, Hanage WP. Horizontal gene transfer and adaptive evolution in bacteria. Nat Rev Microbiol. 2022;20(4):206-18. Epub 20211112. https://doi.org/10.1038/s41579-021-00650-4. PubMed PMID: 34773098.
43. Stalder T, Top E. Plasmid transfer in biofilms: a perspective on limitations and opportunities. NPJ Biofilms Microbiomes. 2016;2(1):16022. Epub 20161019. https://doi.org/10.1038/npjbiofilms.2016.22. PubMed PMID: 28480050; PMCID: PMC5416938
44. Christensen BB, Sternberg C, Molin S. Bacterial plasmid conjugation on semi-solid surfaces monitored with the green fluorescent protein (GFP) from Aequorea victoria as a marker. Gene. 1996;173(1 Spec No):59–65. https://doi.org/10.1016/0378-1119(95)00707-5. PubMed PMID: 8707058.
45. Strugeon E, Tilloy V, Ploy MC, Da Re S. The stringent response promotes antibiotic resistance dissemination by regulating Integron integrase expression in biofilms. MBio. 2016;7(4):Epub 20160816. https://doi.org/10.1128/mBio.00868-16. PubMed PMID: 27531906; PMCID: PMC4992968.
46. Hauryliuk V, Atkinson GC, Murakami KS, Tenson T, Gerdes K. Recent functional insights into the role of (p)ppGpp in bacterial physiology. Nat Rev Microbiol. 2015;13(5):298–309. Epub 20150408. https://doi.org/10.1038/nrmicro3448. PubMed PMID: 25853779; PMCID: PMC4659695.
47. Hobbs JK, Boraston AB. (p)ppGpp and the stringent response: an emerging threat to antibiotic therapy. ACS Infect Dis. 2019;5(9):1505–17. https://doi.org/10.1021/acsinfecdis.9b00204. PubMed PMID: 31287287.
48. Chen Y, Zhang Z, Chen Y, Zhou S, Deng Q, Wang S. Enhancement of inhibition rate of antibiotic against bacteria by molecularly imprinted nanoparticles targeting alarmone nucleotides as antibiotic adjuvants. J Mater Chem B. 2022;10(45):9438-45. Epub 20221123. https://doi.org/10.1039/d2tb00641c. PubMed PMID: 36321529.
49. Colclough A, Corander J, Sheppard SK, Bayliss SC, Vos M. Patterns of cross-resistance and collateral sensitivity between clinical antibiotics and natural antimicrobials. Evol Appl. 2019;12(5):878–87. Epub 20190128. https://doi.org/10.1111/eva.12762. PubMed PMID: 31080502; PMCID: PMC6503891.
50. Sanders CC, Sanders WE Jr, Goering RV, Werner V. Selection of multiple antibiotic resistance by quinolones, beta-lactams, and aminoglycosides with special reference to cross-resistance between unrelated drug classes. Antimicrob Agents Chemother. 1984;26(6):797–801. https://doi.org/10.1128/AAC.26.6.797. PubMed PMID: 6098219; PMCID: PMC180026
51. Herencias C, Rodriguez-Beltran J, Leon-Sampedro R, Alonso-Del Valle A, Palkovicova J, Canton R, San MA. Collateral sensitivity associated with antibiotic resistance plasmids. elife. 2021;10:Epub 20210120. https://doi.org/10.7554/eLife.65130. PubMed PMID: 33470194; PMCID: PMC7837676.
52. Maltas J, Wood KB. Pervasive and diverse collateral sensitivity profiles inform optimal strategies to limit antibiotic resistance. PLoS Biol. 2019;17(10):e3000515. Epub 20191025. https://doi.org/10.1371/journal.pbio.3000515. PubMed PMID: 31652256; PMCID: PMC6834293.

53. Waller NJE, Cheung C-Y, Cook GM, McNeil MB. The evolution of antibiotic resistance is associated with collateral drug phenotypes in Mycobacterium tuberculosis. Nat Commun. 2023;14:1517. https://doi.org/10.1038/s41467-023-37184-7.
54. Marcelino VR, Wille M, Hurt AC, González-Acuña D, Klaassen M, Schlub TE, Eden J-S, Shi M, Iredell JR, Sorrell TC, Holmes EC. Meta-transcriptomics reveals a diverse antibiotic resistance gene pool in avian microbiomes. BMC Biol. 2019;17(1):31. https://doi.org/10.1186/s12915-019-0649-1.
55. Aguiar-Pulido V, Huang W, Suarez-Ulloa V, Cickovski T, Mathee K, Narasimhan G. Metagenomics, Metatranscriptomics, and metabolomics approaches for microbiome analysis. Evol Bioinformatics Online. 2016;12(Suppl 1):5–16. https://doi.org/10.4137/EBO.S36436.
56. Terrón-Camero LC, Gordillo-González F, Salas-Espejo E, Andrés-León E. Comparison of metagenomics and Metatranscriptomics tools: a guide to making the right choice. Genes. 2022;13(12):2280. https://doi.org/10.3390/genes13122280.
57. Lim YW, Schmieder R, Haynes M, Willner D, Furlan M, Youle M, Abbott K, Edwards R, Evangelista J, Conrad D, Rohwer F. Metagenomics and metatranscriptomics: windows on CF-associated viral and microbial communities. J Cyst Fibros. 2013;12(2):154–64. Epub 20120828. https://doi.org/10.1016/j.jcf.2012.07.009. PubMed PMID: 22951208; PMCID: PMC3534838.
58. Korry BJ, Cabral DJ, Belenky P. Metatranscriptomics reveals antibiotic-induced resistance gene expression in the Murine gut microbiota. Front Microbiol. 2020;11:322.
59. Ma P, Amemiya HM, He LL, Gandhi SJ, Nicol R, Bhattacharyya RP, Smillie CS, Hung DT. Bacterial droplet-based single-cell RNA-seq reveals antibiotic-associated heterogeneous cellular states. Cell. 2023;186(4):877–91 e14. Epub 20230127. https://doi.org/10.1016/j.cell.2023.01.002. PubMed PMID: 36708705; PMCID: PMC10014032.
60. Xu Z, Wang Y, Sheng K, Rosenthal R, Liu N, Hua X, Zhang T, Chen J, Song M, Lv Y, Zhang S, Huang Y, Wang Z, Cao T, Shen Y, Jiang Y, Yu Y, Chen Y, Guo G, Yin P, Weitz DA, Wang Y. Droplet-based high-throughput single microbe RNA sequencing by smRandom-seq. Nat Commun. 2023;14(1):5130. Epub 20230823. https://doi.org/10.1038/s41467-023-40137-9. PubMed PMID: 37612289; PMCID: PMC10447461.
61. Jain P, Weinrick BC, Kalivoda EJ, Yang H, Munsamy V, Vilcheze C, Weisbrod TR, Larsen MH, O'Donnell MR, Pym A, Jacobs WR Jr. Dual-reporter Mycobacteriophages (Phi2DRMs) reveal preexisting mycobacterium tuberculosis persistent cells in human sputum. MBio. 2016;7(5):Epub 20161025. https://doi.org/10.1128/mBio.01023-16. PubMed PMID: 27795387; PMCID: PMC5080378.
62. Jones C, Holland IB. Role of the SulB (FtsZ) protein in division inhibition during the SOS response in Escherichia coli: FtsZ stabilizes the inhibitor SulA in maxicells. Proc Natl Acad Sci USA. 1985;82(18):6045–9. https://doi.org/10.1073/pnas.82.18.6045. PubMed PMID: 2994059; PMCID: PMC390696.
63. Zou J, Peng B, Qu J, Zheng J. Are bacterial Persisters dormant cells only? Front Microbiol. 2021;12:708580. Epub 20220202. https://doi.org/10.3389/fmicb.2021.708580. PubMed PMID: 35185807; PMCID: PMC8847742.
64. Chatterjee N, Walker GC. Mechanisms of DNA damage, repair, and mutagenesis. Environ Mol Mutagen. 2017;58(5):235–63. Epub 20170509. https://doi.org/10.1002/em.22087. PubMed PMID: 28485537; PMCID: PMC5474181.
65. Mamber SW, Brookshire KW, Forenza S. Induction of the SOS response in Escherichia coli by azidothymidine and dideoxynucleosides. Antimicrob Agents Chemother. 1990;34(6):1237–43. https://doi.org/10.1128/AAC.34.6.1237. PubMed PMID: 2118327; PMCID: PMC171791.
66. Oda Y. Induction of SOS responses in Escherichia coli by 5-fluorouracil. Mutat Res. 1987;183(2):103–8. https://doi.org/10.1016/0167-8817(87)90051-4. PubMed PMID: 3102958.

67. Crane JK, Alvarado CL, Sutton MD. Role of the SOS response in the generation of antibiotic resistance in vivo. Antimicrob Agents Chemother. 2021;65(7):e0001321. Epub 20210617. https://doi.org/10.1128/AAC.00013-21. PubMed PMID: 33875437; PMCID: PMC8373240.
68. Pribis JP, Zhai Y, Hastings PJ, Rosenberg SM. Stress-induced mutagenesis, gambler cells, and stealth targeting antibiotic-induced evolution. MBio. 2022;13(3):e0107422. Epub 20220606. https://doi.org/10.1128/mbio.01074-22. PubMed PMID: 35658528; PMCID: PMC9239211.
69. Maslowska KH, Makiela-Dzbenska K, Fijalkowska IJ. The SOS system: a complex and tightly regulated response to DNA damage. Environ Mol Mutagen. 2019;60(4):368–84. Epub 20190107. https://doi.org/10.1002/em.22267. PubMed PMID: 30447030; PMCID: PMC6590174.
70. Ragheb MN, Thomason MK, Hsu C, Nugent P, Gage J, Samadpour AN, Kariisa A, Merrikh CN, Miller SI, Sherman DR, Merrikh H. Inhibiting the evolution of antibiotic resistance. Mol Cell. 2019;73(1):157–65 e5. Epub 20181115. https://doi.org/10.1016/j.molcel.2018.10.015. PubMed PMID: 30449724; PMCID: PMC6320318.
71. Kisker C, Kuper J, Van Houten B. Prokaryotic nucleotide excision repair. Cold Spring Harb Perspect Biol. 2013;5(3):a012591. Epub 20130301. https://doi.org/10.1101/cshperspect.a012591. PubMed PMID: 23457260; PMCID: PMC3578354.
72. Ragheb MN, Merrikh C, Browning K, Merrikh H. Mfd regulates RNA polymerase association with hard-to-transcribe regions in vivo, especially those with structured RNAs. Proc Natl Acad Sci USA. 2021;118(1) https://doi.org/10.1073/pnas.2008498118. PubMed PMID: 33443179; PMCID: PMC7817204.
73. Adefisayo OO, Dupuy P, Nautiyal A, Bean JM, Glickman MS. Division of labor between SOS and PafBC in mycobacterial DNA repair and mutagenesis. Nucleic Acids Res. 2021;49(22):12805–19. https://doi.org/10.1093/nar/gkab1169. PubMed PMID: 34871411; PMCID: PMC8682763.
74. Bar-Yaacov D, Mordret E, Towers R, Biniashvili T, Soyris C, Schwartz S, Dahan O, Pilpel Y. RNA editing in bacteria recodes multiple proteins and regulates an evolutionarily conserved toxin-antitoxin system. Genome Res. 2017;27(10):1696–703. Epub 20170901. https://doi.org/10.1101/gr.222760.117. PubMed PMID: 28864459; PMCID: PMC5630033.
75. Nie W, Wang S, He R, Xu Q, Wang P, Wu Y, Tian F, Yuan J, Zhu B, Chen G. A-to-I RNA editing in bacteria increases pathogenicity and tolerance to oxidative stress. PLoS Pathog. 2020;16(8):e1008740. Epub 20200821. https://doi.org/10.1371/journal.ppat.1008740. PubMed PMID: 32822429; PMCID: PMC7467310.
76. Verstraeten N, Knapen WJ, Kint CI, Liebens V, Van den Bergh B, Dewachter L, Michiels JE, Fu Q, David CC, Fierro AC, Marchal K, Beirlant J, Versees W, Hofkens J, Jansen M, Fauvart M, Michiels J. Obg and membrane depolarization are part of a microbial bet-hedging strategy that leads to antibiotic tolerance. Mol Cell. 2015;59(1):9–21. Epub 20150604. https://doi.org/10.1016/j.molcel.2015.05.011. PubMed PMID: 26051177.
77. Bar-Yaacov D, Pilpel Y, Dahan O. RNA editing in bacteria: occurrence, regulation and significance. RNA Biol. 2018;15(7):863–7. Epub 20180802. https://doi.org/10.1080/15476286.2018.1481698. PubMed PMID: 30071181; PMCID: PMC6161690.
78. Moghal A, Mohler K, Ibba M. Mistranslation of the genetic code. FEBS Lett. 2014;588(23):4305–10. Epub 20140916. https://doi.org/10.1016/j.febslet.2014.08.035. PubMed PMID: 25220850; PMCID: PMC4254111.
79. Schuntermann DB, Fischer JT, Bile J, Gaier SA, Shelley BA, Awawdeh A, Jahn M, Hoffman KS, Westhof E, Soll D, Clarke CR, Vargas-Rodriguez O. Mistranslation of the genetic code by a new family of bacterial transfer RNAs. J Biol Chem. 2023;299(7):104852. Epub 20230522. https://doi.org/10.1016/j.jbc.2023.104852. PubMed PMID: 37224963; PMCID: PMC10404621
80. Su HW, Zhu JH, Li H, Cai RJ, Ealand C, Wang X, Chen YX, Kayani MU, Zhu TF, Moradigaravand D, Huang H, Kana BD, Javid B. The essential mycobacterial amidotransferase GatCAB is a modulator of specific translational fidelity. Nat Microbiol. 2016;1(11):16147. Epub 20160826. https://doi.org/10.1038/nmicrobiol.2016.147. PubMed PMID: 27564922.

81. Li YY, Cai RJ, Yang JY, Hendrickson TL, Xiang Y, Javid B. Clinically relevant mutations of mycobacterial GatCAB inform regulation of translational fidelity. MBio. 2021;12(4):e0110021. Epub 20210706. https://doi.org/10.1128/mBio.01100-21. PubMed PMID: 34225495; PMCID: PMC8406222
82. Wong SY, Javid B, Addepalli B, Piszczek G, Strader MB, Limbach PA, Barry CE 3rd. Functional role of methylation of G518 of the 16S rRNA 530 loop by GidB in Mycobacterium tuberculosis. Antimicrob Agents Chemother. 2013;57(12):6311–8. Epub 20131007. https://doi.org/10.1128/AAC.00905-13. PubMed PMID: 24100503; PMCID: PMC3837903.
83. Javid B, Sorrentino F, Toosky M, Zheng W, Pinkham JT, Jain N, Pan M, Deighan P, Rubin EJ. Mycobacterial mistranslation is necessary and sufficient for rifampicin phenotypic resistance. Proc Natl Acad Sci USA. 2014;111(3):1132–7. Epub 20140106. https://doi.org/10.1073/pnas.1317580111. PubMed PMID: 24395793; PMCID: PMC3903211.
84. Chaudhuri S, Li L, Zimmerman M, Chen Y, Chen YX, Toosky MN, Gardner M, Pan M, Li YY, Kawaji Q, Zhu JH, Su HW, Martinot AJ, Rubin EJ, Dartois VA, Javid B. Kasugamycin potentiates rifampicin and limits emergence of resistance in Mycobacterium tuberculosis by specifically decreasing mycobacterial mistranslation. elife. 2018;7:Epub 20180828. https://doi.org/10.7554/eLife.36782. PubMed PMID: 30152756; PMCID: PMC6160228.
85. Munita JM, Arias CA. Mechanisms of antibiotic resistance. Microbiol Spectr. 2016;4:2. https://doi.org/10.1128/microbiolspec.VMBF-0016-2015. PubMed PMID: 27227291; PMCID: PMC4888801.
86. Dzyubak E, Yap MN. The expression of antibiotic resistance Methyltransferase correlates with mRNA stability independently of ribosome stalling. Antimicrob Agents Chemother. 2016;60(12):7178–88. Epub 20161121. https://doi.org/10.1128/AAC.01806-16. PubMed PMID: 27645242; PMCID: PMC5118997.
87. Spigaglia P, Barbanti F, Mastrantonio P. Horizontal transfer of erythromycin resistance from Clostridium difficile to Butyrivibrio fibrisolvens. Antimicrob Agents Chemother. 2005;49(12):5142–5. https://doi.org/10.1128/AAC.49.12.5142-5145.2005. PubMed PMID: 16304188; PMCID: PMC1315942.
88. Stokes JM, Lopatkin AJ, Lobritz MA, Collins JJ. Bacterial metabolism and antibiotic efficacy. Cell Metab. 2019;30(2):251–9. Epub 20190703. https://doi.org/10.1016/j.cmet.2019.06.009. PubMed PMID: 31279676; PMCID: PMC6990394.
89. Kohanski MA, Dwyer DJ, Hayete B, Lawrence CA, Collins JJ. A common mechanism of cellular death induced by bactericidal antibiotics. Cell. 2007;130(5):797–810. https://doi.org/10.1016/j.cell.2007.06.049. PubMed PMID: 17803904.
90. Cho H, Uehara T, Bernhardt TG. Beta-lactam antibiotics induce a lethal malfunctioning of the bacterial cell wall synthesis machinery. Cell. 2014;159(6):1300–11. https://doi.org/10.1016/j.cell.2014.11.017. PubMed PMID: 25480295; PMCID: PMC4258230
91. Conlon BP, Rowe SE, Gandt AB, Nuxoll AS, Donegan NP, Zalis EA, Clair G, Adkins JN, Cheung AL, Lewis K. Persister formation in Staphylococcus aureus is associated with ATP depletion. Nat Microbiol. 2016;1(5):1–7. https://doi.org/10.1038/nmicrobiol.2016.51.
92. Saito K, Mishra S, Warrier T, Cicchetti N, Mi J, Weber E, Jiang X, Roberts J, Gouzy A, Kaplan E, Brown CD, Gold B, Nathan C. Oxidative damage and delayed replication allow viable Mycobacterium tuberculosis to go undetected. Sci Transl Med. 2021;13(621):eabg2612. Epub 20211124. https://doi.org/10.1126/scitranslmed.abg2612. PubMed PMID: 34818059; PMCID: PMC8903021.
93. Saito K, Warrier T, Somersan-Karakaya S, Kaminski L, Mi J, Jiang X, Park S, Shigyo K, Gold B, Roberts J, Weber E, Jacobs WR Jr, Nathan CF. Rifamycin action on RNA polymerase in antibiotic-tolerant Mycobacterium tuberculosis results in differentially detectable populations. Proc Natl Acad Sci USA. 2017;114(24):E4832–E40. Epub 20170530. https://doi.org/10.1073/pnas.1705385114. PubMed PMID: 28559332; PMCID: PMC5474769.
94. Das B, Bhadra RK. (p)ppGpp metabolism and antimicrobial resistance in bacterial pathogens. Front Microbiol. 2020;11:563944. Epub 20201009. https://doi.org/10.3389/fmicb.2020.563944. PubMed PMID: 33162948; PMCID: PMC7581866.

95. Arenz S, Abdelshahid M, Sohmen D, Payoe R, Starosta AL, Berninghausen O, Hauryliuk V, Beckmann R, Wilson DN. The stringent factor RelA adopts an open conformation on the ribosome to stimulate ppGpp synthesis. Nucleic Acids Res. 2016;44(13):6471–81. Epub 20160525. https://doi.org/10.1093/nar/gkw470. PubMed PMID: 27226493; PMCID: PMC5291266.
96. Danchik C, Wang S, Karakousis PC. Targeting the Mycobacterium tuberculosis stringent response as a strategy for shortening tuberculosis treatment. Front Microbiol. 2021;12:744167. Epub 20211007. https://doi.org/10.3389/fmicb.2021.744167. PubMed PMID: 34690990; PMCID: PMC8529327.
97. Zainabadi K, Walsh KF, Vilbrun SC, Mathurin LD, Lee MH, Saito K, Mishra S, Ocheretina O, Pape JW, Nathan C, Fitzgerald DW. Characterization of differentially detectable mycobacterium tuberculosis in the sputum of subjects with drug-sensitive or drug-resistant tuberculosis before and after two months of therapy. Antimicrob Agents Chemother. 2021;65(8):e0060821. Epub 20210716. https://doi.org/10.1128/AAC.00608-21. PubMed PMID: 34060896; PMCID: PMC8284451
98. Zhang J, Yang H, Li J, Hu J, Lin G, Tan BK, Lin S. Current perspectives on viable but non-culturable foodborne pathogenic bacteria: a review. Food Secur. 2023;12(6):Epub 20230310. https://doi.org/10.3390/foods12061179. PubMed PMID: 36981106; PMCID: PMC10048424.
99. Ramamurthy T, Ghosh A, Pazhani GP, Shinoda S. Current perspectives on Viable but Non-Culturable (VBNC) pathogenic bacteria. Front Public Health. 2014:2. https://doi.org/10.3389/fpubh.2014.00103. PubMed PMID: 25133139; PMCID: PMC4116801
100. Fernandez-Garcia L, Blasco L, Lopez M, Bou G, Garcia-Contreras R, Wood T, Tomas M. Toxin-antitoxin systems in clinical pathogens. Toxins (Basel). 2016;8(7):Epub 20160720. https://doi.org/10.3390/toxins8070227. PubMed PMID: 27447671; PMCID: PMC4963858.
101. Harms A, Maisonneuve E, Gerdes K. Mechanisms of bacterial persistence during stress and antibiotic exposure. Science. 2016;354(6318) https://doi.org/10.1126/science.aaf4268. PubMed PMID: 27980159.
102. Hino M, Zhang J, Takagi H, Miyoshi T, Uchiumi T, Nakashima T, Kakuta Y, Kimura M. Characterization of putative toxin/antitoxin systems in Vibrio parahaemolyticus. J Appl Microbiol. 2014;117(1):185–95. https://doi.org/10.1111/jam.12513.
103. Leszczynska D, Matuszewska E, Kuczynska-Wisnik D, Furmanek-Blaszk B, Laskowska E. The formation of Persister cells in stationary-phase cultures of Escherichia Coli is associated with the aggregation of endogenous proteins. PLoS One. 2013;8(1):e54737. https://doi.org/10.1371/journal.pone.0054737.
104. Dewachter L, Bollen C, Wilmaerts D, Louwagie E, Herpels P, Matthay P, Khodaparast L, Khodaparast L, Rousseau F, Schymkowitz J, Michiels J. The dynamic transition of persistence toward the viable but nonculturable state during stationary phase is driven by protein aggregation. MBio. 2021;12(4):e00703–21. https://doi.org/10.1128/mBio.00703-21.
105. Nowakowska J, Oliver JD. Resistance to environmental stresses by Vibrio vulnificus in the viable but nonculturable state. FEMS Microbiol Ecol. 2013;84(1):213–22. Epub 20130102. https://doi.org/10.1111/1574-6941.12052. PubMed PMID: 23228034.
106. Liao XY, Liu DH, Ding T. Nonthermal plasma induces the viable-but-nonculturable state in via metabolic suppression and the oxidative stress response. Appl Environ Microb. 2020;86(5):ARTN e02216–19. https://doi.org/10.1128/AEM.02216-19. PubMed PMID: WOS:000519018300008.
107. Pu Y, Li Y, Jin X, Tian T, Ma Q, Zhao Z, Lin SY, Chen Z, Li B, Yao G, Leake MC, Lo CJ, Bai F. ATP-dependent dynamic protein aggregation regulates bacterial dormancy depth critical for antibiotic tolerance. Mol Cell. 2019;73(1):143–56 e4. Epub 20181121. https://doi.org/10.1016/j.molcel.2018.10.022. PubMed PMID: 30472191.
108. Keim KC, George IK, Reynolds L, Smith AC. The clinical significance of Staphylococcus aureus small Colony variants. Lab Med. 2023;54(3):227–34. https://doi.org/10.1093/labmed/lmac101. PubMed PMID: 36226897.

109. Melter O, Radojevic B. Small colony variants of Staphylococcus aureus--review. Folia Microbiol (Praha). 2010;55(6):548–58. Epub 20110121. https://doi.org/10.1007/s12223-010-0089-3. PubMed PMID: 21253898.
110. Guo H, Tong Y, Cheng J, Abbas Z, Li Z, Wang J, Zhou Y, Si D, Zhang R. Biofilm and small Colony variants-an update on Staphylococcus aureus strategies toward drug resistance. Int J Mol Sci. 2022;23(3):Epub 20220122. https://doi.org/10.3390/ijms23031241. PubMed PMID: 35163165; PMCID: PMC8835882.
111. Zhou S, Rao Y, Li J, Huang Q, Rao X. Staphylococcus aureus small-colony variants: formation, infection, and treatment. Microbiol Res. 2022;260:127040. Epub 20220420. https://doi.org/10.1016/j.micres.2022.127040. PubMed PMID: 35468490.
112. Edwards AM. Phenotype switching is a natural consequence of Staphylococcus aureus replication. J Bacteriol. 2012;194(19):5404-12. Epub 20120803. https://doi.org/10.1128/JB.00948-12. PubMed PMID: 22865841; PMCID: PMC3457229.
113. Safi H, Gopal P, Lingaraju S, Ma S, Levine C, Dartois V, Yee M, Li L, Blanc L, Ho Liang HP, Husain S, Hoque M, Soteropoulos P, Rustad T, Sherman DR, Dick T, Alland D. Phase variation in Mycobacterium tuberculosis glpK produces transiently heritable drug tolerance. Proc Natl Acad Sci USA. 2019;116(39):19665–74. https://doi.org/10.1073/pnas.1907631116. PubMed PMID: 31488707.
114. Garcia LG, Lemaire S, Kahl BC, Becker K, Proctor RA, Denis O, Tulkens PM, Van Bambeke F. Antibiotic activity against small-colony variants of Staphylococcus aureus: review of in vitro, animal and clinical data. J Antimicrob Chemother. 2013;68(7):1455–64. Epub 20130313. https://doi.org/10.1093/jac/dkt072. PubMed PMID: 23485724.
115. Ventura M, Rieck B, Boldrin F, Degiacomi G, Bellinzoni M, Barilone N, Alzaidi F, Alzari PM, Manganelli R, O'Hare HM. GarA is an essential regulator of metabolism in Mycobacterium tuberculosis. Mol Microbiol. 2013;90(2):356–66. Epub 20130908. https://doi.org/10.1111/mmi.12368. PubMed PMID: 23962235.
116. Wolff KA, Nguyen HT, Cartabuke RH, Singh A, Ogwang S, Nguyen L. Protein kinase G is required for intrinsic antibiotic resistance in mycobacteria. Antimicrob Agents Chemother. 2009;53(8):3515–9. Epub 20090615. https://doi.org/10.1128/AAC.00012-09. PubMed PMID: 19528288; PMCID: PMC2715596.
117. Lobritz MA, Belenky P, Porter CB, Gutierrez A, Yang JH, Schwarz EG, Dwyer DJ, Khalil AS, Collins JJ. Antibiotic efficacy is linked to bacterial cellular respiration. Proc Natl Acad Sci USA. 2015;112(27):8173–80. https://doi.org/10.1073/pnas.1509743112. PubMed PMID: 26100898; PMCID: PMC4500273.
118. Lapinska U, Voliotis M, Lee KK, Campey A, Stone MRL, Tuck B, Phetsang W, Zhang B, Tsaneva-Atanasova K, Blaskovich MAT, Pagliara S. Fast bacterial growth reduces antibiotic accumulation and efficacy. elife. 2022;11:Epub 20220607. https://doi.org/10.7554/eLife.74062. PubMed PMID: 35670099; PMCID: PMC9173744.
119. Fange D, Nilsson K, Tenson T, Ehrenberg M. Drug efflux pump deficiency and drug target resistance masking in growing bacteria. Proc Natl Acad Sci USA. 2009;106(20):8215–20. Epub 20090501. https://doi.org/10.1073/pnas.0811514106. PubMed PMID: 19416855; PMCID: PMC2688851.
120. Rolfe MD, Rice CJ, Lucchini S, Pin C, Thompson A, Cameron AD, Alston M, Stringer MF, Betts RP, Baranyi J, Peck MW, Hinton JC. Lag phase is a distinct growth phase that prepares bacteria for exponential growth and involves transient metal accumulation. J Bacteriol. 2012;194(3):686–701. Epub 20111202. https://doi.org/10.1128/JB.06112-11. PubMed PMID: 22139505; PMCID: PMC3264077.
121. Allison KR, Brynildsen MP, Collins JJ. Metabolite-enabled eradication of bacterial persisters by aminoglycosides. Nature. 2011;473(7346):216–20. https://doi.org/10.1038/nature10069. PubMed PMID: 21562562; PMCID: PMC3145328.
122. Meylan S, Porter CBM, Yang JH, Belenky P, Gutierrez A, Lobritz MA, Park J, Kim SH, Moskowitz SM, Collins JJ. Carbon sources tune antibiotic susceptibility in Pseudomonas aeruginosa via tricarboxylic acid cycle control. Cell Chem Biol. 2017;24(2):195–206. Epub

20170119. https://doi.org/10.1016/j.chembiol.2016.12.015. PubMed PMID: 28111098; PMCID: PMC5426816.
123. Peng B, Su YB, Li H, Han Y, Guo C, Tian YM, Peng XX. Exogenous alanine and/or glucose plus kanamycin kills antibiotic-resistant bacteria. Cell Metab. 2015;21(2):249–62. https://doi.org/10.1016/j.cmet.2015.01.008. PubMed PMID: 25651179.
124. Kussell E, Kishony R, Balaban NQ, Leibler S. Bacterial persistence: a model of survival in changing environments. Genetics. 2005;169(4):1807–14. Epub 20050131. https://doi.org/10.1534/genetics.104.035352. PubMed PMID: 15687275; PMCID: PMC1449587.
125. Gutierrez A, Jain S, Bhargava P, Hamblin M, Lobritz MA, Collins JJ. Understanding and sensitizing density-dependent persistence to Quinolone antibiotics. Mol Cell. 2017;68(6):1147–54 e3. Epub 20171207. https://doi.org/10.1016/j.molcel.2017.11.012. PubMed PMID: 29225037.
126. Murima P, McKinney JD, Pethe K. Targeting bacterial central metabolism for drug development. Chem Biol. 2014;21(11):1423–32. Epub 20141016. https://doi.org/10.1016/j.chembiol.2014.08.020. PubMed PMID: 25442374.
127. Mislin GL, Schalk IJ. Siderophore-dependent iron uptake systems as gates for antibiotic Trojan horse strategies against Pseudomonas aeruginosa. Metallomics. 2014;6(3):408–20. Epub 20140131. https://doi.org/10.1039/c3mt00359k. PubMed PMID: 24481292.
128. Kinzel O, Tappe R, Gerus I, Budzikiewicz H. The synthesis and antibacterial activity of two pyoverdin-ampicillin conjugates, entering Pseudomonas aeruginosa via the pyoverdin-mediated iron uptake pathway. J Antibiot (Tokyo). 1998;51(5):499–507. https://doi.org/10.7164/antibiotics.51.499. PubMed PMID: 9666179.
129. Zhanel GG, Golden AR, Zelenitsky S, Wiebe K, Lawrence CK, Adam HJ, Idowu T, Domalaon R, Schweizer F, Zhanel MA, Lagace-Wiens PRS, Walkty AJ, Noreddin A, Lynch Iii JP, Karlowsky JA. Cefiderocol: a Siderophore Cephalosporin with activity against carbapenem-resistant and multidrug-resistant gram-negative bacilli. Drugs. 2019;79(3):271–89. https://doi.org/10.1007/s40265-019-1055-2. PubMed PMID: 30712199.
130. Ventola CL. The antibiotic resistance crisis: part 1: causes and threats. P T. 2015;40(4):277–83. PubMed PMID: 25859123; PMCID: PMC4378521.
131. van Opijnen T, Bodi KL, Camilli A. Tn-seq: high-throughput parallel sequencing for fitness and genetic interaction studies in microorganisms. Nat Methods. 2009;6(10):767–72. https://doi.org/10.1038/nmeth.1377. PubMed PMID: 19767758; PMCID: PMC2957483.
132. Cain AK, Barquist L, Goodman AL, Paulsen IT, Parkhill J, van Opijnen T. A decade of advances in transposon-insertion sequencing. Nat Rev Genet. 2020;21(9):526–40. Epub 20200612. https://doi.org/10.1038/s41576-020-0244-x. PubMed PMID: 32533119; PMCID: PMC7291929.
133. Larson MH, Gilbert LA, Wang X, Lim WA, Weissman JS, Qi LS. CRISPR interference (CRISPRi) for sequence-specific control of gene expression. Nat Protoc. 2013;8(11):2180–96. Epub 20131017. https://doi.org/10.1038/nprot.2013.132. PubMed PMID: 24136345; PMCID: PMC3922765.
134. Li S, Poulton NC, Chang JS, Azadian ZA, DeJesus MA, Ruecker N, Zimmerman MD, Eckartt KA, Bosch B, Engelhart CA, Sullivan DF, Gengenbacher M, Dartois VA, Schnappinger D, Rock JM. CRISPRi chemical genetics and comparative genomics identify genes mediating drug potency in Mycobacterium tuberculosis. Nat Microbiol. 2022;7(6):766–79. Epub 20220530. https://doi.org/10.1038/s41564-022-01130-y. PubMed PMID: 35637331; PMCID: PMC9159947.
135. Carey AF, Rock JM, Krieger IV, Chase MR, Fernandez-Suarez M, Gagneux S, Sacchettini JC, Ioerger TR, Fortune SM. TnSeq of Mycobacterium tuberculosis clinical isolates reveals strain-specific antibiotic liabilities. PLoS Pathog. 2018;14(3):e1006939. Epub 20180305. https://doi.org/10.1371/journal.ppat.1006939. PubMed PMID: 29505613; PMCID: PMC5854444.

136. Rego EH, Audette RE, Rubin EJ. Deletion of a mycobacterial divisome factor collapses single-cell phenotypic heterogeneity. Nature. 2017;546(7656):153–7. Epub 20170531. https://doi.org/10.1038/nature22361. PubMed PMID: 28569798; PMCID: PMC5567998.
137. Xu W, DeJesus MA, Rucker N, Engelhart CA, Wright MG, Healy C, Lin K, Wang R, Park SW, Ioerger TR, Schnappinger D, Ehrt S. Chemical genetic interaction profiling reveals determinants of intrinsic antibiotic resistance in Mycobacterium tuberculosis. Antimicrob Agents Chemother. 2017;61(12) https://doi.org/10.1128/AAC.01334-17. PubMed PMID: 28893793; PMCID: PMC5700314.
138. Bosch B, DeJesus MA, Poulton NC, Zhang W, Engelhart CA, Zaveri A, Lavalette S, Ruecker N, Trujillo C, Wallach JB, Li S, Ehrt S, Chait BT, Schnappinger D, Rock JM. Genome-wide gene expression tuning reveals diverse vulnerabilities of M. Tuberculosis. Cell. 2021;184(17):4579–92 e24. Epub 20210722. https://doi.org/10.1016/j.cell.2021.06.033. PubMed PMID: 34297925; PMCID: PMC8382161.
139. Coe KA, Lee W, Stone MC, Komazin-Meredith G, Meredith TC, Grad YH, Walker S. Multi-strain Tn-Seq reveals common daptomycin resistance determinants in Staphylococcus aureus. PLoS Pathog. 2019;15(11):e1007862. Epub 20191118. https://doi.org/10.1371/journal.ppat.1007862. PubMed PMID: 31738809; PMCID: PMC6934316.
140. Bellerose MM, Proulx MK, Smith CM, Baker RE, Ioerger TR, Sassetti CM. Distinct bacterial pathways influence the efficacy of antibiotics against Mycobacterium tuberculosis. mSystems. 2020;5(4):Epub 20200804. https://doi.org/10.1128/mSystems.00396-20. PubMed PMID: 32753506; PMCID: PMC7406225.
141. Leshchiner D, Rosconi F, Sundaresh B, Rudmann E, Ramirez LMN, Nishimoto AT, Wood SJ, Jana B, Bujan N, Li K, Gao J, Frank M, Reeve SM, Lee RE, Rock CO, Rosch JW, van Opijnen T. A genome-wide atlas of antibiotic susceptibility targets and pathways to tolerance. Nat Commun. 2022;13(1):3165. Epub 20220607. https://doi.org/10.1038/s41467-022-30967-4. PubMed PMID: 35672367; PMCID: PMC9174251.
142. Jensen PA, Zhu Z, van Opijnen T. Antibiotics disrupt coordination between transcriptional and phenotypic stress responses in pathogenic bacteria. Cell Rep. 2017;20(7):1705–16. https://doi.org/10.1016/j.celrep.2017.07.062. PubMed PMID: 28813680; PMCID: PMC5584877.
143. Ma S, Morrison R, Hobbs SJ, Soni V, Farrow-Johnson J, Frando A, Fleck N, Grundner C, Rhee KY, Rustad TR, Sherman DR. Transcriptional regulator-induced phenotype screen reveals drug potentiators in Mycobacterium tuberculosis. Nat Microbiol. 2021;6(1):44–50. Epub 2020/11/18. https://doi.org/10.1038/s41564-020-00810-x. PubMed PMID: 33199862; PMCID: PMC8331221.
144. Jenior ML, Papin JA. Computational approaches to understanding Clostridioides difficile metabolism and virulence. Curr Opin Microbiol. 2022;65:108–15. Epub 20211125. https://doi.org/10.1016/j.mib.2021.11.002. PubMed PMID: 34839237; PMCID: PMC8792252.
145. Dunphy LJ, Papin JA. Biomedical applications of genome-scale metabolic network reconstructions of human pathogens. Curr Opin Biotechnol. 2018;51:70–9. Epub 20171207. https://doi.org/10.1016/j.copbio.2017.11.014. PubMed PMID: 29223465; PMCID: PMC5991985.
146. Peterson EJR, Ma S, Sherman DR, Baliga NS. Network analysis identifies Rv0324 and Rv0880 as regulators of bedaquiline tolerance in Mycobacterium tuberculosis. Nat Microbiol. 2016:16078. https://doi.org/10.1038/nmicrobiol.2016.78. http://www.nature.com/articles/nmicrobiol201678#supplementary-information
147. Yang JH, Wright SN, Hamblin M, McCloskey D, Alcantar MA, Schrübbers L, Lopatkin AJ, Satish S, Nili A, Palsson BO, Walker GC, Collins JJ. A white-box machine learning approach for revealing antibiotic mechanisms of action. Cell. 2019;177(6):1649–61.e9. https://doi.org/10.1016/j.cell.2019.04.016.
148. Immanuel SRC, Arrieta-Ortiz ML, Ruiz RA, Pan M, Garcia L, de Lomana A, Peterson EJR, Baliga NS. Quantitative prediction of conditional vulnerabilities in regulatory and metabolic networks using PRIME. NPJ Syst Biol Appl. 2021;7(1):43. Epub 20211206. https://doi.org/10.1038/s41540-021-00205-6. PubMed PMID: 34873198; PMCID: PMC8648758.

149. Orth JD, Thiele I, Palsson BO. What is flux balance analysis? Nat Biotechnol. 2010;28(3):245–8. Epub 2010/03/10. nbt.1614 [pii]. https://doi.org/10.1038/nbt.1614. PubMed PMID: 20212490.
150. Shlomi T, Cabili MN, Herrgard MJ, Palsson BO, Ruppin E. Network-based prediction of human tissue-specific metabolism. Nat Biotechnol. 2008;26(9):1003–10. Epub 2008/08/20. nbt.1487 [pii]. https://doi.org/10.1038/nbt.1487. PubMed PMID: 18711341.
151. Chandrasekaran S, Price ND. Probabilistic integrative modeling of genome-scale metabolic and regulatory networks in Escherichia coli and Mycobacterium tuberculosis. Proc Natl Acad Sci USA. 2010;107(41):17845–50. Epub 2010/09/30. https://doi.org/10.1073/pnas.1005139107. PubMed PMID: 20876091.
152. Ma S, Minch KJ, Rustad TR, Hobbs S, Zhou SL, Sherman DR, Price ND. Integrated modeling of gene regulatory and metabolic networks in Mycobacterium tuberculosis. PLoS Comput Biol. 2015;11(11):e1004543. https://doi.org/10.1371/journal.pcbi.1004543. PubMed PMID: 26618656; PMCID: PMC4664399.
153. Patel P, Wermuth HR, Calhoun C, Hall GA. Antibiotics. Treasure Island: StatPearls; 2023.
154. Kaufmann SHE, Dorhoi A, Hotchkiss RS, Bartenschlager R. Host-directed therapies for bacterial and viral infections. Nat Rev Drug Discov. 2018;17(1):35–56. Epub 20170922. https://doi.org/10.1038/nrd.2017.162. PubMed PMID: 28935918; PMCID: PMC7097079.
155. Furfaro LL, Payne MS, Chang BJ. Bacteriophage therapy: clinical trials and regulatory hurdles. Front Cell Infect Microbiol. 2018;8:376. Epub 20181023. https://doi.org/10.3389/fcimb.2018.00376. PubMed PMID: 30406049; PMCID: PMC6205996.
156. Worthington RJ, Melander C. Overcoming resistance to beta-lactam antibiotics. J Org Chem. 2013;78(9):4207–13. Epub 20130328. https://doi.org/10.1021/jo400236f. PubMed PMID: 23530949; PMCID: PMC3644377.
157. Tooke CL, Hinchliffe P, Bragginton EC, Colenso CK, Hirvonen VHA, Takebayashi Y, Spencer J. beta-Lactamases and beta-Lactamase inhibitors in the 21st century. J Mol Biol. 2019;431(18):3472–500. Epub 20190405. https://doi.org/10.1016/j.jmb.2019.04.002. PubMed PMID: 30959050; PMCID: PMC6723624.
158. Khanna NR, Gerriets V. Beta-Lactamase Inhibitors. Treasure Island: StatPearls; 2023.
159. Shrestha J, Zahra F, Cannady JP. Antimicrobial Stewardship. Treasure Island: StatPearls; 2023.
160. Imamovic L, Sommer MO. Use of collateral sensitivity networks to design drug cycling protocols that avoid resistance development. Sci Transl Med. 2013;5(204):204ra132. https://doi.org/10.1126/scitranslmed.3006609. PubMed PMID: 24068739.
161. Berenbaum MC. What is synergy? Pharmacol Rev. 1989;41(2):93–141. PubMed PMID: 2692037.
162. Cheng YS, Williamson PR, Zheng W. Improving therapy of severe infections through drug repurposing of synergistic combinations. Curr Opin Pharmacol. 2019;48:92–8. Epub 20190824. https://doi.org/10.1016/j.coph.2019.07.006. PubMed PMID: 31454708; PMCID: PMC6858965.
163. Zheng W, Sun W, Simeonov A. Drug repurposing screens and synergistic drug-combinations for infectious diseases. Br J Pharmacol. 2018;175(2):181–91. Epub 20170709. https://doi.org/10.1111/bph.13895. PubMed PMID: 28685814; PMCID: PMC5758396.
164. Ocampo PS, Lazar V, Papp B, Arnoldini M, Abel Zur Wiesch P, Busa-Fekete R, Fekete G, Pal C, Ackermann M, Bonhoeffer S. Antagonism between bacteriostatic and bactericidal antibiotics is prevalent. Antimicrob Agents Chemother. 2014;58(8):4573–82. https://doi.org/10.1128/AAC.02463-14. PubMed PMID: 24867991; PMCID: PMC4135978.
165. Chait R, Craney A, Kishony R. Antibiotic interactions that select against resistance. Nature. 2007;446(7136):668–71. https://doi.org/10.1038/nature05685. PubMed PMID: 17410176.
166. Latter G, Grice JE, Mohammed Y, Roberts MS, Benson HAE. Targeted topical delivery of Retinoids in the management of Acne Vulgaris: current formulations and novel delivery systems. Pharmaceutics. 2019;11(10):Epub 20190924. https://doi.org/10.3390/pharmaceutics11100490. PubMed PMID: 31554188; PMCID: PMC6835300.

167. Wong D, Bach H, Sun J, Hmama Z, Av-Gay Y. Mycobacterium tuberculosis protein tyrosine phosphatase (PtpA) excludes host vacuolar-H+-ATPase to inhibit phagosome acidification. Proc Natl Acad Sci USA. 2011;108(48):19371–6. Epub 20111115. https://doi.org/10.1073/pnas.1109201108. PubMed PMID: 22087003; PMCID: PMC3228452.
168. Iqbal N, Iqbal N. Imatinib: a breakthrough of targeted therapy in cancer. Chemother Res Pract. 2014;2014:357027. Epub 20140519. https://doi.org/10.1155/2014/357027. PubMed PMID: 24963404; PMCID: PMC4055302.
169. Jeong EK, Lee HJ, Jung YJ. Host-directed therapies for tuberculosis. Pathogens. 2022;11(11):Epub 20221103. https://doi.org/10.3390/pathogens11111291. PubMed PMID: 36365041; PMCID: PMC9697779.
170. Bruns H, Stegelmann F, Fabri M, Dohner K, van Zandbergen G, Wagner M, Skinner M, Modlin RL, Stenger S. Abelson tyrosine kinase controls phagosomal acidification required for killing of Mycobacterium tuberculosis in human macrophages. J Immunol. 2012;189(8):4069–78. Epub 20120917. https://doi.org/10.4049/jimmunol.1201538. PubMed PMID: 22988030; PMCID: PMC3684563.
171. Napier RJ, Rafi W, Cheruvu M, Powell KR, Zaunbrecher MA, Bornmann W, Salgame P, Shinnick TM, Kalman D. Imatinib-sensitive tyrosine kinases regulate mycobacterial pathogenesis and represent therapeutic targets against tuberculosis. Cell Host Microbe. 2011;10(5):475–85. https://doi.org/10.1016/j.chom.2011.09.010. PubMed PMID: 22100163; PMCID: PMC3222875.
172. Cheng C, Sun J, Yu H, Ma T, Guan C, Zeng H, Zhang X, Chen Z, Song H. Listeriolysin O pore-forming activity is required for ERK1/2 phosphorylation during listeria monocytogenes infection. Front Immunol. 2020;11:1146. Epub 20200603. https://doi.org/10.3389/fimmu.2020.01146. PubMed PMID: 32582211; PMCID: PMC7283531.
173. Zumla A, Rao M, Wallis RS, Kaufmann SH, Rustomjee R, Mwaba P, Vilaplana C, Yeboah-Manu D, Chakaya J, Ippolito G, Azhar E, Hoelscher M, Maeurer M. Host-directed therapies network c. Host-directed therapies for infectious diseases: current status, recent progress, and future prospects. Lancet Infect Dis. 2016;16(4):e47–63. https://doi.org/10.1016/S1473-3099(16)00078-5. PubMed PMID: 27036359; PMCID: PMC7164794.
174. Garcia-Fernandez E, Koch G, Wagner RM, Fekete A, Stengel ST, Schneider J, Mielich-Suss B, Geibel S, Markert SM, Stigloher C, Lopez D. Membrane microdomain disassembly inhibits MRSA antibiotic resistance. Cell. 2017;171(6):1354–67 e20. Epub 20171102. https://doi.org/10.1016/j.cell.2017.10.012. PubMed PMID: 29103614; PMCID: PMC5720476.
175. Chanishvili N. Phage therapy--history from Twort and d'Herelle through Soviet experience to current approaches. Adv Virus Res. 2012;83:3–40. https://doi.org/10.1016/B978-0-12-394438-2.00001-3. PubMed PMID: 22748807.
176. Wright A, Hawkins CH, Anggard EE, Harper DR. A controlled clinical trial of a therapeutic bacteriophage preparation in chronic otitis due to antibiotic-resistant Pseudomonas aeruginosa; a preliminary report of efficacy. Clin Otolaryngol. 2009;34(4):349–57. https://doi.org/10.1111/j.1749-4486.2009.01973.x. PubMed PMID: 19673983.
177. Loc-Carrillo C, Abedon ST. Pros and cons of phage therapy. Bacteriophage. 2011;1(2):111–4. https://doi.org/10.4161/bact.1.2.14590. PubMed PMID: 22334867; PMCID: PMC3278648.
178. Favor AH, Llanos CD, Youngblut MD, Bardales JA. Optimizing bacteriophage engineering through an accelerated evolution platform. Sci Rep. 2020;10(1):13981. Epub 20200819. https://doi.org/10.1038/s41598-020-70841-1. PubMed PMID: 32814789; PMCID: PMC7438504.
179. Schooley RT, Biswas B, Gill JJ, Hernandez-Morales A, Lancaster J, Lessor L, Barr JJ, Reed SL, Rohwer F, Benler S, Segall AM, Taplitz R, Smith DM, Kerr K, Kumaraswamy M, Nizet V, Lin L, McCauley MD, Strathdee SA, Benson CA, Pope RK, Leroux BM, Picel AC, Mateczun AJ, Cilwa KE, Regeimbal JM, Estrella LA, Wolfe DM, Henry MS, Quinones J, Salka S, Bishop-Lilly KA, Young R, Hamilton T. Development and use of personalized bacteriophage-based therapeutic cocktails to treat a patient with a disseminated resistant Acinetobacter baumannii infection. Antimicrob Agents Chemother. 2017;61(10):Epub

20170922. https://doi.org/10.1128/AAC.00954-17. PubMed PMID: 28807909; PMCID: PMC5610518.
180. Ferry T, Kolenda C, Laurent F, Leboucher G, Merabischvilli M, Djebara S, Gustave CA, Perpoint T, Barrey C, Pirnay JP, Resch G. Personalized bacteriophage therapy to treat pandrug-resistant spinal Pseudomonas aeruginosa infection. Nat Commun. 2022;13(1):4239. Epub 20220722. https://doi.org/10.1038/s41467-022-31837-9. PubMed PMID: 35869081; PMCID: PMC9306240.
181. Oechslin F. Resistance development to bacteriophages occurring during bacteriophage therapy. Viruses. 2018;10(7):Epub 20180630. https://doi.org/10.3390/v10070351. PubMed PMID: 29966329; PMCID: PMC6070868.
182. Chan BK, Sistrom M, Wertz JE, Kortright KE, Narayan D, Turner PE. Phage selection restores antibiotic sensitivity in MDR Pseudomonas aeruginosa. Sci Rep. 2016;6:26717. Epub 20160526. https://doi.org/10.1038/srep26717. PubMed PMID: 27225966; PMCID: PMC4880932.
183. Chan BK, Abedon ST, Loc-Carrillo C. Phage cocktails and the future of phage therapy. Future Microbiol. 2013;8(6):769–83. https://doi.org/10.2217/fmb.13.47. PubMed PMID: 23701332.
184. Baym M, Stone LK, Kishony R. Multidrug evolutionary strategies to reverse antibiotic resistance. Science. 2016;351(6268):aad3292. https://doi.org/10.1126/science.aad3292. PubMed PMID: 26722002; PMCID: PMC5496981.
185. Doern CD. When does 2 plus 2 equal 5? A review of antimicrobial synergy testing. J Clin Microbiol. 2014;52(12):4124–8. Epub 20140611. https://doi.org/10.1128/JCM.01121-14. PubMed PMID: 24920779; PMCID: PMC4313275.
186. Yilancioglu K, Cokol M. Design of high-order antibiotic combinations against M. tuberculosis by ranking and exclusion. Sci Rep. 2019;9(1):11876. Epub 20190815. https://doi.org/10.1038/s41598-019-48410-y. PubMed PMID: 31417151; PMCID: PMC6695482.
187. Larkins-Ford J, Degefu YN, Van N, Sokolov A, Aldridge BB. Design principles to assemble drug combinations for effective tuberculosis therapy using interpretable pairwise drug response measurements. Cell Rep Med. 2022;3(9):100737. Epub 20220908. https://doi.org/10.1016/j.xcrm.2022.100737. PubMed PMID: 36084643; PMCID: PMC9512659.
188. Chung CH, Chandrasekaran S. A flux-based machine learning model to simulate the impact of pathogen metabolic heterogeneity on drug interactions. PNAS Nexus. 2022;1(3):pgac132. Epub 20220722. https://doi.org/10.1093/pnasnexus/pgac132. PubMed PMID: 36016709; PMCID: PMC9396445
189. Ma S, Jaipalli S, Larkins-Ford J, Lohmiller J, Aldridge BB, Sherman DR, Chandrasekaran S. Transcriptomic signatures predict regulators of drug synergy and clinical regimen efficacy against tuberculosis. MBio. 2019;10(6):Epub 20191112. https://doi.org/10.1128/mBio.02627-19. PubMed PMID: 31719182; PMCID: PMC6851285.
190. Allen RC, Pfrunder-Cardozo KR, Hall AR. Collateral sensitivity interactions between antibiotics depend on local abiotic conditions. mSystems. 2021;6(6):e0105521. Epub 20211130. https://doi.org/10.1128/mSystems.01055-21. PubMed PMID: 34846167; PMCID: PMC8631318.
191. Barrena N, Valcarcel LV, Olaverri-Mendizabal D, Apaolaza I, Planes FJ. Synthetic lethality in large-scale integrated metabolic and regulatory network models of human cells. NPJ Syst Biol Appl. 2023;9(1):32. Epub 20230715. https://doi.org/10.1038/s41540-023-00296-3. PubMed PMID: 37454223; PMCID: PMC10349875.
192. Kaelin WG Jr. Synthetic lethality: a framework for the development of wiser cancer therapeutics. Genome Med. 2009;1(10):99. Epub 20091027. https://doi.org/10.1186/gm99. PubMed PMID: 19863774; PMCID: PMC2784312.
193. Navare AT, Mast FD, Olivier JP, Bertomeu T, Neal ML, Carpp LN, Kaushansky A, Coulombe-Huntington J, Tyers M, Aitchison JD. Viral protein engagement of GBF1 induces host cell vulnerability through synthetic lethality. J Cell Biol. 2022;221(11):Epub 20221028. https://doi.org/10.1083/jcb.202011050. PubMed PMID: 36305789; PMCID: PMC9623979.

194. Mast FD, Navare AT, Van Der Sloot AM, Coulombe-Huntington J, Rout MP, Baliga NS, Kaushansky A, Chait BT, Aderem A, Rice CM, Sali A, Tyers M, Aitchison JD. Crippling life support for SARS-CoV-2 and other viruses through synthetic lethality. J Cell Biol. 2020;219(10) https://doi.org/10.1083/jcb.202006159. PubMed PMID: 32785687; PMCID: PMC7659715.
195. Pratapa A, Balachandran S, Raman K. Fast-SL: an efficient algorithm to identify synthetic lethal sets in metabolic networks. Bioinformatics. 2015;31(20):3299–305. Epub 20150617. https://doi.org/10.1093/bioinformatics/btv352. PubMed PMID: 26085504.
196. Antimicrobial Resistance C. Global burden of bacterial antimicrobial resistance in 2019: a systematic analysis. Lancet. 2022;399(10325):629–55. Epub 20220119. https://doi.org/10.1016/S0140-6736(21)02724-0. PubMed PMID: 35065702; PMCID: PMC8841637.
197. O'Neill J, Resistance RoA, Trust W. Tackling drug-resistant infections globally: final report and recommendations: review on antimicrobial resistance; 2016.
198. Schaberle TF, Hack IM. Overcoming the current deadlock in antibiotic research. Trends Microbiol. 2014;22(4):165–7. https://doi.org/10.1016/j.tim.2013.12.007. PubMed PMID: 24698433.
199. Merrikh H, Kohli RM. Targeting evolution to inhibit antibiotic resistance. FEBS J. 2020;287(20):4341–53. Epub 20200608. https://doi.org/10.1111/febs.15370. PubMed PMID: 32434280; PMCID: PMC7578009.

Chapter 4
Unveiling the Proteomic Landscape of Bacterial Virulence and Antibiotic Resistance Mechanisms

Adriana Canedo Miranda and Cristiano Valim Bizarro

4.1 Introduction

Antimicrobial resistance (AMR) refers to the ability of a microorganism to resist the effects of an antimicrobial drug (AMD) that was initially effective in treating infections caused by that microorganism [61]. Standard treatments become ineffective against infections caused by resistant microorganisms, worsening clinical outcomes, and increasing the risk of disease spread to others. As a trait acquired by previously susceptible microorganisms, AMR occurs naturally over time as a consequence of genetic changes. It mainly involves the occurrence and selection of spontaneous mutations or the acquisition of resistance genes through horizontal gene transfer (HGT) [11, 30].

In addition to acquired resistance, microorganisms can have natural resistance to an AMD. Natural resistance can be further divided into intrinsic and induced resistance [47]. Intrinsic resistance is an innate ability to resist a chemotherapeutic agent, with all or almost all isolates of a given microbial species being able to resist that antimicrobial agent independent of previous exposure to it. As an example, the outer membrane (OM) of Gram-negative bacteria is a cellular barrier that makes them naturally resistant to many compounds that are effective antibiotics against Gram-positive bacteria [11, 28, 47]. The lack of a target site can also provide intrinsic resistance. For example, the absence of a cell wall makes *Mycoplasma* species resistant to glycopeptides and β-lactam antibiotics [35]. In the case of induced

A. C. Miranda · C. V. Bizarro (✉)
Centro de Pesquisas em Biologia Molecular e Funcional (CPBMF) and Instituto Nacional de Ciência e Tecnologia em Tuberculose (INCT-TB), Escola de Ciências da Saúde e da Vida, Pontifícia Universidade Católica do Rio Grande do Sul (PUCRS), Porto Alegre, Brazil

Programa de Pós-Graduação em Biologia Celular e Molecular, Pontifícia Universidade Católica do Rio Grande do Sul, Porto Alegre, Brazil
e-mail: miranda.adriana@edu.pucrs.br; cristiano.bizarro@pucrs.br

V. Soni, A. S. Akhade (eds.), *Antimicrobial Resistance: Factors to Findings*,
https://doi.org/10.1007/978-3-031-65986-7_4

resistance, although the genes conferring resistance are also naturally occurring in the resistant organism, they are only expressed to levels that afford resistance after exposure to AMD. Increased expression of multidrug-efflux pumps after drug exposure is a classic example of induced resistance [47].

Evolutionarily speaking, exposure to antimicrobials could also "induce" resistance, through the so-called increased evolvability hypothesis, put forward in the context of antibiotics and the evolution of antibiotic resistance. According to this hypothesis, antibiotic exposure would lead to an increase in the background level of spontaneous mutations, and, consequently, the rate of resistant bacteria being selected [46]. This antibiotic-induced increase in mutation rates could be the result of the selection of hypermutator phenotypes: cells with constitutively high mutation rates [45]. These cells usually possess loss-of-function mutations in proofreading or mismatch repair enzymes [46]. Alternatively, drug exposure could lead to a "transient hypermutation" phenotype by inducing the expression or activity of error-prone DNA polymerases [45, 46]. Additionally, recent studies suggest a link between persistence and resistance. For example, upon exposure to lethal doses of ciprofloxacin or rifampicin, the minor population of mycobacterial persisters that remain alive upregulate the error-prone DNA polymerase DnaE2 in a pathway dependent on SOS response activation, leading to higher mutation rates and rapid evolution of antibiotic resistance [49].

4.1.1 Mechanisms of Resistance

Mechanisms of resistance to AMDs can be broadly classified into four categories (Fig. 4.1): (i) drug uptake limitation, (ii) drug target modification, (iii) drug inactivation, and (iv) increase in efflux pump activity.

4.1.1.1 Drug Uptake Limitation

As already mentioned, the presence of the outer membrane makes Gram-negative bacteria intrinsically resistant to many compounds that are effective against Gram-positive microorganisms [11, 28, 47]. The presence of this structure makes them more impermeable to compounds of different chemical classes, limiting drug uptake.

Additionally, there are three main mechanisms for limiting the amount of drug uptake by bacteria: (i) reduction of porin levels, (ii) mutations that alter the transport of molecules through the porin channel, changing its selectivity, and (iii) production of alternative porins [43].

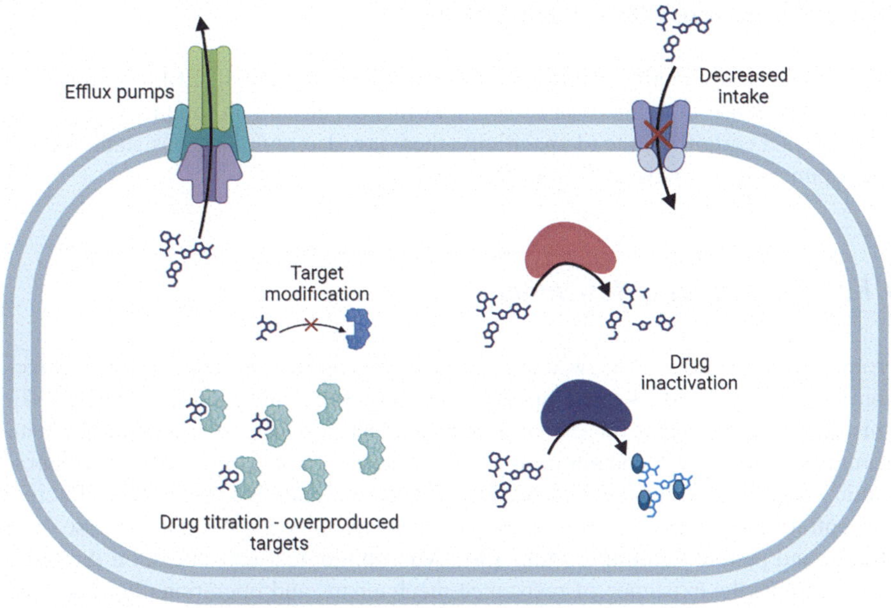

Fig. 4.1 Mechanisms of antimicrobial resistance. The five different mechanisms of resistance within the bacterial cell. See text for more details. (Created with BioRender.com)

4.1.1.2 Drug Target Modification

Many enzymes serve as direct targets of AMDs [4]. Therefore, mutations that affect their structure, particularly on their drug-binding site, reducing drug affinity or even preventing binding at all, are a common resistance mechanism. These modifications allow the target to continue functioning even in the presence of the AMD, making it inefficient. There can also be an increase in target expression, which titrates out the effect of the antimicrobial agent.

Besides drug target modification as a result of mutations occurring directly on their coding sequences, structural changes on drug targets leading to resistance can be accomplished by modifying enzymes [50]. For example, the activity of methyltransferases on the 23S rRNA of bacterial ribosomes can lead to resistance to macrolides [12].

4.1.1.3 Drug Inactivation

One of the most common mechanisms of resistance involves the direct modification of AMD structures. A diverse set of enzyme activities, including hydrolases, transferases, and oxidoreductases, are known to change the structure and in that way the activity of AMDs. These include β-lactamases, esterases, acetyltransferases, glycosyltransferases, phosphotransferases, monooxygenases, and lyases, among others [4].

4.1.1.4 Increase in Efflux Pump Activity

This allows for the active ejection of the AMD out of the cell. Efflux pumps can suffer an increase in expression to augment this expel, and although some pumps can present a certain degree of specificity, there are also multidrug-resistant (MDR) efflux pumps that behave in the opposite manner.

4.1.2 Quantifying Resistance

From the clinical point of view, a microorganism is resistant to an AMD if an infection with this agent does not respond to therapy using this particular AMD [57]. This depends not only on the antimicrobial activity per se, quantifiable with Minimum Inhibitory Concentration (MIC) determinations (Fig. 4.2), since therapeutic success also depends on AMD's pharmacokinetic properties, its effective concentration on the site of infection, patient tolerability to potential adverse drug effects, and patient's immune status [30]. Microbiological resistance, on the other hand, implies the existence of resistance mechanisms and operationally is defined by means of MIC data [57]. The latter definition is particularly useful in studies of ecology and evolution of resistance, where no clinical breakpoints can be defined.

4.1.3 Threat of AMR Dissemination

AMR stands as a prominent global health menace, driven by the ever-growing evolution and emergence of disease-causing agents. This relentless process complicates treatment efficacy, as the development of drug resistance renders AMDs less effective, ultimately contributing to treatment challenges and failures [30]. Currently, there is a wide misuse and overuse of AMDs, such as the inadequate use to treat parasitic or viral infections or the premature withdrawal of antibiotic use due to clearance of symptoms. These misuses overexpose pathogens to these drugs, promoting the development and dissemination of AMR. One common example is the misuse of antibiotics during colds, which promotes the selection and spread of antibiotic-resistant bacteria. Another example is recurrent urinary tract infection, a condition in which patients tend to stop taking antibiotics once the symptoms are gone, yet sometime later, the remaining colonies of bacteria proliferate once again [61]. Depending on the case, the individual may have to call upon other drugs since the infection can already present resistance to the previously used treatment. There is also extensive use of antibiotics in the animal food industry, either in a preventative manner [47] or as growth promoters [26], which exposes these animals to antimicrobial agents constantly, which may in turn cause pathogens to develop resistance that can then be passed on to humans consuming these animals. A transfer of resistance genes from different bacterial species may also occur. This leads to the

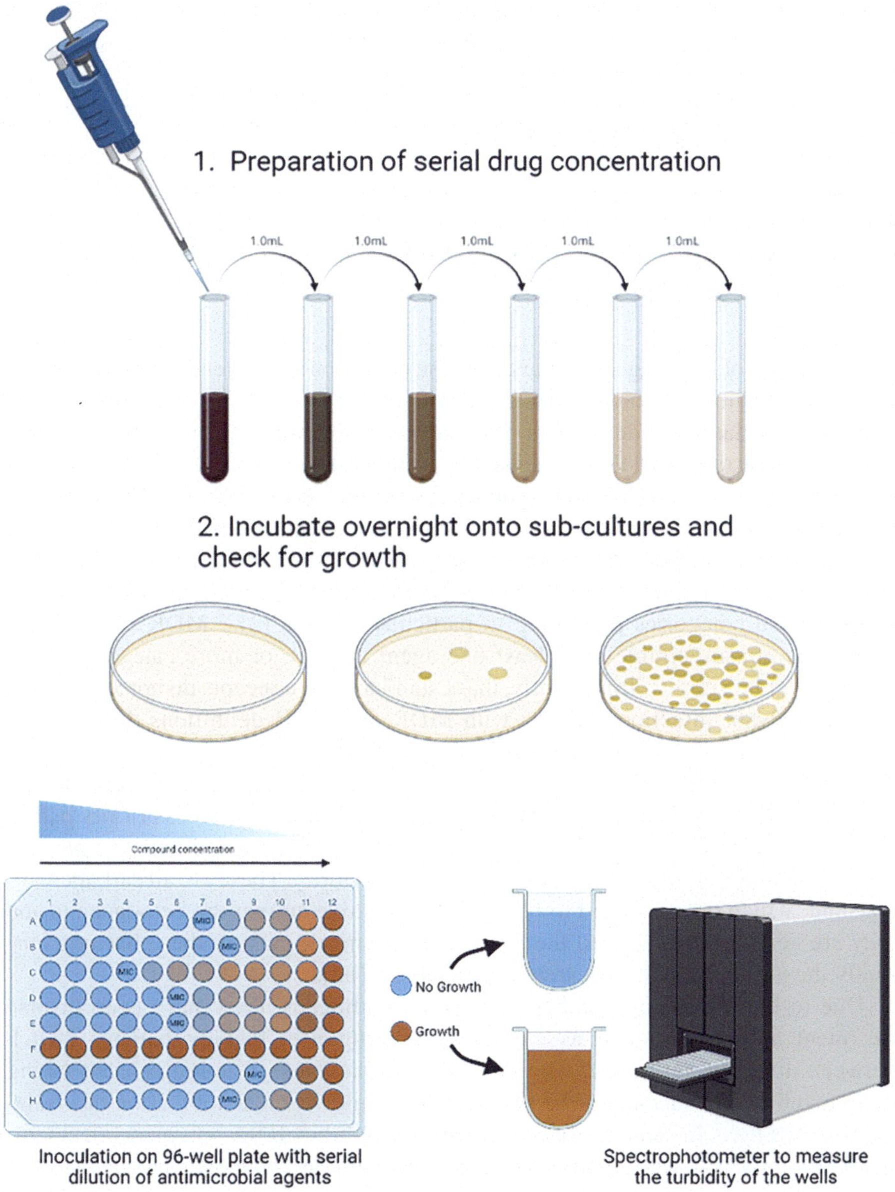

Fig. 4.2 Broth microdilution assay for measuring Minimum Inhibitory Concentration. First, a serial dilution of the antimicrobial agents is prepared to be then sub-cultured onto agar plates and determine the lowest concentration in which bacteria are prevented from growing. This approach can also be inoculated in a microplate to evaluate diverse antimicrobial agents with more dilutions in which the MIC is visually determined from the turbidity of the well using a spectrophotometer. (Created with BioRender.com)

ineffectiveness of available treatments, making these diseases more difficult or even impossible to treat until new drugs are developed. This does not only cause a great impact on treatment, making it more extensive, but also increases the rate of mortality from untreatable diseases as well as increasing the number of hospitalizations, which in turn, increases healthcare costs.

4.1.4 Defining Patterns of Resistance Among Clinically Relevant Bacteria

The necessity of ongoing surveillance for the evolution and spread of diverse bacterial strains exhibiting complex patterns of multidrug resistance prompted the creation of standardized definitions for acquired resistance [27]. Definitions for multidrug-resistant (MDR), extensively drug-resistant (XDR), and pandrug-resistant (PDR) were standardized jointly for *Staphylococcus aureus*, *Enterococcus* spp, *Enterobacteriaceae*, *Pseudomonas aeruginosa,* and *Acinetobacter* spp. [27]. PDR isolates are nonsusceptible to all agents in all antimicrobial categories considered, while XDR isolates are nonsusceptible to at least 1 agent in all but 2 or fewer antimicrobial categories pertaining to particular subsets [27]. MDR strains were defined as nonsusceptible to at least one agent in three or more categories from these particular subsets. Importantly, these standardized concepts do not apply to all clinical-relevant infectious agents, with MDR and XDR definitions differing for organisms such as *Mycobacterium tuberculosis* [14].

There are several pathogens that portray a higher demand for new treatment due to their emerging and stronger resistant strains. According to WHO's priority pathogen list for the development of new antibiotics in 2017, bacteria such as *Acinetobacter baumannii* and *Enterobacteriaceae* are of critical priority. These are all carbapenem-resistant, which are supposed to be the most effective among the beta-lactams and they are generally considered the last-resort treatment, making these bacteria currently the top priority for research of new targets [62].

Due to the increasing number of resistant pathogens, there have been diverse treatment approaches such as variability in drug use, making patients switch between different antibiotics with different mechanisms of action or even taking them together. For instance, when treating tuberculosis, the combination of isoniazid (inhibits mycolic acid synthesis), rifampicin (inhibits RNA polymerase), ethambutol (inhibits cell wall synthesis), and pyrazinamide (unknown mechanism), is carried out to obstruct three pathways at once in case the clinical isolate presents any kind of resistance mechanism toward any of them [55].

4.2 Proteomics

Proteomics is the analysis of proteomes within organisms or systems. These proteomes are the entire protein content within these compartments under determined conditions [29]. A wide range of biological systems and processes can be studied using proteomics approaches, such as single organisms, groups of organisms, individual organs, or even single cells under different physiological conditions or drug treatments. Considering the involvement of numerous proteins in both antimicrobial resistance and the mechanisms of action of AMDs, adopting proteome-wide approaches is indispensable for gaining a comprehensive understanding of these intricate processes [47]. There is a wide variety of techniques used to study these conditions, including mass spectrometry technologies and bioinformatics (Table 4.1).

4.2.1 Initial Approaches

4.2.1.1 Gel Based

It is a widely used technique using proteins' size and charge properties to separate and analyze them (Fig. 4.3). Firstly, a sample preparation is carried out in which the proteins must be extracted from the biological material of interest. Techniques may vary depending on the type of the sample and may include protein extraction and purification, or even reduction or alkylation to guarantee the adequate solubilization and removal of other unwanted substances. This is then followed by gel electrophoresis, which is the step that appropriately separates proteins based on their size and charge [9, 15, 44, 63].

Table 4.1 Comparison of different proteomic approaches and their methodological divergencies

	Approach	Identification
Shotgun proteomics	Discovery-based to analyze a protein complex aiming to identify as many proteins as possible inside the given sample	Proteins are digested into peptides for MS analysis. This data is then matched against a database to identify the diversity of proteins within the sample
Targeted proteomics	Measure the abundance of determined proteins of interest inside a complex	Specific protein targets must be previously determined. Peptide biomarkers can be employed to identify proteins of interest. Selected reaction monitoring or parallel reaction monitoring may be used for the quantification of these peptides.
Top-down proteomics	Direct analysis of intact proteins to identify species, including PTMs and sequence variants	High-resolution MS and fragmentation methods
Bottom-up proteomics	Characterizes complex protein samples	Prior digestion of proteins into smaller peptides to be then separated and identified with MS

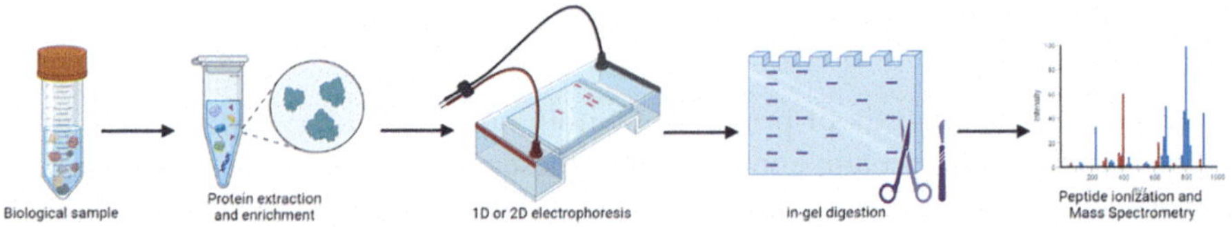

Fig. 4.3 Gel-based proteomics timeline. Starting out with the extraction and enrichment of proteins within the biological sample in question. Proteins are then subjected to electrophoresis for their proper size-based separation to be further digested, making them appropriate for mass spectrometry analyses, in which peptide spectra are matched against their sequences. (Created with BioRender.com)

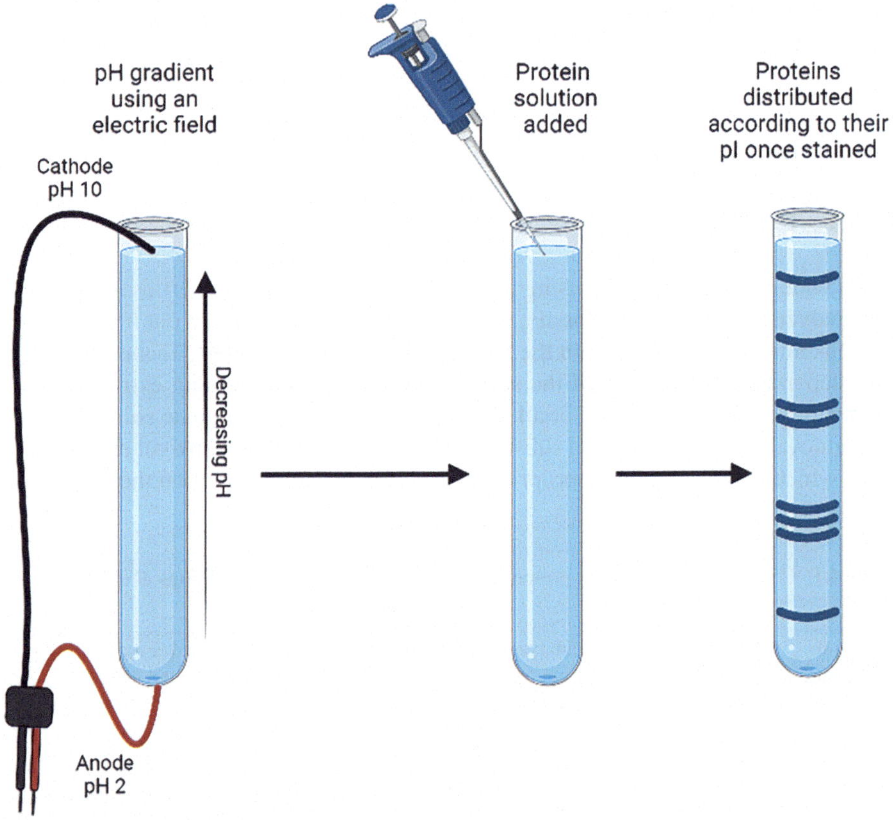

Fig. 4.4 Isoelectric Focusing. Technique used for protein separation based on their isoelectric point. (Created with BioRender.com)

The most used techniques are SDS-PAGE and Isoelectric Focusing (Fig. 4.4). SDS-PAGE (Sodium Dodecyl Sulfate Polyacrylamide Gel Electrophoresis) separates proteins based on their molecular weight. The proteins are first denatured using Sodium Dodecyl Sulfate (SDS) detergent and are then loaded into the

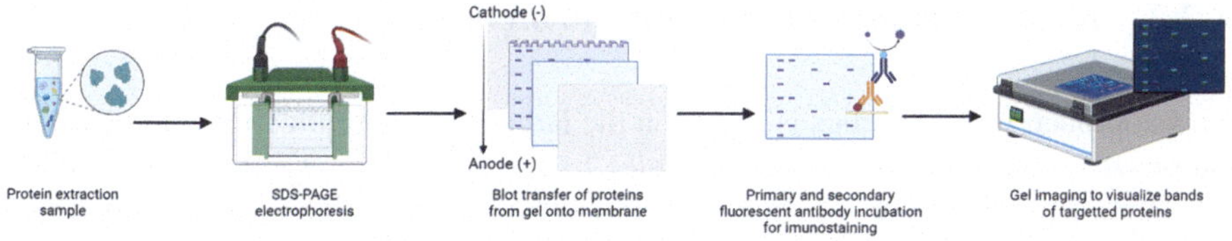

Fig. 4.5 Western blotting technique procedure timeline. This procedure is widely used for the separation of proteins based on their size. See text for more information. (Created with BioRender.com)

polyacrylamide gel, in which the proteins will migrate once an electric current is applied. Smaller proteins will have a faster migration.

Moreover, Isoelectric Focusing (IEF) uses proteins' isoelectric point, which is the pH at which it carries no net charge, to separate them. In this case, an electric field is also applied in a gel matrix with a pH gradient in which proteins will be loaded and migrate toward their corresponding isoelectric point. Once they are separated, protein bands can be visualized using stains such as Coomassie Brilliant Blue or silver stains [38].

Other visualization methods, like immunoblotting (Western Blotting) (Fig. 4.5) can be employed to target specific proteins with antibodies. Finally, mass spectrometry can be used for further peptide identification once the bands are obtained.

4.2.1.2 Peptide Mass Fingerprinting

This technique uses proteins' unique mass profiles for identification. The proteins are first enzymatically digested into smaller peptides by proteolytic enzymes like trypsin. This mixture of peptides is then analyzed with mass spectrometry, measuring the mass-to-charge ratio (m/z) of ions. Matrix-assisted laser desorption/ionization (MALDI) and electrospray ionization (ESI) are commonly employed techniques for the ionization of the peptides to be subjected to MS. Once all the mass spectra are obtained, a database search is performed in which the spectra are compared against theoretical peptide masses from a protein sequence database. This database can be general or customized, containing only the proteins of interest. The theoretical and experimental peptide masses are then compared with each other to finally identify the best matches and use them to infer the presence and identity of the proteins in the sample. Finally, the identified peptides will collectively form a peptide mass fingerprint, which provides a unique pattern that will be used to properly identify the protein in question [28, 29].

4.2.2 High-Throughput Shotgun Proteomics

This approach in proteomics is based on the large-scale identification and quantification of proteins in a biological sample. To do so, proteins go through standard digestion, followed by mass spectrometry analysis to infer the presence and quantify the corresponding proteins [29].

1. Proteins of interest must be extracted from the biological samples according to their following extraction procedure.
2. Once the proteins have been extracted, enzymatic digestion is carried out to generate a peptide mixture out of these proteins.
3. The peptides will be subjected to liquid chromatography (LC) in order to separate them based on their physicochemical properties. Some of these properties are hydrophobicity and/or charge. Reverse-phase chromatography, strong cation exchange chromatography, and high-pH reverse-phase chromatography are a few examples of this methodology.
4. The separated peptides are then subjected to mass spectrometry analysis. There are two main approaches for shotgun proteomics data acquisition (Fig. 4.6):

   (a) Data-dependent acquisition (DDA)

   In this method, only a fixed number of the highest abundance precursor ions from the first stage of each tandem MS cycle are fragmented during the second stage, limiting the identification of low-intensity peptides.

   (b) Data-independent acquisition (DIA)

   In this approach, all the peptides from the first stage of each tandem MS cycle will be fragmented in the second stage. In the broadband DIA, the fragmentation is performed simultaneously on all precursor ions from each cycle. Alternatively, the scanning range (typically 400–1200 m/z) can be divided into a series of intervals that are scanned sequentially, the so-called Sequential Window Acquisition of all Theoretical Mass Spectra (SWATH-MS).
5. Database search and peptide identification: Once the spectra have been detected, these are then compared to protein sequence databases in order to detect and identify the corresponding peptides based on the best experimental spectrum match. Algorithms such as SEQUEST, Mascot, and MaxQuant can be employed for this.
6. Quantitative analysis: The peptides may be then quantified to visualize differential expression patterns at the proteome level. This elucidates the abundance of determined proteins within the sample. A few common quantitative approaches include:

   (a) Label-free quantification: This approach compares the peak intensity or the number of spectral counts for each peptide between different samples.

   (b) Isotope labeling: Isotopic tags can be used among samples. Some samples may also be left untagged. Therefore, a comparison between both tagged and untagged samples may be done with an accurate quantification of both.

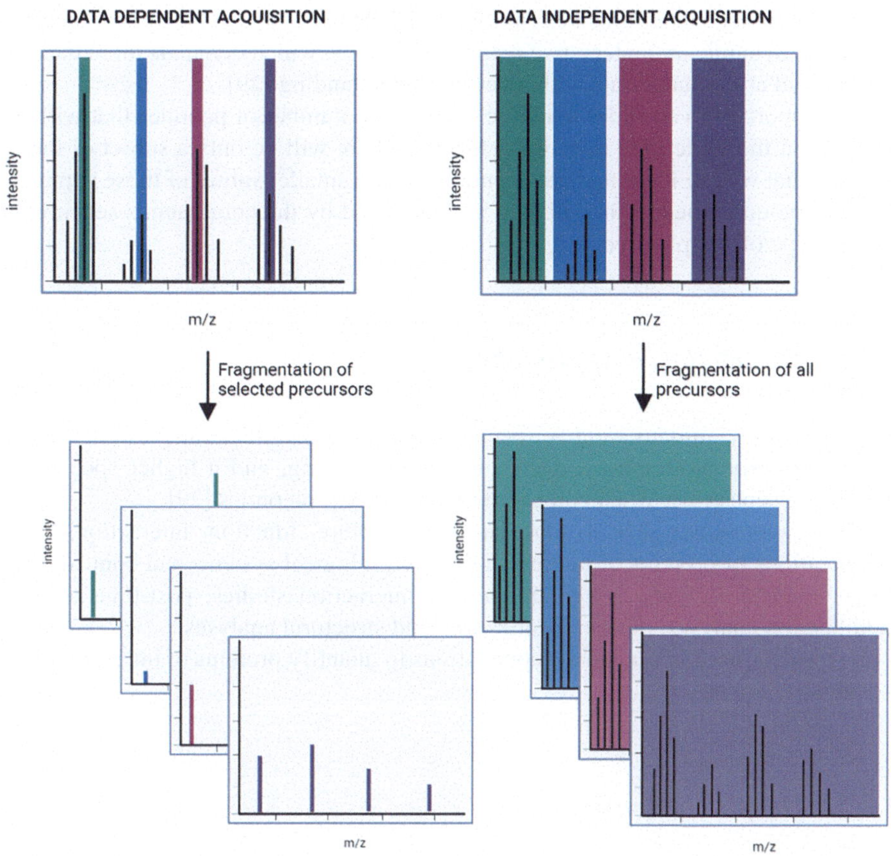

Fig. 4.6 Graphical representation of data-dependent and data-independent acquisition approaches. (Created with BioRender.com)

(c) Tandem mass tags or Isobaric Tags (iTRAQ): This allows for specific peptide labeling in diverse samples.

7. Data analysis: Bioinformatics tools are used to analyze the large datasets obtained across the experiments. Statistical analyses, quality control, protein clustering, and other downstream analyses are included in this step.

4.2.3 Quantitative Methods

This approach is specific for quantifying the amount of determined proteins in a sample of interest. There is a need to quantify proteins rather than just identify them to elucidate biological processes at a cellular level. These methods may, for instance, contribute to understanding why there may be an increased number of certain types of proteins under a determined circumstance. This kind of analysis is more complex due to the dynamic and interactive nature of proteins, which demands more data to be analyzed at the same time with a deeper understanding [29].

It is important to consider that, due to the large number of peptides that will be analyzed at the same time in a single MS run, there will be only a subset of these peptides that will be identified and from this only a smaller subset of those peptides that will be quantified. This will also be influenced by the equipment's sensitivity and the type of sample used.

4.2.4 Target-Based Approaches

When the study aims to identify specific target proteins, the strategies must have optimized chromatography, a deeper instrument tuning, and a higher specificity rather than sensibility, when compared to discovery proteomics [29].

These approaches seek to elucidate the structure, function, interactions, and modifications of proteins of interest in diverse biological systems and contexts. To do so, there are plenty of protein–protein interaction studies, post-translational modification analyses, protein localization, and structural analyses.

It is widely used in diagnostic applications to quantify proteins of interest within a complex sample.

4.2.5 Deep Proteomes

This is a challenging task in the area since these proteomes are present in limited quantities and tend to be in complex biological samples. Because of this, there must be an increase in sensitivity in which enrichment techniques such as subcellular fractionation, immunoprecipitation, affinity-based enrichment, and PTM enrichment techniques may be used. In order to reduce the sample complexity, prefractionation techniques like gel-based separation or liquid chromatography may be applied to then carry out enhanced mass spectrometry techniques: Orbitrap and Fourier-transform ion cyclotron resonance (FT-ICR) analyzers, which present improved sensitivity and accuracy; and selected reaction monitoring or parallel reaction monitoring, which allow for the specific detection and quantification of predefined peptides [29]. Finally, signal amplification methods may be carried out to have a further refinement of the deep proteome analysis like antibody-based signal amplification, in which secondary antibodies conjugated with enzymes are used, as well as isotope tagging.

4.3 Comparative Proteomics

This branch of research compares protein profiles or differences in protein expression among different biological samples or organisms and under different biological contexts. This allows for the understanding of differences in protein abundance and their possible reasons, post-translational modifications, protein–protein interactions, or even protein localization as a response to diverse mechanisms, developmental stages, or environmental conditions.

4.3.1 Membrane and Cell-Surface Proteomics

Membrane Proteomics

1. Protein extraction must be performed due to the cell membrane's hydrophobic nature.
2. Enrichment techniques.
3. Protein separation methods such as electrophoresis or liquid chromatography.
4. MS analysis: shotgun or targeted proteomics.
5. Data analysis.

Cell-Surface Proteomics

1. Cell-surface biotinylation to isolate and label surface proteins using biotin for the cell to be lyzed afterward
2. Mass spectrometry analysis
3. Spatial proteomics to visualize a more precise localization of the cell-surface proteins
4. Functional characterization by investigating different interactions or the cell-surface protein expression dynamics when exposed to different stimuli

4.3.2 Interactome Proteomics

Protein–Protein

1. Antibodies to specific target proteins may be adopted to identify coupled partners.
2. Identification and characterization using affinity purification coupled with mass spectrometry.
3. Stable isotope labeling or label-free quantification can be used to quantify differences in protein interactions under different conditions.

4. Identified interactions can be further validated with co-immunoprecipitation, co-localization, or functional assays.

Protein–Metabolite

1. Proteomics is combined with metabolomics.
2. Affinity capture to enrich and identify interactions.
3. Comparison of proteomic profiles when treated with different metabolites: changes in protein abundance and post-translational modifications.
4. Functional analysis of interactions: subcellular localization, structural, and binding affinity analyses.
5. Integration with pathway analysis tools.

4.4 Technical Advances

4.4.1 Single-Cell Proteomics

This involves the proteomic analysis of individual cells. This allows for a more detailed overview of cellular heterogeneous functioning and identification of the different protein expression patterns at the single-cell level. In this manner, this technique can potentially unveil insights into cellular dynamics, variability, and diversity among different cells within a biological system [21].

4.4.2 Metaproteomics

In contrast to single-cell proteomics, this approach studies collective proteins or diverse proteomes within the same community in an ecosystem or environment. This allows for a wider understanding of the ecosystem as a whole, being able to identify different roles within the community.

4.4.3 Proteogenomics

This involves the integration of both genomic and transcriptomic data along with proteomics to provide a more comprehensive view of the expression patterns and functioning of that organism. Database construction is involved and allows for the prediction of novel proteins [37].

These different approaches lead to both the discovery of new functional peptides and the deeper understanding ranging from single cells up to entire communities. These analyses lead to both the elucidation of antimicrobial resistance mechanisms, which may involve diverse newly discovered peptides through proteogenomics, for instance, or also virulence factors. Many studies have shown strong relationships between virulence factors and resistance phenotypes, meaning they can complement each other instead of acting individually [17, 54].

4.5 Techniques

4.5.1 Whole Genome Sequencing

This process involves the sequencing of the entire DNA genetic information of an individual, composed of coding and noncoding regions. There are many different approaches such as Sanger sequencing and next-generation sequencing.

4.5.2 Mass Spectrometry

Lipidomics: This branch of metabolomics embraces the large-scale study of diverse pathways involved in lipid metabolism. There are four main lipid profiling technologies:

GC-MS (Gas chromatography): Ideal for free fatty acids and steroids. The lipids are initially transformed into volatile esters.

LC-MS: This type of profiling is used for phospholipids, neutral lipids, and sphingolipids since they are less volatile.

Direct infusion-MS: This analysis skips the step of chromatography in order to avoid its negative effects, making it faster but does not separate those compounds that present the same *m/z*. Therefore, it is used for certain lipid types.

Imaging-MS: This technique gathers both molecular information and spatial resolution and is basically the identification of biomarkers in the present tissues or cells via imaging mass spectrometry by ionizing those lipid biomarkers or metabolites in the samples.

There are diverse applications in metabolic, dermatologic, and neurologic diseases. Diseases such as diabetes, hypertension, and obesity require precise detection methods for an effective treatment to be carried out. Also, psoriasis and atopic dermatitis are two dermatologic diseases that require the analysis of lipid metabolites [16, 59, 60]. Finally, lipids play an important role in neurological functioning and diseases, considering neurons' myelin sheath's critical role in neurodegeneration and neurodegenerative diseases such as Alzheimer's disease [64].

Proteomics: different from lipidomics, this involves the large-scale analysis of proteins, including their identities, structure, and chemical properties.

Differential Gene Expression Analyses: This branch is mainly the in silico analysis of gene expression patterns, focusing mainly on comparing significant phenotypic differences among different environments.

Interactome: Besides the identification of both proteins and lipids themselves in different situations with different techniques, there are many interactions involving these molecules that may serve as biomarkers themselves too. The interactome aims to study these interactions involving macromolecules within a cell to identify functional properties as well as their physical interactions.

MLVA Analysis: (Multiple Locus Variable-Number Tandem Repeat Analysis)

This technique generates a DNA fingerprint by analyzing the number of short tandem repeat sequences. This molecular typing technique is used to characterize bacterial strains for epidemiological research. By analyzing the repeat sequences among the genome, allowing for the detection of different strains among homogeneous species, researchers can identify these variants which may also aid in the detection of resistant strains, for instance.

4.6 Multi-omics in AMR Research

Given that AMR is a growing public health concern, there is a high demand for the discovery of new molecular targets, the elucidation of unknown mechanisms of action, and the development of new drugs. To do so, firstly basic research must be done to understand the biological functioning of bacteria and how these organisms are developing strategies to escape antimicrobial action as well as which mutations are being selected during resistance development. Multi-omics analyses integrating genomics, transcriptomics, and proteomics can yield invaluable information to effectively address this formidable challenge [10] (Fig. 4.7). Comparative genomic analyses of susceptible and resistant clinical isolates can facilitate the identification of single nucleotide polymorphisms (SNPs) and other genetic changes related to the resistance trait [24, 25]. Besides the possibility of pinpointing the genetic basis of the mechanisms of resistance at play, genetic changes associated with resistance

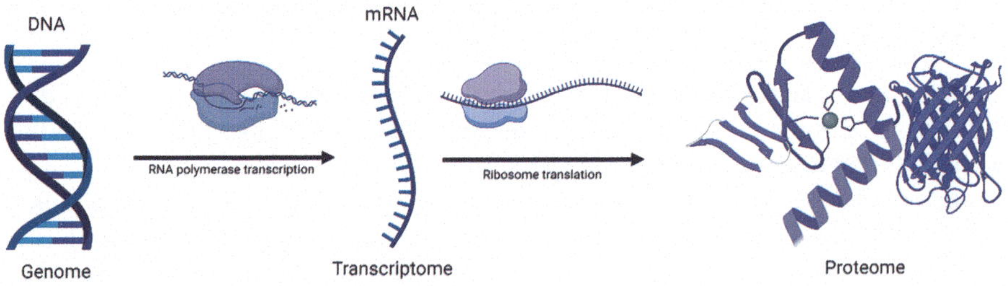

Fig. 4.7 The different analysis levels within an organism. Genomic analyses allow for the precise identification of single nucleotide polymorphisms. Transcriptomic data, such as RNA-seq, can be used to quantify expression levels among different transcripts, as well as the detection of operons, among others. Ribosome profiling can provide a direct overview of the "translatome", the subset of transcripts that are being translated in a particular biological context. Finally, proteomic analyses take a closer look into the protein's peptides within a biological sample. (Created with BioRender.com)

may reveal compensatory adaptations that reduce the fitness cost of resistance mutations for clinical isolates that are not under selective pressure of the AMD [51]. These compensatory adaptations may underpin alternate patterns of gene expression that compensates for enzymes underproduced or inactivated in resistant isolates. A classic example is the hyperexpression of the alkyl hydroperoxidase AhpC in isoniazid (INH)-resistant clinical isolates of *M. tuberculosis* deficient in the activity of the catalase-peroxidase KatG. INH is a prodrug that must be activated by KatG inside mycobacterial cells to give rise to the pharmacologically active INH-NAD adduct. Mutations that eliminate KatG activity prevent the activation of INH and are the major mechanism of INH resistance found in clinical isolates. Nevertheless, this resistance mechanism comes with a fitness cost, as KatG is a key enzyme in the metabolism of organic peroxides. Secondary mutations promoting the hyperexpression of AhpC compensate for KatG deficiency, protecting the bacilli against the toxic effects of these organic species [52]. The systematic integration of transcriptomic and genomic data has the potential to reveal evolutionary adaptations in terms of coordinated gene expression responses that compensate for the fitness costs of resistance traits. This approach was applied to study the changes in transcriptional levels related to compensatory mutations found in rifampicin-resistant *M. tuberculosis* [65].

Transcriptome analysis using RNA-seq data as well as Ribo-seq data is appropriate since the mRNA levels will reflect the transcriptional profile of that microorganism in a particular environment. However, even though ribosome profiling gives more evidence of a peptide being translated, this approach only visualizes those fragments inside the ribosome, being up to 30 nucleotides in length and making them harder to map with their corresponding mRNAs. Now RNA-seq data targets all the different RNA molecules within a sample, being up to 90 nucleotides in length, depending on the molecule type. Finally, proteome analysis helps elucidate both the quantitative and qualitative profile of the bacteria of interest. In this manner, peptide-spectrum matches (PSMs) reflect which proteins are found within a sample. In other words, PSMs provide experimental proof that those mRNA sequences are being translated and their corresponding proteins are present in a particular sample. Some mechanisms involved in antimicrobial resistance, such as changes in expression of the drug targets by either upregulation, to titrate the effect of the drug, or downregulation, to avoid the drug encountering its target at all, may be analyzed with proteomic approaches as well as RNA-seq data. If the gene is being transcribed, the mRNA transcripts can then be quantified, and once it is translated, its peptides can also be identified with mass spectrometry. The expression of beta-lactamases, which degrade a specific class of antibiotics, along with changes in toxin–antitoxin systems or even changes in membrane proteins, can also be quantified with proteomic approaches [58]. A decrease in metabolic activity is also a phenotypic profile bacterial cells adopt to survive antimicrobial activity. This decrease is caused by a downregulation of a wide variety of genes, and protein-metabolite analyses as well as common proteomic approaches and RNA-seq data can help confirm this dormant state, as well as identify which specific proteins are

Table 4.2 Databases used in diverse omics approaches for the characterization of resistance mechanisms

	Tool	Function
Databases	STRING	Known and predicted protein–protein interactions
	DAVID	Gene ontology terms, a comprehensive set of functional annotation tools to understand the biological meaning behind large lists of genes
	Pfam	Protein families
	InterPro	Protein domains and superfamilies
	SMART	Simple Modular Architecture Research Tool: used in the identification and analysis of protein domains within protein sequences
	KEGG	Collection of databases dealing with genomes, biological pathways, diseases, drugs, and chemical substances
	PDB	(Protein DataBase) 3D structural data of large biological molecules, such as proteins and nucleic acids
	PATRIC	Bacterial infectious diseases via integration of vital pathogen information with rich data and analysis tools
	CARD	Comprehensive Antibiotic Resistance Database collects and organizes reference information on antimicrobial resistance genes, proteins, and phenotypes. The database covers all types of drug classes and resistance mechanisms and structures its data based on an ontology
	Victors	Genes were experimentally observed to be necessary for virulence. Included are virulence factors for many different bacteria, viruses, parasites, and fungi, which are pathogenic to animals and humans
	VFDB	Virulence Factor Database for virulence factors in bacterial pathogens
	NDARO	National Database of Antibiotic Resistant Organisms (NDARO) to access AMR data to facilitate real-time surveillance of pathogenic organisms

involved in this process. Examples of databases that can be integrated into multi-omics approaches to study resistance mechanisms are provided in Table 4.2.

4.7 Demonstrative Published Analyses

4.7.1 Techniques Used for the Characterization of Resistance

To identify potential mutations in various strains, it is essential to obtain the reference genome of the wild-type strain, presumed to be the susceptible strain. This reference genome serves as a basis for comparing a known sequence with the whole-genome sequencing data of clinical isolates, enabling the potential identification of point mutations associated with resistance.

A popular example of a point mutation, which is a single nucleotide polymorphism (SNP), is *Mycobacterium tuberculosis*'s *inhA*(S94A) allele. Vilchèze and colleagues [56] were able to identify that this specific nucleotide substitution was sufficient to confer resistance to both Isoniazid and Ethionamide. This amino acid substitution of a serine for an alanine ends up compromising the protein's structure which leads to a decrease in the binding affinity of the activated drugs to their target in *M. tuberculosis*, allowing the bacteria to survive due to the drugs' inefficiency. To evaluate this, they introduced the *inhA*(S94A) point mutation by specialized linkage transduction, which involves the use of a bacteriophage to transfer this genetic material, into *M. tuberculosis*. Then they confirmed the presence of the mutated alleles by sequencing and evaluated the minimum inhibitory concentration (MIC) by treating the strains with increasing concentrations of both ETH and INH to analyze the phenotype of the mutated strains. They also evaluated the overexpression of the target using RT-PCR to be able to determine whether resistance was due to titration or the S94A mutation. X-ray crystallographic analyses were also executed to visualize the target's inhibition. This combination of techniques gives a broad overview of the possible mechanisms of resistance causing an ineffective antimicrobial treatment, but other approaches can be used in parallel to further elucidate the actual means of resistance within a microorganism.

Foudraine and colleagues [13] performed a multi-omics approach by combining susceptibility testing, whole genome sequencing (WGS), and also mass spectrometry to be able to detect the mechanisms of resistance at the proteomic level in both *E. coli* and *K. pneumoniae*. They aimed to verify whether known resistance mechanisms can explain resistance to beta-lactams, aminoglycosides, and fluoroquinolones. In this manner, they identified proteins that are associated with resistance and showed that liquid chromatography-mass spectrometry (LC-MS/MS) can be used as a complementary approach to further characterize resistance mechanisms. In this case, the researchers gathered bacterial isolates that were already known as resistant with prior knowledge of some phenotypic characteristics that could be contributing to resistance such as the hyperproduction of a beta-lactamase. Antimicrobial susceptibility testing was carried out using broth microdilution to determine the MIC of all isolates followed by whole genome sequencing to compare the genotypic and phenotypic resistant characteristics. Finally, LC-MS/MS was employed in Data-Dependent Acquisition to detect and quantify the specific peptides associated with the phenotypic profiles. Once all this experimental data was obtained, they carried out a COG (Clusters of Orthologous Groups) analysis with WebMGA to verify to what extent bacterial isolates were characterized in comparison to the information obtained with WGS. They also used SeqSphere to characterize and compare the isolates based on their core genome sequences, which evaluates the conserved genes that are supposed to be present in most strains of a species. Finally, BioNumerics was used to portray clusterization using minimum spanning trees. This represents the degree of genetic relatedness between the different isolates of each bacterium as well as clusters of those who are closely related, allowing for the detection of potential transmission chains and evolutionary relationships among them.

Another study performed by Barlow and colleagues [7] also highlights the importance of quantifying protein production when analyzing MICs. Their purpose was to investigate the impact of the membrane fusion protein AdeT1, which is the primary resistance mechanism found in *Acinetobacter baumannii*, on resistance in

E. coli. This protein is known to be a component of multidrug efflux pumps which convey resistance to diverse pathogens. The study included evaluating MICs after plasmid construction containing the protein-encoding gene *adeT1* and transferring it to *E.* coli to be then induced by IPTG. In other words, the authors attempted to characterize mechanisms of resistance in *E.* coli by inducing the genes responsible for the protein's transcription and inducing heterologous expression. They found through agar-based MIC assay, however, that the wild-type protein did not confer resistance, but it did when one of its amino acids was propionylated. This study confirms the importance of analyzing protein production as well when performing MIC assays.

When considering the diverse antimicrobial resistance mechanisms portrayed by different microbes, either blocking the entrance or expelling the antimicrobial agent is a widely adopted mechanism. This involves the porins or transporters within membranes and cell walls, in which cell-surface proteomics takes place. Azam and colleagues proposed a strategy for both the expression and purification of these proteins [3]. Due to hydrophobicity, membrane proteins possess characteristics that make them challenging to extract and purify for structural analyses and this could potentially explain why only 2% of the data in the Protein Data Bank are from membrane proteins. Today, roughly 60% of antimicrobial agents target these structures, highlighting the importance of the development of new techniques to be able to accurately characterize them. In this study, they aimed to study and compare their suggested method to characterize VanS, which is both an integral membrane protein and a histidine protein kinase sensor that binds to vancomycin and triggers the expression of resistance genes. To do so, they applied analytical ultracentrifugation (AUC), which has already been used for membrane protein analyses such as those with high alpha-helical content. This technique is used to measure the sedimentation of a molecule in a centrifugal field. Interactions between the sensory protein and vancomycin were elucidated in this manner. They also used spectroscopy to further visualize the structural and binding characteristics of the membrane proteins. More specifically, the authors used near-UV circular dichroism to investigate the binding of proteins to their ligands and far-UV CD spectroscopy for the secondary structure of the purified proteins, which also allowed them to study the denaturation properties of the purified proteins.

Furthermore, there are many other studies that besides developing or suggesting new techniques for proteomic studies, are evaluating the reliability of already existing and widely used approaches. James and colleagues tested the toxicity of the *E. coli* pET vector system, which is widely used for protein over-expression in many fields [20]. Surprisingly, the authors found that protein over-expression in *E. coli* is bactericidal, leading to more than 99.9% cell death, independently of the recombinant protein overproduced being a soluble or integral membrane protein. Based on the mutations detected in over-expressing cells, the bactericidal activity, and other observations, it is suggested that the mechanism of protein overproduction in *E. coli* using the pET vector system is similar to the development of antimicrobial resistance. Examples of techniques employed to study AMR are provided in Table 4.3.

Table 4.3 Examples of different techniques commonly used in research of antimicrobial resistance (AMR)

	Technique	Aim
Techniques to study AMR	Susceptibility testing	Measure a microorganism's vulnerability once exposed to specific concentrations of an antibiotic
	Fitness assays	Determine the microorganism's capability of survival in challenging situations. This may allow for the quantification of biofilm and planktonic growth rates and survival in environmental conditions (water, heat, starvation)
	Molecular dynamics	For the in silico prediction of physical interactions amongst atoms and molecules. A simulation of diverse protein–protein or protein–ligand interactions may be applied
	Cut-well agar diffusion assay	To evaluate the extracts or compounds derived from microorganisms as antimicrobial activity
	GC-MS	Gas chromatography is used for complex compounds
	GC-FID	Gas chromatography with flame ionization detection for the analysis of chemical compounds in a mixture
	OPTA method	Biochemical assay used to measure the activity of certain enzymes—commonly used to assess the activity of DD-Cpases (involved in cell wall metabolism)
	MTT assay	Colorimetric assay which measures cell viability and metabolic activity
	Crystal violet staining	Aims to quantify DNA in relation to the number of cells by staining the nuclei of adherent cells. This could help assess biofilm formation, for instance
	Immunomagnetic separation	Separates targeted organisms from food. Allows for the detection of these organisms to determine the pathogenicity of food, blood, or feces
	Docking simulation	Binding affinity prediction
	Circular dichroism spectroscopy	This absorption spectroscopy evaluates structures using circularly polarized light. Used for structural analyses and interactions with other molecules
	Fourier-transform infrared spectroscopy	It measures the infrared absorption or emission from a solid, liquid, or gas. Can be used to analyze the secondary structure of proteins

(continued)

Table 4.3 (continued)

	Technique	Aim
	Genome-wide transposon mutagenesis	It involves the insertion of transposons that carry a selectable marker, such as an antibiotic resistance gene, which allows researchers to easily identify and isolate the cells or organisms in which the transposons have been inserted. Used to study the function and role of individual genes in an organism's genome
	Phage display	Used for the high-throughput screening of protein interactions
	Two-hybrid	Allows for the discovery of protein interactions amongst them as well as with DNA
	Gene interaction networks	Identification of functional interactions among pairs of genes to understand the impact they have on the organism's phenotype. Cytoscape is an example of a software used for this
	Clustering analysis	To identify and visualize genes in different clusters of co-expression networks. MCODE is one of the available programs to do this
	Topological parameters	Quantitative measurements that characterize the structural properties of a network or graph. Provides insights into the organization, connectivity, and importance of nodes (genes) within the network. NetworkAnalyzer and Cytohubba may be used for this kind of analysis
	Genome mining	Identifies and analyzes the biosynthetic gene cluster region in the bacteria

4.7.2 Diagnostics and Resistance

Given the rapid dissemination and spread of resistant strains, there is an increasing demand for more accurate, fast, and sensitive diagnostic tests that enable earlier detection of resistant infections and help avoid the inefficient use of antibiotics and unnecessary patient exposure. Mkony and colleagues aimed to determine the reliability of Rubarth's newborn scale (RNOS) and C-reactive protein assessment (CRP) [32]. Today, the accurate and immediate diagnosis of neonatal sepsis is of extreme relevance because a delay in treatment can eventually lead to the mortality of neonates. While some may require the use of antibiotics, others do not, and this unnecessary use must be avoided to reduce the risk of resistance. To verify the accuracy of these tests, they combined the analyses with blood culture, in which diverse media, such as blood and chocolate agar as well as MacConkey, were used due to the possibility of the presence of different pathogens in the samples. To further characterize the sepsis-causing bacteria, they also performed biochemical tests and Gram identification tests, for which they compared all results. From the combined analyses, they found that both RNOS and CRP present a good sensitivity but low specificity. This is important when differentiating the pathogens involved to adopt the most appropriate treatment. They also found that only 19.2% of neonates with a sepsis suspect actually required antibiotics, meaning that roughly a fifth of those patients may be subjected to unnecessary antibiotic treatment and increase the risk of resistance. The authors finally suggest a second CRP 12 h later in order to exclude patients without sepsis and highlight the need for a more specific diagnostic procedure. This can also be seen in plenty of urinary tract infections, requiring a disk diffusion test to treat properly since they tend to be recurrent.

But the relevance of having an accurate diagnosis and treatment is not only relevant in bacterial-infected suspects but also in the context of viral infections. The recent COVID-19 pandemic is a great example of this, in which many infected patients were not only inefficiently treated with antibiotics when they only had a viral infection, increasing both the exposure and risk for resistance, but also the inadequate treatment for those with secondary infections. Pink and colleagues specifically worked on how to determine this differentiation with CRP and procalcitonin (PCT) levels, to be able to apply adequate treatment for each individual case [42]. CRP is elevated during inflammatory processes and because of this, patients with higher levels are often prescribed with antibiotics despite not having any secondary infections. Moreover, PCT is used as a biomarker in bacterial infections to differentiate etiology. To do so, they implemented a retrospective single-center study in which hospitalized adult patients with confirmed COVID-19 pneumonia, using RT-PCR and radiological chest examination, were included. Levels of both biomarkers were obtained on patient admission and after defined intervals by the authors. The researchers executed a variety of bacterial detection examinations from different sample types on different agar media such as the lateral flow antigen test for *L. pneumophila* and blood cultures for both aerobic and anaerobic bacteria to differentiate those patients with secondary infections into a single group from those exclusive to SARS-CoV-2. They found that patients had higher levels of both biomarkers during the entire study when they had secondary infections and that PCT levels usually were normal in most patients with only COVID-19. The authors indicate those patients without secondary infection should be exempted from antibiotic prescription since most of the patients included in the study did not present this during hospitalization. On the other hand, a significant portion of patients did develop a secondary infection over time, which is quite common given the fact that this kind of environment gathers infected patients with compromised immune systems and opportunistic infections become inevitable.

Although hospitals execute extensive cleaning protocols with diverse disinfectants to avoid the spread of diseases, other pathogens can further develop resistance to disinfectants and cross-resistance to other antibiotics. This was shown by Barakat and colleagues, who evaluated the relation between resistance development and two different disinfectants commonly used in this environment [6]. They analyzed both povidone-iodine (PVP-I), which is one of the most used antiseptics in hospitals as pre-operative scrub and to avoid post-operative infections, and propanol-based mecetronium ethyl sulfate (PBM), which is also widely used as a hand disinfectant among healthcare providers. The authors used *S. aureus* as the reference strain and determined the MIC using both biocides. To test whether there was any correlation between repeated exposure and resistance, they exposed the microbe to subinhibitory concentrations repeatedly until they finally obtained a mutant-resistant strain. They also subjected this mutant to antibiotic treatment to verify any cross-resistance mechanisms in addition to biofilm formation assays, considering this contributes to the development of resistance. Finally, the researchers executed an alpha-hemolysin activity assay on both strains since it is commonly found in the bacterium and related to its pathogenesis, and a qRT-PCR to analyze virulence gene expression with biocide subinhibitory concentrations. The authors found no resistance developed and a decrease in virulence gene expression when using PVP-I, whereas PBM did induce resistance development in *S. aureus* after repeated exposure. The same resistant strain also showed cross-resistance against widely used antibiotics such as ciprofloxacin and found that biofilm formation as well as hemolysin activity on the resistant strain was almost the same as the parent strain. This makes PBM discouraged when treated with resistant pathogens and PVP-I ends up being the most recommended antiseptic to be used in healthcare. Similar cases have been reported in the past with triclosan triggering cross-resistance to antibiotics, which led to its use being suspended by the US FDA.

4.7.3 Bacterial Virulence

Proteomics and other omics techniques may also be used to explore bacterial virulence, which can be explained as the degree of damage caused to a host by an infective bacterial pathogen. There are diverse virulence factors with distinct functions such as endo and exotoxins, capsules to protect against phagocytosis, siderophores to retrieve host iron, invasion, and adherence factors to colonize host tissues. Proteomics has been widely used to identify these mechanisms, which may involve moonlight proteins, toxin–antitoxin systems, and type IV secretion systems, among others. All these mechanisms have a metabolic price, which may also lead to changes within the pathogen's functioning as well as the host's physiology in order for the bacterium to promote pathogenesis. To do so, these factors and virulence genes can be both up or downregulated in different situations or clinical isolates, as seen in Barakat's research on *S. aureum* [6], and these modifications can be detected with WGS, PCR, and proteomics.

Secretome studies and surface proteomics are also interesting approaches in this field given that many of these virulence factors are found within the surface components of bacterial cells. A particular study conducted by Naha and Ramaiah used an interactomic approach to elucidate relations between resistance and virulence genes in MDR *Shigella flexneri* [36]. Through this topological and clustering approach, they aimed to identify potential biomarkers. The authors obtained both virulence and resistance genes from diverse repositories such as PATRIC and CARD, from which then they gathered their interaction properties from STRING and DAVID databases. They obtained the domain and gene ontology terms for all to then construct gene interaction networks, which were visualized in Cytoscape, and clusters were analyzed with MCODE. The researchers also used the microarray expression profiles obtained from the NCBI-GEO database to conduct a differential gene expression analysis with both virulence and resistance genes and then validated their results with qRT-PCR. They also performed structural and molecular dynamics analyses of hub protein histidine kinase EvgS, identified in the previous analysis, along with its more than 50 functional ligands. The authors found EvgS's interaction partners were related to resistance and virulence mechanisms, with the potential to burden the PhoPQ system, which is a two-component system responsible for response regulation to magnesium starvation, among other mechanisms, and has been previously identified as a potential molecular target. This study is a great example of how the interactomics approach can identify proteins involved in diverse mechanisms and influence both virulence and resistance in an organism.

Moreover, besides showing a relationship with resistance, virulence genes can also be related to biofilm formation as well as persistence, the latter referring to the existence of a subset of bacterial cells, within a genetically susceptible population, which is phenotypically tolerant to cidal doses of an AMD. This phenotypic state is usually associated with dormant cells that have a very low level of metabolic activity and can be induced under extreme conditions, such as starvation. A great example of this is a study carried out by Bahador and colleagues, in which the authors gathered samples from burn patients and, given that they are susceptible to infections, analyzed *A. baumannii* isolates [5]. They wanted to research any relationships between colistin resistance, which is a last resort treatment option for MDR *A. baumannii* infections, and virulence genes to be able to break the barrier between in vivo and in vitro discrepancies. To identify the isolates, they used the API20NE system, which are a series of biochemical tests that analyze the metabolic properties of the isolates to identify the species, and later confirmed the results using the *gyrB* multiplex PCR. MLVA analysis was adopted to explore the isolates' genotypic diversity and compared gene expression levels using quantitative real-time and reverse transcription PCR when exposed to antibiotics both in vivo collecting specimens from infected mice and in vitro. They also performed antimicrobial susceptibility testing and biofilm formation assays. In this manner, the authors found colistin resistance is associated with a *downregulation* of *recA,* which was previously thought to increase resistance with its expression, along with the *upregulation* of other virulence genes. This result highlights the divergence between in vivo and in vitro environments and how different genes may be involved in both resistance and virulence mechanisms.

Other studies involving the same bacterium, such as that of Kim and colleagues, show that a specific gene, *zrlA*, involved in zing uptake regulation, also influences the formation of outer membrane vesicles (OMV), when deleted, and contributes to resistance [22]. This was executed with a combination of antimicrobial susceptibility tests and qRT-PCR to quantify gene expression. They also performed biochemical DD-CPase assays to assess enzymatic activity, OMV isolation, and MTT assay to assess cell viability. Likewise, Momtaz and colleagues identified both virulence and resistance genes in *E. coli* isolate from urinary tract infection (UTI) patients by combining antimicrobial susceptibility testing and PCR assays [33]. Once again, the misuse of antibiotics, especially in recurrent UTIs, contributes to the development of resistance, which may in turn contribute to an increase in virulence and pathogenesis. Additionally, another localized research from Al-Ajmi and colleagues found the most prevalent *E. coli* serotype in human disease, O157, in ruminants with resistance to several common antibiotics due to their constant exposure to antibiotics [1]. This is of significant concern since resistant strains can potentially transfer these genes from animals to humans and the environment.

Lastly, another study by Ioannidis and colleagues also investigated the relationship between virulence genes, biofilm formation, and resistance in *S. enterica* clinical isolates using antimicrobial susceptibility testing on both planktonic bacteria and biofilm forms and further PCR to identify virulence genes associated with resistance [19]. They found all of the virulence genes to be present in up to 74.3% of the strains and that some strains carrying a single virulence gene were still capable of pathogenesis, highlighting the biological importance of these genes. Furthermore, they found that some strains lacking two specific virulence genes had more resistant phenotypes, once again portraying the relationship between virulence and resistance.

4.7.4 Antimicrobial Peptides

Moving on, in another study by Roca-Pinilla and colleagues, new strategies for recombinant proteins were explored for the research of antimicrobial peptides (AMPs) [48]. AMPs are naturally occurring, and highly conserved peptides found across all organisms. Most of them present amphiphilic properties that combine with their diverse biological functions along with multiple attacking target sites.

These are natural peptides presented by all organisms that play important roles in the innate immune system, having significant effects against pathogens. Due to their toxicity, AMPs are challenging for recombinant protein synthesis since they can affect the producer cell, so the authors attempted to overcome this issue by developing a multidomain antimicrobial protein to increase the overall production and effectiveness as well as decrease the toxicity of AMPs. The authors were able to successfully develop this new strategy with recombinant protein nanoclusters and obtained stable antimicrobial agents following this new approach.

Antimicrobial peptides can be seen across nature and can be expressed by bacteria themselves. This is the case of *Paenibacillus peoriae* IBSD35, in which Ngashangva and colleagues attempted to combine genomics and proteomics to further study this strain's antimicrobial properties [39]. Genome sequencing, analysis of extracellular metabolites with LC-MS/MS to evaluate antimicrobial peptides present, genome mining, and electron microscopy were used for this. In this manner, the authors analyzed those proteins being differentially expressed under different conditions, but also the peptides being released as a response to different stimuli using mass spectrometry. When taking a deeper look into gene clusters, the authors found nonribosomal peptide synthases were responsible for both the production of AMPs as well as other metabolites.

Moreover, as previously mentioned, these AMPs can be found all over nature and have successful antimicrobial effects among other organisms. There are plenty of studies analyzing the effect of AMPs from a determined bacterium against another pathogen, as well as the effect of AMPs from a plant extract against another bacteria. This is the case of garlic *Allium sativum* and *V. cholerae.* Bruns and colleagues carried out an investigation using this medicinal plant's extract against the pathogen responsible for cholera, which presents a previously described resistance mechanism within its multidrug efflux pump to diverse antimicrobial agents [8]. They showed that allyl sulfide, the active component within the extract, has the potential to modulate the multidrug efflux pump in *V. cholerae*. To do so, they executed classical analyses such as susceptibility testing using the disk diffusion method as well as MIC assays to assess the effect of the extract on the efflux pump. A fluorescence-based assay was implemented to measure the activity of the latter, in which the accumulation of a fluorescent dye was inversely proportional to the activity of the pump. The activity was also measured along with MIC assays once again, in which a decrease in MIC values indicated an increase in susceptibility and therefore, an effective treatment with the extract.

Another study with a similar hypothesis on humans was carried out by Mulpuru and Mishra [34]. They explored efflux pump inhibitors from the microbiome and their effect on multidrug efflux pumps of *P. aeruginosa*. A previous study had determined other AMPs from the microbiome that were able to confer resistance to *P. aeruginosa* and evaluated their effect on three outer membrane proteins that were previously identified as responsible for the resistance mechanisms of the pathogen. A virtual screening of the AMPs was carried out from the Antimicrobial Peptide Database, and the structure of the multidrug efflux pumps was obtained from the Protein Data Bank to be then analyzed with docking and molecular dynamic simulations for binding interactions and stability with the hypothesized pathogen's proteins. To closely study the residue interaction networks, the authors implemented RINalyzer and Cytoscape to generate the residue interaction networks and visualize them, respectively. They confirmed the potential of the human microbiome's AMP inhibitory effect on multidrug efflux pumps from *P. aeruginosa.*

In the same direction, another study by Pfeil and colleagues showed that some host defense peptides may portray an important role in membrane disruption and the authors evaluated this hypothesis using spectroscopy, imaging, and molecular dynamics simulations [41]. In other words, some host defense peptides disrupt phospholipid microbial bilayers in a carpet-like manner. This makes them interesting candidates for new therapeutic strategies. The authors argue this is due to the diverse mechanisms of action presented by these peptides, besides the elevated cost it would take to develop resistance against them. To evaluate this, they synthesized and purified peptides and confirmed their identities using analytical RP-HPLC and MALDI-ToF mass spectrometry. Molecular dynamics simulations and diverse spectroscopy techniques, such as circular dichroism spectroscopy for the secondary structure of peptides in the presence of synthetic membranes, were implemented to further characterize the specific antimicrobial mechanisms portrayed by the host peptides.

Furthermore, antimicrobial peptides may also exhibit regulatory properties as shown by Stincone and colleagues. They specifically aimed to analyze the effects of AMPs nisin and fengycin on *L. monocytogenes* [53]. Nisin is already used in the food industry for pathogenic control and Fengycin is also known to portray antifungal and antibacterial characteristics and the authors researched their effects on the pathogen using both mass-spectrometry lipidomics and proteomics. They also used the LIPID MAPS Structure Database's raw data to analyze and identify the compounds, as well as the String v11 Database and Cytoskape to characterize protein-protein interactions. They found that the two AMPs regulate expression on biofilm formation in different manners: fengycin inhibited formation while nisin promoted assembly. Fengycin also induced lipid profile modifications and proteomic data showed a decrease in cell wall protein genes involved in biofilm formation. These intriguing results provide new insights on AMPs effects on other pathogens and suggest new alternatives to combat *L. monocytogenes* and other multi-resistant strains.

4.8 Future Prospects

Both computational and technological advances are driving the field of AMR research. Advances in both liquid chromatography (LC) and mass spectrometry (MS) technologies are allowing greater proteomic coverages, enhanced sensitivities, and accurate quantitations at a fast pace [18]. Changes in proteome composition and abundance, in the set or level of post-translational modifications (PTMs), in the network of both protein–protein and protein–metabolite interactions can now be scrutinized in greater detail than ever. When applied to the study of AMD mechanisms of action (MoA) and mechanisms of AMR, these strategies unveil richer and previously unanticipated complex pictures [31, 40].

Additionally, the development of multiplexing strategies together with lower turnaround times are bringing proteomics to clinics [2]. These high-throughput capabilities are allowing the use of proteome-wide approaches in larger sample sizes, which will provide a deeper understanding of AMR dynamics and mechanisms in real life. With greater proteomic coverages, improved sensitivities, and higher throughput comes a remarkable increase in data availability. The number and size of datasets of proteomics data are overwhelming and it seems that this trend will continue for the years to come. We expect newer specialized databases and novel tools in this field to pop up even more in the next years. Data science and analytics skills are already important assets for anyone interested in AMD discovery and AMR research, and they will become an essential part of any endeavor in these fields. Importantly, multi-omics approaches, integrating data from genomics, transcriptomics, proteomics, and metabolomics will be pivotal to pushing forward AMD and AMR research, and indeed it is a research trend that is already taking place [10, 23].

In the long term, we expect that robust, well-grounded mechanistic explanations for both AMD MoAs and AMR mechanisms will open new avenues for the development of novel and effective therapies against diseases caused by drug-resistance strains, reducing the present high-attrition rates of drug development pipelines.

4.9 Concluding Remarks

Antimicrobial resistance is a public health concern that will continuously grow over the years. This is heavily influenced by the extensive inadequate use of antibiotics along with the constant evolution of microorganisms in search of increasing survival rates. The use of antimicrobial agents has been widely adopted for health as well as in the food industry, in order to avoid the spread of infections from the consumption of contaminated animals. However, this has also been related to the spread of resistance due to the excessive use of antibiotics in animals to previously avoid infection. Furthermore, many antibiotics are poorly used due to their adverse effects, having people prematurely abandon treatment once symptoms are gone, and have also been misused to treat viral diseases, for instance. All of these factors thoroughly contribute to the development of resistance among a wide variety of pathogens, and due to their constant expansion, there are still many mechanisms of resistance that have not been elucidated yet. Understanding these mechanisms is vital to be able to fight against this public matter by the development of new drugs and the development of new treatments through the combination of drugs with different targets.

Earlier detection of resistant phenotypes also holds a special degree of relevance to applying adequate treatment before unnecessary drug exposure. There are many techniques used for the detection of phenotypic resistance, such as antimicrobial susceptibility testing widely used in clinical settings, and other approaches used in research for deeper analysis and understanding of these resistance mechanisms. Given the fact that these processes involve genetic mutations, which then lead to altered protein structures or protein expression, proteomics altogether with genomic and structural analyses have a wide variety of techniques to research mechanisms of antimicrobial resistance. Analyzing the proteome can uncover insights into the detailed understanding of microorganisms' biological responses to determined environmental conditions, such as survival under antimicrobial treatment.

There are many different proteomic approaches that cover different research objectives, such as targeted proteomics being based on the analysis of specific previously known proteins, whereas shotgun proteomics aims to identify all proteins present in a given sample. These different methodologies can be applied for various research purposes, allowing for both the detection and quantification of annotated and novel proteins. The latter can play an important role in elucidating certain mechanisms, having the potential to fill gaps in previously known pathways. Quantitative results may also be compared under different experimental conditions, subjecting samples to different antimicrobial treatments, for instance, to be able to determine by which means multi-drug resistance may be acquired. Additionally, phenotypic resistance does not arise solely from antimicrobial exposure, as it can be intrinsic or acquired from different chemical agents or even different microorganism exposure, as seen with antimicrobial peptides. Such chemical agents can be products used in our everyday lives, as seen with disinfectant agents previously used in hospitals, and this highlights the importance of understanding the mechanisms undermining resistant phenotypes to avoid the spread and strengthening of resistance. Because this demand holds a global significance, although there can be different complications such as cost and time, proteomic techniques are under constant development and refinement to further elucidate these processes and therefore, contribute to the fight against the spread of resistance.

References

1. Al-Ajmi D, Rahman S, Banu S. Occurrence, virulence genes, and antimicrobial profiles of Escherichia coli O157 isolated from ruminants slaughtered in Al Ain, United Arab Emirates. BMC Microbiol. 2020;20(1):210. https://doi.org/10.1186/s12866-020-01899-0.
2. Alizadeh M, Yousefi L, Pakdel F, Ghotaslou R, Rezaee MA, Khodadadi E, Oskouei MA, Soroush Barhaghi MH, Kafil HS. MALDI-TOF mass spectroscopy applications in clinical microbiology. Adv Pharmacol Pharm Sci. 2021:9928238. https://doi.org/10.1155/2021/9928238.
3. Azam AA, Kinder JM, Khan GN, Alase A, Ma P, Liu Y, Ault JR, Henderson PJF, Chowdhry BZ, Alexander BD, Harding SE, Phillips-Jones MK. Production of membrane proteins for characterisation of their pheromone-sensing and antimicrobial resistance functions. Eur Biophys J. 2018;47(7):723–37. https://doi.org/10.1007/s00249-018-1325-z.
4. Bacterial enzymes and antibiotic resistance. Retrieved 21 Nov 2023, from https://cyberleninka.ru/article/n/bacterial-enzymes-and-antibiotic-resistance/viewer.
5. Bahador A, Farshadzadeh Z, Raoofian R, Mokhtaran M, Pourakbari B, Pourhajibagher M, Hashemi FB. Association of virulence gene expression with colistin-resistance in Acinetobacter

baumannii: analysis of genotype, antimicrobial susceptibility, and biofilm formation. Ann Clin Microbiol Antimicrob. 2018;17(1):24. https://doi.org/10.1186/s12941-018-0277-6.

6. Barakat NA, Rasmy SA, Hosny AE-DMS, Kashef MT. Effect of povidone-iodine and propanol-based mecetronium ethyl sulphate on antimicrobial resistance and virulence in Staphylococcus aureus. Antimicrob Resist Infect Control. 2022;11(1):139. https://doi.org/10.1186/s13756-022-01178-9.
7. Barlow VL, Lai S-J, Chen C-Y, Tsai C-H, Wu S-H, Tsai Y-H. Effect of membrane fusion protein AdeT1 on the antimicrobial resistance of Escherichia coli. Sci Rep. 2020;10(1):Article 1. https://doi.org/10.1038/s41598-020-77339-w.
8. Bruns MM, Kakarla P, Floyd JT, Mukherjee MM, Ponce RC, Garcia JA, Ranaweera I, Sanford LM, Hernandez AJ, Willmon TM, Tolson GL, Varela MF. Modulation of the multidrug efflux pump EmrD-3 from Vibrio cholerae by Allium sativum extract and the bioactive agent allyl sulfide plus synergistic enhancement of antimicrobial susceptibility by A. sativum extract. Arch Microbiol. 2017;199(8):1103–12. https://doi.org/10.1007/s00203-017-1378-x.
9. Carrette O, Burkhard PR, Sanchez J-C, Hochstrasser DF. State-of-the-art two-dimensional gel electrophoresis: a key tool of proteomics research. Nat Protoc. 2006;1(2):812–23. https://doi.org/10.1038/nprot.2006.104.
10. Chernov VM, Chernova OA, Mouzykantov AA, Lopukhov LL, Aminov RI. Omics of antimicrobials and antimicrobial resistance. Expert Opin Drug Discov. 2019;14(5):455–68. https://doi.org/10.1080/17460441.2019.1588880.
11. Cox G, Wright GD. Intrinsic antibiotic resistance: mechanisms, origins, challenges and solutions. Int J Med Microbiol. 2013;303(6):287–92. https://doi.org/10.1016/j.ijmm.2013.02.009.
12. Dzyubak E, Yap M-NF. The expression of antibiotic resistance Methyltransferase correlates with mRNA stability independently of ribosome stalling. Antimicrob Agents Chemother. 2016;60(12):7178–88. https://doi.org/10.1128/aac.01806-16.
13. Foudraine DE, Strepis N, Stingl C, ten Kate MT, Verbon A, Klaassen CHW, Goessens WHF, Luider TM, Dekker LJM. Exploring antimicrobial resistance to beta-lactams, aminoglycosides and fluoroquinolones in E. coli and K. pneumoniae using proteogenomics. Sci Rep. 2021;11(1):Article 1. https://doi.org/10.1038/s41598-021-91905-w.
14. Global Tuberculosis Report 2023. Retrieved 21 Nov 2023, from https://www.who.int/teams/global-tuberculosis-programme/tb-reports/global-tuberculosis-report-2023.
15. Görg. Current two-dimensional electrophoresis technology for proteomics - PROTEOMICS - Wiley Online Library, 2004. Retrieved 21 Nov 2023, from https://analyticalsciencejournals.onlinelibrary.wiley.com/doi/abs/10.1002/pmic.200401031?casa_token=i4bbE3jzMzEAAAAA:F5V8a_TR_PQ9ZBSCjZj8DQuCoIoHl8mXf8TxLeTisQf-A57LDPCs6gfsyWnMVe80TgkuEMunVZLZjQ.
16. Han X. Lipidomics for studying metabolism. Nat Rev Endocrinol. 2016;12(11):Article 11. https://doi.org/10.1038/nrendo.2016.98.
17. Hasin Y, Seldin M, Lusis A. Multi-omics approaches to disease. Genome Biol. 2017;18(1):83. https://doi.org/10.1186/s13059-017-1215-1.
18. He Y, Shishkova E, Peters-Clarke TM, Brademan DR, Westphall MS, Bergen D, Huang J, Huguet R, Senko MW, Zabrouskov V, McAlister GC, Coon JJ. Evaluation of the Orbitrap Ascend Tribrid mass spectrometer for shotgun proteomics. Anal Chem. 2023;95(28):10655–63. https://doi.org/10.1021/acs.analchem.3c01155.
19. Ioannidis A, Papavasileiou K, Papavasileiou E, Bersimis S, Chatzipanagiotou S. Distribution of six effector protein virulence genes among Salmonellaenterica enterica Serovars isolated from children and their correlation with biofilm formation and antimicrobial resistance. Mol Diagn Ther. 2013;17(5):311–7. https://doi.org/10.1007/s40291-013-0039-2.
20. James J, Yarnall B, Koranteng A, Gibson J, Rahman T, Doyle DA. Protein over-expression in Escherichia coli triggers adaptation analogous to antimicrobial resistance. Microb Cell Factories. 2021;20(1):13. https://doi.org/10.1186/s12934-020-01462-6.
21. Kelly RT. Single-cell proteomics: Progress and prospects. Mol Cell Proteomics. 2020;19(11):1739–48. https://doi.org/10.1074/mcp.R120.002234.

22. Kim N, Kim HJ, Oh MH, Kim SY, Kim MH, Son JH, Kim SI, Shin M, Lee YC, Lee JC. The role of Zur-regulated lipoprotein A in bacterial morphology, antimicrobial susceptibility, and production of outer membrane vesicles in Acinetobacter baumannii. BMC Microbiol. 2021;21(1):27. https://doi.org/10.1186/s12866-020-02083-0.
23. Lee S-Y, Yun SH, Lee YG, Choi C-W, Leem S-H, Park EC, Kim G-H, Lee JC, Kim SI. Proteogenomic characterization of antimicrobial resistance in extensively drug-resistant Acinetobacter baumannii DU202. J Antimicrob Chemother. 2014;69(6):1483–91. https://doi.org/10.1093/jac/dku008.
24. Liu F, Hu Y, Wang Q, Li HM, Gao GF, Liu CH, Zhu B. Comparative genomic analysis of Mycobacterium tuberculosis clinical isolates. BMC Genomics. 2014;15(1):469. https://doi.org/10.1186/1471-2164-15-469.
25. Liu L, Shen P, Zheng B, Yu W, Ji J, Xiao Y. Comparative genomic analysis of 19 clinical isolates of Tigecycline-resistant Acinetobacter baumannii. Front Microbiol. 2020;11. https://www.frontiersin.org/articles/10.3389/fmicb.2020.01321
26. Ma F, Xu S, Tang Z, Li Z, Zhang L. Use of antimicrobials in food animals and impact of transmission of antimicrobial 26. resistance on humans. Biosafety and Health 2021;3(1):32–38. https://doi.org/10.1016/j.bsheal.2020.09.004
27. Magiorakos A-P, Srinivasan A, Carey RB, Carmeli Y, Falagas ME, Giske CG, Harbarth S, Hindler JF, Kahlmeter G, Olsson-Liljequist B, Paterson DL, Rice LB, Stelling J, Struelens MJ, Vatopoulos A, Weber JT, Monnet DL. Multidrug-resistant, extensively drug-resistant and pandrug-resistant bacteria: an international expert proposal for interim standard definitions for acquired resistance. Clin Microbiol Infect. 2012;18(3):268–81. https://doi.org/10.1111/j.1469-0691.2011.03570.x.
28. Mancuso G, Midiri A, Gerace E, Biondo C. Bacterial antibiotic resistance: the most critical pathogens. Pathogens. 2021;10(10):Article 10. https://doi.org/10.3390/pathogens10101310.
29. Marcus K, editor. Quantitative methods in proteomics, vol. 893. Humana Press; 2012. https://doi.org/10.1007/978-1-61779-885-6.
30. Martinez JL. General principles of antibiotic resistance in bacteria. Drug Discov Today Technol. 2014;11:33–9. https://doi.org/10.1016/j.ddtec.2014.02.001.
31. Mijakovic I, Macek B. Impact of phosphoproteomics on studies of bacterial physiology. FEMS Microbiol Rev. 2012;36(4):877–92. https://doi.org/10.1111/j.1574-6976.2011.00314.x.
32. Mkony MF, Mizinduko MM, Massawe A, Matee M. Management of neonatal sepsis at Muhimbili National Hospital in Dar es Salaam: diagnostic accuracy of C – reactive protein and newborn scale of sepsis and antimicrobial resistance pattern of etiological bacteria. BMC Pediatr. 2014;14(1):293. https://doi.org/10.1186/s12887-014-0293-4.
33. Momtaz H, Karimian A, Madani M, Safarpoor Dehkordi F, Ranjbar R, Sarshar M, Souod N. Uropathogenic Escherichia coli in Iran: Serogroup distributions, virulence factors and antimicrobial resistance properties. Ann Clin Microbiol Antimicrob. 2013;12(1):8. https://doi.org/10.1186/1476-0711-12-8.
34. Mulpuru V, Mishra N. Antimicrobial peptides from human microbiome against multidrug efflux pump of Pseudomonas aeruginosa: a computational study. Probioti Antimicrobial Proteins. 2022;14(1):180–8. https://doi.org/10.1007/s12602-022-09910-y.
35. Mycoplasmas and their antibiotic resistance: the problems and prospects in controlling infections. Retrieved 21 Nov 2023, from https://cyberleninka.ru/article/n/mycoplasmas-and-their-antibiotic-resistance-the-problems-and-prospects-in-controlling-infections/viewer.
36. Naha A, Ramaiah S. Structural chemistry and molecular-level interactome reveals histidine kinase EvgS to subvert both antimicrobial resistance and virulence in Shigella flexneri 2a str 301. 3 Biotech. 2022;12(10):258. https://doi.org/10.1007/s13205-022-03325-w.
37. Nesvizhskii AI. Proteogenomics: concepts, applications and computational strategies. Nat Methods. 2014;11(11):Article 11. https://doi.org/10.1038/nmeth.3144.
38. Neuhoff V, Arold N, Taube D, Ehrhardt W. Improved staining of proteins in polyacrylamide gels including isoelectric focusing gels with clear background at nanogram sensitivity using

Coomassie Brilliant Blue G-250 and R-250. Electrophoresis. 1988;9(6):255–62. https://doi.org/10.1002/elps.1150090603.
39. Ngashangva N, Mukherjee PK, Sharma C, Kalita MC, Sarangthem I. Integrated genomics and proteomics analysis of Paenibacillus peoriae IBSD35 and insights into its antimicrobial characteristics. Sci Rep. 2022;12(1):Article 1. https://doi.org/10.1038/s41598-022-23613-y.
40. Park AJ, Krieger JR, Khursigara CM. Survival proteomes: the emerging proteotype of antimicrobial resistance. FEMS Microbiol Rev. 2016;40(3):323–42. https://doi.org/10.1093/femsre/fuv051.
41. Pfeil M-P, Pyne ALB, Losasso V, Ravi J, Lamarre B, Faruqui N, Alkassem H, Hammond K, Judge PJ, Winn M, Martyna GJ, Crain J, Watts A, Hoogenboom BW, Ryadnov MG. Tuneable poration: host defense peptides as sequence probes for antimicrobial mechanisms. Sci Rep. 2018;8(1):Article 1. https://doi.org/10.1038/s41598-018-33289-y.
42. Pink I, Raupach D, Fuge J, Vonberg R-P, Hoeper MM, Welte T, Rademacher J. C-reactive protein and procalcitonin for antimicrobial stewardship in COVID-19. Infection. 2021;49(5):935–43. https://doi.org/10.1007/s15010-021-01615-8.
43. Prajapati JD, Kleinekathöfer U, Winterhalter M. How to enter a bacterium: bacterial Porins and the permeation of antibiotics. Chem Rev. 2021;121(9):5158–92. https://doi.org/10.1021/acs.chemrev.0c01213.
44. Rabilloud T, Chevallet M, Luche S, Lelong C. Two-dimensional gel electrophoresis in proteomics: past, present and future. J Proteome. 2010;73(11):2064–77. https://doi.org/10.1016/j.jprot.2010.05.016.
45. Rees VE, Deveson Lucas DS, López-Causapé C, Huang Y, Kotsimbos T, Bulitta JB, Rees MC, Barugahare A, Peleg AY, Nation RL, Oliver A, Boyce JD, Landersdorfer CB. Characterization of Hypermutator Pseudomonas aeruginosa isolates from patients with Cystic Fibrosis in Australia. Antimicrob Agents Chemother. 2019;63(4) https://doi.org/10.1128/aac.02538-18.
46. Revitt-Mills SA, Robinson A. Antibiotic-induced mutagenesis: under the microscope. Front Microbiol. 2020;11. https://www.frontiersin.org/articles/10.3389/fmicb.2020.585175
47. Reygaert WC. An overview of the antimicrobial resistance mechanisms of bacteria. AIMS Microbiol. 2018;4(3):482–501. https://doi.org/10.3934/microbiol.2018.3.482.
48. Roca-Pinilla R, López-Cano A, Saubi C, Garcia-Fruitós E, Arís A. A new generation of recombinant polypeptides combines multiple protein domains for effective antimicrobial activity. Microb Cell Factories. 2020;19(1):122. https://doi.org/10.1186/s12934-020-01380-7.
49. Salini S, Bhat SG, Naz S, Natesh R, Kumar RA, Nandicoori VK, Kurthkoti K. The Error-Prone Polymerase DnaE2 mediates the evolution of antibiotic resistance in Persister mycobacterial cells. Antimicrob Agents Chemother. 2022;66(3):e01773–21. https://doi.org/10.1128/aac.01773-21.
50. Schaenzer AJ, Wright GD. Antibiotic resistance by enzymatic modification of antibiotic targets. Trends Mol Med. 2020;26(8):768–82. https://doi.org/10.1016/j.molmed.2020.05.001.
51. Schrag SJ, Perrot V, Levin BR. Adaptation to the fitness costs of antibiotic resistance in Escherichia coli. Proc R Soc Lond Ser B Biol Sci. 1997;264(1386):1287–91. https://doi.org/10.1098/rspb.1997.0178.
52. Sherman DR, Mdluli K, Hickey MJ, Arain TM, Morris SL, Barry CE, Stover CK. Compensatory ahpC gene expression in isoniazid-resistant Mycobacterium tuberculosis. Science. 1996;272(5268):1641–3. https://doi.org/10.1126/science.272.5268.1641.
53. Stincone P, Fonseca Veras F, Micalizzi G, Donnarumma D, Vitale Celano G, Petras D, de Angelis M, Mondello L, Brandelli A. Listeria monocytogenes exposed to antimicrobial peptides displays differential regulation of lipids and proteins associated to stress response. Cell Mol Life Sci. 2022;79(5):263. https://doi.org/10.1007/s00018-022-04292-4.
54. Subramanian I, Verma S, Kumar S, Jere A, Anamika K. Multi-omics data integration, interpretation, and its application. Bioinform Biol Insights. 2020;14:1177932219899051. https://doi.org/10.1177/1177932219899051.

55. Treating and managing tuberculosis | American Lung Association. Retrieved 21 Nov 2023, from https://www.lung.org/lung-health-diseases/lung-disease-lookup/tuberculosis/treating-and-managing.
56. Vilchèze C, Wang F, Arai M, Hazbón MH, Colangeli R, Kremer L, Weisbrod TR, Alland D, Sacchettini JC, Jacobs WR. Transfer of a point mutation in Mycobacterium tuberculosis inhA resolves the target of isoniazid. Nat Med. 2006;12(9):Article 9. https://doi.org/10.1038/nm1466.
57. Walsh F. Investigating antibiotic resistance in non-clinical environments. Front Microbiol. 2013;4. https://www.frontiersin.org/articles/10.3389/fmicb.2013.00019
58. Wang P, Li R-Q, Wang L, Yang W-T, Zou Q-H, Xiao D. Proteomic analyses of Acinetobacter baumannii clinical isolates to identify drug resistant mechanism. Front Cell Infect Microbiol. 2021;11. https://www.frontiersin.org/articles/10.3389/fcimb.2021.625430
59. Wenk MR. The emerging field of lipidomics. Nat Rev Drug Discov. 2005;4(7):Article 7. https://doi.org/10.1038/nrd1776.
60. Wenk MR. Lipidomics: new tools and applications. Cell. 2010;143(6):888–95. https://doi.org/10.1016/j.cell.2010.11.033.
61. WHO | Antimicrobial resistance. Retrieved 21 Nov 2023 from https://web.archive.org/web/20150310081111/http://www.who.int/mediacentre/factsheets/fs194/en/.
62. WHO | Antimicrobial resistance 2015, March 10. https://web.archive.org/web/20150310081111/http://www.who.int/mediacentre/factsheets/fs194/en/.
63. Wittmann-Liebold B, Graack H-R, Pohl T. Two-dimensional gel electrophoresis as tool for proteomics studies in combination with protein identification by mass spectrometry. Proteomics. 2006;6(17):4688–703. https://doi.org/10.1002/pmic.200500874.
64. Wood PL. Lipidomics of Alzheimer's disease: current status. Alzheimers Res Ther. 2012;4(1):5. https://doi.org/10.1186/alzrt103.
65. Xu Z, Zhou A, Wu J, Zhou A, Li J, Zhang S, Wu W, Karakousis PC, Yao Y-F. Transcriptional approach for decoding the mechanism of rpoC compensatory mutations for the fitness cost in rifampicin-resistant Mycobacterium tuberculosis. Front Microbiol. 2018;9. https://www.frontiersin.org/articles/10.3389/fmicb.2018.02895

Chapter 5
Microbial Metabolomics to Understand Antimicrobial Resistance and Drug Metabolism

Ramya Venkataraman, Kehilwe Nakedi, Ajay Suresh Akhade, and Vijay Soni

5.1 Introduction

At the dawn of the era of molecular biology, a unidirectional flow of information from DNA to RNA to proteins, which subsequently produced metabolites, was hypothesized. However, with the advent of genomics, proteomics, and more recently metabolomics, a deeper understanding of living systems has been garnered. The flow of biochemical information is no longer thought to be unidirectional; rather, it is a complex web of interaction between the DNA, RNA, proteins, and metabolites [1, 2]. Moreover, no longer are metabolites considered just the "brick and mortar" of the cells, rather they perform a range of important functions as immunomodulators, environmental sensors, signaling molecules, and endogenous toxins [3].

Metabolic systems of organisms are highly dynamic and may be referred to as the "first responders" during any kind of cellular response such as environmental fluctuations or stress responses. Monitoring metabolite dynamics is critical for gathering a deeper understanding of any cellular response [4]. The metabolome, comprising a diverse array of compounds, including lipids, sugars, amino acids, organic acids, and other intermediates, provides a comprehensive profile of the biochemical status and metabolic activities within a biological system. Metabolomics is the large-scale study of these small molecules comprising the metabolome and has emerged as a powerful tool for performing basic research on pathogen metabolism,

R. Venkataraman
Laboratory of Innate Immunity, National Institute of Immunology, New Delhi, India

K. Nakedi · V. Soni (✉)
Division of Infectious Diseases, Weill Department of Medicine, Weill Cornell Medicine, New York, NY, USA
e-mail: vis2032@med.cornell.edu

A. S. Akhade
Institute for Systems Biology, Seattle, WA, USA

V. Soni, A. S. Akhade (eds.), *Antimicrobial Resistance: Factors to Findings*,
https://doi.org/10.1007/978-3-031-65986-7_5

identifying antimicrobial resistance (AMR) mechanisms, biomarker discovery, and drug development [5].

Metabolomics has evolved greatly in the past 25 years and is using advanced high-throughput analytical technologies with computational approaches. Liquid and gas chromatography (LC and GC) are the primary technologies employed for metabolite separation, while mass spectrometry (MS) and nuclear magnetic resonance (NMR) are the key tools dominating metabolite detection. Over time, various adaptations in separation, detection, and computational methodologies have emerged to meet the increasing demands of the field [6].

Bacterial metabolism encompasses a range of pathways critical for the survival of the organism. Metabolomic strategies aim to unravel the intricate metabolic network functioning within the organism, offering valuable insights into both normal physiology and the pathophysiology of bacteria [3]. While metabolites share a high degree of conservation among different organisms, each organism possesses a unique metabolome that undergoes substantial variation due to external and internal factors. Identifying these minor fluctuations by employing metabolomics can be challenging; however, rapid advancement in analytical tools is aiding in overcoming this hurdle [3].

There exists an intimate relationship between a host and a pathogen, wherein each organism is highly sensitive to the presence of the other. This sensitivity leads to changes in the steady-state levels of numerous metabolites within both entities [7–9]. These fluctuations in metabolites play a pivotal role in shaping pathogen establishment, triggering the host immune response, and inducing host cell toxemia, among other effects. Metabolomics is being extensively used in intercepting these signals; which is beneficial in guiding basic research on pathogenesis and performing translational studies [2].

AMR, identified as a global world threat by WHO, claimed the lives of 4.95 million people in 2019 and is predicted to become the primary cause of death worldwide by 2050 [10]. This places significant pressure on the scientific community to devise a solution for this critical problem. Metabolomics has emerged as a highly versatile and valuable technology in addressing this issue. An increasing number of metabolomic studies are being executed not only to identify new drug targets but also to delineate pathogen's mechanisms leading to AMR [5]. Metabolomics is also being employed for developing rapid diagnostic tests (RDT) that play a pivotal role in identifying drug-resistant bacteria and guiding effective treatment strategies [11]. The process of biomarker discovery and drug development has also been greatly accelerated by metabolomic studies [12, 13]. This chapter aims to summarize the key role metabolomics has portrayed in understanding bacterial metabolism, drug metabolism, and the development of AMR. We have described various methods, their challenges, and future direction in metabolomics and microbial drug metabolism. Further, applications of metabolomics in novel drug target identification and biomarker discovery are discussed. We believe that this chapter would serve as a guideline for microbiologists and pharmaceutical experts to understand undercover agents of metabolism and their role in AMR development.

5.2 Understanding Bacterial Metabolism Using Metabolomics

Metabolic reactions fuel all cellular activities occurring in a bacterial system. They are also instrumental in transmitting intracellular signals as a response to environmental cues. Furthermore, there is evidence in the literature demonstrating that in pathogenic strains, metabolism is linked to virulence and invasive capabilities. However, the tightly regulated metabolome is highly complex and intertwined [14]. Metabolomics has been useful in measuring primary metabolites, secondary metabolites, second messengers, quorum sensing compounds, etc., which are involved in bacterial response to changes in environmental cues. The knowledge thus gleaned is of extreme significance in combating bacterial infections at different steps that include identifying metabolic disturbances underlying infections, discovering novel drug targets and biomarkers, and developing therapeutic interventions for combating antimicrobial resistance (AMR) [13, 15].

5.2.1 Methods Used to Study Bacterial Metabolism

Metabolites provide a comprehensive snapshot of the internal dynamics of an organism. Depending upon the component to be analyzed, there exists a vast array of methods for studying bacterial metabolism. The major aspects of cellular metabolism that are commonly investigated include measuring metabolite concentration, metabolic flux, metabolite–protein interactions, metabolite–transcription factor interactions, and metabolite-associated post-translational modifications.

Metabolite concentrations are primarily measured by employing metabolomics approaches. With the recent technological advancement in the field of metabolomics, the accuracy and throughput of metabolite quantitation have improved tremendously. The primary technologies employed for metabolomics include MS and NMR. Each method comes with its specific advantages and drawbacks. When selecting a technique, factors such as sample type, cost, and approach are considered to make an informed choice [16]. Typically, mass spectrometry (MS) is preferred when the sample size is large. NMR detection provides measurement in a nondestructive manner. Furthermore, the metabolomics approach may be broadly differentiated into untargeted or targeted. The untargeted approach aims for broader sample coverage but may not allow the complete identification of molecular structures. The targeted approach, on the other hand, aims for the analysis of a set of metabolites with enhanced resolution [5].

Standard MS instruments are coupled with separation devices via electron ionization (EI) or electrospray ionization (ESI). Depending upon the nature of the samples, separation is performed using gas chromatography (GC), liquid chromatography (LC), or capillary electrophoresis (CE). The past decade has witnessed a massive advancement in MS technologies with respect to chromatographic separation efficiency, resolution, detector sensitivity, data acquisition, and computational power. These advancements have enabled better metabolite quantification and identification [1]. NMR metabolomics is also a powerful tool and invaluable for the field of systems biology. Detailed analysis of NMR spectra provides structural and quantitative information about the metabolites in the sample. NMR involves minimal sample handling and is not dependent on chromatography for the purification or separation of metabolites. Moreover, multiple resonances for a single molecule greatly enhance the accuracy of metabolite identification. However, performing NMR is expensive, less sensitive, and semiquantitative. In comparison, MS-based technologies provide high-throughput accurate mass determination and structural elucidation and are more sensitive. Furthermore, MS instrumentation is available at lower costs, depending on the level of analytical resolution required, making it a popular choice for metabolomic studies [17].

Metabolic flux analysis (MFA) experiments measure the metabolite exchange between cells and the rate of metabolic processes by employing heavy metal labeling. Isotope-labeled reagents are added to microbial cultures, which are taken up by the cells, and the labeled isotopes get incorporated into the metabolites. Carbon-13 [^{13}C] is the most commonly employed stable isotope for performing flux experiments. MFA studies have measured flux in *Escherichia coli* cultured densely on solid agar, co-cultures of multiple species simultaneously without separation, and flux observed under nutrient stress [18–20]. Computational modeling is further used for predicting flux; such experiments have greatly expanded the scope of metabolic analysis. A variant of this approach was recently used to gain insights into the metabolism of intracellular *Salmonella.* Instead of using ^{13}C-labeled glucose in a culture medium which makes it difficult to distinguish host and bacterial metabolites, Mitosch et al. took advantage of the inability of host cells to metabolize mannitol to specifically label and estimate the metabolites of intracellular *Salmonella* during residence in mammalian cell lines RAW264.7 and HeLa [21].

As metabolites are also involved in the allosteric regulation of proteins, recent efforts also aim to identify metabolite–protein interactions. Experiments are primarily designed to infer physical interactions between metabolite and protein or assess functional enzyme activity at different metabolite concentrations during dynamic transition [14].

5.2.2 Bacterial Metabolism Pathways and Metabolomics

The major metabolic pathways that are performed within a bacterium include nutrient uptake, carbon metabolism, energy generation, supply of amino acids, and protein synthesis [14]. The bacterial growth curve provides valuable insights into the metabolic reactions occurring in a cell. Though the importance of bacterial growth is often undervalued while performing metabolic studies, it highlights the dynamic physiological changes observed in response to the environmental cues being evaluated. It is also effective as a quality control tool. The critical regulatory mechanism of carbon catabolite repression (CCR), which determines the order of preference for nutrient uptake in the presence of multiple carbon sources, was first identified by Monad by analyzing growth curves in *Escherichia coli* cultured on multiple sugars [22]. Untargeted metabolic profiling of *Mycobacterium tuberculosis* (*Mtb*) by de Carvalho et al. demonstrated a previously unobserved ability of *Mtb* to catabolize multiple carbon sources simultaneously to achieve enhanced monophasic growth [23]. This phenomenon could be promoting the pathogenicity of *Mtb*. In another study by Mould et al., unbiased metabolomics analysis in *Pseudomonas aeruginosa* wild-type and LasR$^-$ demonstrated broad differences in the intracellular citrate levels. Furthermore, they have shown that cross-feeding between different subpopulations *of P. aeruginosa* enhances LasR$^-$ mutant fitness which is associated with worse health outcomes in patients with cystic fibrosis [24]. A novel study by Zhang et al. analyzed the metabolome of *Streptococcus pneumoniae* under methionine starvation and reported the accumulation of glutamine and lactate with lowered intracellular pH. This build-up switches the metabolism to a low-growth, pro-survival state that helps in prolonged survival under conditions of amino acid starvation [25]. A metabolomics approach has also been useful in demonstrating the critical role of glycogen in increasing competition and survival in *E. coli*, during conditions of rapid nutrient fluctuations [26].

Carbon metabolism pathways play a central role in the sustenance and survival of bacterial species. These pathways are densely interconnected and have tight regulations, which are yet to be fully understood. These regulations occur at the level of transcription, translation, post-translation, or through allosteric regulation [14]. The major central metabolism pathways include glycolysis, the pentose phosphate pathway (PPP), the citric acid cycle, the glyoxylate shunt, the methylcitrate cycle, and gluconeogenesis. Although the CCM pathways have been conserved across various species, different organisms encounter different selective pressures thereby resulting in different metabolic needs [27]. The bacterial metabolic states play a crucial role in determining antibiotic resistance as well [28–30]. However, there is an expanse of knowledge about bacterial metabolism that is largely lacking even now. Being integral to the sustenance of cells, comprehending CCM pathways would also be critical for developing novel methods to target antibiotic-resistant strains. Several reports have demonstrated that metabolomic approaches have been useful in delving deeper into this aspect. For example, a study by Chen et al. has utilized GC–MS-based metabolomics to demonstrate repressed CCM in cefoperazone-sulbactam (SCF)-resistant *Pseudomonas aeruginosa,* which is also suggested to be driving SCF resistance [31].

Other metabolic pathways such as amino acid and fatty acid metabolism pathways also play a critical role in regulating bacterial pathogenesis. Reports have demonstrated that perturbations in fatty acid biosynthesis pathways could also be determining antibiotic resistance. For example, it has been reported that elevated biosynthesis of fatty acids and decreased pyruvate and energy metabolism in *Edwardsiella tarda* mediate ciprofloxacin resistance [30]. In another study, Hao et al. have demonstrated that virulence in *Brucella* sp. is modulated via an amino acid metabolism pathway [32].

5.2.3 Metabolomics in Bacterial Pathogenesis and Host–Microbes Interaction

The interaction between the host and microbe defines the progression of any infection. The host senses the pathogen and elicits an immune response, while the pathogen devises strategies to evade the host and exploit its machinery. In addition to being a nutrient source, host metabolites can in turn also be sensed by the pathogen facilitating adaption to the metabolic shift of the intracellular host milieu. The accumulation of succinate in *Salmonella*-infected macrophages was shown to be sensed by intracellular *Salmonella* through direct uptake leading to induction of key virulence genes of the *Salmonella* pathogenicity island-2 (SPI-2) and antimicrobial resistance [33]. Increased levels of pyruvate and lactate in infected macrophages were also shown to enhance expression of SPI-2 genes of intracellular *Salmonella* [34]. Host metabolites are attractive targets for the pathogen to modify as they serve as a direct link for inducing biochemical and phenotypic changes in the host. Host and pathogen compete for metabolic precursors which could cause perturbations in both the host and pathogen responses [9]. Recent studies have identified modulation in several host metabolome components including glucose, fatty acids, and amino acids such as arginine, tryptophan, and asparagine as a consequence of host–pathogen interactions. Studying host–pathogen interaction-associated metabolome changes is critical as they would help develop novel therapeutic approaches to prevent the progression of infection [2].

Metabolomics has played a significant role in gaining a deeper understanding of the changes induced in the host metabolome as a result of pathogen invasion. Metabolomic studies can also help shed light on the metabolic changes favoring pathogen establishment in the host and variations in the host environment in response to infection [2]. Kentner et al. performed NMR-based metabolomics in *Shigella* infected cell culture supernatant and demonstrated rerouting of host cell metabolic products by the pathogen to enable rapid bacterial expansion [35]. Metabolomic studies have also been useful in identifying essential nutrients acquired by the pathogen from the host for proliferation and survival. For example, the study by Zimmermann et al. identifies 33 nutrients consumed by *Mtb* during initial macrophage infection for establishing intracellular growth [36] (Fig. 5.1a).

5.2.4 Combinatorial Approaches for Studying Host–Pathogen Interactions

To further broaden the scope of our understanding of host–pathogen interactions, metabolomics is being performed alongside technologies such as molecular cartography and 16S rRNA sequencing. Using a combination of strategies helps in gaining a deeper insight into the host–pathogen system as has been illustrated in Fig. 5.1 (Figure republished from [2]). The molecular cartography approach helps provide a

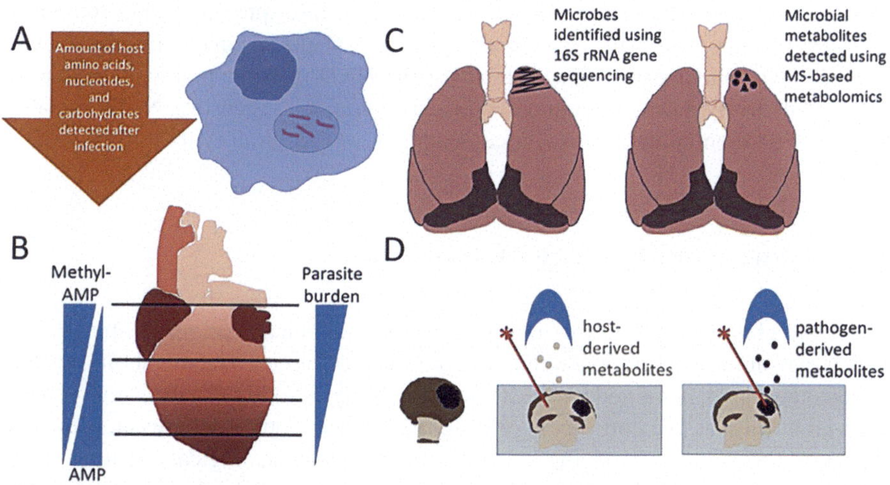

Fig. 5.1 Representative host–microbe metabolomics studies. (**a**) Rerouting of host metabolism observed in vitro in *M. tuberculosis*-infected macrophages. (**b**) Trypanosoma cruzi tropism correlated with host metabolite distribution. (**c**) Multi-omics approach to study host–pathogen interactions in a diseased human lung. (**d**) Identification of pathogen-derived molecules using MALDI–MS imaging of mushroom tissue within a Janthinobacterium agaricidamnosum-infected region. MALDI, matrix-assisted laser desorption ionization, MS, mass spectrometry. (Figure and caption are republished from [2])

spatial distribution of the metabolites and pathogens in the host system. In a study by McCall et al., the authors employed an LC–MS-based untargeted metabolomic approach with molecular cartography on *Trypanosoma cruzi* infected and uninfected heart tissue samples in mice and observed for the first time preferential parasite localization to the base of the heart region and specific distributions of nucleoside derivatives and eicosanoids, which has been correlated to tissue-damaging immune responses. Their study has also been successful in detecting novel cardiac chemical signatures related to the disease severity and outcome [37] (Fig. 5.1b). Another study used a similar approach for assessing the differential distribution of specific sugars in the cystic fibrosis lung model (Fig. 5.1c) [38]. In one other report, MALDI–MS imaging was employed for identifying pathogen-derived metabolites in host mushroom tissue (Fig. 5.1d) [39].

Using a multi-omics approach has also helped deepen our understanding of the dynamic host–pathogen system. In a study by Wang et al., metabolomic and transcriptomic analysis was employed for elucidating a novel strategy utilized by *Salmonella Typhi* for replication in host cells. They demonstrate that *S. Typhi* enhances its pathogenicity by increasing glucose uptake in human macrophages [40]. An increase in macrophage glycolysis upon *S. Typhimurium* infection is accompanied by abrogation of serine synthesis resulting in the accumulation of 2-phosphoglycerate (2PG), 3PG, and phosphoenolpyruvate (PEP), which are then used as carbon sources for intracellular replication [34].

Klabunde et al. have employed metabolomics, transcriptomics, and proteomics to gain insights into the bronchial epithelial response upon infection with *Streptococcus pneumoniae*. They identified the NAD^+ salvage pathway as an antibacterial pathway during infection conditions; therefore, suggesting NAD^+ metabolism to be a potential therapeutic target against lower respiratory tract infection.

5.3 Understanding Drug Metabolism

Drugs are chemical substances that result in phenomenal changes in the system, which may or may not be beneficial after they are consumed. The metabolic system, on the other hand, is trained to transform foreign, hazardous substances into readily excretable wastes. The entire field of drug metabolism comes in between these two processes and is precisely trying to understand the pharmacological and toxicological consequences of a drug [41]. Metabolism plays a pivotal role in mediating the activation, deactivation, toxification, and detoxification of drugs [42]. Understanding drug metabolism is critical to the process of drug development as it establishes the safety and efficacy of the drug being investigated. Also identifying the major enzymes acting on the drug assists in predicting drug–drug interaction, which is another important facet that needs to be analyzed during the process of drug development.

5.3.1 Drug Metabolism Reactions

Drug metabolism reactions are categorized as Phase I (modification), Phase II (conjugation), and in certain cases Phase III (additional modification and excretion) reactions.

Phase I reactions primarily modify the chemical structure of drugs through oxidation, hydrolysis, and cyclization/decyclization reactions. In the case of polar molecules, Phase I reactions involve the removal of hydrogen molecules or the addition of oxygen molecules. Phase I reactions can also alter an inactive compound, also referred to as a prodrug, into an active drug. Prodrugs can be defined as drugs that have little to no active pharmacological activity and undergo biotransformation to form a therapeutically active structure. They are very frequently encountered in the field of drug development [41]. An example of a Phase I modification reaction is the conversion of diazepam, commonly prescribed for anxiety disorders, into desmethyldiazepam and then to oxazepam. Both of these products have physiological and psychological responses similar to that of diazepam. The cytochrome P 450 system is involved in the catalyzes of a majority of Phase I reactions [43].

Phase II modifications involve conjugation reactions where a drug molecule couples with another molecule forming a pharmacologically inert and water-soluble compound that can be easily excreted. Conjugation mechanisms encompass a range of processes like methylation, acetylation, sulphation, glucuronidation, and glycine or glutathione conjugation. These processes take place in different bodily sites, including the liver, kidneys, lungs, intestines, and other organ systems. For instance, in Phase II metabolism, oxazepam undergoes conjugation with a molecule known as glucuronide, rendering the drug biologically inactive and facilitating its excretion without undergoing additional chemical alterations [43].

In case a drug molecule undergoes Phase III metabolism, additional conjugation reactions may occur finally culminating in the excretion of the metabolites. Enzyme catalysis occurring during the Phase I, II, and III reactions is critical for the efficient metabolism of the drugs that depends on the type and concentration of enzymes. Monoamine oxidase and cytochrome P450 are responsible for metabolizing several drug compounds [43].

5.3.2 Methods Used to Study Drug Metabolism

The study of drug metabolism involves assessing the physiological response of a candidate drug, its target engagement, and its mechanism of action. Traditionally, enzymatic assays, electrophoresis, and spectroscopy were employed for performing drug metabolism studies [6]. Presently, a combination of in vitro, in silico, and in vivo methods are being employed to efficiently and reliably predict drug metabolism as a single approach does not provide all the needed information. The simplest in vitro approaches involve the use of recombinant enzyme systems with the drug candidate, other in vitro approaches involve incubation with microsomal preparations (membrane vesicles of the endoplasmic reticulum with membrane-bound enzymes) and hepatocytes or hepatocyte-derived cell lines. For performing in vivo metabolism studies, new preclinical humanized mice models either having human drug metabolizing genes or human hepatocytes have been developed which may help in drug metabolism studies [42].

Over the past decade, metabolomics technology has evolved extensively for performing quantitative biochemical analysis. Presently, liquid and gas chromatography (LC and GC) are primary technologies for separating metabolites, while MS and NMR-based metabolomics have become standard detection tools for identifying metabolites while performing drug metabolomics studies. Further advancements are continuously happening in improving metabolite separation and identification. NMR does not require prior analyte separation, whereas MS is usually coupled to GC or LC. As mentioned in the previous section, both NMR and MS have their advantages and disadvantages and the appropriate platform must be selected based on the requirement. The most recently developed metabolomic approaches currently being employed have been detailed in Table 5.1 as published by Alarcon-Barrera et al. (Table republished from [6]).

Metabolomic approaches further coupled with in silico tools such as IsoScore or Mass-MetaSite guide the detection and identification of metabolite, helping in the prediction of metabolic fate. In silico tools allow planning for a strategy with shows

Table 5.1 Examples of industrially applied metabolomics approaches

Approach	Advantages	Disadvantages	Future directions (examples)
Untargeted LC-MS/MS metabolomics	Discovery of unknowns Broadest possible coverage	Hardly quantitative Metabolite identification remains a bottleneck	Chiral metabolomics [44] Advanced informatics for metabolite identification [45]
Targeted LC-MS-based metabolomics	Known metabolite identities (semiquantitative)	Agnostic to unknowns and restricted to panel of compounds	Chiral metabolomics [46] Isotopically labeled extracts for quantitation MFA [47]
Flow injection-based platforms	No separation column; (semi-)quantitative	Matrix effects Analyte overlap	Increasing coverage and structural details [48]
NMR-based metabolomics	Nondestructive quantitative analysis	Lower sensitivity compared with LC-MS	MFA [49]
MS Imaging (MSI)	Direct on-tissue analysis for spatially resolved metabolomics	Complex and time-consuming Fresh frozen materials required	On-tissue derivatization [50] Multi-omics integration [51]

Table republished from [6]

maximum promise and conserves resources as well as time for optimizing the metabolic stability of drug candidates a priori [42].

5.3.3 *Metabolomics in Drug Metabolism*

Metabolomics plays an extremely significant role in the field of drug research and has particularly been of use in early drug development. The primary applications of metabolomics in drug research and development include identifying novel drug targets and delineating the mode of action of drugs. As metabolomics assists in predicting drug pharmacokinetics, pharmacodynamics, and drug toxicity, it accelerates the entire process of drug discovery. Metabolomics is also useful in exploring novel disease biomarkers, drug–drug interaction, and drug repurposing [52].

A study by Yang et al. has employed metabolomic analysis to demonstrate antibiotics-induced change in the host metabolic environment that led to decreased efficacy and altered immune response [53]. Another study by Wang et al. employs LC-MS-based metabolomics *M. tuberculosis* treated with bedaquiline to demonstrate a secondary inhibition of the glutamine biosynthesis pathway along with its primary inhibition of the ATP synthase pathway. The inhibition of glutamine biosynthesis pathway renders the bacteria hypersensitive to inhibitors of glutamine synthesis [54]. Metabolomic analysis has been employed for investigating the mechanism of action when a combination of β-lactam and aminoglycoside antibiotics are administered against multidrug-resistant *Pseudomonas aeruginosa* strains. Studies revealed alterations encompassing various metabolic pathways, including fatty acid metabolism, peptidoglycan synthesis, and lipopolysaccharide metabolism [55]. Such metabolomic analysis reveals mechanistic details that can be of great value in devising antimicrobial drug combinations [54].

Metabolomics has proven to be very useful in identifying novel drug targets as well. It utilizes a mechanism-based approach for discovering novel antimicrobial drugs, opening new avenues for managing and controlling bacterial infections [56]. A shift toward mechanism-based screening techniques allows more novel sources of drug candidates to be explored as compared to classical screening techniques.

5.3.4 Metabolomics in Drug Resistance and Biomarkers Associated with AMR

Metabolomics is proving to be of immense significance in deciphering AMR-related metabolism and identifying novel biomarkers. The metabolic state of a bacterial cell could be a major factor contributing to AMR. Decreased metabolic rates could lead to decreased antibiotic uptake by the pathogen resulting in AMR. Conversely, an increase in cell metabolism could also be due to an increase in the metabolic activity of the bacteria for supporting energy-demanding mechanisms such as cell wall modifications and overexpression of efflux pumps. Therefore, analyzing the metabolic state plays a critical role in understanding AMR [5]. For example, the study by Mok et al. employs targeted metabolomics in a multi-omics approach to highlight K-13 mutations in *Plasmodium* rewired the metabolic program involving the TCA cycle, purine, glutamate, and pyruvate metabolism driving AMR artemisinin [57]. Kok et al. published a review in 2022 on unraveling antimicrobial resistance using metabolomics and summarized an exhaustive overview of studies researching the role of metabolism in AMR using metabolomics techniques detailed in Table 5.2 (republished from [5]).

Another application of metabolomics is in the identification of biomarkers that include measurable and quantifiable metabolites induced by a specific target or drug candidate [6, 82]. The biomarkers assayed can be of a dynamic or static nature. Dynamic biomarkers are widely employed for observing disease progression in patients and drug development. Static biomarkers are of significance as a prognostic tool, assisting in the prediction of clinical response. Biomarker discovery also aids pharmacokinetic and pharmacodynamic studies. For example, metabolomics was used for analyzing the volatile organic metabolome of *Steptococcus pyrogenes,* and 27 candidate biomarkers were identified that distinguish *S. pyrogenes* from other species [83]. In another study, metabolomics and 16S rDNA sequencing were employed for identifying biomarkers of drug-induced liver injury [84]. Metabolomics has proven to be an extremely useful tool for discovering infection-associated

Table 5.2 Overview of studies researching the role of metabolism in AMR using metabolomics techniques

AMR mechanism	Metabolic pathways	Analytical approach	Antibiotics	Species	References
Metabolic adaptation in energy environment					
Energy metabolism influences antibiotic efficacy	Nucleotides	Colorimetric[a]	Ampicillin[b], carbenicillin[b], gentamicin[c], kanamycin[c], streptomycin[c], ciprofloxacin[d], levofloxacin[d], norfloxacin[d], cefsulodin[e]	*E. coli*	[58]
	Amino acids, nucleotides	Untargeted LC and GC–MS	Ampicillin[b], gentamicin[c], levofloxacin[d], norfloxacin[d], daptomicin[e], rifampin[e]	*E. coli*, *Staphylococcus aureus*	[59]
Nutrient supplementation	Glycolysis, TCA cycle	Untargeted GC–MS	Kanamycin[c]	*Edwardsiella tarda*	[60, 61]
	TCA cycle	Untargeted LC and GC–MS	Tobramycin[c], chloramphenicol[e], linezolid[e], rafampin[e]	*E. coli*	[62]
Respiration and secondary antibiotic effects	Amino acids, glycolysis, lipids, TCA cycle	Untargeted LC-MS	Chloramphenicol	*E. tarda*	[63]
	Amino acids, glutathione, glycolysis, pentose phosphate, TCA cycle	Untargeted GC–MS	Gentamicin	*Vibrio alginolyticus*	[29]
	Glycolysis, TCA cycle	(Non) targeted LC–MS	Streptomycin, isoniazid, rifampicin	*M. tuberculosis*	[64]
Biofilm heterogeneity	Nucleotides, pentose phosphate	Targeted NMR and CE–MS	N.A.	*Burkholderia glumae*	[65]
	Lactate, TCA cycle	Targeted NMR	N.A.	*Shewanella oneidensis*	[66]
Metabolic adaptation to the microenvironment					

(continued)

Table 5.2 (continued)

AMR mechanism	Metabolic pathways	Analytical approach	Antibiotics	Species	References
Metabolic adaptation in antibiotic-resistant cells	Nutrient consumption	Targeted LC–UV and LC–MS	N.A.	*P. aeruginosa*	[67]
	Lipids, glycolysis, TCA cycle	Untargeted MS	Ampicillin, norfloxacin, chloramphenicol	*E. coli*	[68]
	Amino acids, glycolysis, nucleotides, TCA cycle	Untargeted MS	Kanamycin, nalidixic acid, norfloxacin, ofloxacin, chloramphenicol	*E. coli*	[69]
	Glycolysis, lipids, nucleotides, pentose phosphate, TCA cycle	Untargeted and targeted LC–MS	Methicillin	*S. aureus*	[70]
Antibiotic-induced metabolic adaptation	Amino acids, nucleotides, TCA cycle	Untargeted NMR	Ampicillin, carbenicillin, ciprofloxacin, ofloxacin, streptomycin, cefalexin, doxycycline, tetracycline	*E. coli*	[71]
	Amino acids, glycolysis, lipids, nucleotides, TCA cycle	Untargeted LC–MS	Ceftazidime, fosmidomycin, triclosan	*E. coli*	[72]
	Nucleotides	Colorimetric[a]	Ciprofloxacin, streptomycin, bedaquilline, isoniazid, rifampicin	*M. smegmatis*	[73]
	Amino acids, nucleotides, TCA cycle	Targeted LC–MS	Ampicillin, ciprofloxacin, kanamycin	*S. aureus*	[74]
	Amino acids, glycolysis, nucleotides, pentose phosphate	Untargeted LC–MS	Ciprofloxacin	*E. coli*	[53]

(continued)

Table 5.2 (continued)

AMR mechanism	Metabolic pathways	Analytical approach	Antibiotics	Species	References
Biofilm formation	Quorum sensing	Targeted MALDI–SIMS	N.A.	*P. aeruginosa*, *S. aureus*	[75]
	Quorum sensing	Targeted MALDI–SIMS	N.A.	*P. aeruginosa*, *S. aureus*	[76]
Cell wall modifications					
Cell wall disruption and synthesis	Amino acids, lipids, pentose phosphate, TCA cycle	Untargeted LC–MS	Colistin	*Klebsiella pneumoniae*	[77]
	Lipids, pentose phosphate, TCA cycle	Untargeted LC–MS	Colistin	*Acinetobacter baumannii*	[78]
	Glycolysis, lipids, TCA cycle	Untargeted GC–MS	Colistin	*M. tuberculosis*	[79]
	Amino acids, glutathione, lipids, nucleotides, pentose phosphate	Untargeted LC–MS	Colistin, doripenem	*A. baumannii*	[80]
	Lipids	Targeted MALDI–MS	Colistin	*E. coli*	[81]

Table republished from [5]
Abbreviations: *CE–MS* capillary electrophoresis–mass spectrometry, *GC–MS* gas chromatography with mass spectrometry, *LC* liquid chromatography, *LC–UV* liquid chromatography with UV detection, *MALDI–SIMS* matrix-assisted laser desorption ionization–secondary ion mass spectrometry, *NMR* nuclear magnetic resonance, PMF proton motive force, ROS reactive oxygen species, TCA cycle tricarboxylic acid cycle

biomarkers and is being identified for developing models to predict infection, causative agent, severity of disease and outcome, and for monitoring antimicrobial therapy. However, a majority of the studies conducted so far are still at the testing stage and need to be advanced to the clinical level [8, 12].

5.3.5 Metabolomics in Drug Toxicity

Drug metabolism may result in the formation of beneficial metabolites (detoxification) or detrimental (toxification), this is referred to as the toxicological consequence of a drug compound. The toxicological consequence of a drug has become a high priority during development as new drugs with a complete focus on producing drugs that form products with minimal toxicity. Once in the system, drugs undergo biotransformation and may result in metabolites with altered properties. Based on the pharmacological and toxicological properties of the drug metabolites, they can be classified into active, inactive, or reactive metabolites. As the reactive metabolites can bind to cellular macromolecules, such as proteins and DNA, they disrupt normal cellular functions thereby leading to toxicity. Hence, it is crucial to identify the reactive metabolites during drug metabolism studies.

A myriad of enzymes exist in the system that are involved in drug metabolism, including Cytochrome P450 (CYP) enzymes, dehydrogenases, flavin-containing monooxygenases, hydrolases, peroxidases, UDP-glucuronosyltransferases (UGTs), sulfotransferases, and glutathione S-transferases. Out of these, cytochrome P450, peroxidases, and transferases are primarily responsible for the generation of toxic products. The enzymes not only have different levels of expression between species, but their expression levels also vary from one organ to another. Different factors such as diet, gender, age, biological rhythms, and lifestyle also need to be considered that might be altering the drug metabolism. Drug toxicity studies are extremely challenging and meticulous analysis of drug metabolism is therefore warranted, to ensure no toxic metabolite formed gets missed. Metabolomics has gained increasing significance as a tool for assessing drug-induced toxification [41–43].

5.4 Metabolomics in Antimicrobial Resistance Diagnostics

Rapid antimicrobial sensitivity testing (AST) is key and essential in curbing the widespread misuse of antibiotics, which dramatically accelerates the frequency of antibiotic resistance. Rapid diagnostic tests that identify drug-resistant bacteria determine antimicrobial susceptibility can guide effective treatment strategies and support clinical trials for new antimicrobial drugs. Appropriate use of antimicrobials is also a key in the management of antimicrobial resistance, as it limits the exposure of patients and environments to antibacterial under only those circumstances in which the drugs are likely to improve the health status of an individual.

Metabolomic approaches, both targeted and untargeted, have been used to interrogate the changes in the metabolome during bacteria's exposure to antibiotics. These metabolite profiling strategies are crucial in identifying metabolomic biomarkers of resistance antibiotics. Using advanced analytical instrumentation, data processing, and statistical analysis, the global changes in the metabolome between resistant and susceptible pathogen strains can be compared, and novel biomarkers of resistance can then be identified.

Metabolomics enables metabolite profiling of an organism, and the quantification of the changes in biochemical networks resulting from evolution and phenotypic adaptation are associated with a broad range of AMR mechanisms [5, 85]. One of the successful applications of metabolomics in diagnosing AMR was done in 2012 [86]. Sparbier and colleagues used matrix-assisted laser desorption ionization-time of flight mass spectrometry (MALDI–TOF MS) for rapid detection of β-lactam resistance, which is based on the expression/overexpression of β-lactamases. By measuring the mass shift of resultant hydrolysis of the amide bond in the β-lactam ring after exposure to β-lactam antibiotics, the authors could classify bacteria according to sensitivity and resistance. This technology has since been used successfully to diagnose AMR in clinically relevant bacterial species [87–91], summarized here [11].

Another group used MALDI–TOF MS to detect AMR to carbapenems, which are routinely used to treat infections by multidrug-resistant Gram-negative bacteria. They used the meropenem as a marker of resistance and MALDI–TOF MS to measure the presence or absence of meropenem sodium salt peaks as well as their degradation products. Their method could detect resistance to carbapenems in Enterobacteriaceae and *P. aeruginosa* with a sensitivity of 96.67% and a specificity of 97.87% [92].

5.4.1 Raman Spectroscopy Metabolomics for AST

Other platforms have been used to profile metabolomic resistance markers in clinical isolates. Absorption spectra at specific wavelengths provide important clues crucial for determining the structure of unknown metabolic features, whereas the peak area under the curve of absorption spectra is used for quantitation. Karanja et al. [93] used vibrational spectroscopy based on infrared absorption or stimulated Raman scattering (SRS) imaging identifying the metabolic characteristics of fluconazole-resistant *C. albicans*. Raman scattering is free from water interference, making it ideal for probing biological samples [94, 95]. The authors found that resistant strains accumulated more lipid droplets than the fluconazole-susceptible strains. To further improve the ability of SRS to diagnose AST, they tracked isotope-labeled glucose molecules and found that labeled biomolecules in the lipid droplet were significantly decreased in susceptible and no significant change in the resistant strains. This work proved that alterations in metabolic signatures captured by RSS are an ideal biomarker for fluconazole resistance.

5.4.2 NMR Metabolomics for AST

NMR metabolomics is a reliable and quantitative tool that has been used previously in the antimicrobial research space for metabolomic profiling in response to antibiotics [96]. To study the metabolic changes that result in antimicrobial resistance in *E. coli*, Aries and Cloninger mutated the bacteria in the presence of compounds f C16-DABCO and mannose functionalized dendrimers (DABCOMD). NMR hydrophilic metabolomic workflow was employed for both the wild type and mutant samples. They found the metabolites that changed significantly between those groups were components of peptidoglycan synthesis, which is a critical component of the cell membrane. N-acetylglucosamine had a 25-fold change difference between the wild type and mutant. They concluded that *E. coli* altered its cell membrane composition to protect itself from the large positive charge found on the multivalent C16-DABCO and mannose functionalized PAMAM dendrimers.

5.4.3 Metabolomics and Machine Learning for AST

Even though metabolomics technologies have advanced extensively over the years, analysis of the full of an organism in a single analytical method is still not possible because of the great diversity in the physicochemical characteristics and the broad range of concentrations of the metabolites. A combination of deep learning and metabolomics is where the future of AST is heading. These technologies combined would allow deeper deconvolution of already available metabolomics datasets containing a wealth of information that is still yet to be mined. The studies discussed above used the presence and absence of a few single peaks on MS spectra to identify AMR. However, the conventional approach using few single peaks would result in insufficient prediction power without using complete information of whole MS spectra. While machine learning has been used for other -omics data analysis, it has been recently applied, at a low level to metabolomics data. Mass spectrometry-based metabolomics has relied on machine learning to process spectral peaks, metabolites, and small molecule identification and quantification. In a recent review [97], of the 27 studies they systematically reviewed, only 9 of those used MS and ML to identify antimicrobial susceptibility testing [97]. This highlights the need for more machine learning algorithms to interrogate information-rich mass spectrometry-based metabolomic datasets.

The first work demonstrating the power of combined machine learning and metabolomic profiling to predict drug-resistant cancer phenotypes with high accuracy was as recent as 2019 [98]. They used the single-probe technique to conduct single-cell metabolomics of cultured chronic myelogenous leukemia (CML) cells to obtain metabolomic data that was then analyzed by machine learning algorithm. As compared to typical tests (t-test, PCA ANOVA, etc.) that have been successfully used in the past to predict biomarkers from metabolomics, the authors combined three previous machine learning tools that have been previously used for other -omics datasets. They exploited the predictive power of random forest (RF), penalized logistic regression (LR), and artificial neural network (ANN) models to distinguish and predict drug-resistant phenotypes from metabolomics datasets.

5.4.4 Metabolomics in Identifying Mechanisms of Resistance

Metabolomics is set up to be at the frontier of discovering novel mechanisms of resistance [5]. The metabolome is closely related to biological phenotype and consequently affects their susceptibility to antimicrobials. Metabolomics is an essential tool in unraveling the phenotypical AMR mechanisms and metabolic adaptations that occur during genetic resistance, as demonstrated by Zhou and colleagues [99]. They profiled the metabolome of gentamicin-susceptible and resistant *Salmonella choleraesuis* (*S. choleraesuis*), the causative agent of bacteremia in humans [100]. They assessed the changes in metabolite abundances between the strains and found that oxidative phosphorylation was the most perturbed pathway. The authors concluded that the proton motive force of *S. choleraesuis* was suppressed because of disturbed oxidative phosphorylation, ultimately mediating the resistance of *S. choleraesuis* resistance to gentamicin.

Resistance can be based on a range of genetic and phenotypic resistance mechanisms. These include altered outer membrane permeability preventing the antibiotic from entering the cells. Another resistance mechanism is antibiotic target site alterations often because of spontaneous mutation of a bacterial gene from exposure to drugs [101]. A good example of intrinsic resistance to drugs is demonstrated by *P. aeruginosa*, an opportunistic human pathogen. It has an innate multidrug resistance promoted by broadly specific multidrug efflux systems due to mutational overexpression of the efflux genes [102]. It also has biofilm production that adds a layer to its multidrug resistance. The natural multilayered resistance of *P. aeruginosa* highlights that antibiotic resistance is multifactorial and should be studied at different system levels. Mielko and colleagues studied the metabolome of resistant versus susceptible *P. aeruginosa* strains and found that intracellular amino acids were the most upregulated metabolites in the drug-resistant strains [103]. Metabolomics and lipidomics also revealed novel resistance mechanisms to shed a broader light on the AMR enigma that is *P. aeruginosa*, Han et al. found that resistance to polymyxin was associated with decreased phospholipid levels and they also noted significant changes in amino acid and carbohydrate metabolism, particularly the intermediate metabolites from 4-amino-4-deoxy-l-arabinose synthesis and the methionine salvage cycle pathways [104].

The metabolic status of bacteria significantly affects their relative susceptibility to antibiotics and metabolic pathways related to antibiotic resistance. A few studies have been carried out that compared metabolomic profiles between drug-sensitive and drug-resistant as well as multidrug-resistant strains of different organisms. Peng et al. studied changes in the metabolome of kanamycin-resistant *Edwardsiella tarda*, an opportunistic pathogen in humans [60]. They found that glucose and alanine abundances were depleted in kanamycin-resistant isolates and that the addition of these metabolites restored susceptibility to kanamycin. In another study, lipidomic analyses revealed the change in bacterial membrane phospholipid composition as a metabolic resistance mechanism used by *S. aureus* to prevent membrane lysis by the antibiotic, daptomycin [7]. Rego et al. also used a metabolomics platform to characterize the metabolic landscape of drug-susceptible (DS), multidrug-resistant (MDR), and extensively drug-resistant (XDR) *M. tuberculosis* [105]. They found, as in the case of drug-resistant *P. aeruginosa*, that *Mtb* strains of differing resistance had altered amino acid profiles. Amino acids were also found to be enriched in multidrug-resistant *E. coli* [106], as well as biosynthesis of phenylpropanoids and purine metabolism, making these candidate markers of resistance.

5.4.5 Potential for Integrated Systems and Technologies

There is currently, to our best knowledge, no test on the shelves that uses metabolites to diagnose AMR. A "good" metabolomic drug susceptibility testing (DST) would integrate one or more technologies that are currently available. For example, metabolomic biomarkers identified in in vitro biochemical screening studies can be translated into a product on the shelves on a lateral flow device as a point of care diagnostics. We have discussed metabolomics-based biomarkers of drug resistance in the sections above, now we need to take lessons from transcriptomics and proteomics to figure out how to get these and the ones to follow to the clinic, taking into consideration that the population who needs them is usually resource constrained.

Metabolomic approaches, both targeted and untargeted, have been used to interrogate the changes in the metabolome during bacteria's exposure to antibiotics. These metabolite profiling strategies are crucial in identifying metabolomic biomarkers of resistance antibiotics. Using advanced analytical instrumentation, data processing, and statistical analysis, the global changes in the metabolome between resistant and susceptible pathogen strains can be compared, and novel biomarkers of resistance can then be identified.

A recent study reported that glutamine promotes the function of antibiotics against multidrug-resistant bacteria [107]. This study analyzed the metabolic spectrum of clinically isolated multidrug-resistant and susceptible *E. coli* and identified glutamine, which was inhibited by drug-resistant bacteria, as a biomarker of AMR. Lin et al. validated the candidate biomarkers identified in their study; biosynthesis of amino acid and upregulation of cysteine and methionine metabolism (discussed in other sections of this chapter). They used GC–MS to validate candidate markers and could predict antibiotic drug resistance characteristics and mechanisms in bacteria at the metabolite level [106]. Flonza Isa et al. used Q-TOF to identify candidate urine-based biomarkers in TB patients [108]. These identified molecules were able to differentiate between patients with active TB and healthy controls. An interesting future would be where these metabolomic biomarkers can be integrated into new devices with biosensor arrays for the detection of specific infectious diseases, same as with other -omics platforms [109].

Microbial lipids are in recent years a hot topic and studies have been published that show them to be sensitive and selective biomarkers for antimicrobial resistance. Antimicrobial resistance can be mediated via microbial lipids. For example, antimicrobial peptides (AP) are ubiquitous and are bactericidal through targeting microbial membrane integrity and lipopolysaccharides (LPS) modification [110]. These and other lipid species can be identified via MALDI–TOF MS [88, 111, 112]. However, these studies have noted considerable difficulties in sample preprocessing [113] and quantity in biological samples. The use of other technologies in combination with MS-based metabolomics opens up possibilities that will enable researchers to overcome the lipidomics sample processing pitfalls and finally have an AST test on the shelves. One such technology that offers that possibility is microfluidics, which can be useful to separate hydrophobic signatures such as lipids clinical samples. Microfluidics provides the potential of having a microanalysis system where chromatographic separation, fluorescent tagging reactions, and detection are accomplished with no added sample handling and other applications reviewed by Carvalho et al. [114].

5.5 Challenges and Future Directions

Metabolomics has been instrumental in driving significant advancements across bacterial metabolism, drug metabolism, drug discovery, and antimicrobial resistance studies. The accuracy of metabolomics makes it an ideal tool for physiological and metabolic prediction studies. Nonetheless, some limitations of metabolomics need to be addressed. Awareness of its drawbacks and potential pitfalls empowers researchers to ask pertinent questions and design experiments that efficiently collect relevant information. The following sections highlight the major challenges being confronted in metabolomic research and its future perspectives.

5.5.1 Challenges Faced in Metabolomics Research

Although metabolomics data can predict the metabolism of cells and organisms, it is incapable of providing complete information about physiology. Metabolic circuits functioning at the cellular level are extremely complex due to their interconnectivity and require a combination of various disciplines, including microbiology, chemistry, and statistics, to obtain a comprehensive picture of the process. The complexity of cellular metabolism is further increased because of the ability of the microorganisms to rapidly change their metabolic states in response to internal and external stimuli. The dynamic nature of the metabolome often impacts the reproducibility of the results. Therefore, a combination of metabolomics, transcriptomics, and proteomics is needed to obtain accurate results [22]. Additionally, metabolomics generates large volumes of raw data that require the expertise of computational scientists. Therefore, a close interaction of different disciplines is imperative for deriving meaning from the vast quantity of data generated.

There is a lack of standardization in metabolomic approaches and different laboratories opt for different protocols, this affects the reliability and reproducibility of the experiments. Optimizing microbial growth conditions is of paramount importance in a well-executed microbial metabolomics study. However, establishing a standardized growth curve is often considered trivial. Alterations in growth conditions contribute to minor metabolic changes, which over time accumulate resulting in major changes that are a common cause for errors in microbial metabolomics experiments. Metabolite degradation during sample collection and processing also impacts the accuracy of the experiment [22]. The two major approaches employed for metabolomic studies include NMR and MS. Though NMR is costlier, both these technologies require considerable funding to cover the cost and maintenance of the analytical equipment. Furthermore, NMR lacks sensitivity and is semi-quantitative. MS, on the other hand, is a destructive method and requires a separate instrument for sample separation that increases the expenditure [115]. Therefore, one needs to carefully select a workflow that is appropriate for the study. Another roadblock hindering metabolomics study is the unavailability of open access to relevant data, which are often maintained as trade secrets in companies and research organizations. Open access to data repositories will broaden the available resources and thereby assist the metabolite prediction process. Limited resources, organizational structures, and the limitations of current methods are other bottlenecks that deter metabolomic studies [42].

5.5.2 Future Directions and Advancements in Metabolomics

Although metabolomics technology is expensive and analytically challenging, it is of immense value for investigating a plethora of subjects including host–pathogen interaction, drug development, infectious biomarker identification, and AMR diagnostics. The ongoing rapid advancements in the development of robust and sensitive metabolomics technologies are expected to address the challenges, contributing to smoother operations in metabolomics research. Powerful in vitro systems such as the 3D bioreactor technology, resembling in vivo systems closely, are being developed that would mimic the host system better. As these systems provide a more "in vivo" like environment, they will be extremely beneficial in drug toxicity and drug metabolism studies [116–120]. Analytical technologies are constantly undergoing more sophistication and increase in sensitivity. Miniaturized separation devices are being developed that have helped in drastically reducing the required sample volume for MS. Metabolic flux analysis experiments employing stable isotopes are being increasingly used for analyzing dynamic interactions of metabolic pathways. A novel method known as comprehensive isotope targeted MS (CIT-MS), combining strengths of both MS and NMR, is a promising approach being developed for quantitative metabolomics and stable isotope tracing. With the advancement in microfluidics, single-cell metabolomics studies coupled with other -omics approaches are also gaining momentum in drug discovery studies. The use of single-cell metabolomics is anticipated to increase significantly in the near future [6].

5.6 Conclusion

An exciting and evolving area of research, metabolomics is at the forefront of providing novel insights into the intricacies of pathogen metabolism, host–pathogen interaction, AMR mechanisms, biomarker discovery, and drug development. Targeted approaches, analyzing hundreds of metabolites, are extensively being employed for biomarker identification. Untargeted approaches provide support by identifying as yet unknown metabolites and pathways. Drug metabolomic studies aid in early drug discovery. They help in identifying novel targets, provide a mechanism of action, and assist in detecting toxic metabolite production. Curtailing the rampant administration of antimicrobials is imperative for managing AMR. Metabolomic studies have been beneficial for interrogating the alterations in the bacterial metabolome upon antibiotic administration, thereby helping in identifying antibiotic resistance markers and deciphering AMR pathways.

Despite the massive advancements made in the techniques of metabolomics, numerous challenges still need to be addressed. The number of metabolites in organisms varies from hundreds to hundreds of thousands and simultaneously analyzing the entire set of metabolites using a single method is extremely difficult. Therefore, a combination of approaches is needed to arrive at a valid conclusion. A greater degree of synergy between computational approaches and experimental characterization has the potential to enhance the efficiency of metabolite identification and prediction studies. Metabolism is a dynamic process and its complexity is further influenced by several internal and external factors, which affect the reproducibility of the results. A combination of -omics approaches will be useful in providing a holistic outlook of the intricate cellular pathways. The field of metabolomics is experiencing continuous development and rapid progress. Emerging technologies such as metabolomic flux analysis studies, stable isotope tracing experiments, MS Imaging, and single-cell metabolomics are among the exciting advancements on the horizon.

References

1. Baidoo EEK. Microbial metabolomics: a general overview. In: Baidoo EEK, editor. Microbial metabolomics: methods and protocols. New York: Springer New York; 2019. p. 1–8.

2. Newsom SN, McCall LI. Metabolomics: eavesdropping on silent conversations between hosts and their unwelcome guests. PLoS Pathog. 2018;14(4):e1006926.
3. Wishart DS. Metabolomics for investigating physiological and pathophysiological processes. Physiol Rev. 2019;99(4):1819–75.
4. Link H, et al. Real-time metabolome profiling of the metabolic switch between starvation and growth. Nat Methods. 2015;12(11):1091–7.
5. Kok M, et al. Unraveling antimicrobial resistance using metabolomics. Drug Discov Today. 2022;27(6):1774–83.
6. Alarcon-Barrera JC, et al. Recent advances in metabolomics analysis for early drug development. Drug Discov Today. 2022;27(6):1763–73.
7. Jiang J-H, et al. Antibiotic resistance and host immune evasion in *Staphylococcus aureus* mediated by a metabolic adaptation. Proc Natl Acad Sci. 2019;116(9):3722–7.
8. Tounta V, et al. Metabolomics in infectious diseases and drug discovery. Mol Omics. 2021;17(3):376–93.
9. Zeng J, et al. Metabolomics insights into the interaction between Pseudomonas plecoglossicida and Epinephelus coioides. Sci Rep. 2022;12(1):13309.
10. Murray CJ, et al. Global burden of bacterial antimicrobial resistance in 2019: a systematic analysis. Lancet. 2022;399(10325):629–55.
11. Hrabák J, Chudácková E, Walková R. Matrix-assisted laser desorption ionization-time of flight (maldi-tof) mass spectrometry for detection of antibiotic resistance mechanisms: from research to routine diagnosis. Clin Microbiol Rev. 2013;26(1):103–14.
12. Araújo R, et al. Infection biomarkers based on metabolomics. Metabolites. 2022;12(2):92.
13. Clish CB. Metabolomics: an emerging but powerful tool for precision medicine. Cold Spring Harb Mol Case Stud. 2015;1(1):a000588.
14. Chubukov V, et al. Coordination of microbial metabolism. Nat Rev Microbiol. 2014;12(5):327–40.
15. Soni V, Wang Z, Singh V. Editorial: bacterial metabolomics approach towards antimicrobials and resistance. Front Microbiol. 2023;14:1222594.
16. Grim CM, Luu GT, Sanchez LM. Staring into the void: demystifying microbial metabolomics. FEMS Microbiol Lett. 2019;366(11):fnz135.
17. Halouska S, et al. Revisiting protocols for the NMR analysis of bacterial metabolomes. J Integr OMICS. 2013;3(2):120–37.
18. Gebreselassie NA, Antoniewicz MR. (13)C-metabolic flux analysis of co-cultures: a novel approach. Metab Eng. 2015;31:132–9.
19. Wolfsberg E, Long CP, Antoniewicz MR. Metabolism in dense microbial colonies: (13)C metabolic flux analysis of E. coli grown on agar identifies two distinct cell populations with acetate cross-feeding. Metab Eng. 2018;49:242–7.
20. Ji F, et al. Determination of intracellular metabolites concentrations in Escherichia coli under nutrition stress using liquid chromatography-tandem mass spectrometry. Talanta. 2018;189:1–7.
21. Mitosch K, et al. A pathogen-specific isotope tracing approach reveals metabolic activities and fluxes of intracellular Salmonella. PLoS Biol. 2023;21(8):e3002198.
22. Joshua CJ. Metabolomics: a microbial physiology and metabolism perspective. Methods Mol Biol. 2019;1859:71–94.
23. de Carvalho LP, et al. Metabolomics of *Mycobacterium tuberculosis* reveals compartmentalized co-catabolism of carbon substrates. Chem Biol. 2010;17(10):1122–31.
24. Mould DL, et al. Citrate cross-feeding between *Pseudomonas aerguinosa* genotypes supports *lasR* mutant fitness. bioRxiv. 2023.
25. Zhang C, et al. Amino acid starvation-induced glutamine accumulation enhances pneumococcal survival. mSphere. 2023;8(3):e0062522.
26. Sekar K, et al. Bacterial glycogen provides short-term benefits in changing environments. Appl Environ Microbiol. 2020;86(9):e00049–20.

27. Baughn AD, Rhee KY. Metabolomics of central carbon metabolism in *Mycobacterium tuberculosis*. Microbiol Spectr. 2014;2(3). https://doi.org/10.1128/microbiolspec.mgm2-0026-2013.
28. Jiang M, et al. Na(+)-NQR Confers aminoglycoside resistance via the regulation of l-alanine metabolism. MBio. 2020;11(6):e02086–20.
29. Zhang S, et al. Reduced ROS-mediated antibiotic resistance and its reverting by glucose in Vibrio alginolyticus. Environ Microbiol. 2020;22(10):4367–80.
30. Su YB, et al. Enhanced biosynthesis of fatty acids is associated with the acquisition of ciprofloxacin resistance in *Edwardsiella tarda*. mSystems. 2021;6(4):e0069421.
31. Chen YT, et al. Repressed central carbon metabolism and its effect on related metabolic pathways in cefoperazone/sulbactam-resistant *Pseudomonas aeruginosa*. Front Microbiol. 2022;13:847634.
32. Hao M, et al. Regulation of the gene for alanine racemase modulates amino acid metabolism with consequent alterations in cell wall properties and adhesive capability in *Brucella spp*. Int J Mol Sci. 2023;24(22):16145.
33. Rosenberg G, et al. Host succinate is an activation signal for *Salmonella* virulence during intracellular infection. Science. 2021;371(6527):400–5.
34. Jiang L, et al. Salmonella Typhimurium reprograms macrophage metabolism via T3SS effector SopE2 to promote intracellular replication and virulence. Nat Commun. 2021;12(1):879.
35. Kentner D, et al. Shigella reroutes host cell central metabolism to obtain high-flux nutrient supply for vigorous intracellular growth. Proc Natl Acad Sci USA. 2014;111(27):9929–34.
36. Zimmermann M, et al. Integration of metabolomics and transcriptomics reveals a complex diet of *Mycobacterium tuberculosis* during early macrophage infection. mSystems. 2017;2(4). https://doi.org/10.1128/msystems.00057-17.
37. McCall LI, et al. Mass spectrometry-based chemical cartography of a cardiac parasitic infection. Anal Chem. 2017;89(19):10414–21.
38. Garg N, et al. Three-dimensional microbiome and metabolome cartography of a diseased human lung. Cell Host Microbe. 2017;22(5):705–716.e4.
39. Graupner K, et al. Imaging mass spectrometry and genome mining reveal highly antifungal virulence factor of mushroom soft rot pathogen. Angew Chem Int Ed Engl. 2012;51(52):13173–7.
40. Wang J, et al. *Salmonella enterica* Serovar Typhi induces host metabolic reprogramming to increase glucose availability for intracellular replication. Int J Mol Sci. 2021;22(18):10003.
41. Testa B. Drug metabolism for the perplexed medicinal chemist. Chem Biodivers. 2009;6(11):2055–70.
42. Kirchmair J, et al. Predicting drug metabolism: experiment and/or computation? Nat Rev Drug Discov. 2015;14(6):387–404.
43. Susa ST, Hussain A, Preuss CV. Drug metabolism. StatPearls [Internet]: StatPearls Publishing; 2023.
44. Pandey R, et al. Novel strategy for untargeted chiral metabolomics using liquid chromatography-high resolution tandem mass spectrometry. Anal Chem. 2021;93(14):5805–14.
45. Giera M, Yanes O, Siuzdak G. Metabolite discovery: biochemistry's scientific driver. Cell Metab. 2022;34(1):21–34.
46. Kimura T, et al. Chiral amino acid metabolomics for novel biomarker screening in the prognosis of chronic kidney disease. Sci Rep. 2016;6(1):26137.
47. Hermann G, et al. 13C-labelled yeast as internal standard for LC–MS/MS and LC high resolution MS based amino acid quantification in human plasma. J Pharm Biomed Anal. 2018;155:329–34.
48. Cabruja M, et al. In-depth triacylglycerol profiling using MS3 Q-Trap mass spectrometry. Anal Chim Acta. 2021;1184:339023.
49. Fernández-García J, et al. Stable isotopes for tracing mammalian-cell metabolism in vivo. Trends Biochem Sci. 2020;45(3):185–201.

50. Angelini R, et al. Visualizing cholesterol in the brain by on-tissue derivatization and quantitative mass spectrometry imaging. Anal Chem. 2021;93(11):4932–43.
51. Xu G, Li J. Recent advances in mass spectrometry imaging for multiomics application in neurology. J Comp Neurol. 2019;527(13):2158–69.
52. Pang H, Hu Z. Metabolomics in drug research and development: the recent advances in technologies and applications. Acta Pharm Sin B. 2023;13(8):3238–51.
53. Yang JH, et al. Antibiotic-induced changes to the host metabolic environment inhibit drug efficacy and alter immune function. Cell Host Microbe. 2017;22(6):757–765.e3.
54. Wang Z, et al. Mode-of-action profiling reveals glutamine synthetase as a collateral metabolic vulnerability of *M. tuberculosis* to bedaquiline. Proc Natl Acad Sci. 2019;116(39):19646–51.
55. Yang H, et al. Metabolomics reveals the mechanism of action of meropenem and amikacin combined in the treatment of *Pseudomonas aeruginosa*. Front Cell Infect Microbiol. 2023;13:1327452.
56. Aminov R. Metabolomics in antimicrobial drug discovery. Expert Opin Drug Discov. 2022;17(9):1047–59.
57. Mok S, et al. Artemisinin-resistant K13 mutations rewire *Plasmodium falciparum's* intra-erythrocytic metabolic program to enhance survival. Nat Commun. 2021;12(1):530.
58. Lopatkin AJ, et al. Bacterial metabolic state more accurately predicts antibiotic lethality than growth rate. Nat Microbiol. 2019;4(12):2109–17.
59. Lobritz MA, et al. Antibiotic efficacy is linked to bacterial cellular respiration. Proc Natl Acad Sci. 2015;112(27):8173–80.
60. Peng B, et al. Exogenous alanine and/or glucose plus kanamycin kills antibiotic-resistant bacteria. Cell Metab. 2015;21(2):249–62.
61. Su Y-B, et al. Fructose restores susceptibility of multidrug-resistant *Edwardsiella tarda* to kanamycin. J Proteome Res. 2015;14(3):1612–20.
62. Meylan S, et al. Carbon sources tune antibiotic susceptibility in *Pseudomonas aeruginosa* via tricarboxylic acid cycle control. Cell Chem Biol. 2017;24(2):195–206.
63. Wang C, et al. Metabolites in the TCA cycle promote resistance to chloramphenicol of *Edwardsiella tarda*. J Proteome Res. 2020;20(1):972–81.
64. Nandakumar M, Nathan C, Rhee KY. Isocitrate lyase mediates broad antibiotic tolerance in *Mycobacterium tuberculosis*. Nat Commun. 2014;5(1):4306.
65. Sauer K, et al. *Pseudomonas aeruginosa* displays multiple phenotypes during development as a biofilm. J Bacteriol. 2002;184(4):1140–54.
66. Petrova OE, et al. Microcolony formation by the opportunistic pathogen *Pseudomonas aeruginosa* requires pyruvate and pyruvate fermentation. Mol Microbiol. 2012;86(4):819–35.
67. La Rosa R, Johansen HK, Molin S. Convergent metabolic specialization through distinct evolutionary paths in *Pseudomonas aeruginosa*. MBio. 2018;9(2). https://doi.org/10.1128/mbio.00269-18.
68. Zampieri M, et al. Metabolic constraints on the evolution of antibiotic resistance. Mol Syst Biol. 2017;13(3):917.
69. Hoerr V, et al. Characterization and prediction of the mechanism of action of antibiotics through NMR metabolomics. BMC Microbiol. 2016;16:1–14.
70. Aros-Calt S, et al. Annotation of the Staphylococcus aureus metabolome using liquid chromatography coupled to high-resolution mass spectrometry and application to the study of methicillin resistance. J Proteome Res. 2015;14(11):4863–75.
71. Choi Y, et al. Comparison of fitness cost and virulence in chromosome-and plasmid-mediated colistin-resistant Escherichia coli. Front Microbiol. 2020;11:798.
72. Piddock LJ. Clinically relevant chromosomally encoded multidrug resistance efflux pumps in bacteria. Clin Microbiol Rev. 2006;19(2):382–402.
73. Deter HS, Hossain T, Butzin NC. Antibiotic tolerance is associated with a broad and complex transcriptional response in E. coli. Sci Rep. 2021;11(1):6112.
74. Frydenlund Michelsen C, et al. Evolution of metabolic divergence in *Pseudomonas aeruginosa* during long-term infection facilitates a proto-cooperative interspecies interaction. ISME J. 2016;10(6):1323–36.

75. Vuong C, et al. Impact of the agr quorum-sensing system on adherence to polystyrene in Staphylococcus aureus. J Infect Dis. 2000;182(6):1688–93.
76. Fuhrer T, et al. High-throughput, accurate mass metabolome profiling of cellular extracts by flow injection–time-of-flight mass spectrometry. Anal Chem. 2011;83(18):7074–80.
77. Li B, et al. Colistin resistance gene mcr-1 mediates cell permeability and resistance to hydrophobic antibiotics. Front Microbiol. 2020;10:3015.
78. Pamp SJ, et al. Tolerance to the antimicrobial peptide colistin in *Pseudomonas aeruginosa* biofilms is linked to metabolically active cells, and depends on the pmr and mexAB-oprM genes. Mol Microbiol. 2008;68(1):223–40.
79. Aye SM, et al. Lipid A profiling and metabolomics analysis of paired polymyxin-susceptible and-resistant MDR Klebsiella pneumoniae clinical isolates from the same patients before and after colistin treatment. J Antimicrob Chemother. 2020;75(10):2852–63.
80. Vincent IM, et al. Untargeted metabolomics to ascertain antibiotic modes of action. Antimicrob Agents Chemother. 2016;60(4):2281–91.
81. Koen N, van Breda SV, Loots DT. Elucidating the antimicrobial mechanisms of colistin sulfate on *Mycobacterium tuberculosis* using metabolomics. Tuberculosis. 2018;111:14–9.
82. Wishart DS. Applications of metabolomics in drug discovery and development. Drugs R D. 2008;9(5):307–22.
83. Berna AZ, et al. Volatile profiling distinguishes *Streptococcus pyogenes* from other respiratory streptococcal species. mSphere. 2023;8(5):e00194–23.
84. He K, et al. Combined analysis of 16S rDNA sequencing and metabolomics to find biomarkers of drug-induced liver injury. Sci Rep. 2023;13(1):15138.
85. Vrioni G, et al. MALDI-TOF mass spectrometry technology for detecting biomarkers of antimicrobial resistance: current achievements and future perspectives. Ann Transl Med. 2018;6(12):240.
86. Sparbier K, et al. Matrix-assisted laser desorption ionization-time of flight mass spectrometry-based functional assay for rapid detection of resistance against β-lactam antibiotics. J Clin Microbiol. 2012;50(3):927–37.
87. Clark AE, et al. Matrix-assisted laser desorption ionization-time of flight mass spectrometry: a fundamental shift in the routine practice of clinical microbiology. Clin Microbiol Rev. 2013;26(3):547–603.
88. Leung LM, et al. Identification of the ESKAPE pathogens by mass spectrometric analysis of microbial membrane glycolipids. Sci Rep. 2017;7(1):6403.
89. Oviaño M, et al. Universal protocol for the rapid automated detection of carbapenem-resistant Gram-negative bacilli directly from blood cultures by matrix-assisted laser desorption/ ionisation time-of-flight mass spectrometry (MALDI-TOF/MS). Int J Antimicrob Agents. 2016;48(6):655–60.
90. Carvalhaes CG, et al. Detection of carbapenemase activity directly from blood culture vials using MALDI-TOF MS: a quick answer for the right decision. J Antimicrob Chemother. 2014;69(8):2132–6.
91. Papagiannitsis C, et al. MALDI-TOF MS meropenem hydrolysis assay with NH4HCO3, a reliable tool for the direct detection of carbapenemase activity. J Clin Microbiol. 2015;53(5):1731–5.
92. Hrabak J, et al. Carbapenemase activity detection by matrix-assisted laser desorption ionization-time of flight mass spectrometry. J Clin Microbiol. 2011;49(9):3222–7.
93. Karanja CW, et al. Stimulated Raman imaging reveals aberrant lipogenesis as a metabolic marker for azole-resistant Candida albicans. Anal Chem. 2017;89(18):9822–9.
94. Smith WE, Rodger C. Surface-enhanced raman scattering (SERS), applications*. In: Lindon JC, editor. Encyclopedia of spectroscopy and spectrometry. 2nd ed. Oxford: Academic Press; 1999. p. 2822–7.
95. Zhang C, Aldana-Mendoza JA. Coherent Raman scattering microscopy for chemical imaging of biological systems. J Phys: Photonics. 2021;3(3):032002.
96. Aries ML, Cloninger MJ. NMR metabolomic analysis of bacterial resistance pathways using multivalent quaternary ammonium functionalized macromolecules. Metabolomics. 2020;16(8):82.

97. Weis CV, Jutzeler CR, Borgwardt K. Machine learning for microbial identification and antimicrobial susceptibility testing on MALDI-TOF mass spectra: a systematic review. Clin Microbiol Infect. 2020;26(10):1310–7.
98. Liu R, Zhang G, Yang Z. Towards rapid prediction of drug-resistant cancer cell phenotypes: single cell mass spectrometry combined with machine learning. Chem Commun (Camb). 2019;55(5):616–9.
99. Zhou Y, et al. Exogenous D-ribose promotes gentamicin treatment of several drug-resistant *Salmonella*. Front Microbiol. 2022;13:1053330.
100. Chiu CH, Su LH, Chu C. *Salmonella enterica* serotype Choleraesuis: epidemiology, pathogenesis, clinical disease, and treatment. Clin Microbiol Rev. 2004;17(2):311–22.
101. Lambert PA. Bacterial resistance to antibiotics: modified target sites. Adv Drug Deliv Rev. 2005;57(10):1471–85.
102. Poole K. Multidrug efflux pumps and antimicrobial resistance in *Pseudomonas aeruginosa* and related organisms. J Mol Microbiol Biotechnol. 2001;3(2):255–64.
103. Mielko KA, et al. Metabolomics comparison of drug-resistant and drug-susceptible *Pseudomonas aeruginosa* strain (intra- and extracellular analysis). Int J Mol Sci. 2021;22(19):10820.
104. Han ML, et al. Alterations of metabolic and lipid profiles in polymyxin-resistant *Pseudomonas aeruginosa*. Antimicrob Agents Chemother. 2018;62(6):e02656–17.
105. Rêgo AM, et al. Metabolic profiles of multidrug resistant and extensively drug resistant *Mycobacterium tuberculosis* unveiled by metabolomics. Tuberculosis. 2021;126:102043.
106. Lin Y, et al. Comparative metabolomics shows the metabolic profiles fluctuate in multi-drug resistant Escherichia coli strains. J Proteome. 2019;207:103468.
107. Zhao XL, et al. Glutamine promotes antibiotic uptake to kill multidrug-resistant uropathogenic bacteria. Sci Transl Med. 2021;13(625):eabj0716.
108. Isa F, et al. Mass spectrometric identification of urinary biomarkers of pulmonary tuberculosis. EBioMedicine. 2018;31:157–65.
109. Resurreccion EP, Fong K-W. The integration of metabolomics with other omics: insights into understanding prostate cancer. Metabolites. 2022;12(6):488.
110. Gunn JS. Bacterial modification of LPS and resistance to antimicrobial peptides. J Endotoxin Res. 2001;7(1):57–62.
111. Smith PBW, Snyder AP, Harden CS. Characterization of bacterial phospholipids by electrospray ionization tandem mass spectrometry. Anal Chem. 1995;67(11):1824–30.
112. Ernst RK, et al. Unique lipid a modifications in *Pseudomonas aeruginosa* isolated from the airways of patients with cystic fibrosis. J Infect Dis. 2007;196(7):1088–92.
113. Sorensen M, et al. Rapid microbial identification and colistin resistance detection via MALDI-TOF MS using a novel on-target extraction of membrane lipids. Sci Rep. 2020;10(1):21536.
114. Carvalho BG, et al. Advanced microfluidic technologies for lipid nano-microsystems from synthesis to biological application. Pharmaceutics. 2022;14(1):141.
115. Pinu FR, Goldansaz SA, Jaine J. Translational metabolomics: current challenges and future opportunities. Metabolites. 2019;9(6):108.
116. Goralski TDP, et al. A novel approach to interrogating the effects of chemical warfare agent exposure using organ-on-a-chip technology and multiomic analysis. PLoS One. 2023;18(2):e0280883.
117. Gallagher EM, et al. Normalization of organ-on-a-Chip samples for mass spectrometry based proteomics and metabolomics via Dansylation-based assay. Toxicol In Vitro. 2023;88:105540.
118. Essaouiba A, et al. Analysis of the transcriptome and metabolome of pancreatic spheroids derived from human induced pluripotent stem cells and matured in an organ-on-a-chip. Mol Omics. 2022;18(8):791–804.
119. Dornhof J, et al. Microfluidic organ-on-chip system for multi-analyte monitoring of metabolites in 3D cell cultures. Lab Chip. 2022;22(2):225–39.
120. van Midwoud PM, Verpoorte E, Groothuis GMM. Microfluidic devices for in vitro studies on liver drug metabolism and toxicity. Integr Biol. 2011;3(5):509–21.

Chapter 6
Role and Relation of Microbiome to Address Antimicrobial Resistance

Chandrima Bhattacharya, Mehmed Taha Dinc, Srijani Basu, Riddhi Chaudhuri, Chenlian Fu, Gresia Cervantes, Upasana Das Adhikari, Rupobrata Panja, Jake Qiu, Anusha Aditya, and Christopher E. Mason

Abbreviations

AMP	Antimicrobial peptides
AMR	Antimicrobial resistance
AMU	Antimicrobial use
ARB	Antibiotic-resistant bacteria
ARG	Antimicrobial resistance gene
BGC	Biosynthetic gene cluster
CDI	*Clostridium difficile* infection
CSD	Cesarean delivery
EFSA	European Food Safety Authority
ESAC-Net	European Surveillance of Antimicrobial Consumption Network
FMT	Fecal microbiome transplant

C. Bhattacharya
Tri-Institutional Program in Computational Biology and Medicine, New York, NY, USA

The Bin Talal Bin Abdulaziz Alsaud Institute for Computational Biomedicine, Weill Cornell Medicine, New York, NY, USA

M. T. Dinc
Department of Medicine, Boston Medical Centre, Boston, MA, USA

S. Basu
Department of Medicine, Weill Cornell Medicine, New York, NY, USA

R. Chaudhuri
Purdue Institute for Drug Discovery, West Lafayette, IN, USA

Department of Chemistry, Purdue University, West Lafayette, IN, USA

C. Fu
Tri-Institutional Program in Computational Biology and Medicine, New York, NY, USA

Memorial Sloan Kettering Cancer Center, New York, NY, USA

G. Cervantes
Polytechnic University of Guanajuato, Cortazar, Guanajuato, Mexico

V. Soni, A. S. Akhade (eds.), *Antimicrobial Resistance: Factors to Findings*,
https://doi.org/10.1007/978-3-031-65986-7_6

HGT	Horizontal gene transfer
ISS	International Space Station
Mac	Macrolide-specific efflux
MATE	Multi-antimicrobial extrusion
MDR	Multidrug resistance
MSRA	Methicillin-resistant *Staphylococcus aureus*
PCR	Polymerase chain reaction
SCFA	Short-chain fatty acids
UTI	Urinary tract infection
VD	Vaginal delivery
VRE	Vancomycin-resistant enterococci
WHO	World Health Organization
WWTP	Wastewater treatment plants

6.1 Introduction

6.1.1 Overview of Antimicrobial Resistance and Its Relationship with the Human Microbiome

Over 30,000 mortalities due to drug-resistant infections have been reported per year in the United States, with economic tolls linked to treating bacterial infections amounting to millions of dollars spent on tackling the global threat [1]. The rise of

U. D. Adhikari
Department of Immunology, Harvard University, Boston, MA, USA

Ragon Institute of MGH, Harvard University and Massachusetts Institute of Technology, Boston, MA, USA

R. Panja
Center for Computational and Integrative Biology, Rutgers University-Camden, Camden, NJ, USA

J. Qiu
New York Genome Center, New York, NY, USA

Physiology, Biophysics, and Systems Biology PhD Program, Weill Cornell Medicine, New York, NY, USA

A. Aditya
Irving Medical Centre, Columbia University, New York, NY, USA

C. E. Mason (✉)
The Bin Talal Bin Abdulaziz Alsaud Institute for Computational Biomedicine, Weill Cornell Medicine, New York, NY, USA

Physiology, Biophysics, and Systems Biology PhD Program, Weill Cornell Medicine, New York, NY, USA

World Quant, New York, NY, USA
e-mail: chm2042@med.cornell.edu

resistant strains is a threat to the efficacy of antibiotics associated with treating bacterial infections [2]. Inappropriate use of antimicrobial medications along with the slow development of novel antimicrobials is a predominant factor leading to the rise of AMR or antimicrobial resistance [1]. Irrespective of this, very few studies have focused on the development of resistance in the enteric microbiota for healthy individuals (Fig. 6.1) [3]. The human gut microbiota is most commonly exposed to antibiotics, and the metagenomic analysis from fecal samples in the study by Bag et al. revealed that the enteric microbiota of healthy individuals are enriched with numerous AMR genes [2]. In the gut microbiome, most resistant traits are passed on via horizontal gene transfer (HGT), and based on literature precedence, 20% to >30% of the human enteric microbiome might possess multidrug resistance (MDR) even with less or absence of exposure to antibiotics. Interestingly, a study of 252 human fecal metagenomes from the United States, Denmark, and Spain assessing AMR across 68 different classes and subclasses of antibiotics has shown that long-term availability of antibiotics, as well as antibiotic usage in animals, predominantly contributes to global resistomes [4].

The reservoir of antibiotic resistance gene (ARG) in healthy individuals extends beyond the gut microbiome. Apart from the enteric microbiome, the human oral microbiome harbors the second most significant population of microbes in the body and the presence of ARG has been identified across all ages from neonatal to adult

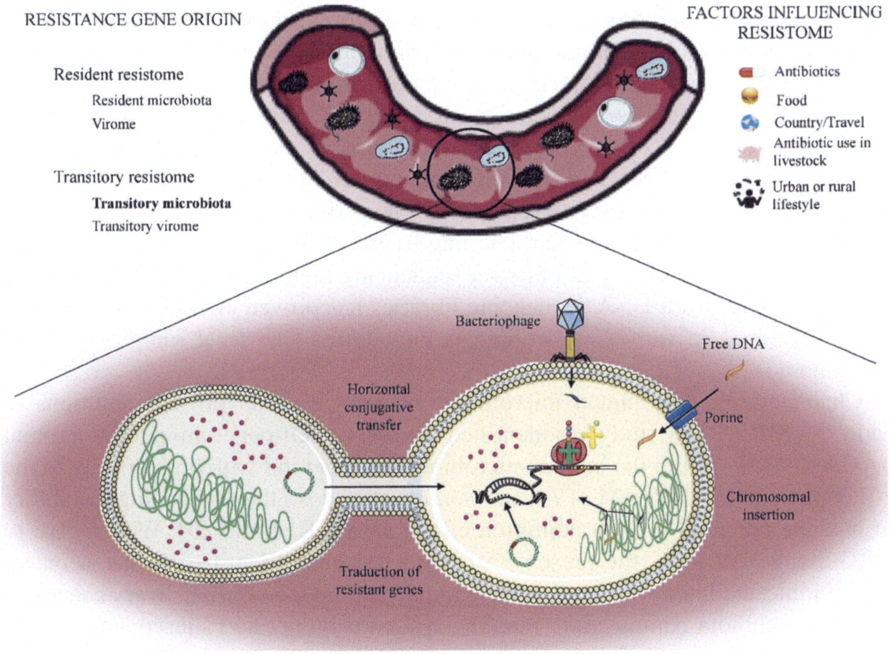

Fig. 6.1 Various modes of dissemination of antibiotic resistance genes (ARG) and predominant factors that impact resistome phenotypes. (Figure and caption are reprinted from the citation [3])

populations [5]. The oral cavity can host around 1000 different microbial taxa, with Streptococci being predominant in children and carrying tet genes, while other taxa such as Veillonella show multidrug resistance (MDR) traits, including resistance to ampicillin and penicillin, even without direct antibiotic selection pressure [4, 6]. The healthy oral microbiota in humans has a wealth of ARG, and dissemination is known to exhibit HGT [5]. With regard to ARG spread via bacterial transmission, it is crucial to note that the buccal cavity cannot be considered a closed system as widespread dissemination of oral microorganisms to the gut region has been observed by Ref. [6], and the transmission frequency is even higher for opportunistic pathogens. Also, oral microbes can spread into other regions of the body via blood vessels, and oral biofilm flocs on hard tissues can be transmitted to other body sites via the sloughing phenomenon [6]. Therefore, there is a need to delve deeper into the mechanisms by which ARG are acquired and transmitted in both gut and oral microbiome thereby contributing to the development of pathogenic infections.

6.1.2 Importance of Microbiome in Addressing Antimicrobial Resistance

By 2050, AMR is estimated to result in around approximately ten million annual deaths; therefore, surveillance for ARG in both the environment and those associated with human microbiomes is important. Currently, various cultures (e.g., rectal sample culture to isolate gut microbes) and non-culture methods (e.g., gut resistome profiling by metagenomic sequencing or identification of novel ARG by functional metagenomics) are used to characterize ARG in the microbiome [7]. However, thorough characterization of microbial resistome remains challenging for researchers due to the diversity of ARG and low ARG abundance in most natural environments, DNA extraction from such environments is non-trivial, and ARGs are linked to mobile elements that comprise around one-fourth of the genetic material of those microbiotas. The development of newer sequencing methods like DARTE-QM that can enrich ARG from DNA sequencing libraries has proven effective in characterizing resistomes in various environments including swine fecal samples and soil microbiota [8].

Characterizing the human microbiome strategically can provide solutions to address AMR. A new study has identified the role of cell-penetrating antimicrobial peptides (AMPs) from the human microbiota that can bind with high affinity to the efflux pump MexAB-OprM of *Pseudomonas aeruginosa* thereby having the ability to mediate potential growth inhibition effects in MDR strains [9]. With regard to ARG characterization, the presence of ARG has been identified in the neonatal microbiome and interesting studies have found that mode of birth delivery can affect enteric microbiota composition, immune system priming, and subsequently AMR in children [10]. In newborns, there is rapid colonization of microbes from the mother and external environment; however, in cesarean delivery (CSD), transmission of Bacteroides strain from the mother is impeded and there is increased colonization of opportunistic pathogens like *Enterococcus*, *Enterobacter*, and *Klebsiella* strains making the individuals with CSD more susceptible to bacterial infections [11]. Functional analysis of the metagenome of 1-year-old infants who have undergone vaginal delivery (VD) indicates an increased gene copy number of carbapenem and phenazine biosynthetic pathways, and interestingly, these two naturally occurring antibiotics are used in the clinic to combat pathogenic infections [10]. Therefore, gut microbiome can play a crucial role in fighting resistance against pathogenic species and the discovery of novel antibacterial molecules (like AMPs and natural antibiotics) can significantly contribute to the development of novel therapy against AMR.

6.1.3 Simple Explanations of the Role of Environmental Microbiome and Its Impact on Bacterial Evolution Against Drug Treatments

The heterogeneity of microbes in the external environment is much greater compared to those present inside humans as it comprises the huge diversity of microbial species in the soil, air, wastewater, agriculture, and livestock. Despite extensive studies, a huge population of environmental microbes has remained underdiscovered; therefore, their impact on human health remains unknown. As human beings are exposed to a variety of environmental microbes that comprises both beneficial and pathogenic microorganisms that have a deep impact on the human microbiome, there is a rising need to study the interaction of the environmental and human microbiome [12].

A major consequence of the indiscriminate use of antibiotics is the rise in the antimicrobial-resistant bacterial (ARBs) population in various environments which is referred to commonly as "antibiotic resistance pollution" (Table 6.1) [13]. For instance, an epidemiological study reported an increased abundance of ARGs since the 1940s in all classes of antibiotics from the soil obtained from various locations in the Netherlands [14]. Resistance against antibiotics can develop by two significant mechanisms, that is, via acquiring de novo mutations or the environment might already harbor ARGs, and their abundance increases in response to selection pressure [13]. Interestingly, the presence of non-antimicrobial agents like metals, pesticides, and biocides also contributes to AMR as zinc, mercury, or copper can promote the selection of resistant microbes. Recent studies have demonstrated ARG can be transmitted to humans via animal products due to indiscriminate usage of antimicrobial agents used for accelerating growth in poultry or livestock, and this includes resistant pathogens like *Salmonella* sp., foodborne *E. coli*, and *Campylobacter* sp. [15]. Also, sewage from hospitals and other environments is passed through wastewater treatment plants (WWTPs) to reduce pathogen numbers and then released into surface water. However, sewage provides a nutrient-rich environment that

Table 6.1 Major categories of MRD bacteria and associated ARGs isolated from environmental samples

Pathogen	Source of isolation	Antimicrobial resistance genes	Responsible for
Escherichia coli	Household water, drinking water, poultry swabs, milk, vegetables, raw meat, and chickens	$bla_{CTX-M-9}$, qnrS, bla_{CTX}, bla_{TEM}, bla_{CTX-M}+ bla_{SHV}, bla_{SHV}, bla_{CTX-M}+ bla_{TEM}, bla_{TEM}+ bla_{SHV}, bla_{CTX-M}+ bla_{SHV}+ bla_{TEM}, bla_{CMY-2}	Urinary tract infections, diarrhea, neonatal meningitis, septicemia, abdominal pain, pneumonia, nosocomial bacteremia
Staphylococcus aureus	Retail meat and broiler chickens	mecA, tetK, blaZ	Bovine mastitis, osteomyelitis, endocarditis, gastroenteritis, toxin syndromes, abscess formation
Klebsiella pneumoniae, *Klebsiella oxytoca*	Milk	aadA22, arr-3, aac(3)-Id, cmlA, bla_{TEM}, ereA2, bla_{SHV}, bla_{OXA}, bla_{CTX-M}	Urinary tract infections, bovine mastitis, septic shock, pulmonary pneumonia, wound infections, nosocomial infections, hepatic and pancreatic disease, septicemia, bacteremia
Enterococcus faecium	Chicken feces and water	tetM, ermB, tetL	Vancomycin-resistant enterococci (VRE) infections, neonatal meningitis nosocomial infections
Salmonella spp.	Retail pork, pigs, broiler chickens	pse-1, tetA, tetB, ant(3″)-Ia	Salmonellosis, bacteraemia
Campylobacter spp.	Chickens	NA	Post-infectious irritable bowel syndrome, campylobacteriosis
Pseudomonas spp.	Untreated and treated water	bla_{AmpC}, bla_{OXA}, bla_{IMP}, bla_{TEM}, tetC, tetA, sul1	Blood-borne infections, central nervous system infections, urinary tract infections, wound infections, endocarditis, nosocomial infections
Vibrio spp.	Urban water, shellfish	dfrA15, bla_{P1}	Cholera, diarrhea, profuse watery stools

Table adapted from Ref. [1]
NA not available

supports rapid microbial growth therefore eventually harboring novel ARGs, ARBs, and antibiotics thus turning the bed for the evolution and transmission of AMR [13].

Over the years, the upsurge of superbugs that are gradually acquiring resistance against all known natural or synthetically derived antimicrobials is a major concern to human health [16]. Superbugs like methicillin-resistant *Staphylococcus aureus* (MRSA) and vancomycin-resistant enterococci (VRE) are hotspots of ARGs against multiple antibiotic classes. Such ARG panels are often associated with integrons, and the amalgamation of various ARGs into a single integron results in patterns of cross-selection that lead to ARG maintenance in the gene loci. For instance, MDR strains of Enterobacteria isolated from hospitals have ARGs against antibiotics such as chloramphenicol, spectinomycin, and streptomycin that have not been used in the context of hospital environments over decades [13]. Apart from "established" ARGs that have been well characterized, around 95% of the pan-resistome of the environmental microbial population is dominated by "latent" ARGs that are typically unknown and largely neglected in genome high-throughput sequence-based studies. Analysis of 10,000 metagenomic samples revealed that numerous latent ARGs are present in the wastewater which acts as large reservoirs of latent ARGs thus posing as a major health hazard [17].

6.2 The Microbiome and Antimicrobial Resistance

6.2.1 Definition and Composition of Microbiome

The gut harbors the largest population of microbiota, with each individual hosting between 10 and 50 trillion microbial cells. The genes of these cells are collectively referred to as the microbiome, while the various taxa of microbes associated with humans, including bacteria, bacteriophages, archaea, fungi, eukaryotic viruses, and protozoa, are collectively known as the microbiota [18–21].

A significant number of microorganisms that make up the gut microbiota in adult humans, consisting of approximately 1000 bacterial species, can be broadly classified into two bacterial groups: Bacteroidetes (comprising Gram-negative, nonmotile forms) and Firmicutes, primarily composed of Gram-positive bacteria [19–21].

The concept of metagenomics refers to the total description of DNA from all microbial samples, but it also applies to the intricate mixtures of gut microbes. In some studies, it can also refer to specific marker genes like the 16s rRNA gene, which is utilized to investigate the phylogeny and taxonomy of bacteria. Notably, the number of named bacterial species has risen to over 8000, representing a remarkable increase of more than 450% in the number of described taxa since 1980.

The gastrointestinal tract microbiota holds a determining role in maintaining our digestive system, serving various functions such as food fermentation, protection against pathogens, stimulation of the immune system, and synthesis of essential vitamins. Typically, the gut microbiota comprises six main phyla: Bacteroidetes, Firmicutes, Proteobacteria, Actinobacteria, Fusobacteria, and Verrucomicrobia, with Firmicutes and Bacteroidetes being the predominant types. Furthermore, when exploring fungal populations in the gut (known as gut mycobiota), extensively studied species include *Saccharomyces*, *Candida*, *Cladosporium*, and *Malassezia*. Besides bacteria and fungi, the microbiota of the gut also encompasses bacteriophages, viruses, and archaea, with *Methanobrevibacter smithii* being a prominent member. These viruses, collectively known as the virome, can infect human cells or other microorganisms, influencing bacterial community composition within the microbiome. Additionally, less common microorganisms like protists and helminths (parasitic worms) can also be part of the human microbiome [22, 23].

The structure of the human microbiota can vary significantly among individuals and can be influenced by various factors, such as genetics, diet, age, and environmental exposures [8, 9]. The gut microbiota has a profound effect on health and drives multiple physiological processes like immune system function, digestion, and metabolism. Disruptions in the balance of the microbiota, also called dysbiosis, have been linked to conditions such as inflammatory bowel disease, obesity, and autoimmune disorders [24].

Research into the human microbiota is ongoing, and scientists continue to uncover its complex interactions with the human body and its potential implications for health and disease. A well-balanced gut microbiota commonly displays traits like extensive species variety, plentiful microbial genetic diversity, and a consistent core microbiome. However, it is crucial to note that the relative distribution of microorganisms varies from person to person and can even fluctuate within the same individual. Variables like age and environmental factors, such as medication usage, can cause fluctuations in the structure of the human gut microbiota [25].

Furthermore, gut microbiota composition differs along the gastrointestinal tract. For example, Proteobacteria members like Enterobacteriaceae are abundant in the small bowel but are less frequently encountered in the colon. In contrast, the colon often harbors Bacteroidetes, including Bacteroidaceae, Prevotellaceae, and Rikenellaceae. These variations are primarily attributed to the distinct environmental conditions in these regions. The small intestine exhibits a quicker transit time and elevated bile concentration, whereas the colon features slower flow rates, a more moderate pH, and a greater abundance of anaerobic microorganisms [26].

Age is another factor influencing gut microbiota composition. Generally, microbiota diversity increases from childhood to adulthood and decreases in older age, typically over 70. In pre-adulthood, a child' microbiota is dominated by species like *Veillonella*, *Clostridium coccoides* spp., *Akkermansia muciniphila*, *Bacteroides*, and *Clostridium botulinum* spp. By the time a child turns three, its gut microbiota begins to look like that of adults, with three primary microbial phyla—Bacteroidetes, Firmicutes, and Actinobacteria—becoming predominant. As individuals age, modifications in diet and the immune system can modify the structure and composition of the gut microbiota, frequently resulting in reduced numbers of Bifidobacterium and heightened numbers of Proteobacteria and Clostridium. The decrease in anaerobic bacteria such as Bifidobacterium might exacerbate inflammatory conditions owing to its involvement in regulating the immune system [27, 28].

Acknowledging the substantial impact of the gut microbiota on human health and its contribution to diverse biological processes and disease progression, research on the human microbiota has advanced beyond basic compositional investigations. There is now a heightened focus on understanding the causal relationships of microbiota functions, aided by advancements in high-throughput sequencing techniques, microbiota interaction modeling, and simulation (Fig. 6.2).

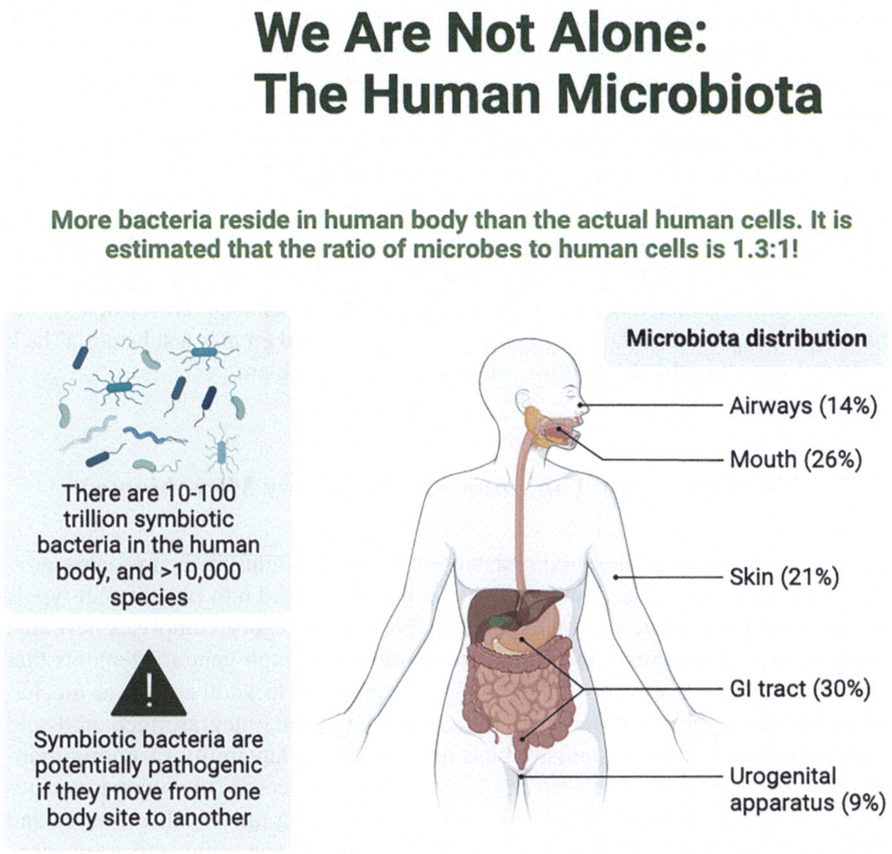

Fig. 6.2 The human body houses trillions of microbes. (Created by Biorender)

6.2.2 Role of the Microbiome in Antibiotic Resistance

Antimicrobial resistance (AMR) is a critical global health issue posing a significant threat to the effectiveness of commonly used medications against a widening spectrum of infections caused by microbes. The emergence of AMR in bacteria is an issue of great concern. Over the last few decades, bacteria that cause deadly infections have progressively developed resistance to newly introduced antibiotics (Coman and Vodnar, 2020).

The World Health Organization (WHO) has long stressed the need for a better and globally coordinated endeavor to tackle AMR. The World Health Organization (WHO) initiated the Global Strategy for Containment of Antimicrobial Resistance in 2001, presenting a thorough guideline of interventions designed to curb the dissemination of resistant microbes. WHO released "The Evolving Threat of Antimicrobial Resistance – Options for Action," which advocated for a multifaceted approach to address this issue. This approach encompasses strengthening healthcare systems and surveillance, enhancing the prudent utilization of antimicrobials in healthcare facilities and communities, bolstering infection prevention and control protocols, advocating for the development of novel drugs and vaccines, and securing political dedication.

Recognizing the primary role of surveillance in addressing AMR, the WHO in April 2014 released the first global report on AMR surveillance. This report compiled insights from global surveillance networks, demonstrating the valuable role of data obtained through surveillance where available. Such data can guide treatment decisions, elucidate AMR trends, pinpoint areas requiring immediate intervention, and monitor the effectiveness of strategies to combat resistance. Nevertheless, the absence of proper surveillance in many parts of the world creates a substantial lack of clarity regarding the distribution and scope of this problem.

6.2.3 Mechanisms of Antibiotic Production by Microbiome

The progression of antimicrobial resistance (AMR) is a multifaceted process governed by a spectrum of mechanisms that can be categorized into two distinct types: heritable and phenotypic mechanisms [29]. The first category comprises heritable mechanisms, wherein resistance traits are conferred through genetic elements that can be passed from one generation of microorganisms to another. These mechanisms encompass plasmid-mediated, transposon-mediated, integrase-mediated, and intricate toxin-antitoxin systems, which underpin the adaptability of microorganisms to antibiotic-induced selective pressures [30]. Conversely, the secondary category encompasses phenotypic mechanisms, orchestrating microbial responses and attributes in reaction to environmental cues. These mechanisms encompass the construction of biofilms, cooperative interactions through quorum sensing, dynamic adaptations within swarming, and the emergence of persister cells [31]. Together, these mechanisms unveil the ingenious strategies employed by microorganisms to elude the impact of antimicrobial agents. In understanding the interplay between heritable and phenotypic AMR mechanisms, we delve into the complex landscape of bacterial adaptations, revealing their remarkable strategies for evading antimicrobial challenges. A table is given with a few common microorganisms, antimicrobials, and the associated molecular resistance mechanisms (Table 6.2).

Inherent antimicrobial resistance is a distinguishing feature observed in both pathogenic and non-pathogenic microorganisms, as it plays an indispensable role in bacterial survival and adaptation within dynamic environments [29]. Innate bacterial resistance arises from bacterial cell membrane's impermeability to antibiotics, influenced by the drugs' physicochemical properties and efflux pumps, the lack of specific target molecules, and enzymes that deactivate antibiotic compounds [42]. These mechanisms result from the orchestrated interplay of genetic and inheritable components alongside diverse phenotypic traits involved in a broad spectrum of metabolic functions. Only in recent times has the significance of these intricate factors been acknowledged as critical in instigating and propagating antimicrobial resistance (AMR).

Table 6.2 Few microorganisms and their method for gaining AMR

Microorganisms	Antimicrobial compound	Resistance mechanism	Reference
Campylobacter spp.	Chloramphenicol	Point mutation on *gyrA* gene	[32]
	Aminoglycosides	Methylation of ribosomal target	[33]
	Tetracycline	Multidrug efflux pump (CmeABC)	[34]
	β-Lactams		[35]
	Cotrimoxazole		
	Tylosin		
	Lincosamides		
Staphylococcus spp.	Penicillin	Staphylococcal cassette chromosome-*mec* genes	[36]
	Methicillin		[37]
	Vancomycin		
Salmonella spp.	Kanamycin	AcrAB-TolC: multidrug efflux pumps	[38]
	Amoxicillin	*qnr* genes-containing plasmid	[39]
	Ceftriaxone	Multiple point mutations on *gyrA*, *gyrB*, *parC*, *parE* genes	
	Nalidixic acid		
Enterococcus spp.	Erythromycin	*lsa*(E) genes	[40]
	Tetracyclines		[41]
	Aminoglycosides		
	Chloramphenicol		
	Vancomycin		

6.3 Microbiome as a Reservoir of Antimicrobial Resistance Genes

The coevolution of humans with microbial communities is integral to understanding the thorough impact of past antibiotic usage. Antibiotics, celebrated as a foundation of modern medicine in the mid-twentieth century, have ushered in an era where infectious diseases seemed on the brink of eradication [43]. These drugs revolutionized medical practices, making complex surgeries, immunomodulatory treatments, and organ transplants possible. The balance of microbial species is often thrown into disarray which results in dysbiosis with frequent overuse disrupting the finely tuned interplay between human physiologies and their commensal microbiota [44]. The impact of past antibiotic uses in the host, and a few molecular perspectives of acquired antimicrobial resistance are two aspects that we will discuss.

6.3.1 Impact of Past Antibiotics Uses in a Host

Antibiotic usage in host organisms has complex impacts that extend beyond the immediate treatment of microbial infections. Antibiotics, despite being largely effective, have the potential to effectuate disrupting consequences within the host

[45]. Increased susceptibility to infections, emerging resistant pathogens, long-term alterations in the microbiome, and potential impacts on host metabolism and immunity are a few of the after-effects. The inappropriate use of antibiotics can have community-level implications, contributing to antimicrobial resistance [46]. Here, we discuss the impact of past antibiotic uses in hosts and a few other molecular perspectives of acquired antimicrobial resistance as markers of the aforementioned topic.

Understanding these specific impacts of antibiotic usage is essential for healthcare professionals, as it informs clinical decisions, antibiotic stewardship efforts, and research to lessen the adverse implications of antibiotic treatment [47]. Furthermore, investigations into approaches aimed at reinstating and sustaining a well-balanced microbiome post-antibiotic therapy hold the potential to mitigate the persistent occurrence and dissemination of antimicrobial resistance genes among the microbial communities that reside within our bodies. In essence, acknowledging the microbiome's characteristic as a reservoir for antimicrobial resistance genes underscores the imperative for a comprehensive approach to combat antimicrobial resistance, one that considers not only the host but also the interconnected microbial communities [48].

One of the foremost concerns arising from historical antibiotic utilization resides in the genesis and dispersal of antibiotic-resistant bacteria. Antibiotics impose selective pressure on the host organism, fostering the development of bacterial strains that exhibit resistance to these drugs [49]. These antibiotic-resistant bacteria persist within the host's microbiome, effectively serving as a reservoir for antimicrobial resistance genes. The presence of resistant genes among commensal bacteria elevates the potential for horizontal gene transfer, facilitating the transmission of resistance traits to pathogens [50]. Moreover, the global proliferation of multidrug-resistant bacteria has become a critical quandary in contemporary healthcare, partly attributable to the sluggish pace of new antimicrobial drug development compared to the rapid development of resistance to antibiotics. Consequently, the pervasive use and, at times, inappropriate application of antibiotics have emerged as substantial catalysts for this burgeoning public health crisis [51]. Consequently, addressing the aftermath of preceding antibiotic employment is not merely a concern confined to individual well-being but represents a collective endeavor to ameliorate the ever-mounting quandary of antibiotic resistance while preserving the intricate equilibrium of the human microbiome.

A table is prepared to provide more insight (Table 6.3):

6.3.2 Various Methods of Acquired AMR

In affluent countries, over-prescription of antibiotics by general practitioners, even when not medically warranted, contributes significantly to their inappropriate use. Diagnostic uncertainty often leads to unnecessary prescriptions, especially when

Table 6.3 Current hypothesis on impact of antibiotics in host

Cause	Effect	Reference
Microbiome disruption	Antibiotics can disrupt the stability of beneficial and harmful bacteria in the host's microbiome, negatively affecting microbial diversity	[52]
Proliferation of resistant strains	Antibiotics exert selective pressure, favoring the survival and proliferation of antibiotic-resistant bacteria within the host	[53]
Increased vulnerability	An altered microbiome can make the host more susceptible to opportunistic infections	[54]
Increased risk of infections	Prolonged antibiotic use may increase the risk of secondary infections by antibiotic-resistant pathogens	[55]
Resistance gene transfer	The host's microbiome can accumulate antibiotic-resistant genes potentially facilitating bacterial horizontal transfer, including gene transfer to pathogens	[56]
Metabolic effects	Antibiotics may alter the host's metabolic processes through changes in gut bacteria, potentially affecting nutrient absorption and metabolism	[57]
Allergic reactions	Antibiotics may affect the immune system resulting in allergic reactions (mild rashes to severe anaphylaxis) in some individuals	[58]
Developmental impacts	Antibiotic use during critical periods of development, such as in infants, may have long-term impacts on immune system development and overall health	[59]
Community-level impact	Antibiotic use within a population can lead to the emergence of antibiotic-resistant strains, affecting public health on a larger scale	[60]
Superinfections	Disruption of the microbiome can create opportunities for superinfections, where new pathogens colonize the host after the initial infection has been treated with antibiotics	[61]
Gut-brain axis	Alterations in the gut microbiome due to antibiotics may influence neurological conditions through the gut-brain axis	[62]

distinguishing between viral and bacterial infections is challenging. Self-medication also plays a role in this issue.

In many developing countries, the excessive use of antibiotics is due to their easy availability without a prescription. In both situations, there is a widespread belief that antibiotics are miraculous cures for various ailments.

In hospitals, prolonged and intensive use of antibiotics is a major factor in the emergence and spread of highly antibiotic-resistant nosocomial infections. Other factors include the presence of immunosuppressed patients, invasive surgical procedures, lengthy hospital stays, and inadequate infection control measures.

- *Antibiotic Consumption in Communities and Hospitals*

According to the European Surveillance of Antimicrobial Consumption Network (ESAC-Net), antibiotic usage in the community across Europe in 2012 varied between 11.3 and 31.9 defined daily doses (DDD) per 1000 individuals per day [49]. Beta-lactams/penicillin emerged as the predominant antibiotics prescribed. In hospitals, ESAC-Net estimated the average antibiotic consumption for systemic use in the EU/EEA to be 2.0 DDD per 1000 individuals in 2012, with beta-lactams/penicillin also being frequently administered. A point prevalence survey conducted in Europe in 2011 revealed that 35.0% of hospitalized patients were undergoing antibiotic treatment. In the United States, healthcare practitioners prescribed a total of 258.0 million antibiotic courses in 2010, with penicillin and macrolides constituting the most commonly prescribed categories [63].

- *Misconceptions About Antibiotics and Self-Medication*

Numerous studies highlight a deficiency in antibiotic knowledge among the general population, encompassing misconceptions regarding their efficacy against viruses and insufficient awareness concerning antibiotic resistance and potential side effects. A 2009 survey in Europe revealed that 20% of respondents took antibiotics for flu-like symptoms and 14% for common colds, despite knowing antibiotics do not work against viruses. In the UK, 38% of adults did not know that antibiotics do not work for most coughs or colds, while in Sweden, knowledge about antibiotics and resistance was better [64]. The inappropriate use of antibiotics is associated with behaviors like not completing prescribed treatments and self-medication, which is common in both developing and affluent countries.

- *Antibiotic Utilization in Agriculture and Livestock Farming*

A substantial portion of antibiotic consumption is attributed to food-producing animals and aquaculture, thereby exacerbating antibiotic resistance. While growth-promoting antibiotic usage is prohibited in Europe, it persists in the United States and other nations. According to the European Surveillance of Veterinary Antimicrobial Consumption (ESVAC), considerable antibiotic sales for food-producing animals were recorded in 2012, with tetracyclines, penicillin, and sulphonamides being the most commonly sold classes.

- *Environmental Factors and Antibiotic Resistance*

The environment plays a crucial role in antibiotic resistance. Soil acts as a reservoir of resistance genes, and water contaminated with fecal microorganisms and organic fertilizers can spread drug-resistant bacteria. Wastewater treatment plants (WWTPs) can harbor resistant bacteria, with high concentrations of resistant genes found in some studies.

- *Emergence of Resistance in Pathogenic Bacteria*

Increasing antibiotic resistance poses challenges in treating common infections like pneumonia and cystitis. These infections now require more complex and costly treatments. Several bacterial species, including *S. aureus*, non-typhoidal Salmonella (NTS), *K. pneumoniae*, and drug-resistant tuberculosis, exemplify antibiotic-resistant pathogens causing infections in various settings [65].

The diverse methods through which antibiotic resistance (AMR) emerges pose a severe threat to global public health. Addressing this crisis necessitates multifaceted strategies, but the development of new antibiotics remains an urgent and vital component of this response. The exploration of the microbiome for novel antibiotics represents a promising frontier in the fight against drug-resistant infections. As research advances, we can anticipate the discovery of new antibiotics and therapeutic strategies that leverage the unique capabilities of our microbial companions.

6.4 Various Methodologies Used to Understand the Microbiome-AMR Relationships

Antibiotics and microbiome research have been co-existing since the dawn of microbiome research: the humans' use of antibiotics increases natural selection that gives selective advantages to microbial populations that have antibiotic resistance against the popularly used antibiotics, which can render certain antibiotic drugs ineffective [66]. To study this relationship and help maintain the effectiveness of relevant drugs, numerous experimental and computational methods were developed to identify AMR genes in microbiome populations and colonies. In the advent of the age of high-throughput sequencing and metagenomics, the accuracies and throughput of these methods have also increased. This section discusses both the experimental and computational methods to identify antibiotic-resistant microbiome populations and/or AMR genes.

6.4.1 Experimental Methods

The experimental methods of identifying antibiotic resistance have aimed to identify specific colonies with resistance as well as paving the way for finding the genes that render such resistance [67]. These methods have been nicely categorized experimental methods to identify microbial populations resistant to antibiotics or AMR genes into culture-based analysis and molecular identification of AMR genes. We adapt this framework to categorize the experimental methods.

- *Culture-Based Analysis*

Culture-based analysis is by far the most used experimental method for screening resistant bacterial strains [67]. As the name suggests, it involves the culturing of collected microbial samples in one or multiple different cultures to allow for the growth of some of the collected bacteria species into colonies. The decision on the cultures is key: it not only provides the necessary nutrients that the species of interest will be able to survive and reproduce but also provides a base on which one introduces antibiotic resistance to select for resistant colonies. This enables an accurate and sensitive detection of resistant strains [68, 69] for select, well-defined antibiotics. Culture-based analysis is the gold standard for identifying resistance in certain microbial species due to its intuitive and thorough nature for hypothesis-driven questions [69]: if a colony can reproduce on the media with the selectively introduced antibiotics, then it poses resistance toward it.

The experimental and selective nature of this technique also leads to the downsides of this method, the main ones being low throughput and relatively low generalizability. The time required for culturing is long, and the same experiment has to be done for each studied antibiotic and each collected sample [70], leading to limited throughput of experiments. Additionally, some bacteria cannot survive on the commonly used media but are viable bacteria species that can carry antibiotic resistance. This can be due to inevitable human factors such as temperature and oxygen not being enough for the species to reproduce effectively [71], or can be because the media themselves were not suitable enough for the species [72]. This leads to relatively low generalizability as one must ensure the rigor of the setup before further screening. Finally, culture-based analysis is meant only for identifying resistant colonies, but not for identifying the genes responsible for such resistance. It must be coupled with additional molecular and computational methods to identify the bacterial species in the colonies or to find the AMR genes. Some such techniques are also discussed in the next subsection *Molecular Identification of AMR genes.*

- *Molecular Identification of AMR Genes*

A more direct approach to identifying antimicrobial resistance is through molecular identification of AMR genes. Such methods usually consist of two goals: to identify what AMR genes exist and which microbial species within one's samples possess such AMR genes. Broadly, one is also usually interested in the microbial taxonomy distribution landscape in the sample to see what species are generally survivable in an environment with antibiotics of interest, which can also help identify new AMR genes.

Polymerase chain reaction (PCR) and sequencing are likely the most commonly used molecular identification methods. PCR is also quite a specific method as it requires the design of primers to define genetic regions for identification. Different PCR techniques exist to achieve different cost, time, and effectiveness goals of researchers [69]. However, one thing of note is that conventional PCR techniques cannot differentiate genes from viable microbial cells versus non-viable ones, the latter of which is undesirable. This can be overcome by real-time PCR [73] or multiplex PCR techniques [74]. In addition to PCR, direct sequencing of the samples also enables direct identification of potential functional regions of interest. Different sequencing techniques possess different advantages: while this is not the focus of this, it is worth mentioning that the development of long-read sequencing such as nanopore enables state-of-the-art high-quality genome reconstruction [75], posing the potential of bettering the downstream taxonomy calling (species identification) [76] and AMR identification [77]. The algorithms for species calling and AMR identification will be discussed in Sect. 6.4.2.

It is also worth mentioning that mass spectrometry (MS) is a potential new avenue for identifying AMR species. Different from genetic sequencing-based identification, MS, typically MALDI MS, identifies AMR species from environmental samples with antibiotics based on the proteins and peptides derived from the microbial cells, which can accurately identify unknown bacteria on multiple levels including genus, species, and even subspecies level [78, 79]. Combining PCR and MS technologies also provides the possibility of species and even strain identification [80, 81]. While MS is not as commonly used as sequencing to identify AMR species and genes, one can consider this as one additional goal while conducting MS for other functional characterization.

6.4.2 Computational Methods

While it is necessary to conduct experimental methods outlined above for AMR species and gene identification, the actual species and gene identification usually require computational methods to deconvolve the data generated from the experiments. For example, one can directly sequence environmental samples or the cultured isolates of environmental samples but cannot speedily identify AMR genes and what species carry such genes without the support of computational tools. Here, we outline some computational methods for taxonomy calling as well as AMR gene identification.

Taxonomy calling or characterizing the presence and absence as well as abundance of different species in given samples is one of the most common goals of microbiologists. In the context of AMR research, one usually is given a sequenced direct or cultured sample of environmental microbes and needs to identify the species abundance in the sample. Such methods usually work by aligning certain biomarkers from sequencing data with their annotated species databases. The most commonly used methods are the MetaPhlAn [82–85] and Kraken [86, 87], (Braken [88]) families of taxonomy callers, among others such as PathSeq [89], MegaBLAST [90], taxMaps [91], and DIAMOND [92]. It is worth noting that different taxonomy callers use different biomarkers and biomarker databases to identify species. For example, the Kraken family, PathSeq, and taxMaps use DNA sequencing data, DIAMOND uses protein data, whereas MetaPhlAn family uses clade-specific markers. These differences render different taxonomy callers' strengths and weaknesses in identifying different species. A review by Ye et al. comprehensively benchmarks and summarizes the performance of these taxonomy callers, among others, in different areas ranging from the ability to correctly classify species, to correctly compute abundance, and to computational burden [93]. With one's choice of taxonomy caller, one can directly apply these methods to sequencing data to identify the available bacterial species and abundance.

AMR gene identification works similarly to taxonomy calling in that it usually identifies AMR genes by aligning genetic regions in sequencing data with databases storing such information. While there are many existing methods, with the commonly used ones being abricate [94], Ariba [95], ResFinder [96], RGI [97], AMRFinderPlus [98], and SRST2 [99], there has not been comprehensive benchmarking on the effectiveness and time efficiency of these computational methods. Some differences exist between these methods, however. For one, different methods support different kinds of data: SRSR2 supports sequencing data, abricate, and ARIBA supports assembly data, whereas AMRFinderPlus supports protein data; RGI supports both nucleotide and protein data. For two, they align with different kinds of databases for reference genes: abricate, Ariba, SRST2, and AMRFinderPlus all support multiple databases commonly including the NCBI and CARD databases [97, 100], whereas ResFinder uses its own ResFinder and PointFinder databases and RGI supports the CARD database.

In conclusion, there exist many experimental and computational methods to identify resistant bacterial colonies, species, and AMR genes. One may choose one or a combination of methods based on their cost and effectiveness requirements as well as the type of data they have at hand. With the advent of metagenomics and the steady improvement of sequencing technologies, lower amounts of samples will be required to achieve highly accurate species and AMR gene identification, especially with increasingly comprehensive computational databases and methods. There still exists an urgent need for benchmarking computational methods that conduct AMR gene identification and determine which methods work the best in which scenario.

6.5 Role of Environmental Microbial Reservoir AMR Development and Spread

AMRs are those microbes that do not respond to any agent to eliminate them, multiplying significantly through different transmission routes. These microorganisms constitute from bacteria to fungi or viruses. Since its existence became known, it has generated concern in world-class institutions and organizations such as the World Health Organization, which created a database of recurrent pathogenic AMR, attributing its resistance as acquired or natural [101].

Certainly, there is evidence that allows us to know the role of environmental reservoirs in the spread of AMR. This section will analyze their role as a transmission that has been reported in research. In this section, we address the role that environmental microbial reservoirs have had in AMR development and resistance. Spread transmission can be divided into two highly common cases; one is the incorrect use of antibiotics in various industries that allow human-animal contact; this refers to the agricultural and food chain; subsequently environmental interactions due to specific changes [15, 102].

The processes of evolution cannot be invaded by organisms. All of them have genetic mutations, which can sometimes be difficult to handle; the prevalence of AMR is a result of evolution [103]. On another hand, the AMR genes can be acquired on two levels according to the literature. The first is when the genes move within the genome, which is called the intracellular level, while the inter-cellular level consists of a horizontal spread [104].

6.5.1 Some Reports and Case Studies

Microorganisms such as *Staphylococcus aureus*, *Neisseria gonorrhoeae*, and *Clostridium difficile* have been classified as resistant to antibiotics; annually, there is a high mortality rate due to resistance acquisition in bacteria. European countries and some on the American continent, especially the United States and Canada, have struggled to address the problem through their research. Countries with deplorable economic conditions have primary consequences on their population, in addition, causing a lack of research and reports that provide an accurate database. This section exemplifies some cases that have arisen in regions around the world.

In 2022, a published study exposed wastewater representing repercussions on human health. Countries that lack management systems present AMR spread. A precedent of it happened in the city of Montevideo in Uruguay where currently, only 90% of houses are connected to a sewage system, and water represented a potential vehicle for AMR spread because hospitals do not have a controlled water waste system. Hospital water represents a high danger as it circulates a large number of pathogens that produce spectrum-beta-lactamase and carbapenem, being common in microorganisms that represent resistance, concluding that infected humans correlate with the abundance of AMR in a vital element found in the environment, which unfortunately may increase the mortality and disease cases.

Furthermore, from 2008 to 2014, Japan experienced genetic diversity in *Campylobacter coli* in animals and humans; being reported antimicrobial resistance of *Campylobacter coli*; the study indicates that precisely water, land, and other environmental factors may be responsible for the spread of pathogens; especially if there is environmental contamination; attributed to urbanization, agriculture, and waste management activities, also suggesting the poultry and cattle as major reservoirs. The transmission is given in foodborne; foods that come from animals, although this is a chain, where animals are affected by environmental conditions, thus leading to human transmission by consuming derived products [105].

A country that is annexed to this list of cases is regions of Canada, where broiler chicken farming faced antimicrobial resistance in *Campylobacter*, a bacteria associated with stomach infections. The objective of the study consisted of comparing regulations for the treatment of said food, confirming that the unregulated use of antibiotics can affect the aliment industry, having direct repercussions on human health. In addition, a disparity was highlighted between areas of Canada, associating this difference with how antimicrobials are used in veterinary medicine [106].

In the region of New Zealand, another similar study was conducted on agricultural systems with contrasting systems and practices, from irrigation systems to the elimination of pesticide use, where *E. coli* bacteria isolates were screened, having the purpose of detecting antimicrobial resistance and showing evidence that the excessive and unregulated use of antimicrobials and other agrochemicals can increase the spread of AMR from soil to other environments [107].

Although various cases have occurred in Asia, Africa, and America during the last decades, Europe has been one of the continents that has implemented successful policies and practices; such is the case in Sweden; which emphasized the importance of responsible antibiotic use in agriculture and veterinary medicine to safeguard both animal and public health [108]. In this case, the acquisition and spread of AMR are diverse, so countries must provide the conditions and regulations that reduce cases in any sector.

6.5.2 *Various Routes of AMR Spread and Their Transfer to Humans and Animals*

AMR transmission routes are varied, so it represents a big problem when dealing with them, thanks to the reported cases; it is possible to classify the most common routes. Here, we fully explain the diversity of AMR spread pathways and their transfer between human, environmental, and animal interactions.

Routes of transmission:

- Human-human
- Animal-to-human transmission (zoonotic spread)
- Spread via animal-derived food products
- Environment
- Travel and global movement
- Wildlife as reservoirs

Human-Human One of the last pandemics that shook the world COVID-19 confirmed that throughout history, it has been proven that humans can transmit endless diseases through contact with contaminated fluids or spaces. In healthcare settings, this transmission route is of particular concern, as resistant infections can be contracted during hospital stays or medical procedures [109]. Methicillin-resistant *Staphylococcus aureus* (MRSA) is a clear example that by having contact from one person to another, they can become infected [110, 111]. However, the transmission in individuals is a result of exposure to natural environments [112, 113]; it must be remembered that humans are continually in constant change and exposed to diverse environments [114]; in this case, to mitigate AMR, the main sources of acquisition must be identified. Nowadays, the specific characterization of AMR acquisition in humans remains unknown.

Animal-to-Human Transmission (Zoonotic Spread) Contact between humans and animals leads to zoonotic diseases. In a portion of cases, animals can transmit pathogens with antimicrobial resistance. This occurs in different scenarios, for instance, through the consumption of contaminated animal products, direct contact with animals in farming activities, and some wildlife interactions [115]. However, there is a slow nature in the transfer of antimicrobial resistance specifically between animals and humans. High-risk consequences usually occur transversally, meaning there is no instantaneous transmission [116]. Transmission in animals is also due to various aspects. For instance, during transportation and shipping, there is a high risk of transmission of antibiotic-resistant bacteria, given the unhygienic measures in vehicles. In addition, foodborne transmission, agricultural exposure, and environmental interactions contribute to the spread of antibiotic resistance [117].

Spread Via Animal-Derived Food Products The food industry prevails because it is considered an elemental key to the life of the human being; however, it has been reported that the food chain can negatively impact health due to its contamination. Antimicrobial resistance can infiltrate the food chain when antibiotics are used in agriculture or aquaculture. In particular, poultry where antimicrobials like tetracyclines, sulfonamides, and penicillin are widely used [118], developing resistance to certain microorganisms. The consumption of meat, dairy, or seafood from animals treated with antibiotics may introduce resistant bacteria into the human digestive system; including in livestock animals, which can also be affected by contaminated foods.

Environment The environment acts as a substantial reservoir of antimicrobial resistance acquisition. Contaminated water bodies, soil, and air can harbor resistant microorganisms. Humans and animals can be exposed to these reservoirs through various means, leading to potential infections. Poor wastewater treatment, the use and management of pesticides in soils, and even animal and human feces facilitate the resistance of bacteria. It is worth mentioning that certain environments, due to their unique conditions, tend to provide genetic characteristics associated with the transmission of antibiotic-resistant genes (ARGs) [119]. The environment will continue to suffer constant interactions and changes, especially due to urbanization; therefore, the measures must be considered with the objective of reducing risks in the population.

Travel and Global Movement International travel involves the movement of millions of people through more than 100 countries. Therefore, people can be vectors of transmission, considering other factors such as the transportation of food and objects.

Wildlife as Potential Reservoirs Without surprise, the origin of many diseases has been investigated in wildlife reservoirs, which can carry AMR bacteria and spread it quickly to humans or some domestic animals; clearly, insects and bugs represent potential carriers of antimicrobial resistance.

In summary, due to the current situation and the problems faced due to AMR in the health field, it must be considered that the environment is a determining aspect for its evolution and spreading; being a constant and urgent fight. This remains a constant and urgent battle. While some studies have highlighted this phenomenon, the mechanisms and role of AMR remain unclear. However, advances in technology and research can provide tools to combat it.

6.6 Human Resistome and Its Relationship with AMR Evolution and Spread

6.6.1 Sources and Factors

Deep within each of us lies a silent defense system, a potent arsenal: the human resistome. The term encompasses all identified genes for antibiotic resistance as well as their precursors, and potential pathways for resistance found in the microbial community residing in our bodies [120]. Antibiotic resistance has become more prevalent, and the study of the resistome is gaining more importance.

Year after year, it is becoming more common to encounter pathogens that are displaying resistance, limiting our treatment options in various settings, and creating challenges for us. Hence, it is of utmost importance to understand the driving forces of this challenge. That is why we must listen very carefully to the human resistome, because while our human genome sheds light on our evaluation, the resistome provides insight into our interaction with microbes and the use and misuse of medications.

Although it may seem that the human body possesses a singular resistome entity, it is a complex assembly of unique resistomes. Various studies have characterized resistomes within the human body. These include the gut resistome, skin resistome, nares resistome, vaginal resistome, oral cavity resistome, and nasopharyngeal resistome [121–124]. Each of these resistomes has its unique characteristics and a specific set of antibiotic resistance genes [122]. This diversity highlights the complexity of the resistome.

We may likely identify more resistomes within the human body in the future, but, as of now, the human gut resistome stands as the most thoroughly examined among the currently defined resistomes as most metagenomic studies focus on the human gut [125–129]. As noted by Singh et al. (2019), antimicrobial resistance genes can be transferred between species of commensal bacteria and transient pathogens in the human gut due to the combination of factors that create an ideal environment for ARG transfer in the gut, thus highlighting the paramount role of the human resistome in the spread of ARGs.

In this chapter, we will explore various sources influencing the establishment of the human resistome and, in turn, its role in the evolution of antimicrobial resistance (AMR). Three primary domains will be categorized to understand the interconnected factors shaping the human resistome. The list starts with the human host itself, which serves a dual role. Humans not only act as the habitat of the human resistome but are also one of the principal sources. Next, we will review the critical role of animals in shaping human resistome. Far from being passive participants, animals act as vital channels for the transfer and evolution of ARGs, directly and indirectly contributing to the human resistome. Finally, environmental factors will be examined. We aim to provide a comprehensive understanding of the three crucial interconnected spheres contributing to the human resistome: humans, animals, and the environment.

6.6.1.1 Humans as a Source of the Human Resistome and the Related Factors

Various sources and factors contribute to the formation of the human resistome, as it is dynamic and ever-changing. One of the main contributors to human resistome is antibiotic use in healthcare. Several studies have demonstrated that antibiotic use in clinical practice contributes to the spread of ARGs, particularly within the human gut microbiome. Early research focusing on infants revealed that antibiotic exposure is correlated with long-lasting increases in the gut resistome with extended carriage of multidrug-resistant bacteria [130]. Reanalyzing the human microbiome project highlighted the most abundant ARGs in the project, which were against beta-lactams, MLS, tetracycline, multidrug resistance genes, and fluoroquinolone. These findings align with the antibiotics most extensively prescribed during the project's timeline, suggesting a link between antibiotic resistance and their consumption [122, 131].

Building on the role of antibiotics, it is essential to delve into the intrinsic reservoirs of ARGs within individuals. The gut microbiome inherently acts as a reservoir for ARGs. This process begins as early as infancy, and ARGs are present even in antibiotic-free infants. This is likely due to mother-to-child transmission [132]. The initial gut resistome established in infancy accumulates more ARGs through later antibiotic exposure and other factors. Bacteria that are moving through the gut may interact with existing commensal bacteria. Such interactions allow for the exchange of ARGs, contributing to the emergence of resistance in the transient species [133]. This interaction amplifies resistance in the transient bacteria, which can re-enter the human population through environmental exposure, contributing to the human resistome [122].

Although antimicrobial use (AMU) is deemed a main contributor to the spread of ARGs, multiple studies have highlighted the role of several factors, such as sanitation, travel, diet, access to clean water, and infection control [134, 135]. Exemplary traveling, as a potential factor, is associated with gut colonization by ESβL-producing *Enterobacteriaceae* [135].

Human-to-human transmission is another way for the bacteria to spread resistance. Especially in places where sanitation is not particularly good, oral-fecal transmission might spread resistant bacteria within the community. Similarly, sexual encounters could also serve as a transmission method for certain resistant bacteria. In this regard, the well-studied historical case of sexually transmitted *Neisseria gonorrhoeae* can be used as an example. Initially, in the 1930s, *N. gonorrhoeae* was being treated primarily with sulfonamides. In 15 years, the resistance rate exceeded 75%, and penicillin became the treatment of choice. However, after the introduction of penicillin, reports on the development of resistance in *N. gonorrhoeae* began. A similar pattern was observed with tetracycline. Over several decades, *N. gonorrhoeae* has gained resistance to three major groups of antibiotics [136]. Although it is unclear to what extent ARGs are found in asymptomatic careers, multiple studies have shown that asymptomatic *N. gonorrhoeae* infections are common in men and women [137], making them a part of the human resistome.

6.6.1.2 Animals as a Source of the Human Resistome and the Related Factors

While humans themselves play a significant role in shaping the human resistome, animals also contribute substantially to this equation, by being a major source of ARGs and antibiotic-resistant bacteria. Factors such as antibiotic use in animals and poor husbandry practices contribute to animal resistome. Various studies have shown that ARGs have shown to be transmitted between both animals and humans through a combination of direct and indirect interactions, such as through contaminated food, soil, or water. One of the pioneering studies demonstrating the transfer of resistance from animals to humans was conducted in 1969. Although the study concluded that the transfer did not persist in the human gut, they showed the possibility and potential for ARG transfer between humans and animals [138, 139].

One of the major factors leading to increased ARG abundance is antibiotic usage in livestock and poultry. Although the relative benefit of antibiotic use is still an area of research, only in the United States, did the utilization of antibiotics for animal farming increase more than 100 times between 1950 and 2004 [140]. There are several incentives for the use of antibiotics in animals. One primary use area is the treatment and prevention of disease. The broad use of antibiotics in this sector acts as a significant catalyst for the evolution of antimicrobial resistance within the animal population, which may result in the selection of ARGs that may ultimately be transferred to humans [120].

Another type of use is a low-dose antibiotic treatment for growth promotion [141]. This type of use exposes animals that are otherwise healthy to antibiotics. Although these are primarily low-dose antibiotics, there is evidence that even low doses can also contribute to the expansion of the resistome in animals [120].

Antibiotics not only expand the animal resistome but also result in significant excretion of these drugs to the environment. Depending on the antibiotic and the route given, up to 90% of the delivered dose was found to be excreted as an active metabolite or in its original non-metabolized form. Hence, animal feces and urine may contain high concentrations of antibiotics [142]. This raises a significant concern, especially considering that some animal waste is used as manure.

A recent study characterized the role of manure treatment in ARG transfer to the environment. Soil treatment with manure led to increased ARG abundance and decreased microbial diversity in the soil. Interestingly, the study also discussed that although the quantity of live antibiotic-resistant bacteria is low in composted manure, it still plays a role in ARG spread. Apart from the live bacteria, the manure treatment is thought to be responsible for the spread of ARGs through mobile ARGs and due to the presence of residual antibiotics [143].

As early as 1976, Levy and his colleagues were among the first to report the transfer of resistant organisms from animals to humans [144]. This study provides critical insight into the relationship between animal and human resistomes. This transmission can occur in several ways, one of which involves direct contact with animals. This pathway particularly places individuals such as farm and slaughterhouse workers at elevated risk due to their close interactions with animals [145].

Another important mode of transmission is the consumption of food products carrying resistant bacteria and ARGs. Strong evidence has shown that food products, at each point during processing, can harbor a significant number of antimicrobial-resistant bacteria along with their resistance genes [145]. This prevalence of antimicrobial resistance is not confined to a single stage or type of food product. A myriad of studies have reported the presence of ARGs in various food products at multiple stages from bulk milk to cooked meats [146]. These food items not only harbor the resistance genes, but they also function as carriers, passing these genes to humans when consumed [147]. The transfer is not just straightforward, but also occurs indirectly, as these food items mix with other environmental factors, promoting the broader distribution of the resistance genes. This shows that the food industry serves as a crucial bridge facilitating the migration of ARGs from animals to humans, contributing to human resistome.

6.6.1.3 Environment as a Source of the Human Resistome and Related Factors

The environment is a potent shaper of the human resistome. The relationship is dynamic, attributable to our historical and ongoing encounters with microorganisms, and our antibiotic use patterns, including misuse. The repercussions of such misuse are now apparent, necessitating a deeper exploration into the complexities of the environmental resistome.

It is crucial to acknowledge the inherent ability of the environment to facilitate its resistome development. Although antibiotics are undeniably important catalysts and pressure points for the selection and emergence of antibiotic resistance, they are not the sole contributors. Resistance existed within the environment even before the widespread usage of antibiotics, as evidenced by historical data [148]. This also highlights that the environment is a natural source of antibiotics and their resistance genes and is mainly seen as the primary reservoir for ARGs [149–151].

The environment plays a critical role as a natural reservoir for ARGs [147]. This role is influenced by a wide range of factors. Although ARGs are naturally present in minimal quantities across diverse ecosystems, certain environmental conditions can increase their abundance. Notably, the presence of heavy metals, crude oil, and sewage has been linked to a surge in ARG prevalence [152].

Rivers and other surface waters play a pivotal role in the dynamics of the human resistome and act as reservoirs for ARGs [120, 153–156]. They are the ultimate recipients of various pollutants from multiple sources, such as agricultural, industrial, and domestic sources. In this capacity, they function as a collective "sink" for ARGs, accumulating them and potentially spreading them to distant environments. Numerous studies have validated that rivers close to human habitation harbor a significantly larger pool of ARGs than pristine, untouched rivers [120]. One study underscored the impact of human presence on ARGs dissemination by demonstrating a positive correlation between the concentration of resistant bacteria and the size of the nearby human population [157]. These findings clearly illustrate the role of human-related factors in the proliferation of ARGs and the pivotal role of surface waters.

A major study on resistomes in soil was conducted in 2006. This study included 480 soil-derived bacterial isolates and screened them against 21 different classes of antibiotics. The results revealed a startling level of resistance with most bacteria exhibiting resistance to seven to eight antibiotic classes [158]. There have since been studies indicating that soil is the origin of ARGs. One study revealed that permafrost, which can preserve microorganisms and their DNA for thousands of years and provides a unique record of life in the past, contains ARGs [159]. Even isolated environments such as pristine caves are not exempt from the presence of ARGs. A study examining bacteria in a cave isolated from the surface for millions of years revealed a multidrug-resistant genotype [160]. This suggests that antibiotic resistance is not a recent phenomenon but has been an integral part of microbial life for thousands of years in the environment.

As summarized in the review article by Carvalho and Santos, the antibiotic burden in rivers can circle back to humans through various pathways. These pathways include water treatment plants and drinking water. Additionally, wildlife inhabiting these aquatic environments, where antibiotics and resistant bacteria thrive, can acquire antibiotic resistance genes (ARGs) through their feeding habits and living conditions [161]. When these animals are consumed by humans, they consequently contribute to human resistome. Similarly, crops represent another significant means of transferring antibiotics. When irrigated with water from these contaminated sources or grown in soils enriched with these resistant bacteria, crops can take up and carry antibiotics [147, 161, 162]. Upon consumption, these compounds can contribute to human resistome.

6.6.2 Evidence of Asymptomatic Carriage of AMR Genes

Encountering pathogens does not always cause an aggravated response in a host. Similarly, through evaluation, animals have developed two distinct defensive mechanisms to deploy, when faced with invaders. During an encounter with a foreign pathogen, the animal may choose a strategic adaptation response over an extensive lethal reaction. This requires a physiological modulation of the response by turning the invader into a silent passenger [163]. This phenomenon is called asymptomatic carriage, which is when an individual carries a foreign pathogen that has resistance genes but does not cause signs or symptoms of an infection. Making them practically invisible, this phenomenon results in the concealment of the true extent of AMR in the population, which in turn hinders our ability to effectively address them. Hence, we must develop our tools to recognize asymptomatic passengers, even though they do not show any overt signs of presence.

Although initially thought to be a small contributor, it seems that this phenomenon is more prevalent than initially thought. Multiple studies investigated its prevalence. In one study, the AMR profile of *E. coli* from symptomatic diarrhea cases displayed a remarkable resemblance to the control group with no clinical symptoms [164]. Similarly, the case of MRSA carriage is another example. This bacterium can inhabit the nasal and throat areas without causing any symptoms [165]. This was highlighted in another study showing a significant correlation between prior clinical experience among dental students and a higher incidence of MRSA carriage [166]. Our understanding of this issue is extensive, ranging from the silent carriage of *Streptococcus pyogenes* in the throat to the presence of various bacteria, including *Ureaplasma urealyticum*, in the urogenital area. Even more alarmingly, extensively drug-resistant bacteria have been found in asymptomatic individuals [167–170].

This described hidden and concealed aspect of the resistome is a major player in the narrative of AMR. In this complex story, every chapter matters, including each quiet carrier and each AMR gene adding to the overall story. This is a story that cannot be ignored, this is a story that tells us about our past, and present and, if listened to carefully, could tell us about our future.

6.6.3 Strategies to Control the Resistome: Limitations and Challenges

Understanding different resistomes is a complex endeavor, as they display a diverse and ever-evolving nature, but a crucial one, since a change in one setting, can have a profound effect on the others, given the interconnected nature of all resistomes. By understanding these factors, we have a better chance of creating effective solutions.

Detection of the progression of the resistome poses a major challenge in combating AMR. There is a need for timely and accurate detection methods to address this issue effectively. To address this challenge, an integrated approach encompassing techniques such as functional metagenomics, next-generation sequencing, and state-of-the-art computational methods is essential [171]. The focus of these surveillance efforts should target the known areas prone to AMR evolution, as highlighted by Crofts et al. [171]. These locations typically exhibit high antibiotic usage, such as in agriculture and healthcare facilities, or where AMR emergence is more prominent due to high microbial presence, such as in densely populated regions or areas with inadequate sanitation.

The connection between the broad antibiotic use and the increasing prevalence of AMR is becoming increasingly evident as noted by multiple studies [122, 130, 131, 172]. It is essential to implement strategies to reduce antibiotic exposure in all sectors, including healthcare, veterinary care, and environmental management. These strategies involve promoting prescribing practices based on evidence, establishing programs for antimicrobial stewardship, and raising awareness among the public about responsible antibiotic use [173]. By reducing the pressure on bacteria, we can minimize the occurrence and transmission of resistant strains.

Infection control and prevention play a role in reducing the burden of AMR. Vaccination is an effective method for infection control. Not only do they show the ability to control the spread of AMR, but they also effectively reduce the existing burden of AMR. An illustrative example is a study from 2018 that highlighted the impact of the PVC13 vaccine in combating AMR. This study revealed that introducing the PVC13 vaccine led to a decrease in the previously high prevalence of AMR in pneumococcal disease [174]. In terms of their use in animals, various alternative methods have been suggested as substitutes for antibiotics. For instance, practices such as maintaining hygiene, improving the living conditions of animals, using probiotics, phage therapy, and administering vaccines have been proposed for use in animal farming [175, 176].

Among these alternatives, phage therapy stands out as an option. This method uses bacteriophages [177], which are viruses that have evolved to target specific bacteria for elimination [178]. Unlike broad-spectrum antibiotics, this targeted approach prevents the unintended disruption of beneficial bacteria.

The pursuit of new antibiotics has emerged as a natural strategy. Despite efforts to develop new antibiotics, the rapid rise of bacterial resistance has transformed these solutions into temporary fixes. While upgrading the arsenal of available antibiotics is undeniably important, it is essential to develop innovative and long-lasting strategies to address the growing problem of AMR. The ongoing fight against resistance has sparked the development of new and inventive approaches. These methods go beyond conventional methods and involve antimicrobial peptides, probiotics, nanobiotics, the CRISPR/cas9 system, and bacteriophages [179]. As a remarkable example, nanoparticles can effectively treat bacterial infections by suppressing the proliferation of multidrug-resistant bacteria. They can be used to improve the delivery of conventional antibiotics or act as standalone bactericidal agents [152]. Another promising example was shown when bacteriophages were used effectively to treat a patient with multidrug-resistant bacteria who had not responded to conventional antibiotics [180].

Each solution has its own set of limitations. The reduction in antibiotic usage is not different. One obstacle is the creation of universal regulations for antimicrobial therapy. This process is complicated by the diverse responses from different nations, each with its unique healthcare system and policies. In a clinical context, the challenge is even greater. Reducing antimicrobial usage in patient care requires more precise evaluation of patients, since with the current diagnostic tools, determining which patient needs antibiotics is often not a straightforward decision. Implementing a more thorough assessment will inevitably translate into increased time and resources devoted to each patient, adding to the workload of healthcare providers who are already overwhelmed with significant provider shortages. Even the novel methods that have been introduced are not without their limitations. Apart from the logistical and regulatory obstacles, the use of bacteriophages is linked with bacterial resistance and the presence of neutralizing antibodies, which limit their application [180].

In conclusion, the battle against AMR has been escalating and requires a holistic approach. A better understanding of all parts of the resistome, both within and outside the human body, is of paramount importance if we are to gain the upper hand in this race. By recognizing that each one of us has a role to play in this battle against AMR, it is essential that researchers, healthcare professionals, policymakers, and the general public work together to control its spread. Although there are multiple setbacks and challenges along the road to alleviating the burden of AMR, the novel approaches and exploration of new strategies signify our relentless pursuit of combating AMR, working toward a future where it is no longer a looming threat to the general public.

6.7 Impact of Antibiotics on Microbiome

6.7.1 Effect of Antibiotics on Microbiome Diversity

Antibiotics inhibit the growth of bacteria, and before their discovery, infectious diseases including pneumonia and tuberculosis were a major source of mortality [181]. This class of molecules modernized medicine today, saving lives on a massive scale. Yet, the use of antibiotics comes with consequences.

Our gut is intimately linked with our microbial residents. This community of microbes has been tuned by human evolution to comprise specific microbes and optimized abundances that lead to important health implications including the maturation of our immune system and transforming nutrients into metabolites like short-chain fatty acids (SCFA) [23]. Antibiotics perturb this status quo of mutualism between us and microbes. Acute changes occur to the community structure of our microbiome and consequently formulate the evolutionary forces that develop into antimicrobial resistance. Ultimately leading to dysbiotic gut flora, creating large dynamical changes leaving canonically healthy commensal microbes (*Faecalibacterium prausnitzii* and *Bacteroides fragilis*) to become depleted and upregulation of pathogenic microbes [181].

Trillions of microbes comprising 1000+ species live within human gastrointestinal tracts. Due to this magnitude of scale, exercising control of each microbe within each class of species is challenging for the host. The host offers some ability to control system-wide control on the gut microbiome via the immune system and environmental conditions (e.g., pH) to dictate populations of specific commensal microbes or removal of pathogenic microbes [182]. Microbial dynamics is in large part regulated by other interacting microbial members. High-diverse communities constitute a wide range of interactions among microbes, including both competition and cooperative interactions, that ultimately create stability [183].

As a byproduct of these community dynamics, metabolites are produced that are important for the normal homeostasis of host health. One of these byproducts includes SCFA which are known to be involved in both proximal and distal organs from the gut to the brain [184]. The homeostatic role of SCFAs is multifaceted. Butyrate promotes anti-inflammatory signals to colonic macrophages and dendritic cells as well as regulates the inhibition of stem cells residing in the colonic crypts. Interestingly, butyrate is also an inhibitor of histone deacetylase, hinting toward epigenetic roles [185]. Faecalibacterium phylum is one of the main producers of butyrate and is often seen to be downregulated in antibiotic-treated guts [184]. Bacteroidetes phylum is known to break down difficult-to-degrade carbohydrates into substrates which can then be taken as nutrients from the epithelium. One of these substrates includes acetate which is the most common SCFA found in the normal gut [185]. Acetate has been described as not only a substrate for the host but also for other microbial members including Faecalibacterium phylum.

Unfortunately, antibiotics are not microbe-specific but instead target regions that are conserved across the bacterial kingdom. There are many types of antibiotics such as macrolides and tetracyclines that are broad in their targeting, impacting a wide range of microbes. This causes large-scale collateral damage to the commensal microbes that ranges from inhibition to killing [186]. Therefore, sweeping changes occur impacting the diversity of the microbiome that is crucial to the maintenance of health. Diversity can be tracked through metrics like alpha and beta diversity [187]. Not all microbes get depleted after antibiotic treatment, and some survive and find themselves in a new changed environment. Consequently, this loss of diversity opens an opportunistic niche due to the different types of sugars and amino acids within the local environment. This change in resources within the gut allows other normally lower abundance taxa to flourish in dysbiosis as seen in early infants with increased abundances in bacteria including *Staphylococcus epidermis* and *Streptococcus salivarius* [188]. Colonization resistance becomes reduced due to the impacted commensal microbiota allowing for pathogens like *Clostridium difficile* to gain a competitive advantage and flourish in the new antibiotic-treated gut environment [189].

Most of the bacteria that tend to flourish and become pathogenic are bacteria that were already present in the healthy gut microbiota. Frequently, individuals have *Staphylococcus aureus*, *Streptococcus pneumoniae*, *Pseudomonas aeruginosa*, *Klebsiella pneumoniae*, and *Escherichia coli* present in their gut which are also opportunistic pathogens that find themselves upregulated after antibiotic treatment [56]. These microbes frequently reside throughout the gastrointestinal tract in a region-specific manner due to the varied environmental conditions (pH and nutritional content). Interestingly, antibiotics induce the translocation of these microorganisms to a range of body sites, such as the bloodstream which can cause large complications in health [190].

Antibiotics induce selective pressures for microbes to gain antimicrobial resistance genes to survive and foster. This enrichment of antimicrobial resistance genes has been frequently seen in patients receiving antibiotics causing large health implications that frequently lead to death [191]. This gain-of-function adaptation to antibiotics does not only lend itself to evolving a new suite of AMR genes from the microbe itself but rather much of these AMR genes are sourced from mobile genetic elements. This method of horizontal gene transfer allows for easy spread and acquisition of AMR genes within a microbial environment, with many pathogens weaponizing these suites of genes for their proliferation [183].

These evolutionary remodeling events are dependent on the specific class of antibiotics yet gaining resistance from antibiotics comes at a cost. Multi-resistant plasmids have become a challenge in the hospital setting but, in an antibiotic-free environment, these AMR-encoding genes are associated with a fitness cost. For example, blaCTX-M-15 encodes for a signal peptide that allows β-lactamase to be excreted out from the cytoplasm. This affects normal cellular function and constitutes mutated bacteria with a lower competitive advantage compared to its wild-type counterpart [192]. Thus, these new evolutionary forces and microbial dynamics are environmental contexts.

6.7.2 Consequences of Long-Term Antibiotic Use

Development of the microbiota starts after birth when it is thought that an influx of microbes seeds the gastrointestinal tract of infants from the vaginal canal. In caesarian-section-delivered infants, the initial seeding event comes from the skin and hospital-environment microbes [193]. These sequential colonization events throughout childhood to adolescence are important as they dictate the development and trajectory of a healthy microbiota. Colonization events include weaning to solid foods where an influx of *Bacteroidi*a and *Clostridia* taxa starts to develop inside the gastrointestinal tract [24]. The microbiota's development is crucial to train the immune system. As the infant continues to develop, so those the microbes due to a series of changing environments in pH, nutritional content, and the presence of other microbes throughout the proximal and distal lengths of the gastrointestinal tract, ultimately shaping local microbial communities within the gut. These microbial communities start to become stable toward adulthood but before this timeframe, especially within the first 2 years of life, the microbiome is highly susceptible to perturbations that could lead to life-lasting effects.

The microbiome directly interfaces with epithelial tissue and plays a considerable role in feedback systems such as mucosal production but also the maturation of the human immune system. Major stimulators for neutrophil development and production come from microbial signals such as IL-17A leading to an increase in circulation of granulocyte stimulating factor [194]. The absence of microbial signals will cause apoptosis in neutrophils. Just as microbial signals dynamically control the population of neutrophils, they are also just as critical for the development of macrophages. SCFAs produced by commensal microbiota dampen inflammatory response but influence the production of specific cytokines such as IL-36 to regulate other pathogens.

Most importantly, the commensal microbiota establishes the regulation of immunity which is critical to prevent overt inflammation or autoimmunity in several diseases like inflammatory bowel disease. This control of the regulation of homeostasis is critical. Although the microbiota has been seen to induce inflammatory responses, it also has control over the development of Treg cells, critical for reducing inflammation. Therefore, our microbiome is constantly communicating with both innate and adaptive immune cells to create an immune system that is not overtly inflammatory but also not overtly immunosuppressive. Antibiotics are known to perturb a mature and healthy immune system. Antibiotic usage in adolescents is responsible for immune-related diseases like allergies and asthma [28]. Broad-spectrum antibiotics deplete microbe-producing SCFAs and cause hyperactivation of intestinal macrophages and proinflammatory T helper cells [141].

There are increasing implications for the role of the intestinal neuro-immune axis. The gastrointestinal tract is heavily innervated by a network of neurons, with heavy communication between immune, epithelial, and microbial cells. Neurons express a wide range of receptors for cytokines while immune cells produce neurotransmitters, implicating a mechanistic method for direct crosstalk. Antibiotics are known to impact the regulation of this crosstalk. A cocktail of ampicillin, streptomycin, metronidazole, and vancomycin impacted the function of enteric neurons to drive the motility of the gut. Specifically, *Muscularis Macrophages* express BMP2 signals to activate peristalsis by enteric neurons. This interaction between enteric neurons and *Muscularis macrophages* is mutualistic. Enteric neurons will produce Csf1 to promote *Muscularis macrophage* survival. Yet antibiotics decrease *Muscularis macrophages* BMP2 expression and confer a decrease in the number of BMPR+ enteric neurons. Microbes are also likely to regulate neuronal production of Csf1 as microbial signals enhance neuronal expression of Csfl [195]. Other early life interventions can cause developmental issues in the brain. Associates with neuroatypical behavior are tied to the use of antibiotics. Mice treated with antibiotics were found to lose fear extinction learning, the capability to lose fear after multiple exposures. Antibiotic-treated mice were found to have immature states of microglia that impact synaptic pruning and spine formation due to the loss of microbiota-derived metabolites. This study also elucidated that there are critical time frames when microbiota is introduced in germ-free mice, where if microbes were not supplemented during the neonatal period created a permanent deficiency of fear response in adulthood [196].

We are increasingly becoming aware of the consequential health impacts of perturbing the microbial gut community. Although antibiotics have modernized the way we deal with the most serious infectious diseases, there is also a cost to our health that could impact our long-term development. The non-specific elimination of microbes kills off otherwise normal commensal microbes important for the regulation of a wide range of etiologies spanning from gastrointestinal, neurological, and cardiovascular health. Importantly, understanding the microbial dynamics and development that occurs during the first 2 years of life to adulthood will be considerably important and mindful of antibiotic usage.

6.8 Microbiome-Based Therapies for Antibiotic Resistance

Antibiotic resistance poses a worldwide health emergency as bacteria adapt to survive the medications meant to eradicate them. The emergence of superbugs endangers our capacity to treat infections effectively. Microbiome-based therapies offer innovative strategies to tackle antibiotic resistance by harnessing the power of the human microbiome. In this discussion, we explore how microbiome-based approaches can play a pivotal role in this battle.

6.8.1 Fecal Microbiota Transplantation: Restoring the Balance Within

In recent years, the human microbiome has arisen as a central player in our overall health and well-being. Consisting of trillions of microorganisms, such as bacteria, viruses, fungi, and archaea, the microbiome inhabits various niches in and on the human body. Among the most intriguing areas of microbiome research is fecal microbiota transplantation (FMT), a procedure that involves the transfer of healthy donor feces into the gastrointestinal tract of a recipient. This chapter delves into the history, mechanisms, applications, and challenges associated with FMT, exploring its potential as a therapeutic intervention for a diverse range of conditions.

Historical Perspective

FMT is not a novel concept but has a long history, dating back to ancient Chinese medical texts from the fourth century, where "yellow soup" was used as a remedy for food poisoning and diarrhea [197]. However, it was not until the twentieth century that FMT gained recognition in modern medicine. Initially, it was utilized as a final option for patients with severe, recurrent *Clostridium difficile* infection (CDI) that failed to respond to antibiotics [198]. The success rates of FMT in treating CDI were remarkable, sparking interest in its potential applications beyond this specific condition.

The Microbiota and Gut Health

The gut microbiota has a pivotal role in human health, influencing digestion, metabolism, immunity, and even mental well-being [28]. A balanced gut microbiome is characterized by diversity and stability, with various species of bacteria working in harmony to maintain homeostasis. Dysbiosis, an imbalance in the gut microbiota, is associated with a multitude of diseases, including inflammatory bowel disease (IBD), irritable bowel syndrome (IBS), metabolic disorders, and even mood disorders like depression and anxiety [19].

Mechanisms of FMT

FMT works on the principle of microbial restoration. When a donor's stool is transplanted into a recipient's gut, it introduces a diverse array of microorganisms, helping to restore balance within the recipient's microbiome. The exact mechanisms of how FMT exerts its therapeutic effects are still being investigated, but several key mechanisms have been proposed [199]:

1. *Microbial Competition*: The introduction of a healthy microbiome competes with pathogenic organisms, reducing the abundance of harmful bacteria like *Clostridium difficile*.
2. *Immune Regulation*: FMT may modulate the recipient's immune system, helping to quell inflammation and promote immune tolerance.
3. *Metabolic Changes*: Altered microbial composition can impact host metabolism, potentially benefiting conditions like obesity and diabetes.
4. *Production of Beneficial Metabolites*: Certain bacteria in the donor stool may produce beneficial metabolites, such as short-chain fatty acids, which can have positive effects on gut health.

Applications of FMT

While FMT initially gained prominence in the treatment of CDI, its potential applications have expanded considerably [199]:

1. *Inflammatory Bowel Disease (IBD)*: FMT has shown promise in reducing inflammation and improving symptoms in patients with Crohn's disease and ulcerative colitis.
2. *Irritable Bowel Syndrome (IBS)*: Some studies suggest that FMT may help alleviate symptoms in IBS patients, although more research is needed.
3. *Metabolic Disorders*: FMT has been explored as a potential therapy for conditions like obesity and type 2 diabetes, given its impact on host metabolism.
4. *Autoimmune Disorders*: Research is ongoing to assess the potential of FMT in modulating the immune response in autoimmune diseases like multiple sclerosis.
5. *Mental Health*: Emerging evidence suggests a connection between the gut microbiome and mental health. FMT may have a role in the treatment of mood disorders, but this is still relatively uncharted territory.

Challenges and Considerations

Despite its potential, FMT is not without challenges and considerations [200]:

1. *Donor Screening*: Donor selection is critical to ensure the safety and efficacy of FMT. Rigorous screening for infections and other health factors is essential.
2. *Standardization*: There is a need for standardized protocols and guidelines to optimize FMT procedures and outcomes.
3. *Long-Term Effects*: The long-term consequences of FMT are poorly understood, and further research is required to assess its safety over extended periods.
4. *Regulatory Hurdles*: FMT exists in a regulatory gray area in many countries. Regulatory frameworks need to catch up with the rapid advancements in the field.
5. *Ethical and Social Issues*: Ethical concerns, including informed consent and privacy, must be addressed in FMT research and practice.

6.8.2 *Microbial Biotherapy: Harnessing the Power of Microorganisms for Health*

Microbial biotherapy, also known as "microbiome-based therapy" or "microbial therapy," is a leading-edge field of medicine that harnesses the potential of microorganisms to treat a variety of health conditions. In this one-page overview, we will explore the key concepts, applications, and significance of microbial biotherapy.

The Microbiome Revolution

The human microbiome, a vast and diverse community of microorganisms inhabiting our bodies, has been under the scrutiny of intense research in recent years [201]. This ecosystem of bacteria, viruses, fungi, and other microorganisms influences our health in profound ways, from digestion and metabolism to immune function and even mental well-being. Microbial biotherapy leverages this understanding to promote health and treat diseases.

Key Concepts

Microbial biotherapy is based on several fundamental concepts [202]:

1. *Microbial Balance*: The key to health is maintaining a balanced and diverse microbiome. Dysbiosis, an imbalance in the microbiome, is linked with numerous health conditions.
2. *Microbial Modulation*: By introducing specific microorganisms or altering the existing microbial composition, biotherapy seeks to modulate the microbiome to improve health outcomes.
3. *Targeted Treatment*: Unlike traditional pharmaceuticals, microbial biotherapy aims to target the root causes of diseases by addressing microbial imbalances and restoring homeostasis.

Applications

Microbial biotherapy has shown promise in a broad spectrum of applications such as in FMT, probiotics, prebiotics, synbiotics, and developing novel therapies using engineered microorganisms to target specific disease conditions such as cancer or metabolic disorders [203].

Significance

Microbial biotherapy is significant for several reasons [204]:

1. *Precision Medicine*: It offers a personalized approach to healthcare, tailoring treatments to an individual's unique microbiome profile.
2. *Reduced Side Effects*: Microbial therapies often have fewer side effects compared to traditional pharmaceuticals because they work with the body's natural systems.
3. *Treatment of Complex Conditions*: It holds promise in treating complex and chronic diseases that have proven challenging to manage with conventional treatments.
4. *Preventative Potential*: Microbial biotherapy may also have a role in preventing diseases by maintaining a healthy microbiome from an early age.

Challenges and Future Directions

While microbial biotherapy holds immense potential, it faces challenges related to standardization, regulation, and long-term safety assessment. Researchers and healthcare professionals must work collaboratively to address these challenges and unlock the full potential of this revolutionary field.

In conclusion, microbial biotherapy represents a paradigm shift in medicine, where we view microorganisms not as enemies to be eradicated but as allies in promoting health and combating disease. As our comprehension of the microbiome continues to deepen, microbial biotherapy is poised to revolutionize healthcare by offering safe, effective, and personalized treatments that target the root causes of diseases.

6.8.3 Probiotics and Prebiotics: Nurturing Your Gut Health

Probiotics and prebiotics are two essential components of a healthy gut, each playing a unique role in maintaining the delicate balance of your gut microbiome [205]. In this one-page exploration, we will delve into what probiotics and prebiotics are, their significance for your health, and how you can incorporate them into your daily life.

Probiotics: The Beneficial Microbes
Probiotics are live microorganisms, primarily bacteria and some yeasts, that provide health benefits when ingested in sufficient quantities. These beneficial bacteria have gained considerable attention for their role in supporting gut health and overall well-being. Probiotics can be found in various foods and supplements.

Key Points

- *Gut Health*: Probiotics contribute to a balanced gut microbiome by increasing the population of beneficial bacteria and helping to control harmful microorganisms. This balance is crucial for digestion, nutrient absorption, and immune function.
- *Diverse Strains*: Different probiotic strains offer various health benefits. Common strains include *Lactobacillus* and *Bifidobacterium* species, which are known for their digestive and immune support.
- *Applications*: Probiotics are used to relieve digestive issues like diarrhea, irritable bowel syndrome, and lactose intolerance. They may also support immune function and have potential roles in managing conditions like eczema and urinary tract infections.
- *Sources*: Probiotics are found in fermented foods such as yogurt, kefir, sauerkraut, kimchi, and in supplement form. It is important to choose products with well-defined strains and sufficient live cultures.

Prebiotics: Food for Your Gut Microbes
Prebiotics are non-digestible fibers and compounds found in certain foods that nourish and stimulate the growth and functioning of beneficial gut bacteria. Essentially, they serve as food for your gut microbiome [206].

Key Points

- *Gut Microbiome Support*: Prebiotics play a vital role in maintaining a diverse and healthy gut microbiome. They selectively stimulate the growth of beneficial bacteria while discouraging harmful ones.
- *Digestive Health*: Consumption of prebiotics can improve bowel regularity, reduce constipation, and enhance overall digestive health.
- *Sources*: Common sources of prebiotics include foods like garlic, onions, leeks, asparagus, chicory root, and whole grains. Dietary fibers, particularly soluble fiber, are rich in prebiotics.

- *Synergistic with Probiotics*: Prebiotics and probiotics work synergistically. Combining them in your diet can enhance the effectiveness of probiotics by providing the necessary sustenance for beneficial bacteria.

Incorporating Probiotics and Prebiotics into Your Diet
To nurture your gut health, consider these practical tips:

- *Fermented Foods*: Incorporate probiotic-rich foods into your diet, such as yogurt, kefir, kombucha, and fermented vegetables.
- *Dietary Fiber*: Consume a diet rich in fruits, vegetables, whole grains, and legumes to increase your intake of prebiotic fibers.
- *Supplements*: If you have specific health concerns or dietary restrictions, consider probiotic and prebiotic supplements, but consult with a healthcare professional for guidance.
- *Variety*: A diverse diet ensures you receive a wide range of nutrients and fibers, which can promote microbial diversity in your gut.

In conclusion, probiotics and prebiotics are integral components of a balanced and healthy diet. They support the thriving ecosystem of microorganisms in your gut, which, in turn, contributes to overall well-being. By making mindful food choices and incorporating probiotic and prebiotic-rich foods into your daily routine, you can take proactive steps toward maintaining a happy and harmonious gut microbiome. Your gut health, after all, is intimately connected to your overall health and vitality.

6.9 Microbiome and Antibiotic Stewardship

6.9.1 Role of the Microbiome in Optimizing Antibiotic Use

The development of a myriad range of antibiotics over the decades has also resulted in the global emergence of several antimicrobial-resistant strains. According to Centre for Disease Control and Prevention (CDC) report in 2022, more than 2.8 million infection cases occur every year worldwide due to antimicrobial resistance (AMR), leading to 700,000 cases of death (https://www.cdc.gov/drugresistance/index.html). The major risk of resistance has been posed by gram-negative bacteria out of which the list of the highest occurring ones published by WHO has been listed as acronym ESKAPE (*Enterococcus faecium*, *Staphylococcus aureus*, *Klebsiella pneumoniae*, *Acinetobacter baumannii*, *Pseudomonas aeruginosa*, and *Enterobacter* species). Due to rising cases of antimicrobial resistance, there is an immediate necessity for responsible usage of antibiotics and taking advantage of stewardship approaches to minimize these effects. Extensive usage of over-the-counter antibiotics in the past years has also been reported as a major cause of the emergence of antimicrobial strains. This in turn results in perturbation of the healthy microbiome of the gut and intestine in most cases. The necessary approaches at this moment can be enlisted as follows:

Limited Usage of Antibiotics

1. *Determination of Optimal Route of Delivery*: The administration of antibiotics through the intravenous route is a proposed mechanism for circumventing the process of direct exposure of antibiotics to the gut microbiota. A recent report by Zhou et al. shows oral administration of Ampicillin can lead to disruption of the normal microbiome and lead to the rise of common opportunistic pathogens [207].
2. *Use of Alternative Therapies*: The use of antimicrobial peptides or analogs in whichever cases possible is one of the alternative approaches to excessive usage of antibiotics. Cationic, hydrophilic, short peptides of length 15–150 amino acids have been characterized as antimicrobial peptides that hold important roles as Host Defense peptides (HDPs) and exhibit a broad range of antibacterial properties. Their mechanism of action is usually via the formation of transient gaps or ionic pores in the bacterial membranes. In a study by Fuente-Nunez et al., D-amino acid analog of HDP LL-37 emerged as a potent antibacterial agent against multidrug-resistant *P. aeruginosa* biofilms [208]. In yet another study, Brevinin-1, an AMP derived from the skin secretion of frogs resulted in successful clearance of Methicillin-resistant *S. aureus* (MRSA) and *Enterococcus faecalis* [209, 210].
3. *Use of Bacteriophage or Nanomedicine-Based Products to Disrupt Resistance and Biofilm Formation*: A very recently exploited technology is plant-based nanoparticle synthesis for neutralization of several bacterial species. Different kinds of nanoparticles developed through these methods are gold nanoparticles (AuNPs), silver nanoparticles (AgNPs), and zinc oxide nanoparticles (ZnO NPs). A safe method of preparation of AgNP nanoparticles from *Carissa carandas* L. leaf extract has been demonstrated by Singh et al. leading to effective bacterial clearance against *Shigella flexneri* [211, 212].

Targeted Therapy

1. Usage of fecal microbiota restoration (FMT) and probiotics can be seen as effective therapies to mitigate the effects of excessive antibiotic usage on the microbiome [213].
2. Modulation of microbiome composition to regulate the immune system of the body, majorly the innate immune system or the first line of defense.

6.9.2 *Microbiome-Guided Therapy*

Several studies have shown the complex interplay of the microbiome and the immune system of the body. The first line of defense from foreign bodies consists of mechanical barriers like the integumentary system, mucous membranes, and sweat/tear glands. The skin and the gastrointestinal system have a complex ecosystem of commensal microbes that contribute to the acidic pH environment, which results in a hostile environment for pathogens [214].

One of the most popular examples includes that of the colonization of *K. pneumoniae* in the gastrointestinal and upper respiratory tract, more prevalently in neonates, elderly, and immune-compromised individuals. Excessive usage of antibiotics has increased the number of carbapenem-resistant resistant *K. pneumoniae* (CR-KP), whose percentage has drastically increased over the past decade. At the same time, the indiscriminate use of antibiotics leads to dysbiosis of the commensal Bacteroidetes, thus resulting in further colonization of *K. pneumoniae* in this favorable microenvironment [215].

Another very prominent example of microbiome-guided therapy is that of the enterococci family which are responsible for many urinary tract infections (UTI), endocarditis or other systemic infections. These strains spontaneously acquire multidrug resistance, out of which the most common category is the emergence of vancomycin-resistant enterococci (VRE). Studies on murine models have suggested intestinal microbiota consisting of specific anaerobes like *Barnesiella* species successfully diminish the percentage colonization of Vancomycin-resistant *Enterococcus faecium*. Of note, these commensals are also responsible for favorably modulating the immune microenvironment by activating the Toll-like receptors or TLRs on the residing stromal cells, which in turn stimulates the production of AMP REG 3¥. Some of the residing enterococci directly compete with resistant pathogenic strains by producing neutralizing molecules like bacteriocins.

The other most studied strain in terms of regulating the microbiome is *Clostridium difficile*. A thorough meta-analysis study has suggested these strains spontaneously grow resistance to fluoroquinolones, clindamycin, and cephalosporin antibiotics. However, a carbohydrate-rich diet can present a conducive microenvironment for the growth of particular microbes like *Lachnospiraceae* that can metabolize rare amino acids, thus restricting the growth of *C. difficile*. There are also implications of reports of duodenal infusion of donor feces for the treatment of *C. difficile* infections with a success rate of ~90% [189].

In conclusion, the restoration of the original commensal microbiome can prove to be beneficial for protection against as well as restriction of colonization of disease-causing strains (Fig. 6.3).

6.9.3 One Health Perspective and AMR Development

The recent emergence of AMR strains has led to concern of its overarching effect on not only human health but its larger impact on the environment. The One Health approach takes into account the communion of humans, closely related animals (livestock and domesticated animals), and the surrounding ecosystem [141]. Extensive usage of antibiotics for the welfare of both humans and animals has led to the rise of new "zoonotic" strains which include diseases like rabies, anthrax, and

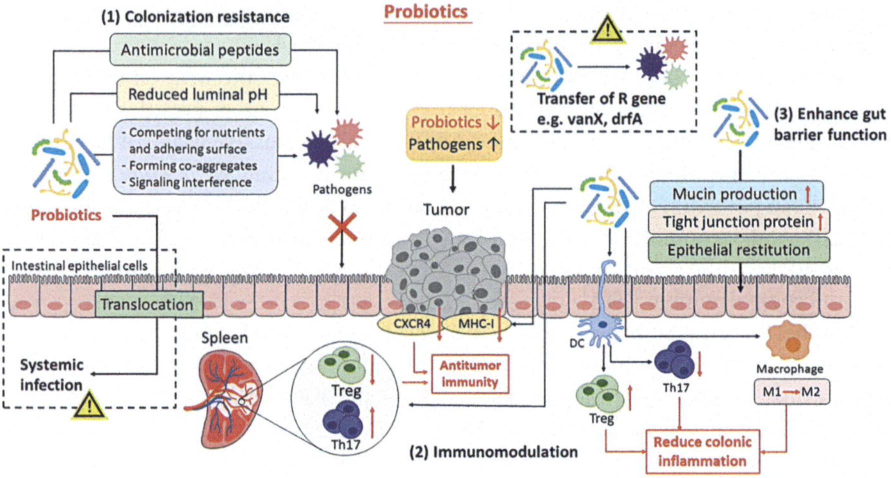

Fig. 6.3 Putative mechanisms of actions of probiotics and their associated risks. (Figure and caption are reprinted from citation [216])

Lyme disease [217]. These hybrid strains are arguably passed from animals to humans or vice-versa due to the following reasons:

1. *Excessive Use of Antibiotics*: The indiscriminate usage of antibiotics in both animals and humans has posed a global health problem. In the case of livestock, there is an abundance of long-term, low-dose medication to prevent prophylaxis from fruits or vegetables. Irresponsible usage of antibiotics for maintaining animal health can lead to rewiring of gut commensals which in turn may result in a cascade effect of changing the entire ecological niche. To circumvent this problem, WHO has proposed the termination of antimicrobials like colistin, fluoroquinolones, and macrolides to livestock, which is otherwise more relevant in the context of human diseases.
2. *Lack of Cleanliness and Proper Sanitation for Livestock Maintenance*: In the existing scenario, there is a lack of awareness and surveillance among farmers and animal owners regarding the caging and maintenance of farm animals and the implications they can have on the ecosystem. Overcrowded cages, unsanitary practices, or lack of maintenance for sick animals may lead to the horizontal transfer of AMR genes.
3. *Lack of Proper Guidelines in the Wet Lab Market*: There is ample discussion among policymakers and non-profit organizations around regulations for wet markets and their implications on transmitted diseases. Dense markets, interspecies mixing, the presence of live animals and long supply chains lead to the emergence of new AMR species as well as major disease outbreaks [218].

The One Health approach advocates for a collaborative effort involving law enforcement, policymakers, the agricultural sector, communities, and pet owners to share the responsibility of reducing the negative impacts of antibiotic use and preventing

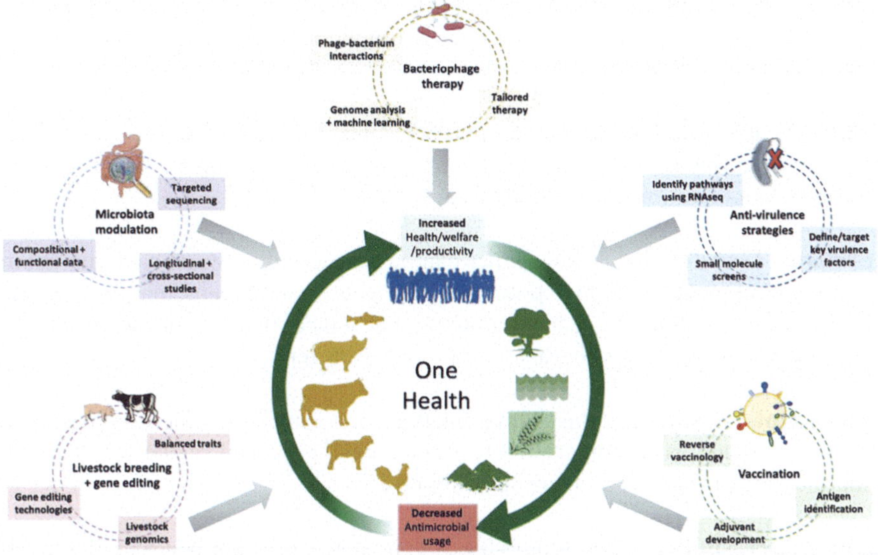

Fig. 6.4 One Health perspective. (Figure and caption are reprinted from citation [219])

the development and dispersion of antimicrobial-resistant strains. A more stringent restriction on over-the-counter drugs, recognition of AMR spread from foodborne or water-borne sources as well as livestock, and strong enforcement of rules and guidelines regarding checking the most common sources of AMR emergence can prove to be powerful tools for antimicrobial stewardship (Fig. 6.4) (https://www.cdc.gov/onehealth/in-action/combating-ar-in-people-and-animals.html).

6.10 Limitations and Challenges and Future Perspectives

6.10.1 Limitations of Microbiome-Based Approaches to Address Antimicrobial Resistance

The human microbiota acts as a shield against various harmful bacterial pathogens, but it is highly sensitive to small imputations and disruptions. Even brief disruptions, like the usage of antibiotics, can diminish their protective effects [214]. Post-antibiotic treatment, pathogens can colonize and multiply, posing serious risks. Investigating how these pathogens can exploit microbiota's defenses can help uncover new infection prevention methods [220]. Excessive antibiotic use consistently leads to patient issues, despite their clear value. Microbiome-based approaches have been proposed as a potential strategy to address antimicrobial resistance. However, there exist several limitations to these approaches [220], some detailed below:

Lack of Understanding There is a lot yet to be learned about the human microbiome and its interaction with antibiotics and antibiotic-resistant bacteria. This lack of understanding makes it difficult to develop effective microbiome-based strategies.

Limited Scope Microbiome-based approaches may only be effective against certain types of antibiotic-resistant bacteria, and may not be a comprehensive solution to the problem of antimicrobial resistance.

Resistance Transfer Resistance genes can be transferred between bacteria within the gut microbiome, including potential pathogens. This means that even if the microbiome is modified to reduce antibiotic resistance, there is an increased risk that resistance genes could be transferred to other bacteria.

Inconsistent Results Studies investigating the use of microbiome-based approaches to address antimicrobial resistance have produced mixed results, with some showing promise and others showing no effect.

Limited Accessibility Microbiome-based approaches may not be accessible to all populations, particularly those in countries with low income where access to healthcare and advanced technologies is limited.

In summary, while microbiome-based approaches have potential as a strategy to address antimicrobial resistance, several limitations need to be addressed before they can be widely implemented.

6.10.2 Challenges to Overcome for Successful Implementation

While microbiome-based therapies hold great promise in the fight against antimicrobial resistance (AMR), several challenges must be surmounted to ensure their successful implementation. These challenges are intricately linked to the complexity of the microbiome and the multifaceted nature of AMR. Here, we explore the obstacles that need to be addressed:

Limited Availability of New Antimicrobial Agents One of the foremost challenges is the scarcity of novel antimicrobial agents. The ongoing quest for new antibiotics remains a pressing need to effectively combat AMR. Microbiome-based therapies often rely on the synergy between the microbiome and traditional antibiotics. A shortage of new agents hinders the development of innovative treatment approaches [221].

Accessibility Disparities Microbiome-based approaches may not be readily accessible to all populations. Particularly, low-income countries face disparities in healthcare access, especially with the limited availability of advanced technologies.

Overcoming these accessibility issues is paramount to ensure that microbiome-based therapies can benefit a broader spectrum of patients worldwide.

Weak Laboratory Capacity Multiple middle- and low-income countries struggle with inadequate laboratory capacity, suboptimal governance of their health system, limited resources, and insufficient health information systems. This weak infrastructure poses a significant hurdle in adopting microbiome-based therapies effectively, especially for those countries bearing a heavy health and economic burden of communicable diseases [222].

Resistance Transfer within the Microbiome Within the complex microbiome, ARGs can be transferred between different bacteria, including potential pathogens. Even if efforts to modify the microbiome to reduce antibiotic resistance show promise, there remains a risk that resistance genes could be disseminated to other bacteria, undermining the therapy's intended effects [223].

Inconsistent Research Outcomes Research studies investigating microbiome-based approaches have produced inconsistent results. While some show promise, others exhibit no discernible effect. This variability underscores the intricate interplay between the microbiome and AMR, necessitating a deeper understanding to achieve more consistent and predictable outcomes [224].

Limited Scope Microbiome-based approaches may primarily target specific types of antibiotic-resistant bacteria, potentially not providing a comprehensive solution to the complex issue of AMR. This limited scope calls for a multifaceted approach that combines microbiome-based therapies with other strategies to address resistance more comprehensively.

Challenges in Diagnosis and Tailoring Therapy Rapid identification of resistant infections and the subsequent adaptation of therapy present formidable challenges. To fully harness microbiome-based strategies, it is essential to overcome these difficulties and explore novel antimicrobial stewardship strategies capable of minimizing resistance more effectively (Fig. 6.5).

In summary, the successful implementation of microbiome-based therapies to combat AMR requires addressing an array of challenges. These encompass the development of novel antimicrobial agents, the enhancement of laboratory capacity, the mitigation of accessibility issues, and the management of resistance transfer within the microbiome. Overcoming these hurdles is pivotal to harnessing the full potential of microbiome-based interventions in the fight against antimicrobial resistance.

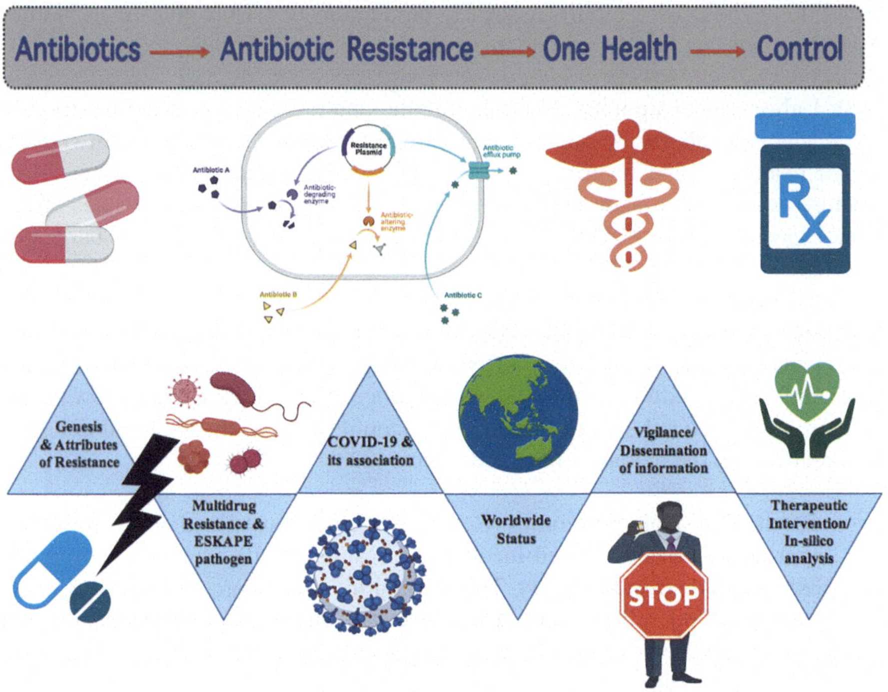

Fig. 6.5 Improved diagnostics and drugs, rigorous clinical testing, and strategic antimicrobial use are vital for curbing resistance. (Figure and caption are reprinted from citation [225])

6.10.3 *Future Applications of Microbiome Research in Addressing Antimicrobial Resistance*

6.10.3.1 Novel Nontraditional Antimicrobial Therapies

Novel nontraditional antimicrobial therapies present a promising frontier in combating bacterial infections and reducing antimicrobial resistance. These unconventional approaches encompass phage therapies, monoclonal antibodies, virulence disruptors, and microbiome-based treatments discussed in Table 6.4 [224]. While offering distinct advantages, they encounter significant hurdles. Successful development necessitates a paradigm shift in clinical trial protocols for narrow-spectrum agents, changes in diagnostic deployment, operationalization, and reimbursement, and the reevaluation of in vitro and in vivo efficacy assays. Effective progress requires close collaboration among industry, government, and regulatory bodies to navigate these unique approaches through their respective challenges.

Table 6.4 Selected advantages and disadvantages of nontraditional approaches

Method	Advantages	Disadvantages
Bacteriophage	Targeted to infect specific bacteria Historically proven efficacy Applicable for topical use Considered generally safe for use in food Preserves the microbiome, causing no collateral damage Acts as a natural predator to the targeted microorganism Evolves concurrently with the target microbe	Cannot be used for broad-spectrum application Swift elimination by the immune system causing challenge in administration Constraints on intracellular delivery Uncertain regulatory pathway Combination use might be challenging Potential susceptibility to resistance
Lysins (Endolysins)	Specific targets Rapid bactericidal action Systemically administration Lower probability of resistance	Delivery is challenging Stability issues Can have off-target effects Potential immunogenicity
Virulence factor disruptors	Reduced selective pressure No collateral damage Combination therapy possible Uses a host-centric approach	Pathogen modification will reduce functionality Limited efficacy
Microbiome therapies	Microbial balance is restored High specificity Reduced risk of resistance	Lack of complete knowledge of optimal microbiome Unclear regulatory pathway Standardization is challenging. Variable responses among individual
Monoclonal antibodies	High specificity Limited impact on commensal microbiota Precise target Limited impact on commensal microbiota Already in use	Expensive Potential immunogenicity Potential for resistance Challenge in administration

6.10.3.2 Genome Mining for Drug Discovery

Since the momentous discovery of penicillin, microbial natural products and their derivatives have remained invaluable in the realm of drug development. The advent of genome mining techniques has substantially expanded our ability to uncover secondary metabolites produced by biosynthetic gene clusters (BGCs). This opens the door to discovering novel natural products with pharmaceutical potential. However, it is worth noting that many BGCs remain dormant or fail to express in their native hosts under laboratory conditions. This challenge has led to the enticing strategy of employing genetically manipulable hosts for heterologous BGC expression, thereby activating these dormant clusters and revealing their potential products.

Table 6.5 highlights recent breakthroughs in capturing intact BGCs and discovering natural product biosynthetic pathways. In essence, the pharmaceutical industry continues to draw from the wellspring of natural products, bolstered by cutting-edge genome mining techniques and creative approaches to awaken silent gene clusters.

Table 6.5 Various approaches for the cloning of complete natural product biosynthetic gene cluster

Class	Strategies	Principles	Capacity	Advantages	Disadvantages	GGCs
Sequence independent libraries cloning	Cosmid/fosmid libraries	Fragmentation, gel-fractionated, ligation and phage packaging	<50 kb	Not requiring genome sequence data Capable of generating natural product with novel structure Capable of covering the complete genetic material Suitable for cloning environmental DNA	Untargeted Laborious and time-consuming Packaging: Large BGCs may be spanned into separate clones	Omnipeptin Anisomycin Ashimides Frigocyclinone Locillomycins
	PAC/BAC libraries	Fragmentation, gel-fractionated and ligation	<300 kb	Not requiring genome sequence data Capable of generating natural product with novel structure Capable of covering the complete genetic material	Untargeted Laborious and time-consuming Technically challenging for large fragment cloning and DNA extraction	Atratumycin Neoabyssomicin/abyssomicin Avermectins Murayaquinone
	FAC libraries	Random fragmentation, adaptor ligation, gel-fractionated and ligation	10–200 kb	Unbiased library Not requiring genome sequence data Capable of generating natural product with novel structure Suitable for fungal BGCs cloning	Untargeted Laborious and time-consuming	Sesterterpenoid Benzomalvin A/D Diketomorpholines Plipastatin Scleric acid Brasiliquinones

Direct cloning	TAR	In vivo homologous recombination of *Saccharomyces cerevisiae*	<100 kb	Cas9-facilitated high efficiency cloning Suitable for cloning large genomic regions	Technically challenging to use yeast spheroplasts for highly transformation efficient Some false-positives Requires careful preparation and/or manipulation of gDNA	Plipastatin Scleric acid Brasiliquinones
	LLHR	RecET-mediated linear-plus-linear homologous recombination in *E. coli*	<~52 kb	Technically easier Suitable for cloning small- and mid-BGCs Simply for using short recombination homologous arms	False-positive Require highly specialized capturing vectors Multi-round selection	Luminmide A/B Bacillomycin Streptoketides
	ExoCET	Cas9 digestion, T4 polymerase for in vitro annealing and RecET-mediated homologous recombination	<~102 kb	Technically easier Simply for using short recombination homologous arms	Low efficiency for clone large-size BGCs Require pathway-specialized vectors False-positive	Salinomycin Spinosad
	CATCH	Cas9-assisted site-specific cleavage and Gibson assembly	<~150 kb	Suitable for cloning large genomic regions	Require carefully preparing the target DNA in gel	Bacillaene Mutanocyclin/SNC1-465

(continued)

Table 6.5 (continued)

Class	Strategies	Principles	Capacity	Advantages	Disadvantages	GGCs
	DiPaC	Q5 high-fidelity PCR amplification and Gibson assembly	<~22 kb per round	Technically easier Extremely efficient for cloning small- to mid-size BGCs	Introduction of new mutations during PCR Impractical for large BGCs	Phenazine fontizine A5 Sodorifen Hapalosin Napsamycin, daptomycin
	Site-specific recombinase	Homologous recombination and recombinase circularizing the plasmid in vivo or in vitro with recombinase	<~200 kb	Effectively avoid the introduction of new mutations Suitable for high-frequency recombination hosts Suitable for cloning large BGCs	Time-consuming Only use in the host with high-frequency natural homologous recombination Usually multi-round recombination	Siderophore Erythromycin
	iCatch	Homologous recombination and in vitro self-ligation	<~20 kb	Suitable for high-frequency recombination hosts	Time-consuming Only use in the host with high-frequency natural homologous recombination Usually need multi-round recombination Carefully prepare the genome DNA	Actinorhodin

Table and caption are reprinted from citation [227]

In 2023, the discovery of clovibactin was made possible through the fusion of biosynthetic gene clusters (BGCs) and genome mining. Clovibactin's emergence represents a significant stride toward developing antibiotics that can combat bacterial pathogens without engendering resistance [226]. By amalgamating BGC and genome mining techniques, this discovery offers hope in the war against antimicrobial resistance and paves the way for future antibiotic innovation.

6.11 Conclusion: A Holistic Approach to Confronting Antibiotic Resistance

The history of antimicrobials traces back to ancient folk medicine [228]. But it was not until the twentieth century, which saw the advent of antibiotics. This started an era that revolutionized medicine by controlling infectious diseases, but alongside we saw the rise of resistance to these compounds, turning into a significant health crisis [228]. Methicillin-resistant *Staphylococcus aureus* (MRSA) is a prominent example. Today, MRSA is resistant to an entire class of antibiotics and has evolved into multidrug-resistant variants, associated with increased virulence and transmission capabilities, leading to higher morbidity and mortality. In general, the emergence of AMR has outpaced our ability to combat infections, thus necessitating advancing novel antibiotic therapies. Alarming statistics reveal that in the United States, around two million people annually contract infections from antibiotic-resistant bacteria, which results in approximately 23,000 deaths, while Europe sees about 25,000 such deaths each year. Resistance rates include 3–6% for azole against Aspergillus, around 7% for Candida isolates against fluconazole, and increasing echinocandin resistance; additionally, the WHO identified artemisinin resistance in five countries of the Greater Mekong subregion in 2015, indicating a growing global challenge of antimicrobial resistance (AMR) as resistance mechanisms evolve and spread [229].

AMR has escalated into a worldwide issue, further intensified by the COVID-19 pandemic, heightened international travel, the transmission of infections in settings like hospitals and nursing homes, and the widespread overuse of antibiotics in human and animal health, often exacerbated by incomplete adherence to prescribed treatments. The situation is aggravated by a limited grasp of the mechanisms and environmental contributors to AMR, alongside a significant delay in creating new antimicrobials over the past two decades. Consequently, the rise of AMR is outpacing the discovery and development of more effective antibiotic therapies. More studies like that done by the MetaSUB Consortium, including StuckOnU, are required as an effort to increase global surveillance [230, 231].

Not only does antimicrobial resistance pose health risks on Earth, but a recent study has elucidated issues in space, where conditions such as microgravity, radiation, and confinement amplify concerns. This can pose issues, especially during long-distance space travel. This study utilized whole-genome sequencing data from the International Space Station (ISS), collected during the Microbial Tracking-1 project, to identify AMR genes in 226 microbes from ISS environmental samples, revealing dominant AMR genes in certain bacteria associated with human infections. The findings, validated through computational and experimental methods, highlight the utility of machine learning in identifying hidden AMR genes in complex datasets, enhancing our understanding of space environmental microbiomes and their potential human health impacts [232, 233].

Thus, as we move forward, we need to focus on modern approaches that are required to complement antibiotic therapies. Modern approaches, including One Health perspective, are pertinent to address AMR, especially given the lack of new antimicrobial discoveries in recent years. This complex and evolving problem requires international collaboration and concerted efforts to develop innovative solutions.

References

1. Karanja CW, et al. Isoquinoline antimicrobial agent: activity against intracellular bacteria and effect on global bacterial proteome. Molecules. 2022;27:5085.
2. Bag S, et al. Molecular insights into antimicrobial resistance traits of commensal human gut microbiota. Microb Ecol. 2019;77:546–57.
3. Baron SA, Diene SM, Rolain J-M. Human microbiomes and antibiotic resistance. Hum Microb J. 2018;10:43–52.
4. Brinkac L, Voorhies A, Gomez A, Nelson KE. The threat of antimicrobial resistance on the human microbiome. Microb Ecol. 2017;74:1001–8.
5. Sukumar S, et al. Development of the oral resistome during the first decade of life. Nat Commun. 2023;14:1291.
6. Anderson AC, et al. The oral microbiota is a reservoir for antimicrobial resistance: resistome and phenotypic resistance characteristics of oral biofilm in health, caries, and periodontitis. Ann Clin Microbiol Antimicrob. 2023;22:37.
7. Singh NK, et al. Multi-drug resistant Enterobacter bugandensis species isolated from the International Space Station and comparative genomic analyses with human pathogenic strains. BMC Microbiol. 2018;18:175.
8. Smith SD, et al. Diversity of Antibiotic Resistance genes and Transfer Elements-Quantitative Monitoring (DARTE-QM): a method for detection of antimicrobial resistance in environmental samples. Commun Biol. 2022;5:216.
9. Mulpuru V, Mishra N. Antimicrobial peptides from human microbiome against multidrug efflux pump of Pseudomonas aeruginosa: a computational study. Probiotics Antimicrob Proteins. 2022;14:180–8.
10. Busi SB, et al. Persistence of birth mode-dependent effects on gut microbiome composition, immune system stimulation and antimicrobial resistance during the first year of life. ISME Commun. 2021;1:8.
11. Shao Y, et al. Stunted microbiota and opportunistic pathogen colonization in caesarean-section birth. Nature. 2019;574:117–21.
12. Panthee B, Gyawali S, Panthee P, Techato K. Environmental and human microbiome for health. Life. 2022;12:456.
13. Kraemer SA, Ramachandran A, Perron GG. Antibiotic pollution in the environment: from microbial ecology to public policy. Microorganisms. 2019;7:180.
14. Knapp CW, Dolfing J, Ehlert PAI, Graham DW. Evidence of increasing antibiotic resistance gene abundances in archived soils since 1940. Environ Sci Technol. 2010;44:580–7.

15. Samreen, Ahmad I, Malak HA, Abulreesh HH. Environmental antimicrobial resistance and its drivers: a potential threat to public health. J Glob Antimicrob Resist. 2021;27:101–11.
16. Baral B, Mozafari MR. Strategic moves of "superbugs" against available chemical scaffolds: signaling, regulation, and challenges. ACS Pharmacol Transl Sci. 2020;3:373–400.
17. Inda-Díaz JS, et al. Latent antibiotic resistance genes are abundant, diverse, and mobile in human, animal, and environmental microbiomes. Microbiome. 2023;11:44.
18. Ursell LK, Metcalf JL, Parfrey LW, Knight R. Defining the human microbiome. Nutr Rev. 2012;70:S38–44.
19. Shreiner AB, Kao JY, Young VB. The gut microbiome in health and in disease. Curr Opin Gastroenterol. 2015;31:69–75.
20. Proctor LM. The human microbiome project in 2011 and beyond. Cell Host Microbe. 2011;10:287–91.
21. Jandhyala SM. Role of the normal gut microbiota. World J Gastroenterol. 2015;21:8787.
22. Milani C, et al. The first microbial colonizers of the human gut: composition, activities, and health implications of the infant gut microbiota. Microbiol Mol Biol Rev. 2017;81:e00036-17.
23. Fan Y, Pedersen O. Gut microbiota in human metabolic health and disease. Nat Rev Microbiol. 2021;19:55–71.
24. Rodríguez JM, et al. The composition of the gut microbiota throughout life, with an emphasis on early life. Microb Ecol Health Dis. 2015;26. https://doi.org/10.3402/mehd.v26.26050.
25. Bäckhed F, et al. The gut microbiota as an environmental factor that regulates fat storage. Proc Natl Acad Sci. 2004;101:15718–23.
26. Kennedy MS, Chang EB. The microbiome: composition and locations. In: Progress in molecular biology and translational science, vol. 176. Elsevier; 2020. p. 1–42.
27. Coman V, Vodnar DC. Gut microbiota and old age: modulating factors and interventions for healthy longevity. Exp Gerontol. 2020;141:111095.
28. Sekirov I, Russell SL, Antunes LCM, Finlay BB. Gut microbiota in health and disease. Physiol Rev. 2010;90:859–904.
29. Reygaert WC. An overview of the antimicrobial resistance mechanisms of bacteria. AIMS Microbiol. 2018;4:482–501.
30. Yang QE, Walsh TR. Toxin–antitoxin systems and their role in disseminating and maintaining antimicrobial resistance. FEMS Microbiol Rev. 2017;41:343–53.
31. Guzmán-Soto I, et al. Mimicking biofilm formation and development: recent progress in in vitro and in vivo biofilm models. iScience. 2021;24:102443.
32. Gharbi M, et al. Distribution of virulence and antibiotic resistance genes in Campylobacter jejuni and Campylobacter coli isolated from broiler chickens in Tunisia. J Microbiol Immunol Infect. 2022;55:1273–82.
33. Hull DM, Harrell E, Van Vliet AHM, Correa M, Thakur S. Antimicrobial resistance and interspecies gene transfer in Campylobacter coli and Campylobacter jejuni isolated from food animals, poultry processing, and retail meat in North Carolina, 2018–2019. PLoS One. 2021;16:e0246571.
34. Qin X, Wang X, Shen Z. The rise of antibiotic resistance in Campylobacter. Curr Opin Gastroenterol. 2023;39:9–15.
35. Darby EM, et al. Molecular mechanisms of antibiotic resistance revisited. Nat Rev Microbiol. 2023;21:280–95.
36. Venugopal N, et al. Molecular detection and typing of methicillin-resistant Staphylococcus aureus and methicillin-resistant coagulase-negative staphylococci isolated from cattle, animal handlers, and their environment from Karnataka, Southern Province of India. Vet World. 2019;12:1760–8.
37. Sadidi R, Azimian A. Prevalence, antimicrobial resistance, and molecular characteristics of coagulase-negative staphylococci isolated from children's blood cultures in Northeastern Iran within 2013–2019. Int J Inf Secur. 2023;9:e135630.
38. Hasan M, Wang J, Ahn J. Ciprofloxacin and tetracycline resistance cause collateral sensitivity to aminoglycosides in Salmonella typhimurium. Antibiotics. 2023;12:1335.

39. Al-Ansari MM, Aljubali MM, Somily AM, Albarrag AM, Masood A. Isolation and molecular characterization of multidrug-resistant Salmonella enterica serovars. J Infect Public Health. 2021;14:1767–76.
40. Sabença C, et al. Next-generation sequencing and MALDI mass spectrometry in the study of multiresistant processed meat vancomycin-resistant enterococci (VRE). Biology. 2020;9:89.
41. Zaheer R, et al. Surveillance of Enterococcus spp. reveals distinct species and antimicrobial resistance diversity across a One-Health continuum. Sci Rep. 2020;10:3937.
42. Munita JM, Arias CA. Mechanisms of antibiotic resistance. Microbiol Spectr. 2016;4(2):1–24.
43. Pancu DF, et al. Antibiotics: conventional therapy and natural compounds with antibacterial activity—A pharmaco-toxicological screening. Antibiotics. 2021;10:401.
44. Belkaid Y, Hand TW. Role of the microbiota in immunity and inflammation. Cell. 2014;157:121–41.
45. Patangia DV, Anthony Ryan C, Dempsey E, Paul Ross R, Stanton C. Impact of antibiotics on the human microbiome and consequences for host health. MicrobiologyOpen. 2022;11:e1260.
46. Llor C, Bjerrum L. Antimicrobial resistance: risk associated with antibiotic overuse and initiatives to reduce the problem. Ther Adv Drug Saf. 2014;5:229–41.
47. Ha DR, Haste NM, Gluckstein DP. The role of antibiotic stewardship in promoting appropriate antibiotic use. Am J Lifestyle Med. 2019;13:376–83.
48. Anthony WE, Burnham C-AD, Dantas G, Kwon JH. The gut microbiome as a reservoir for antimicrobial resistance. J Infect Dis. 2021;223:S209–13.
49. Prestinaci F, Pezzotti P, Pantosti A. Antimicrobial resistance: a global multifaceted phenomenon. Pathog Glob Health. 2015;109:309–18.
50. Jian Z, et al. Antibiotic resistance genes in bacteria: occurrence, spread, and control. J Basic Microbiol. 2021;61:1049–70.
51. Ventola CL. The antibiotic resistance crisis: part 1: causes and threats. P T. 2015;40:277–83.
52. Dahiya D, Nigam PS. Antibiotic-therapy-induced gut dysbiosis affecting gut microbiota—brain axis and cognition: restoration by intake of probiotics and synbiotics. Int J Mol Sci. 2023;24:3074.
53. Vats P, Kaur UJ, Rishi P. Heavy metal-induced selection and proliferation of antibiotic resistance: a review. J Appl Microbiol. 2022;132:4058–76.
54. Sonnenburg JL, Sonnenburg ED. Vulnerability of the industrialized microbiota. Science. 2019;366:eaaw9255.
55. Morris S, Cerceo E. Trends, epidemiology, and management of multi-drug resistant gram-negative bacterial infections in the hospitalized setting. Antibiotics. 2020;9:196.
56. Lamberte LE, Van Schaik W. Antibiotic resistance in the commensal human gut microbiota. Curr Opin Microbiol. 2022;68:102150.
57. Mu C, Zhu W. Antibiotic effects on gut microbiota, metabolism, and beyond. Appl Microbiol Biotechnol. 2019;103:9277–85.
58. Arıkoğlu T, Kuyucu S, Caubet J. New diagnostıc perspectives in the management of pediatrıc beta-lactam allergy. Pediatr Allergy Immunol. 2022;33:e13745.
59. Aversa Z, et al. Association of infant antibiotic exposure with childhood health outcomes. Mayo Clin Proc. 2021;96:66–77.
60. Malik B, Bhattacharyya S. Antibiotic drug-resistance as a complex system driven by socio-economic growth and antibiotic misuse. Sci Rep. 2019;9:9788.
61. De Nies L, Kobras CM, Stracy M. Antibiotic-induced collateral damage to the microbiota and associated infections. Nat Rev Microbiol. 2023;21:789. https://doi.org/10.1038/s41579-023-00936-9.
62. Socała K, et al. The role of microbiota-gut-brain axis in neuropsychiatric and neurological disorders. Pharmacol Res. 2021;172:105840.
63. Vander Stichele RH, et al. European surveillance of antimicrobial consumption (ESAC): data collection performance and methodological approach. Br J Clin Pharmacol. 2004;58:419–28.
64. Gunasekera YD, Kinnison T, Kottawatta SA, Silva-Fletcher A, Kalupahana RS. Misconceptions of antibiotics as a potential explanation for their misuse. A survey of the general public in a rural and urban community in Sri Lanka. Antibiotics. 2022;11:454.

65. Kaye KS, Engemann JJ, Fraimow HS, Abrutyn E. Pathogens resistant to antimicrobial agents: epidemiology, molecular mechanisms, and clinical management. Infect Dis Clin N Am. 2004;18:467–511.
66. Santos-Lopez A, et al. The roles of history, chance, and natural selection in the evolution of antibiotic resistance. eLife. 2021;10:e70676.
67. Penders J, Stobberingh E, Savelkoul P, Wolffs P. The human microbiome as a reservoir of antimicrobial resistance. Front Microbiol. 2013;4:87.
68. Ekong PS, et al. Bayesian estimation of true prevalence, sensitivity and specificity of three diagnostic tests for detection of Escherichia coli O157 in cattle feces. Prev Vet Med. 2017;148:21–7.
69. Galhano BSP, Ferrari RG, Panzenhagen P, de Jesus ACS, Conte-Junior CA. Antimicrobial resistance gene detection methods for bacteria in animal-based foods: a brief review of highlights and advantages. Microorganisms. 2021;9:923.
70. Gheyas AA, Burt DW. Microarray resources for genetic and genomic studies in chicken: a review. Genesis. 2013;2000(51):337–56.
71. Na SH, et al. Detection of oxazolidinone and phenicol resistant enterococcal isolates from duck feces and carcasses. Int J Food Microbiol. 2019;293:53–9.
72. Oliver JD. The viable but nonculturable state in bacteria. J Microbiol. 2005;43 Spec No:93–100.
73. Sails AD, Bolton FJ, Fox AJ, Wareing DRA, Greenway DLA. A reverse transcriptase polymerase chain reaction assay for the detection of thermophilic Campylobacter spp. Mol Cell Probes. 1998;12:317–22.
74. Adzitey F, Huda N, Ali GRR. Molecular techniques for detecting and typing of bacteria, advantages and application to foodborne pathogens isolated from ducks. 3 Biotech. 2013;3:97–107.
75. Liu L, Yang Y, Deng Y, Zhang T. Nanopore long-read-only metagenomics enables complete and high-quality genome reconstruction from mock and complex metagenomes. Microbiome. 2022;10:209.
76. Govender KN, Eyre DW. Benchmarking taxonomic classifiers with Illumina and Nanopore sequence data for clinical metagenomic diagnostic applications. Microb Genom. 2022;8:mgen000886.
77. Viehweger A, et al. Nanopore-based enrichment of antimicrobial resistance genes – a case-based study. GigaByte. 2023;2023:gigabyte75.
78. Karas M, Hillenkamp F. Laser desorption ionization of proteins with molecular masses exceeding 10,000 daltons. Anal Chem. 1988;60:2299–301.
79. Braga PAC, et al. Bacterial identification: from the agar plate to the mass spectrometer. RSC Adv. 2013;3:994–1008. https://pubs.rsc.org/en/content/articlelanding/2013/ra/c2ra22063f
80. Bacconi A, et al. Improved sensitivity for molecular detection of bacterial and Candida infections in blood. J Clin Microbiol. 2014;52:3164–74.
81. Wolk DM, et al. Pathogen profiling: rapid molecular characterization of Staphylococcus aureus by PCR/electrospray ionization-mass spectrometry and correlation with phenotype. J Clin Microbiol. 2009;47:3129–37.
82. Segata N, et al. Metagenomic microbial community profiling using unique clade-specific marker genes. Nat Methods. 2012;9:811–4.
83. Truong DT, et al. MetaPhlAn2 for enhanced metagenomic taxonomic profiling. Nat Methods. 2015;12:902–3.
84. Beghini F, et al. Integrating taxonomic, functional, and strain-level profiling of diverse microbial communities with bioBakery 3. eLife. 2021;10:e65088.
85. Blanco-Míguez A, et al. Extending and improving metagenomic taxonomic profiling with uncharacterized species using MetaPhlAn 4. Nat Biotechnol. 2023;41:1633–44. https://doi.org/10.1038/s41587-023-01688-w.
86. Wood DE, Salzberg SL. Kraken: ultrafast metagenomic sequence classification using exact alignments. Genome Biol. 2014;15:R46.

87. Wood DE, Lu J, Langmead B. Improved metagenomic analysis with Kraken 2. Genome Biol. 2019;20:257.
88. Lu J, Breitwieser FP, Thielen P, Salzberg SL. Bracken: estimating species abundance in metagenomics data. PeerJ Comput Sci. 2017;3:e104.
89. Kostic AD, et al. PathSeq: a comprehensive computational tool for the identification or discovery of microorganisms by deep sequencing of human tissue. Nat Biotechnol. 2011;29:393–6.
90. Morgulis A, et al. Database indexing for production MegaBLAST searches. Bioinformatics. 2008;24:1757–64.
91. Corvelo A, Clarke WE, Robine N, Zody MC. taxMaps: comprehensive and highly accurate taxonomic classification of short-read data in reasonable time. Genome Res. 2018;28:751–8.
92. Buchfink B, Xie C, Huson DH. Fast and sensitive protein alignment using DIAMOND. Nat Methods. 2015;12:59–60.
93. Ye SH, Siddle KJ, Park DJ, Sabeti PC. Benchmarking metagenomics tools for taxonomic classification. Cell. 2019;178:779–94.
94. Seemann T, Abricate, Github https://github.com/tseemann/abricate
95. Hunt M, et al. ARIBA: rapid antimicrobial resistance genotyping directly from sequencing reads. Microb Genom. 2017;3:e000131.
96. Bortolaia V, et al. ResFinder 4.0 for predictions of phenotypes from genotypes. J Antimicrob Chemother. 2020;75:3491–500.
97. Alcock BP, et al. CARD 2023: expanded curation, support for machine learning, and resistome prediction at the Comprehensive Antibiotic Resistance Database. Nucleic Acids Res. 2023;51:D690–9.
98. Feldgarden M, et al. AMRFinderPlus and the Reference Gene Catalog facilitate examination of the genomic links among antimicrobial resistance, stress response, and virulence. Sci Rep. 2021;11:12728.
99. Inouye M, et al. SRST2: rapid genomic surveillance for public health and hospital microbiology labs. Genome Med. 2014;6:90.
100. Klimkaitė L, et al. Novel antibiotic resistance genes identified by functional gene library screening in Stenotrophomonas maltophilia and Chryseobacterium spp. bacteria of soil origin. Int J Mol Sci. 2023;24:6037.
101. Despotovic M, De Nies L, Busi SB, Wilmes P. Reservoirs of antimicrobial resistance in the context of One Health. Curr Opin Microbiol. 2023;73:102291.
102. Singh V, Chibale K. Strategies to combat multi-drug resistance in tuberculosis. Acc Chem Res. 2021;54:2361–76.
103. Uddin TM, et al. Antibiotic resistance in microbes: history, mechanisms, therapeutic strategies and future prospects. J Infect Public Health. 2021;14:1750–66.
104. Boerlin P, Reid-Smith RJ. Antimicrobial resistance: its emergence and transmission. Anim Health Res Rev. 2008;9:115–26.
105. Asakura H, Sakata J, Nakamura H, Yamamoto S, Murakami S. Phylogenetic diversity and antimicrobial resistance of Campylobacter coli from humans and animals in Japan. Microbes Environ. 2019;34:146–54.
106. Dramé O, et al. Antimicrobial resistance of *Campylobacter* in broiler chicken along the food chain in Canada. Foodborne Pathog Dis. 2020;17:512–20.
107. Amoafo OY, Malekar V, Jones E, On SLW. Antibiotic resistance and phylogenetic profiling of Escherichia coli from dairy farm soils; organic versus conventional systems. Curr Res Microb Sci. 2022;3:100088.
108. Wierup M, Wahlström H, Bengtsson B. Successful prevention of antimicrobial resistance in animals—A retrospective country case study of Sweden. Antibiotics. 2021;10:129.
109. Elbehiry A, et al. The development of technology to prevent, diagnose, and manage antimicrobial resistance in healthcare-associated infections. Vaccines (Basel). 2022;10:2100.
110. Ferreira JP, et al. Transmission of methicillin-resistant Staphylococcus aureus between human and hamster. J Clin Microbiol. 2011;49:1679–80.
111. Turner NA, et al. Methicillin-resistant Staphylococcus aureus: an overview of basic and clinical research. Nat Rev Microbiol. 2019;17:203–18.

112. Huijbers PMC, et al. Role of the environment in the transmission of antimicrobial resistance to humans: a review. Environ Sci Technol. 2015;49:11993–2004.
113. Lechner I, Freivogel C, Stärk KDC, Visschers VHM. Exposure pathways to antimicrobial resistance at the human-animal interface—a qualitative comparison of Swiss expert and consumer opinions. Front Public Health. 2020;8:345.
114. Wee BA, Muloi DM, van Bunnik BAD. Quantifying the transmission of antimicrobial resistance at the human and livestock interface with genomics. Clin Microbiol Infect. 2020;26:1612.
115. Dafale NA, Srivastava S, Purohit HJ. Zoonosis: an emerging link to antibiotic resistance under "one health approach". Indian J Microbiol. 2020;60:139–52.
116. Parkhill J. Antimicrobial resistance exchange between humans and animals: why we need to know more. Engineering. 2022;15:11–2.
117. EFSA Panel on Biological Hazards (BIOHAZ), et al. Transmission of antimicrobial resistance (AMR) during animal transport. EFSA J. 2022;20:e07586.
118. Hedman HD, Vasco KA, Zhang L. A review of antimicrobial resistance in poultry farming within low-resource settings. Animals. 2020;10:1264.
119. Larsson DGJ, Flach C-F. Antibiotic resistance in the environment. Nat Rev Microbiol. 2022;20:257–69.
120. Kim D-W, Cha C-J. Antibiotic resistome from the One-Health perspective: understanding and controlling antimicrobial resistance transmission. Exp Mol Med. 2021;53:301–9.
121. Forslund K, Sunagawa S, Coelho LP, Bork P. Metagenomic insights into the human gut resistome and the forces that shape it: prospects & overviews. BioEssays. 2014;36:316–29.
122. Maestre-Carballa L, Navarro-López V, Martinez-Garcia M. A resistome roadmap: from the human body to pristine environments. Front Microbiol. 2022;13:858831.
123. O'Connor L, Heyderman R. The challenges of defining the human nasopharyngeal resistome. Trends Microbiol. 2023;31:816–31.
124. Roachford OSE, Alleyne AT, Kuelbs C, Torralba MG, Nelson KE. The cervicovaginal microbiome and its resistome in a random selection of Afro-Caribbean women. Hum Microb J. 2021;20:100079.
125. Hu Y, et al. Metagenome-wide analysis of antibiotic resistance genes in a large cohort of human gut microbiota. Nat Commun. 2013;4:2151.
126. Rose G, et al. Antibiotic resistance potential of the healthy preterm infant gut microbiome. PeerJ. 2017;5:e2928.
127. Feng J, et al. Antibiotic resistome in a large-scale healthy human gut microbiota deciphered by metagenomic and network analyses. Environ Microbiol. 2018;20:355–68.
128. Forslund K, et al. Country-specific antibiotic use practices impact the human gut resistome. Genome Res. 2013;23:1163–9.
129. Shuai M, et al. Human gut antibiotic resistome and progression of diabetes. Adv Sci. 2022;9:e2104965.
130. Gasparrini AJ, et al. Metagenomic signatures of early life hospitalization and antibiotic treatment in the infant gut microbiota and resistome persist long after discharge. Nat Microbiol. 2019;4:2285–97.
131. Hicks LA, Taylor TH, Hunkler RJUS. Outpatient antibiotic prescribing, 2010. N Engl J Med. 2013;368:1461–2.
132. Saturio S, et al. Old folks, bad boon: antimicrobial resistance in the infant gut microbiome. Microorganisms. 2023;11:1907.
133. Salyers AA, Gupta A, Wang Y. Human intestinal bacteria as reservoirs for antibiotic resistance genes. Trends Microbiol. 2004;12:412–6.
134. Davies J, Davies D. Origins and evolution of antibiotic resistance. Microbiol Mol Biol Rev. 2010;74:417–33.
135. Holmes AH, et al. Understanding the mechanisms and drivers of antimicrobial resistance. Lancet. 2016;387:176–87.
136. da Costa-Lourenço APR, Barros dos Santos KT, Moreira BM, Fracalanzza SEL, Bonelli RR. Antimicrobial resistance in Neisseria gonorrhoeae: history, molecular mechanisms and epidemiological aspects of an emerging global threat. Braz J Microbiol. 2017;48:617–28.

137. Quillin SJ, Seifert HS. Neisseria gonorrhoeae host-adaptation and pathogenesis. Nat Rev Microbiol. 2018;16:226–40.
138. Singh S, Verma N, Taneja N. The human gut resistome: current concepts & future prospects. Indian J Med Res. 2019;150:345–58.
139. Smith HW. Transfer of antibiotic resistance from animal and human strains of escherichia coli to resident E. coli in the alimentary tract of man. Lancet. 1969;293:1174–6.
140. Massé D, Saady N, Gilbert Y. Potential of biological processes to eliminate antibiotics in livestock manure: an overview. Animals. 2014;4:146–63.
141. McEwen SA, Collignon PJ. Antimicrobial resistance: a One Health perspective. Microbiol Spectr. 2018;6(2). https://doi.org/10.1128/microbiolspec.arba-0009-2017.
142. Patyra E, Kwiatek K, Nebot C, Gavilán RE. Quantification of veterinary antibiotics in pig and poultry feces and liquid manure as a non-invasive method to monitor antibiotic usage in livestock by liquid chromatography mass-spectrometry. Molecules. 2020;25:3265.
143. Abbas F, Thomas P, Cully-Duse B, Andronicos NM, Winter G. Cattle–compost–soil: the transfer of antibiotic resistance in livestock agriculture. MicrobiologyOpen. 2023;12:e1375.
144. Levy SB, Fitzgerald GB, Macone AB. Spread of antibiotic-resistant plasmids from chicken to chicken and from chicken to man. Nature. 1976;260:40–2.
145. Marshall BM, Levy SB. Food animals and antimicrobials: impacts on human health. Clin Microbiol Rev. 2011;24:718–33.
146. Founou LL, Founou RC, Essack SY. Antibiotic resistance in the food chain: a developing country-perspective. Front Microbiol. 2016;7:1881.
147. Iwu CD, Korsten L, Okoh AI. The incidence of antibiotic resistance within and beyond the agricultural ecosystem: a concern for public health. MicrobiologyOpen. 2020;9:e1035.
148. Abraham EP, Chain E. An enzyme from bacteria able to destroy penicillin. Nature. 1940;146:837.
149. Martínez JL. Antibiotics and antibiotic resistance genes in natural environments. Science. 2008;321:365–7.
150. Allen HK, et al. Call of the wild: antibiotic resistance genes in natural environments. Nat Rev Microbiol. 2010;8:251–9.
151. Finley RL, et al. The scourge of antibiotic resistance: the important role of the environment. Clin Infect Dis. 2013;57:704–10.
152. Himanshu, et al. Nanobiotics and the one health approach: boosting the fight against antimicrobial resistance at the nanoscale. Biomol Ther. 2023;13:1182.
153. Gordon L, et al. Antimicrobial resistance survey in a river receiving effluents from freshwater fish farms. J Appl Microbiol. 2007;102:1167–76.
154. Hanna N, et al. Monitoring of water quality, antibiotic residues, and antibiotic-resistant Escherichia coli in the Kshipra River in India over a 3-year period. Int J Environ Res Public Health. 2020;17:7706.
155. Lata P, Ram S, Agrawal M, Shanker R. Enterococci in river Ganga surface waters: propensity of species distribution, dissemination of antimicrobial-resistance and virulence-markers among species along landscape. BMC Microbiol. 2009;9:140.
156. Ram S, Vajpayee P, Shanker R. Prevalence of multi-antimicrobial-agent resistant, shiga toxin and enterotoxin producing Escherichia coli in surface waters of river Ganga. Environ Sci Technol. 2007;41:7383–8.
157. Sweattatut R, Boontanon SK, Piyaviriyakul P, Harada H, Fujii S. Antimicrobial resistant Escherichia coli distribution along the lower part of the Chao Phraya River, Thailand. IOP Conf Ser Earth Environ Sci. 2022;973:012005.
158. D'Costa VM, McGrann KM, Hughes DW, Wright GD. Sampling the antibiotic resistome. Science. 2006;311:374–7.
159. D'Costa VM, et al. Antibiotic resistance is ancient. Nature. 2011;477:457–61.
160. Pawlowski AC, et al. A diverse intrinsic antibiotic resistome from a cave bacterium. Nat Commun. 2016;7:13803.
161. Carvalho IT, Santos L. Antibiotics in the aquatic environments: a review of the European scenario. Environ Int. 2016;94:736–57.

162. Kumar K, Gupta SC, Baidoo SK, Chander Y, Rosen CJ. Antibiotic uptake by plants from soil fertilized with animal manure. J Environ Qual. 2005;34:2082–5.
163. Chen GY, Thorup NR, Miller AJ, Li Y-C, Ayres JS. Cooperation between physiological defenses and immune resistance produces asymptomatic carriage of a lethal bacterial pathogen. Sci Adv. 2023;9:eadg8719.
164. Ingle DJ, Levine MM, Kotloff KL, Holt KE, Robins-Browne RM. Dynamics of antimicrobial resistance in intestinal Escherichia coli from children in community settings in South Asia and sub-Saharan Africa. Nat Microbiol. 2018;3:1063–73.
165. Bagge K, Benfield T, Westh H, Bartels MD. Eradicating MRSA carriage: the impact of throat carriage and Panton-Valentine leukocidin genes on success rates. Eur J Clin Microbiol Infect Dis. 2019;38:683–8.
166. Baek YS, Baek S-H, Yoo Y-J. Higher nasal carriage rate of methicillin-resistant Staphylococcus aureus among dental students who have clinical experience. J Am Dent Assoc. 2016;147:348–53.
167. Dumre SP, et al. Asymptomatic throat carriage rate and antimicrobial resistance pattern of Streptococcus pyogenes in Nepalese school children. Kathmandu Univ Med J. 2009;7:392–6.
168. Farfour E, et al. Asymptomatic carriage of extensively drug-resistant bacteria (eXDR), a simple way to assess spontaneous clearance. J Hosp Infect. 2020;104:503–7.
169. Horner P, et al. Should we be testing for urogenital Mycoplasma hominis, Ureaplasma parvum and Ureaplasma urealyticum in men and women? – a position statement from the European STI Guidelines Editorial Board. J Eur Acad Dermatol Venereol. 2018;32:1845–51.
170. Oliver J, et al. Group A Streptococcus pharyngitis and pharyngeal carriage: a meta-analysis. PLoS Negl Trop Dis. 2018;12:e0006335.
171. Crofts TS, Gasparrini AJ, Dantas G. Next-generation approaches to understand and combat the antibiotic resistome. Nat Rev Microbiol. 2017;15:422–34.
172. Acharya KP, Wilson RT. Antimicrobial resistance in Nepal. Front Med. 2019;6:105.
173. Pieri A, et al. Country income is only one of the tiles: the global journey of antimicrobial resistance among humans, animals, and environment. Antibiotics. 2020;9:473.
174. Klugman KP, Black S. Impact of existing vaccines in reducing antibiotic resistance: primary and secondary effects. Proc Natl Acad Sci. 2018;115:12896–901.
175. Sang Y, Blecha F. Alternatives to antibiotics in animal agriculture: an ecoimmunological view. Pathogens. 2015;4:1–19.
176. Tasho RP, Cho JY. Veterinary antibiotics in animal waste, its distribution in soil and uptake by plants: a review. Sci Total Environ. 2016;563–564:366–76.
177. Kahn LH, et al. From farm management to bacteriophage therapy: strategies to reduce antibiotic use in animal agriculture. Ann N Y Acad Sci. 2019;1441:31–9.
178. Putra RD, Lyrawati D. Interactions between bacteriophages and eukaryotic cells. Scientifica. 2020;2020:e3589316.
179. Mancuso G, De Gaetano S, Midiri A, Zummo S, Biondo C. The challenge of overcoming antibiotic resistance in carbapenem-resistant gram-negative bacteria: "attack on titan". Microorganisms. 2023;11:1912.
180. Little JS, et al. Bacteriophage treatment of disseminated cutaneous Mycobacterium chelonae infection. Nat Commun. 2022;13:2313.
181. Fishbein SRS, Mahmud B, Dantas G. Antibiotic perturbations to the gut microbiome. Nat Rev Microbiol. 2023;21:772–88.
182. Weis AM, Round JL. Microbiota-antibody interactions that regulate gut homeostasis. Cell Host Microbe. 2021;29:334–46.
183. Coyte KZ, Schluter J, Foster KR. The ecology of the microbiome: networks, competition, and stability. Science. 2015;350:663–6.
184. Ferreira-Halder CV, Faria AVDS, Andrade SS. Action and function of Faecalibacterium prausnitzii in health and disease. Best Pract Res Clin Gastroenterol. 2017;31:643–8.
185. Van Der Hee B, Wells JM. Microbial regulation of host physiology by short-chain fatty acids. Trends Microbiol. 2021;29:700–12.

186. Maier L, et al. Unravelling the collateral damage of antibiotics on gut bacteria. Nature. 2021;599:120–4.
187. Palleja A, et al. Recovery of gut microbiota of healthy adults following antibiotic exposure. Nat Microbiol. 2018;3:1255–65.
188. Reyman M, et al. Effects of early-life antibiotics on the developing infant gut microbiome and resistome: a randomized trial. Nat Commun. 2022;13:893.
189. Wang R. Clostridioides difficile infection: microbe-microbe interactions and live biotherapeutics. Front Microbiol. 2023;14:1182612.
190. Ubeda C, et al. Vancomycin-resistant Enterococcus domination of intestinal microbiota is enabled by antibiotic treatment in mice and precedes bloodstream invasion in humans. J Clin Invest. 2010;120:4332–41.
191. Murray CJL, et al. Global burden of bacterial antimicrobial resistance in 2019: a systematic analysis. Lancet. 2022;399:629–55.
192. Rajer F, Sandegren L. The role of antibiotic resistance genes in the fitness cost of multiresistance plasmids. mBio. 2022;13:e03552-21.
193. Robertson RC, Manges AR, Finlay BB, Prendergast AJ. The human microbiome and child growth – first 1000 days and beyond. Trends Microbiol. 2019;27:131–47.
194. Deshmukh HS, et al. The microbiota regulates neutrophil homeostasis and host resistance to Escherichia coli K1 sepsis in neonatal mice. Nat Med. 2014;20:524–30.
195. Miller WR, Munita JM, Arias CA. Mechanisms of antibiotic resistance in enterococci. Expert Rev Anti-Infect Ther. 2014;12:1221–36.
196. Chu C, et al. The microbiota regulate neuronal function and fear extinction learning. Nature. 2019;574:543–8.
197. Zhang F, Luo W, Shi Y, Fan Z, Ji G. Should we standardize the 1,700-year-old fecal microbiota transplantation? Am J Gastroenterol. 2012;107:1755.
198. Eiseman B, Silen W, Bascom GS, Kauvar AJ. Fecal enema as an adjunct in the treatment of pseudomembranous enterocolitis. Surgery. 1958;44:854–9.
199. Cammarota G, et al. International consensus conference on stool banking for faecal microbiota transplantation in clinical practice. Gut. 2019;68:2111–21.
200. Kelly CR, et al. Update on fecal microbiota transplantation 2015: indications, methodologies, mechanisms, and outlook. Gastroenterology. 2015;149:223–37.
201. Lloyd-Price J, Abu-Ali G, Huttenhower C. The healthy human microbiome. Genome Med. 2016;8:51.
202. Kho ZY, Lal SK. The human gut microbiome – a potential controller of wellness and disease. Front Microbiol. 2018;9:1835.
203. Hsu BB, et al. Dynamic modulation of the gut microbiota and metabolome by bacteriophages in a mouse model. Cell Host Microbe. 2019;25:803–814.e5.
204. Reid G, Gadir AA, Dhir R. Probiotics: reiterating what they are and what they are not. Front Microbiol. 2019;10:424.
205. Hill C, et al. The International Scientific Association for Probiotics and Prebiotics consensus statement on the scope and appropriate use of the term probiotic. Nat Rev Gastroenterol Hepatol. 2014;11:506–14.
206. Gibson GR, Roberfroid MB. Dietary modulation of the human colonic microbiota: introducing the concept of prebiotics. J Nutr. 1995;125:1401–12.
207. Zhou Y, et al. Antibiotic administration routes and oral exposure to antibiotic resistant bacteria as key drivers for gut microbiota disruption and resistome in poultry. Front Microbiol. 2020;11:1319.
208. De La Fuente-Núñez C, Cardoso MH, De Souza Cândido E, Franco OL, Hancock REW. Synthetic antibiofilm peptides. Biochim Biophys Acta. 2016;1858:1061–9.
209. Mhlongo JT, Waddad AY, Albericio F, De La Torre BG. Antimicrobial peptide synergies for fighting infectious diseases. Adv Sci. 2023;10:2300472.
210. Zhou X, et al. Enhanced antimicrobial activity of N-terminal derivatives of a novel brevinin-1 peptide from the skin secretion of Odorrana schmackeri. Toxins. 2020;12:484.

211. Cesaro A, Lin S, Pardi N, De La Fuente-Nunez C. Advanced delivery systems for peptide antibiotics. Adv Drug Deliv Rev. 2023;196:114733.
212. Zou W, McAdorey A, Yan H, Chen W. Nanomedicine to overcome antimicrobial resistance: challenges and prospects. Nanomedicine. 2023;18:471–84.
213. Jamshidi P, et al. Fecal microbiota transplantation in irritable bowel syndrome: a systematic review and meta-analysis of randomized controlled trials. Int J Mol Sci. 2023;24:14562.
214. Panwar RB, Sequeira RP, Clarke TB. Microbiota-mediated protection against antibiotic-resistant pathogens. Genes Immun. 2021;22:255–67.
215. Sequeira RP, McDonald JAK, Marchesi JR, Clarke TB. Commensal Bacteroidetes protect against Klebsiella pneumoniae colonization and transmission through IL-36 signalling. Nat Microbiol. 2020;5:304–13.
216. Fong W, Li Q, Yu J. Gut microbiota modulation: a novel strategy for prevention and treatment of colorectal cancer. Oncogene. 2020;39:4925–43.
217. Aslam B, et al. Antibiotic resistance: one health one world outlook. Front Cell Infect Microbiol. 2021;11:771510.
218. Lin B, Dietrich ML, Senior RA, Wilcove DS. A better classification of wet markets is key to safeguarding human health and biodiversity. Lancet Planet Health. 2021;5:e386–94.
219. Pollock J, et al. Alternatives to antibiotics in a One Health context and the role genomics can play in reducing antimicrobial use. Clin Microbiol Infect. 2020;26:1617–21.
220. Schwartz DJ, Langdon AE, Dantas G. Understanding the impact of antibiotic perturbation on the human microbiome. Genome Med. 2020;12:82.
221. Chinemerem Nwobodo D, et al. Antibiotic resistance: the challenges and some emerging strategies for tackling a global menace. J Clin Lab Anal. 2022;36:e24655.
222. Iskandar K, et al. Surveillance of antimicrobial resistance in low- and middle-income countries: a scattered picture. Antimicrob Resist Infect Control. 2021;10:63.
223. Barlow G. Clinical challenges in antimicrobial resistance. Nat Microbiol. 2018;3:258–60.
224. Tse BN, et al. Challenges and opportunities of nontraditional approaches to treating bacterial infections. Clin Infect Dis. 2017;65:495–500.
225. Saha M, Sarkar A. Review on multiple facets of drug resistance: a rising challenge in the 21st century. J Xenobiot. 2021;11:197–214.
226. Shukla R, et al. An antibiotic from an uncultured bacterium binds to an immutable target. Cell. 2023;186:4059–4073.e27.
227. Lin Z, Nielsen J, Liu Z. Bioprospecting through cloning of whole natural product biosynthetic gene clusters. Front Bioeng Biotechnol. 2020;8:526.
228. Aminov RI. A brief history of the antibiotic era: lessons learned and challenges for the future. Front Microbiol. 2010;1:134.
229. Dadgostar P. Antimicrobial resistance: implications and costs. Infect Drug Resist. 2019;12:3903–10.
230. Afshinnekoo E, et al. Geospatial resolution of human and bacterial diversity with city-scale metagenomics. Cell Syst. 2015;1:72–87.
231. Danko DC, Mason C. The MetaSUB microbiome core analysis pipeline enables large scale metagenomic analysis. arXiv200912338 Q-Bio. 2020.
232. Madrigal P, et al. Machine learning algorithm to characterize antimicrobial resistance associated with the International Space Station surface microbiome. Microbiome. 2022;10:134.
233. Butler D, et al. Shotgun transcriptome, spatial omics, and isothermal profiling of SARS-CoV-2 infection reveals unique host responses, viral diversification, and drug interactions. Nat Commun. 2021;12(1):1660.

Chapter 7
Environmental Reservoirs, Genomic Epidemiology, and Mobile Genetic Elements

Thato Motlhalamme, Lynthia Paul, and Vinayak Singh

Abbreviations

AMR	Antimicrobial resistance
ARB	Antimicrobial-resistant bacteria
ARG	Antimicrobial resistance genes
AST	Antimicrobial susceptibility testing
BAS	Biocidal active substances
CA-MRSA	Community-associated methicillin-resistant *S. aureus*
CIA	Critically important antibiotics
CMT	Complex mutant TEM
CRE	Carbapenem-resistant *Enterobacterales*
EEA	European Economic Area
EOC	Emerging organic contaminants
ESBL	Extended-spectrum β-lactamases
EU	European Union

T. Motlhalamme
South African Medical Research Council Drug Discovery and Development Research Unit, Department of Chemistry, University of Cape Town, Rondebosch, South Africa
e-mail: thato.motlhalamme@uct.ac.za

L. Paul
Division of Medical Microbiology, Department of Pathology, University of Cape Town, Observatory, South Africa
e-mail: lynn.paul@uct.ac.za

V. Singh (✉)
South African Medical Research Council Drug Discovery and Development Research Unit, Department of Chemistry, University of Cape Town, Rondebosch, South Africa

Holistic Drug Discovery and Development (H3D) Centre, University of Cape Town, Rondebosch, South Africa

Institute of Infectious Disease and Molecular Medicine (IDM), University of Cape Town, Observatory, South Africa
e-mail: vinayak.singh@uct.ac.za

V. Soni, A. S. Akhade (eds.), *Antimicrobial Resistance: Factors to Findings*,
https://doi.org/10.1007/978-3-031-65986-7_7

GIT	Gastrointestinal tract
HGT	Horizontal gene transfer
IS	Insertion sequence
MIC	Minimum inhibitory concentration
MDR	Multidrug resistance
MRSA	Methicillin-resistant *Staphylococcus aureus*
NDM-1	New Delhi metallobetalactamase-1
PPCP	Pharmaceuticals and personal care products
QAC	Quaternary ammonium compounds
ST	Sequence types
TMP-SMX	Trimethoprim-sulfamethoxazole
VIM-1	Verona integron-encoded metallobetalactamase-1
VRE	Vancomycin-resistant Enterococci
WWTP	Wastewater treatment plant

7.1 Introduction

The discovery of antibiotics improved the success rate of therapies against infectious disease; however, this inadvertently led to the emergence of drug-resistant bacteria against most antibacterials in use. Organisms resistant to at least one antimicrobial, in two or more key antimicrobial categories, are referred to as multidrug-resistant (MDR) bacteria, whereas extremely drug-resistant isolates are susceptible to agents in less than two categories. Pan-drug-resistant isolates are not susceptible to any drug in all key antimicrobial categories [1]. Resistance levels have increased to the extent that AMR is now a global health concern that threatens the effectiveness of antibiotics and compromises the ability to treat infectious diseases. Notably, the ESKAPE group of organisms are a major concern, especially for hospitalized patients. A retrospective analysis over a five-year period (2017–2018), looking at AMR data from the European Union [52], highlights data from specific antibiotics most often used against the ESKAPE microbes. The data reveal a notable increase in combined resistance for *Acinetobacter baumannii* against carbapenems, fluoroquinolones, and aminoglycosides (28.2% in 2017 to 36.8% in 2021), which is a major concern as this organism is already resistant to most antibiotics used to treat it in clinical practice. Another major concern noted was carbapenem-resistant *Enterobacterales* (CRE), notably *Escherichia coli* and *Klebsiella pneumoniae*. The rise in AMR against last-resort drugs such as carbapenems is a major concern. This phenomenon is well studied for nosocomial infection, but community-acquired carbapenem-resistant isolates and their associated transmission routes are less well studied [74].

AMR occurs when a microbe (pathogen or commensal) decreases its susceptibility to an antibiotic, most often following exposure to that antibiotic and becomes unsusceptible (acquired resistance) [31]. This phenomenon predates the selective pressure of clinical antibiotic use [39]. While the development of AMR primarily occurs within clinical settings, it is important to recognize the significant role that environmental reservoirs and natural conditions play in shaping the evolution and spread of AMR. Environmental reservoirs, such as soil, water, wildlife, companion animals/pets, insects, and rodents, serve as crucial sources of resistance genes/ mechanisms and provide ideal conditions for the proliferation and exchange of resistant bacteria [69]. These reservoirs are referred to as the 'resistome' [39]. As AMR occurs in all environments, it requires a One Health approach, with combined efforts to minimize the emergence and selection of drug-resistant bacteria in humans, animals, and other environments. This includes stewardship of clinical, veterinarian, agricultural, and economic activities that contributes to the emergence of AMR [122].

The transmission of AMR determinants between bacteria is enabled by the carriage on mobile elements such as plasmids, phages, and transposons [5, 7, 72, 194, 195]. Additionally, natural factors like climate, pollution, and ecological interactions further influence the emergence and dissemination of AMR. Antimicrobial resistance genes (ARGs) of clinical significance represent a fraction of all AMR elements found among environmental bacteria. Pathogens frequently acquire new genetic traits, such as AMR genes, from other microbes in environmental reservoirs, followed by adaptation and further diversification. These traits are highly problematic in the clinical arena, as the use of antibiotics (whether overuse or incorrect use) creates a selective pressure, driving AMR emergence in host [1, 164, 198]. In pathogens where AMR arises due to mutations on its genome, for example *Mycobacterium tuberculosis*, various studies have shown that microbial evolution occurs in host, resulting from the host biochemical environment as well as exposure to antimycobacterial antibiotics. These mutations include gene variants responsible for resistance to anti-TB antimicrobials [61] which could lead to remodelling of metabolic pathways, contributing to resistance phenotypes, such as the *rpoB* mutations in *M. tuberculosis* [102]. Understanding the intricate relationship between AMR and its environmental context is therefore essential for devising comprehensive strategies to combat this growing public health threat.

7.2 Environmental Reservoirs and AMR

The intricate web of life on Earth comprises invertebrate organisms, reptiles as well as humans and a vast array of aquatic, avian, and terrestrial animals, both domesticated and wild. All these domains of life are inhabited by microorganisms. Domesticated animals have walked alongside humanity for generations, serving diverse roles, from companions and farm labourers to sources of sustenance. These animals have adapted to coexist with humans and evolved unique traits distinct from their wild ancestors, frequently living within households and often receive advanced care to ensure their well-being. However, this care has given rise to concerns, with the widespread use of antimicrobials to treat ailments leading to alarming consequences. Antimicrobial overuse, particularly in food animals, fosters the emergence and spread of antibiotic-resistant bacteria, posing a critical threat to both animal and human populations. This symbiotic relationship between humans and companion animals has unveiled the potential for these animals to harbour antimicrobial-resistant bacteria in their gut, capable of transmission to humans through direct contact or environmental contamination. Furthermore, wildlife, existing at the juncture of ecosystems and human activities, plays a previously underestimated role in the ongoing battle against antimicrobial resistance. Wild animals, whether migrating birds with extensive ranges or creatures foraging in human-affected environments, can serve as conduits for the transfer of antibiotic-resistant bacteria and resistance genes between nature and society. In essence, the narrative of antimicrobial resistance transcends species boundaries, encompassing not only domestic and companion animals but also the intricate tapestry of the natural world, posing significant challenges and necessitating comprehensive solutions.

7.2.1 Domesticated and Companion Animals

Domesticated animals are species of animals that have been selectively bred and kept by humans for various purposes, such as companionship, agricultural work, food production, and other beneficial uses. Over generations, these animals have undergone domestication, which involves adapting to living alongside humans and exhibiting behaviours and characteristics that are distinct from their wild ancestors [99]. Some common examples of domestic animals include dogs, cats, cattle, horses, pigs, sheep, and chickens. Companion animals or pets such as dogs, cats, and rabbits are normally kept within or near households. These animals are more likely to receive better health care compared to their wild counterparts leading to the high usage of antimicrobials to treat diseases and wounds [69]. As a result, the global use of antimicrobials in animal husbandry accounts for 60–80% of the total production of these agents [1]. When it comes to animals raised for food, antimicrobials are often administered to whole groups or enclosures via their feed and water. Furthermore, when food animals are subjected to extended periods of low-dose medication to boost growth, it creates an environment conducive to the emergence and dissemination of antibiotic-resistant bacteria among both animal and human populations [90, 153, 196]. Additionally, the overuse of human critically important antibiotics (CIAs) such as macrolides, (fluoro)quinolones, third- and fourth-generation cephalosporins on dogs, cats, and horses is a public health concern, which has led to higher prevalence of multidrug-resistant microbes in companion animals [46].

Companion animals are frequently exposed to antibiotics and can harbour AMR bacteria in their gut, potentially transmitting these pathogens to humans through direct contact or environmental contamination. Animals may be carriers of extended-spectrum β-lactamases (ESBL), biofilm-producing, multidrug-resistant microbes, as described in a survey of *Klebsiella* on the Indian subcontinent [12]. *E. coli* is another common, commensal bacterium residing in the gastrointestinal tract (GIT) of mammals. However, it is also capable of causing a range of diseases. Given its substantial genetic diversity and capacity to acquire ARGs, particularly through HGT, this species can serve as both a source and reservoir of ARGs [11, 85, 158, 160]. Several observational and surveillance studies have reported MDR in *E. coli* isolates in domestic animals and environments they live in. A systematic review of antimicrobial resistance patterns of *E. coli* in Bangladesh found different degrees of resistance to tetracycline, aminoglycosides, amino-penicillin, macrolides, fluoroquinolone, chloramphenicol, and polypeptides in isolate samples from domestic animals, wildlife, and the environment [90]. Another study in Ecuador reported high levels of resistance to ampicillin, tetracycline, sulfamethoxazole-trimethoprim, and chloramphenicol in *E. coli* isolated from chickens, pigs, and cattle [110].

7.2.2 Wild Animals

The role of human and livestock populations in driving AMR is widely acknowledged, but an overlooked aspect is the contribution of wildlife as a reservoir for antibiotic-resistant bacteria. Wild animals, existing at the interface of ecosystems and human activities, can serve as a bridge for the transmission of AMR between the natural world and human society. Some of these animals, for example, certain bird species characterized by extensive home ranges or migratory patterns, can contribute significantly to the dissemination of antimicrobial-resistant bacteria (ARB) and ARGs [10]. For instance, migratory birds such as gulls and storks are known to cover vast distances during their migrations. Their faecal droppings can contaminate water bodies, potentially transferring resistant bacteria or genes between wildlife species and different water bodies [76, 146]. Wild animals foraging in human-influenced environments may also become colonized by bacteria carrying clinically important antibiotic resistance [22, 35].

Antimicrobial-resistant *E. coli* strains discovered in wild birds were shown to share genomic traits with strains found in hospitals, wastewater, and even human pathogens which strongly suggests zoonotic transmission between humans, domesticated, and wild animals [184]. Notably, ESBL-producing Gram-negative bacteria had been identified in avian species, emphasizing the possibility for these birds to spread pathogenic bacteria across continents during migration [10]. Genetic analysis of ESBL-producing enterobacteria from wild birds revealed diverse lineages and sequence types (STs), of which *E. coli* ST131 (CC131) was frequently identified as a global problem. This is due to its distribution and association with the dissemination of specific resistance genes in various environments [141]. Wang et al. [191] also reported several sequence types (ST10, ST69, ST405, ST410, ST224, ST648) of ESBL-producing *E. coli* in wild birds, indicating their involvement in the intercontinental spread of resistance.

Several studies have isolated carbapenemase-producing bacteria, typically found in clinical settings, from wild animals. An NDM-1 (New Delhi metallobetalactamase-1)-producing *Salmonella* enterica subsp. serovar Corvallis was isolated in black kites in Germany. *S. corvallis* was reported frequently on the Asian subcontinent and parts of Africa, suggesting potential long-distance dissemination through the Africa-Asian migratory cycles of black kites [60]. Vittecoq et al. [187] found carbapenemase-producing *E. coli* (verona integron-encoded metallo-betalactamase-1 (VIM-1) carbapenemase) in yellow-legged gulls and proposed the introduction of these strains from neighbouring countries where VIM-1 was prevalent in human isolates. Vancomycin-resistant enterococci (VREs) in clinical settings are a global issue, also in wildlife. For example, vancomycin-resistant *E. faecium* (VRE) strains were isolated from birds in Portugal [166], Australia [134], and North America [133, 135, 151]. Similar occurrences were described for the pandemic-associated VREs from the *E. faecalis* CC16 and CC21 lineages.

Aquaculture is a hotspot for antimicrobial resistance and a notable contributor to the spread of ARB and ARGs. Intensive aquaculture practices involve the use of large amounts of antimicrobial agents to combat bacterial diseases. These agents, along with agricultural residues and by-products added to fish feed, can contribute to the emergence and spread of resistance [21]. ESBL-producing Enterobacteriaceae were isolated in various fish species residing in Tunisia's Bizerte lagoon area. The presence of CTX-M-1- and CTX-M-15-positive ESBLs was particularly prevalent in these fish [78]. Brahmi et al. [25] isolated ESBL-producing isolates from wild fish species in Algeria, including *E. coli* ST131 and ST410 lineages, suggesting the potential spread of resistance lineages through fish populations. In Hong Kong, ESBL-producing *E. coli* and *K. pneumoniae* were identified in Snakehead fish and black carp [159]. Importantly, frozen freshwater fish imported to Saudi Arabia from countries like Thailand, India, and Vietnam were found to carry ESBL-producing *E. coli*, indicating the potential transmission of resistance through the food trade [54]. Furthermore, aquatic wildlife such as rainbow trout in Lebanon [77], as well as fish from other regions, have been found to harbour resistant *E. coli* strains, often associated with multidrug resistance. The presence of *mcr* genes, which confer resistance to colistin, an antibiotic of last-resort, has raised concerns about resistance dissemination from aquaculture to humans [28, 29]. Studies like Kalová et al. [89] have identified *mcr* genes in various *E. coli* isolates from aquaculture products imported to different countries.

The impact of AMR in wildlife extends beyond birds and fish, having been identified also in mammals, including rabbits, wild boars as well as reptiles such as loggerhead sea turtles [19, 171, 172]. Notably, well-known human high-risk methicillin-susceptible and resistant *Staphylococcus aureus* (MRSA) isolates have been described in faeces of various wild animals such as foxes, fallow deer, ravens, and partridges. Some clonal lineages, such as CC130, have been described in both humans and animals [126]. Studies have found antimicrobial resistance in primates as well. For example, Bornbusch et al. [23] and Chong et al. [34] detected resistance in various primate species. This highlights the potential for resistance transmission in settings like zoos or ecotourism due to close human-animal interactions.

7.2.3 Vermin

Insects, rodents, and companion animals/pets are often overlooked but act as significant reservoirs for AMR, posing a potential risk to human health [69]. These diverse organisms harbour numerous microbes including antibiotic-resistant pathogens. The close interaction of these animals and insects with humans and their environments makes them ideal reservoirs of various antimicrobial-resistant microbes, resistance genes, and mobile elements that have an impact on human health.

Insects, such as flies and cockroaches, are known to frequent areas with high bacterial loads, including hospitals, households, and food preparation areas, facilitating the acquisition and transmission of resistant bacteria. Several studies have highlighted the role of insects, such as flies and cockroaches, in the spread of antibiotic-resistant bacteria. In one study conducted in the UK, bacterial strains resistant to multiple antibiotics were found in flies, indicating their potential as reservoirs for antibiotic resistance [20]. Another study focused on cockroaches and demonstrated the HGT of resistance genes between *E. coli* and *Enterobacteriaceae*. These authors also reported on transfer of antimicrobial resistance plasmids between organisms in their gut, suggesting cockroaches as potential carriers of ARB [6]. Houseflies were also found to facilitate the transfer of ARGs within their intestines, including the transfer of resistance genes to other bacterial species. Vertical transmission of antibiotic-resistant bacteria from one fly generation to another was also observed [64] . Studies have shown that insects harbour antimicrobial-resistant microbiota, ARGs, and mobile genetic elements. In one such study, the multidrug-resistant bacteria *E. faecalis*, along with the associated ARGs, *tet(M), erm(B), gelE, esp, asa1*, and the Tn*916/1545* transposon family, were prevalent in the faeces of German cockroaches and the digestive tracts of houseflies captured from swine farm manure [3]. The use of antibiotics in agriculture also created a selective pressure on the microbiome of insects leading to the emergence of resistant phenotypes. This was seen in honeybees where the use of tetracycline contributed to the development of AMR in the bees' gut microbiome [179].

Rodents, such as rats and mice, can carry AMR bacteria in their faeces and contaminate food and water sources, contributing to the dissemination of resistance. The importance of these small mammals, including those found in households, as a source of AMR is not well recognized, yet they play a vital role as sentinels, reservoirs, and vectors of AMR, particularly in areas where anthropogenic factors such as pollution or waste are involved. Microbes in rodents harbour genes conferring resistance to various antibiotics belonging to different antibiotic classes such as penicillin's, tetracyclines, aminoglycosides, quinolones, and sulphonamides. These mammals can be considered a hotspot for the acquisition and transmission of ARGs which can be transmitted to other organisms including humans [69].

7.3 Anthropogenic Pressures Drive AMR Development

The roots of AMR are complex and multifaceted. Although the occurrence is inherent and existed prior to the contemporary selective influence of clinical antibiotic application, anthropogenically produced or concentrated contaminants have exacerbated the development of AMR. Human activities such as the misuse/overuse of antibiotics in human and veterinary settings, for example the use of antibiotics to improve growth in animal husbandry and aquaculture via growth promotors, or disease prevention as metaphylactics and prophylactics is a concern. Industrial activity and hospital waste have resulted in high levels of these compounds being detected in environments; microbes are found which may increase the prevalence of ARGs and ARB [4, 62, 66, 73]. The half-life of antibiotics can vary significantly, ranging from a few hours to hundreds of days. However, antibiotic residues are widely recognized as persistent contaminants of the environment and can be found in various environmental matrices and are known to enhance the spread of antibiotic resistance [129].

Biocidal active substances (BAS) are compounds that have activity against pathogenic organisms [182]. These include various microbiocidal (such as formaldehyde, chlorhexidine, and quaternary ammonium compounds (QACs)) and microbiostatic chemical compounds with activity against a broad range of microbes [87]. Emerging organic contaminants (EOCs) on the other hand are a diverse group of synthetic or natural organic compounds of anthropogenic origin [66]. These compounds exhibit remarkable durability, the propensity to accumulate within biological systems, and are reintroduced into the ecosystem. EOCs include medicines and commercial formulations that were not developed as antimicrobial agents; therefore, their sub-lethal effects were not considered during drug development or in how the products were used. Some classes of EOCs involved in the promotion of AMR include antibacterial agents triclosan and triclocarban, antifungal agents, herbicides, surfactants, salicylates, fluoxetine, ibuprofen, and carbamazepine.

EOCs and BAS are found in everyday products such as pharmaceuticals and personal care products (PPCPs), cleaning products, pesticides, building materials, veterinary, and industrial products [4, 157]. Antibiotic residues, EOCs, and BAS are typically introduced into environmental compartments through waste effluents. Once released, they come into contact native microbes, which can act as a driving force for microbial evolution and the emergence of more resistant strains (Fig. 7.1 *Creative Commons License CC BY 4*). This can potentially disturb the innate equilibrium of ecosystems [129, 192]. The deterioration of enduring chemical compounds poses a formidable challenge, thereby rendering it intricate to predict the consequences of any ensuing contamination arising from their discharge [4, 36, 98, 157].

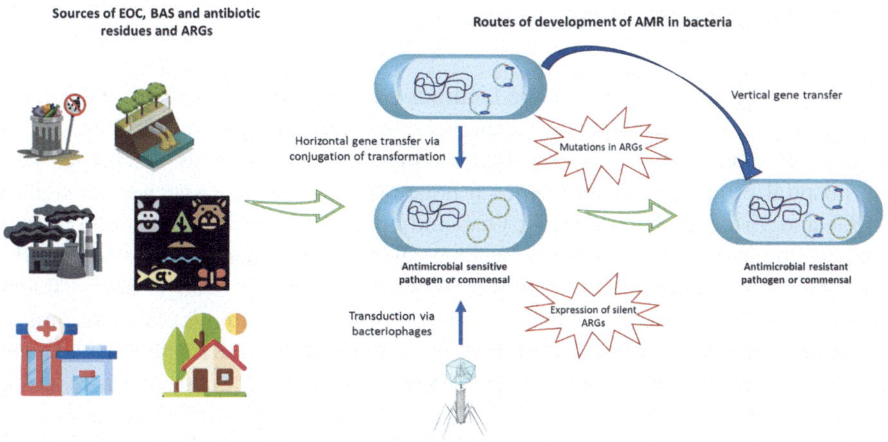

Fig. 7.1 Effects of emerging organic contaminants (EOC), bactericidal active compounds (BAS), and ARGs on the spread of AMR. Exposure to EOC, BAS, and antibiotics is hypothesized to select/enrich for microbes that can defend themselves via acquired ARG elements or intrinsically encoded defence mechanisms. Mobile ARGs are acquired by sensitive bacteria from the surrounding environment or direct interaction with other microbes. This horizontal gene transfer occurs through conjugation with already resistant bacteria, transformation with plasmids or membrane vesicles carrying ARGs, and transduction by phages. Mutations occurring in the bacteria can select for resistance in genes encoding key bacterial proteins which are antibiotic targets. Bacteria can also express silent resistance genes. All or some of these resistance mechanisms can be transferred from parent to progeny (vertical gene transfer). (Creative Commons License CC BY 4 (Some icons are used from https://bioicons.com/))

7.3.1 Household Activities, Medications, and Use of Household Chemicals

EOCs and BAS are found in household products such as personal care products, disinfectants, detergents, cosmetics, pesticides, synthetic musk's, sweeteners, and microplastics which all end up in waste streams generated by the household [4]. However, it is difficult to fully characterize the content of household waste as it differs from home to home. Some ingredients found in these products could also drive AMR. An example is triclosan (C12H17Cl3O2; 2,4,4′-trichloro-2′-hydroxydiphenyl ether) which is a common antiseptic ingredient used in soap as well as other personal care products like toothpaste, lotions, and shampoos [91]. However, IT biocide can also stimulate horizontal transfer of antibiotic resistance at sub-minimum inhibitory concentration (MIC) [87]. The QAC benzalkonium chloride which is found in personal care products like alcohol-based hand rubs and mouthwashes could induce antibiotic resistance against fluoroquinolones [27].

7.3.2 Hospital Waste Management

Hospital waste and other healthcare facilities contain various BAS and EOCs from therapeutic agents, biocide disinfectant, solvents, and antibiotic residues as well as waste from laboratories, patients, and cleaning services. These are discharged into the hospital effluent which becomes a hotspot inter- and intraspecies transfer of ARGs and virulence factors [157]. One study detected high levels of fluoroquinolone, sulphonamide, and tinidazole residues in hospital effluents [51], whereas others have found enrichment of novel ARGs in activated sludge and effluent wastewater. These ARGs conferred resistance to several antibiotics such as aminoglycosides, fluoroquinolones, and β-lactams [113, 199]. ARBs that have been detected in hospital effluent include ESBL-producing *E. coli*, multi-resistant *Pseudomonas,* and MRSA and VREs. Bacteria in hospital effluent have also been shown to carry vancomycin (*vanA*) and methicillin (*mecA*) resistance markers [4, 157]. On the other hand, a South African paper on shared ARGs between natural environment and agricultural settings highlighted the importance of environmental matrices as reservoirs of ARGS that end up in hospital settings. In this study, it was noted that farm settings were one of the major contributors of antibiotic resistance in clinical settings [53]. This highlights the need for robust surveillance systems to track important ARGs not only in clinical setting but also in the environment to limit their spread.

7.3.3 Agricultural Waste

One of the major driving forces contributing to the transmission of antimicrobial resistance to humans through animal, fish, and seafood products is the excessive use of antimicrobial compounds as growth promoters in livestock farming and aquaculture. The increasing consumption of food animal products, particularly in low- and middle-income countries with growing economies, is expected to further drive antibiotic consumption beyond the current levels observed in high-income countries [183]. The largest quantities of antimicrobial agents are utilized in pig (172 mg/kg), poultry (148 mg/kg), cattle (45 mg/kg) farming practices. This consumption is anticipated to double in the BRICS countries (Brazil, Russia, India, China, and South Africa) due to the growing demand for antibiotics to maintain animal health and productivity [183].

Notably, tetracyclines, macrolides, and sulphonamides represent some of the frequently used antimicrobial agents in this context [201]. A significant proportion of these agents are not metabolized by farmed animals, resulting in the release of substances into water and soil through animal excrement, primarily in the form of manure and urine. Studies have indicated that 75–80% of tetracyclines, 50–90% of erythromycin, and 60% of lincomycin doses are excreted in urine and faeces [38]. This results in elevated concentrations of various antimicrobial agents, such as tetracyclines, fluoroquinolones, sulphonamides, and macrolides, identified in animal faeces [38]. This highlights the risk of AMR spreading via animal faeces which are frequently used as fertilizer for agricultural land.

The presence of antibiotics in soil at these high concentrations fosters the preferential growth of ARB, as it drives the change from the antibiotic sensitivity of exposed microbes towards AMR phenotypes [9]. Notable sub-lethal concentrations of antimicrobials in the environment provide pressure for HGT between microbes; ARB and associated ARGs in soils around agricultural facilities are well documented. For instance, a study conducted by Hartmann et al. [75] reported an 18.3% prevalence of the $bla_{\text{CTX-M}}$ gene among soil samples collected from an area with high beef cattle densities in Burgundy, France. These resistance genes are transferred to humans and other environments through the food supply chain and waste materials [95]. Pathogens such as *E. coli* O157:H7 and *Salmonella* spp. are present in compost, soil, and water samples, most likely disseminated from animal faecal origin [17, 82]. Many of these isolates were found to be found to be resistant to multiple antimicrobials of different classes. Direct excretion of antimicrobials from animals in husbandry facilities serves as the primary route of entry into the environment. This is compounded by the application of manure as a fertilizer, runoff from farms, and fish farming practices [66].

7.3.4 Wastewater Treatment

Municipal wastewater systems are significant contributors to antibiotic resistance, acting as direct sources for the dissemination of these resistant elements into the environment [30, 128]. These systems receive wastewater from various origins, including hospitals, households, veterinary facilities, and agriculture, concentrating residual antibiotics and diverse ARBs. Wastewater effluent discharged into the environment contains various microbial communities, mixing with the native microbiota, and often contains sub-inhibitory concentrations of EOCs. Exposure to EOCs at concentrations below the MIC can facilitate the selection of resistance and may also affect various cellular responses, including stress responses and the expression of bacterial virulence genes [4]. Although conventional wastewater treatment plant (WWTP) systems are designed to effectively remove dissolved nutrients and solids, including bacteria, they currently lack specific technologies for reducing ARBs and ARGs [149, 150]. However, researchers have been exploring various technologies to address this issue. These include biological methods (i.e. WWTPs, constructed wetlands, and anaerobic membrane bioreactors) and physical/chemical methods (i.e. adsorption, disinfection, and advanced oxidation processes) [107]. Although studies have reported successful removal of ARBs and ARGs using the above-mentioned technologies [14, 88, 149], there are still considerable gaps to fill before these can be implemented in large-scale WWTPs. As most of the studies conducted thus far have been lab or pilot scale, it is important to evaluate the practicality modifying existing water and wastewater treatment facilities to reduce ARBs and ARGs in the effluent of plants.

7.4 Evolution and Transmission of Antimicrobial Resistance Genes in Environmental Settings

Bengtsson-Palme and coworkers stated that 'The lack of knowledge of how, and under which circumstances, the environment facilitates resistance development makes mitigation of the emergence and dissemination of mobile resistance factors problematic' [16].

Microbes typically share their habitats with other organisms, competing for environmental resources such as nutrients, space, and hydration. In these environments, the fittest competitors predominate, notably those with optimal abilities to occupy and use resources in the specific niche, plus the ability to control the numbers of competitors [67]. All polymicrobial environments, including the GIT of animals and insects, aquatic environment (both free-living microbes and those attached to marine and freshwater life), and soil are niches where interaction between microbes is possible. Microbes use various strategies, including the production of antibiotics, to outcompete any neighbours for a specific environment [67]. It therefore stands to reason that bacteria would develop defence mechanisms against antibiotics. It is also not surprising that ARGs encoding for scavengers of antibiotics are omnipresent in all environments, including the microbiota of animals as well as other environments [84, 127].

Although AMR determinants are present in the environment at low, undetectable levels and microbes produce limited amounts of antibiotics to outcompete other organisms, anthropogenic pressures have introduced large concentrations of abiotic pressures into natural environments where these organisms live. Introduction of these evolutionary pressures elects for microbial diversification and drives the emergence of AMR variants [13]. Various metagenomic studies demonstrated the presence of AMR determinants in nature, ranging from mammalian, avian, insects, and reptiles. At least one study demonstrated enrichment of AMR markers in environments polluted by antibiotics originating from human activity, specifically industrial manufacturing of antibiotics [136].

The use of antibiotics is not the only driver of the evolution of AMR; incorrect use, overuse, the use of old antibiotics, and/or unnecessary or prolonged use of antibiotics drives evolution of bacteria towards drug resistance are some of the factors contributing to AMR evolution [117]. Empirical use of broad-spectrum antibiotics in patients is necessary where the identity of bacteria causing an illness is not known yet. For example, sepsis patients with a clinical picture suggestive of bacterial infection are treated empirically with relevant antibiotic(s) to cover all potential 'suspects' that might cause the condition. Once the causative microbe is identified, antibiotic therapy is de-escalated, meaning antimicrobials are changed to an antibiotic known to be effective against identified organism(s) [148]. Notably, extended hospitalization, prolonged antibiotic use are other known risk factors for AMR in clinical settings [33]. In a study reviewing the risk of AMR development in critically ill patients, an increased risk for AMR development was found for each additional day that β-lactams were administered to treat suspected *Pseudomonas* infection [176]. Antibiotic use as a driver is also evident by the typical emergence of resistant bacteria after antibiotics are introduced into practice [105]. Importantly, drug-resistant bacteria could be disseminated from hospitals to the environment if not managed appropriately which is a major concern [86].

The enrichment of AMR genes in hospital waste and hospitalized patient isolates is a known phenomenon. However, these are not the only risk factors conducive for the development of AMR. Other abiotic stressors include antibiotic exposure of microbes in the environment, chemical disinfectants, fertilizers, and metal-containing pollutants in the environment [4, 157], which selectively enrich the abundance of AMR genes to the extent that it is detectable and a threat to a specific environment/ecosystem. Chemicals such as polyvinyl chloride were also shown to drive the selection/emergence of AMR. A recent report [188] showed that polyvinyl chloride leachate, from plastic pollutants, contributed to enrichment of aminoglycoside resistance, antibacterial peptides, and MDR determinant in a laboratory-based microcosm study. Interestingly, AMR genes encoding macrolides, lincosamides, and streptogramins were significantly lower. This observation suggests that specific abiotic stresses enrich for specific bacterial genera: where AMR determinants are typically associated with Gram-negative bacteria, while Gram-positive bacteria are producers of AMR determinants against aminoglycosides and antibacterial peptides [188]. This further suggests that one could predict/model the emergence and evolution of AMR determinants in specific environments, based on the type of abiotic pressure introduced into an environment. Could this information be useful to tailor intelligent design of preventative or mitigating measures to minimize enrichment of AMR in a specific environmental reservoir?

A recent systematic review showed that anthropogenic pressures on industrialized aquatic spaces drive changes in the sedimentary microbiota towards enriched AMR presence and increased risk of human exposure to AMR determinants [152], while another study showed the feasibility of environmental monitoring to assist plans to mitigate the risk of human exposure to environmental AMR [174]. The systematic review by Rodgers and coworkers [152] delineates the variety of pollutants introduced into estuarine environments where human activity occurs. These include a variety of the major pollutants such as metals and polyaromatic hydrocarbons from terrestrial and water-based industries (fishing, aquaculture, ships, etc.), which were found to modulate the local microbiota and act as a reservoir from where AMR can spread. Studies like these suggest that we should use the knowledge gained to take preventative action to limit the spread of AMR outside of its original environment. This should include monitoring human pollution in spaces and rehabilitating environments to return them to their pre-AMR enriching states. Unfortunately, knowledge in this area is lacking and is a major limiting factor in the fight against AMR from a One Health perspective. Currently, a majority of policies are driven by knowledge of AMR in clinical areas (from the standpoint of antibiotic use as a driver), while studies on exposure of humans to AMR determinants in the environment are neglected [174]. Surveillance and remediation of environmental reservoirs of ARGs and ARB is essential and could better inform policies on combating AMR.

In the clinical arena, differences were observed in different geographical variations with regard to the rate of emergence, incidence, and type of AMR emerging. This suggests that mere use of antibiotics is not the only driver for AMR emergence. For example, a systematic review of AMR in anaerobic bacteria showed that the emergence or lowering of AMR in anaerobe bacteria differs from region to region, depending on the anaerobic bacterial species, strains, geographical locale, antibiotic use in different countries, and specimen type [24]. This suggests that intelligent use of antibiotics should include knowledge of the 'appropriate antibiotic for the specific infection, appropriate dosages to achieve proper blood serum and or tissue levels'. This variation in AMR rates in isolates from various infected sites was also observed in a study of 57,788 *S. aureus* isolates in a specific clinical setting. The study was done to discern the proportion of methicillin, clindamycin, and trimethoprim-sulfamethoxazole (TMP-SMX) resistance in all *S. aureus* isolates identified in a specific outpatient and inpatient setting. The data showed that the number of isolates with either methicillin, clindamycin, and TMP-SMP resistance was associated more with infections in specific bodily sites [186]. With the dawn of methicillin resistance in community-associated methicillin-resistant *S. aureus* (CA-MRSA), clindamycin and TMP-SMX are now used to treat community-acquired *S. aureus* infections. However, clindamycin resistance has now also occurred in this microbe. Therefore, antibiotic choices for treatment for infections should be guided by the site of infection, type of microbe involved, and the length of therapy needed. The differences in antibiotic guidelines and available antibiotics in different locales clearly contribute to observed differences in AMR from different regions. Also, unlawful use of antibiotics (black market acquisition and not via prescription, use of outdated antibiotics, etc.) contributes to emergence of AMR in places where this practice is reported [185].

To understand how AMR evolved and what drives dissemination of AMR determinants, it is imperative that we understand the role of AMR markers in the natural niches of microbes and how it impacts an ecosystem. This knowledge could assist with informed design of therapy regimes tailored to minimize emergence and/or evolution of drug resistance. For example, the perturbation of the GIT microbiota following antibiotic exposure and associated dysbiosis of the human microbiota is well described. Notably, the use of metagenomic sequencing provides good insight of how microbial diversity and numbers are changed due to antibiotic exposure [50]. Yet, a paucity of knowledge exists on how antibiotic use selects for HGT in vivo and how it can be prevented [8]. This knowledge is necessary to assist clinical use of antibiotics without increasing risk for AMR.

Importantly, in the clinical context, we now understand that incorrect or prolonged use of antibiotics enriches bacteria-producing AMR determinants [103]. An example is a study which monitored people on long-term antibiotic, where it was shown that antibiotic use resulted in 56% of the patient cohort ($n = 27$) being colonized by multidrug-resistant commensal Enterococci (vancomycin-resistant), MRSA, and carbapenem-resistant *Enterobacterales* [93]. These bacteria are commensals of the human microbiota, and this implies therefore that prolonged exposure enriched for these organisms. On the other hand, evidence-based studies showed that AMR can be diminished in spaces where antimicrobial stewardship is effectively practised [42]. An emerging thought is that the human GIT microbiota itself must be used as part of the strategy to control emergence of AMR [119, 140]. Most certainly, restoration of gastrointestinal microbiota is highly effective to treat *C. difficile* infection (a typical consequence of dysbiosis) [144]; therefore, it is plausible that gut microbes could be used in a similar manner to limit enrichment of known GIT AMR determinants (via limitation of their producers).

Since microbes in shared environments are most likely in close contact (such as in biofilms or faecal matter), more knowledge is also needed on dynamics in microbiota and biofilms. These polymicrobial environments consist of a multitude of genera, species, and strains, varied phenotypes within the same species, and varied sensitivity to antibiotics. Notably, it also differs in the ability to share genetic information, potentially with different stressors inducing HGT. Antibiotic use is one such stressor, but not the only pressure contributing to sharing of genetic information. Like other environments, the human microbiota is a rich reservoir of ARGs [155]. Oral and GIT microbiomes of healthy adults and babies are rich with antimicrobial resistance determinants against tetracycline, with a large diversity dependent on genera and bacterial phyla [48, 68, 163]. These AMR determinants are readily shared between GIT microbiota and represent a problem for therapy.

Apart from antibiotic-sensitive and antibiotic-resistant microbes, current knowledge indicates that some microbes in an ecosystem are dependent on the presence of antibiotics to improve growth [137], while others are antibiotic-sensitive or drug-resistant. Antibiotic dependence, a phenomenon not widely known or studied, refers to bacteria that requires antibiotics to grow, or has improved growth in the presence of antibiotics. First described in a patient on prolonged streptomycin therapy and a *Brucella* isolate with improved growth in the presence of the antibiotic [71], the phenomenon also had been observed for other antibiotics and bacteria [137, 143]. These antibiotic-dependent microbes are resistant to relevant antibiotics; thus, one should expect enrichment in the host. This is important as overgrowth of some bacteria, usually kept at low numbers by competitors in their microbiota niche, results in disease, for example, *C. difficile* diarrhoea or yeast infections. In the clinical setting, increased numbers of opportunistic pathogens/absence of protective microbes in the microbiota are problematic as it increases patient risk for infection from pathogens that may carry AMR markers. This problem could be exacerbated if antibiotic-dependent opportunistic pathogens carrying other AMR determinants are enriched by antibiotic exposure. Antibiotic dependence had been demonstrated for ESKAPE pathogens, namely *E. coli* and *P. aeruginosa* and *K. pneumoniae*, all showing enhanced growth in the presence of streptomycin and chloramphenicol, respectively [137]. The ability to improve growth when exposed to antibiotics implies that the human microbiota could be enriched for these ESKAPE organisms if growth-promoting antibiotics are ingested. Colistin-dependent *A. baumannii-A. calcoaceticus* complex (ABC) from a patient on extended antibiotic therapy for complicated, antibiotic-resistant diabetic foot infection had also been described; this is a huge concern as this is last-resort drug, mostly used when other treatment options are void due to MDR bacteria [79].

Growth dependence had not been studied in the clinical arena in recent decades, yet improved knowledge of the latter will help with best practice to harness antimicrobials while limiting selection of AMR in our clinical environments. The accelerated growth of one microbiota member will come with cost to the ecosystem and contribute to dysbiosis as the microbial composition, and their biochemical potential (metabolome) will significantly change. Typically, polymicrobial environments drive co-evolution of its members, resulting in a specific metabolome profile of a niche. Antibiotic therapy disturbs this relationship, as a microbial profile change means a change in metabolites, toxins, and AMR markers, with consequences for the specific niche [56]. Therefore, greater understanding of the role of antibiotics and resistance genes in microbiomes (the polymicrobial space shared by microbes) is necessary to counter AMR in the clinical space [43]. At the centre of this, it is important to understand that antibiotics as a mechanism to control bacteria are derived from microbes which predate human use. However, an improved understanding of the human microbiota-containing AMR determinants plus the drivers selected for the evolution of drug resistance in the human microbiota is essential. For example, greater understanding of the mechanisms whereby AMR determinants are spread *in host* and other environments is essential.

7.5 ARG: A Natural, Microbial Defence Against Antimicrobials Produced by Competing Microbes

Antimicrobial compounds are produced by bacteria as a form of biocontrol against microbial competitors in their microenvironment. Most known antimicrobials obtained from microbes had been identified in members of the Actinomycetota (also known as Actinobacteria) [15], and many antibiotics of natural origin (meaning first observed in microbes) are still currently used. These antimicrobials are produced via either a single gene pathway (such as the β-lactamases) or complex pathways (the polyketide pathways and other antimicrobials that are produced as secondary metabolites) [118]. A logical consequence of antimicrobial warfare is the evolution of defence mechanisms against produced antimicrobials; both the antibiotic producer and the target need to protect itself from the action of an antibiotic. In fact, the major antibiotic producer, the genus *Streptomyces*, was the first microbe shown to produce protective factors against the antibiotic it produced. These include a variety of β-lactamases and penicillin-binding proteins, thus proteins involved in cell wall production and modification [131]. The known mechanisms used by microbes as countermeasures include increased expression of genome-encoded factors that limit the entry of antibiotics into the bacterial cell (such as penicillin-binding proteins and efflux pumps), modulation of antimicrobial targets (changing the genetic code e.g. *gyrA*) to evade damage to the target), and production of antibiotic-destroying enzymes (such as β-lactamases and rifamycin-degrading enzymes) [41, 200]. Antibiotic-degrading enzymes are typically encoded either by an ancestral bacterial chromosome or carried on mobilizable genetic elements such as plasmids and transposable elements (transposons, integrons, or gene cassettes) (Table 7.1), while resistance mechanisms countering entry or limit access to a target are usually encoded on the chromosome.

AMR determinants are shared between microbes using mobile genetic determinants as the vehicles for dissemination. This sharing of mobile elements and phages between bacteria is a driver of genetic diversity in microbes, as evidenced by the extensive amount of additional genetic information in bacterial pangenomes. The concept of a pangenome was introduced by Tettelin and coworkers [177] and refers to bacterial evolution through the acquisition of mobilizable genetic material from neighbours. Bacterial core genomes reflect the basic, minimum genetic information contained across a specific species as well as accumulation of mutations (thus vertical evolution). On the other hand, it is now known that the microbial pangenome (core and accessory genome) [177] provides a window on bacterial evolution through the acquisition of mobilizable genetic material. Resistance determinants on the chromosomes could be disseminated via integrated prophages and other mobile elements such as genetic island containing transposons. The excision of prophage DNA from bacteria infected by temperate phages could potentially lead to the inclusion of bacterial DNA during prophage induction, as it gets encapsulated along with the phage DNA. Because of HGT, the microbial pangenome typically contains a variety of transposons, entire mobilizable mobile islands, prophages, and plasmids, driving genetic diversity [26]. In fact, microbial evolution is a continuous dynamic, driven by open pangenomes mostly changing via horizontal exchange between microbes, but also deletion of genetic material from microbes [96]. Two pathogens with a large arsenal of acquired genes in their accessory genomes are *E. coli* and *Clostridium perfringens* [94]. Endogenous infections by opportunistic pathogens like *E. coli* are a known phenomenon that occurs when patient environment/health is conducive to commensals spreading and causing infection outside of its known niche.

7.5.1 Microbes Obtain DNA from the Environment or Directly from Other Microbes

Some microbes are naturally competent where they can take up DNA without chemical induction as done in a laboratory setting. Uptake and natural competence are well-studied in the human pathogen *Haemophilus* and had been shown to be driven by nutrient depletion, for example, if the nucleotide pool is diminished, it

Table 7.1 Characteristics of mobile genetic elements involved in transfer of resistance mechanisms in microbes

Element	Characteristics	Role in spread of resistance genes	Examples and phenotype of specific element	References
Plasmid	Circular, autonomously replicating element; carries genes needed for conjugal DNA transfer	Transfer of resistance genes; mobilization of other elements that carry resistance genes	pBF4, pBFTM10, pBI136-41 kb, 15 kb, 80 kb plasmids that encode high-level MLS resistance conferred by erm genes	[145]
Conjugative transposon	Integrated elements that can excise to form a nonreplicating circular transfer intermediate	Carries genes needed for conjugal DNA transfer	CTnBST-100 kbp conjugative transposon that contains *ermB* gene	[145]
Nonreplicating *Bacteroides* unit (NBU)	Integrated elements that cannot excise and transfer themselves; can be triggered to excise and transfer by conjugative transposons; transfer intermediate is a nonreplicating circle carrying a gene that allows the NBU to take advantage of the conjugal transfer apparatus of a conjugative transposon	Transfer of resistance genes	NBU1—bifunctional protein that binds to the *oriT* region and catalyses the nicking reaction that initiates the transfer process	[106, 145]
Transposon	Can move from one DNA segment to another within the same cell. Very broad host range	Can carry resistance genes from chromosome to plasmid or vice versa	Tn*21*—complex transposon encoding resistance to streptomycin, spectinomycin, sulphonamides, and mercuric ions	[18]
Gene cassette	Circular, nonreplicating DNA segments containing only open reading frames which integrate into integrons	Carry resistance genes		[132]
Integron	Genetic elements that capture promoterless gene cassettes at an attachment site downstream of a promoter, by action of the integrase encoded on the integron; frequently associated with transposons and conjugative plasmids	Forms clusters of resistance genes, all under the control of the integron promoter	Class 1 integrons associated with a variety of resistance gene cassettes, e.g. *aadA* resistance determinant, encoding streptomycin-spectinomycin resistance	[80, 49]

spikes natural competency to obtain relevant material from the environment [147]. The ability to take up DNA has been described for various microbes of clinical relevance. The most widely known example is *A. baumannii*, one of the ESKAPE pathogens. Other pathogens where this was demonstrated include the genera *Burkholderia* [130], *Haemophilus* [40, 114], and *Vibrio* [109, 197]. The messenger molecule cyclic AMP is required for uptake of plasmids in natural competent microbes such as *Haemophilus* [193]. As reviewed by McDonough and coworkers, 'cAMP is widely used by microbial pathogens and their mammalian hosts to regulate cellular processes in response to environmental signals, providing numerous opportunities for pathogens to manipulate their host environments during infection' [121]. Bacteria require the relevant receptor, catabolite activator protein (CRP) as well as the CRP-S (cAMP-CRP complex binding site) promoter where relevant inducer molecule can affect transcription and production of genes involved in the uptake of DNA. Natural competency and DNA uptake are well described in naturally competent human pathogens such as *H. influenza* and *H. parasuis* [40, 114, 193]. Interestingly, while *M. tuberculosis* drug resistance is driven by chromosomal mutations, cyclic AMP has recently been shown to exhibit reduced tolerance to ethambutol, D-cycloserine, and vancomycin (antibiotics targeting the cell wall) when intracellular levels of these messenger molecules were reduced [178].

Both single- and double-stranded DNA are taken up by bacterial cells [175]. Once the genetic material is absorbed into a microbial cell, it either is used as a source of nutrients or incorporated into the genetic makeup (genome) of an organism. Therefore, it is important that we understand how cAMP in the human microbiota is impacted by abiotic exposures, and specifically if/how it contributes to horizontal exchange of genetic material.

7.5.2 Plasmids

Plasmids are circular (with some exceptions in *Streptomyces* [92]), self-replicating pieces of DNA that are extrachromosomal matter inside microbial cells. This extrachromosomal DNA confers various traits, such as virulence and antibiotic resistance factors, not encoded by an ancestral genome to the host microbe. Contrary to popular opinion that antibiotic use is selected for drug resistance and associated AMR markers, the presence of bacteria with self-transmissible, conjugative plasmids (referred to as resistance or R plasmids in older literature) carrying AMR markers had been described in microbes in natural environments [45]. More recent work, using metagenomic analysis, clearly shows that antibiotic resistance genes carried on self-mobilizable plasmids and other mobile elements are present in natural environments, including the mammalian microbiome, and various metagenomic analyses confirm its presence in all ecosystems [84, 136].

Mobilizable plasmids contribute to the dissemination of genes, both intraspecies and between genera. Plasmids are transferred through a conjugative structure called a pilus, associated with the type IV secretion system of bacteria. Extensively reviewed in the literature since its first description in the previous century [124], in laboratory experiments, plasmids can be transferred to antibiotic-sensitive recipient hosts by chemically treating host cells [37], but in nature competency of microbes to take up plasmids is not widely described/understood.

Environments outside of the human body, such as the marine environment, soil, and water, are known to have low numbers of organisms harbouring plasmids encoding resistance to various antibiotics [168]. Evidence for the exchange of AMR-bearing plasmids from marine bacteria to various *E. coli* isolates was first provided in the seminal work of Sizemore and Colwell [168]. Interestingly, in the mentioned study, more antibiotic-resistant marine bacteria were obtained from water than sediment samples. The observation that soil sediment samples contain less AMR markers had also been found in studies using the latest technology [136]. In the Sizemore and Colwell study, multiple attempts to cure marine isolates from plasmids failed; this is interesting considering the concept that extrachromosomal DNA represents a metabolic burden and microbes lose plasmids in the absence of pressure.

The exchange of plasmids between gastrointestinal bacteria of the *Bacteroidetes* phylum is widely reported, with demonstrations of HGT and associated phenotypes using in vitro and in vivo (animals and humans) models. Exchange of the *Bacteroides* plasmid R plasmid, pBF4 confers clindamycin resistance to recipient host, and this plasmid is successfully shared between different species from the genus *Bacteroides* [120]. Furthermore, in vitro experiments confirm that interspecies transfer of clindamycin resistance determinants, conferred by *erm* genes, is shared between gastrointestinal *B. ovatus* and *B. thetaiotaomicron* [112]. AMR factors via plasmids include inhibitor-resistant β-lactamases, 16S rRNA methyltransferases (targeting aminoglycosides) [81], β-lactamases such as the TEM-1-bearing plasmids shared in *Enterobacterales,* and metronidazole resistance-conferring *nim* genes [70, 111, 180]. In vivo exchange of plasmids is plausible and had been illustrated in the anaerobic, spore-forming microorganism *C. perfringens*. This microbe is recognized for its proclivity to readily exchange plasmids, even when some of these plasmids exceed 150 kb in size. In the research conducted by Lacy and colleagues, it was observed that a plasmid was transferred between two strains of *C. perfringens* within the GIT of chickens. Importantly, it is worth noting that no antibiotic pressure was applied in this study, suggesting that this organism either possesses alternative mechanisms driving horizontal gene transfer or has an inherent propensity for effortless plasmid sharing [101].

The spread of antibiotic resistance genes via plasmids is problematic for various organisms currently on the WHO list of bacterial pathogen watchlist. Among CREs, transfer of plasmids with resistance genes is a major concern [97]. Treatment is further complicated if these ESBLs bacteria carry resistance markers such as TEM1 (a class A β-lactamase) on plasmids. This gene and its variants contain mutations that exclude the use of lactamase inhibitors; such complex β-lactamases which contain properties of ESBLs and inhibitor resistance in one gene are widely described.

These complex mutant TEM β-lactamases (CMT), first described by Sirot [167], differ from ancestral TEM-1 (*blaT-1B* gene) through several amino acid substitutions that resembles the IRT-4 TEM-1 variant, plus mutations found in TEM-15, TEM-4, and TEM-3 extended TEM-1 lactamases. Of concern is that the MIC conferred against amoxycillin-clavulanic acid was lower than conferred by IRT-4 alone.

The evolution of TEM-1 into a variety of inhibitor-resistant variants is well described; other CTMs had been described in the literature. Even mutations in non-catalytic regions of the TEM-1 protein were shown to convert variants into ESBLs through conformational changes that increase the active site size, enabling the accommodation of the larger third-generation cephalosporin antibiotics (such as ceftazidime) [32]. The large variety of known *Enterobacterales* plasmids contributing to the multidrug-resistant problem of bacteria in this group has been reviewed and shows that these plasmids are diverse in size and other features such as their origin of replication and the *Mob* factors [154]. Plasmids harbouring genes coding for multiple resistance markers also had been described. A major concern is the emergence of multidrug-resistant ESKAPE organisms harbouring such plasmids. Notably, plasmids encoding for MDR including the last-resort drug tigecycline had been described in *K. pneumoniae*, with resistance to tigecycline encoded by *tet*(X3), *tet*(X4), *tet*(X5), and *tet*(X6). One strain described strain harboured a plasmid with resistance encoded by *bla* NDM-1 and *bla* KPC-2 and *tmexCD2-toprJ2* [108]. Plasmid dissemination in various environments is not well-studied, yet such knowledge is necessary to effectively combat AMR dissemination in vivo and other environments [175].

Antimicrobial markers could also be disseminated via mobile elements present either on the bacterial chromosome or plasmids. Mobile elements such as insertion sequences (IS) contribute to bacterial evolution as it duplicates in a genome (leading to altered genotype and phenotypes), as well as spread between bacteria (thereby disseminating DNA between different bacterial cells) [165]. This is highly relevant for commensal microbes carried in humans and animals. For example, among anaerobic bacteria, dissemination of resistance markers is mainly facilitated via conjugative transposons and large mobile genetic elements. These elements mainly disseminate genes coding for resistance against macrolides (e.g. *ermF*), metronidazole (e.g. *nimA-K*), and tetracycline (*tet*) via conjugative and mobilizable elements [106, 155, 156]. In the case of the anaerobe bacterium *Bacteroides fragilis*, the presence of the *nim* genes (known variants A-K) does not equate to resistance to the drug nitroimidazole. Expression of this gene, which is carried on plasmids or the chromosome, is dependent on the presence of an insertion element upstream of the gene. These transposable elements, such as IS1186 and IS612 [162], contain a sequence resembling the typical promoter found in *Bacteroides*; therefore, the expression of *nim* is only evident when these IS elements are adjacent to the gene. Individuals harbouring *nim*-positive, metronidazole-sensitive *B. fragilis* isolates are at an increased risk of developing metronidazole resistance. Prolonged exposure to metronidazole can exert evolutionary pressure, leading to recombination events that result in strains with IS elements located upstream of the *nim* gene. This genetic alteration ultimately confers metronidazole resistance. Similarly, the gene (*cfiA*) encoding for carbapenem resistance is also dependent on IS for expression [70, 142, 169].

Metronidazole resistance in *B. fragilis* can also be induced by in vitro exposure of the organisms to sub-lethal doses of the drug, in both *nim*-negative and *nim*-positive isolates [65, 161]. Exposure to this drug also resulted in altered metabolic pathways such as those related to oxidative stress, iron metabolism, and DNA repair. Interestingly, the presence/absence of the *nim* gene appears to influence transcriptional responses to metronidazole and the level of observed resistance [47, 100, 138]. These altered metabolic pathways are potentially responsible for the slow growth observed in metronidazole-resistant *B. fragilis* isolates [65].

Notably, screening of wastewater is to be a reservoir of AMR markers coding for resistance to aminoglycosides, beta-lactams, tetracycline, phenicol, and macrolides. Apart from aerobic microbes, the biggest contributor of AMR markers and mobile elements observed were from *Bacteroides* and *Prevotella*, indicating that human and animal faeces can seed the environment with these microbes, from where it can disseminate [173]. Similar mechanisms of resistance, driven by genome alteration mobile elements, had recently been described for the facultative anaerobic gut commensal (and opportunistic pathogen) *Enterococcus* [181]. In this study, blood cultures of *E. faecalis* and *E. faecium* isolates from a patient on extended antibiotic therapy were analysed, revealing evidence of plasmid and interspecies genome recombination events driven by IS. This highlights the potential evolutionary pressure exerted in vivo on commensal microbes via prolonged antibiotic exposure. It further suggests that mobile elements such as plasmids, transposons, and IS elements present in host-microbiota should be studied to understand pressures driving AMR evolution in the host environment.

7.5.3 Phage as Vectors Disseminating AMR Markers

The usefulness of bacteriophages was first described by D'Herelle [63], noting its potential to treat bacterial infections; despite these observations, antibiotics as mode of therapy were adopted across the Western World. However, predation by phage shapes microbial community composition and phenotypic trends in an ecosystem. This is due to lytic phage either destroying the host they infected, or contributing new information to a host by providing genetic material from the previous host (via temperate phage whose DNA integrate into the bacterial chromosome) to new cells it infects.

Some traits acquired by acquisition of phage are as follows:

1. Virulence factors, such as those found disease causing *Vibrio cholerae* 01 and 0139 strains. These strains produce the binary toxin ToxAB that elicits disease development. However, the *CtxAB* locus originates from infecting CTXΦ cholera phage. It is postulated that phage infection occurs in the marine environment,

the natural habitat of non-toxinogenic *V. cholerae* [44, 190]. Environmental factors such as ultraviolet exposure were shown to increase CTXΦ infection.
2. Modulation of microbial transcriptomes and proteomes, leading to modified phenotypes. *Bacillus subtilis* infected by lytic phage ϕ29 produced [125] exhibited altered transcriptome patterns, with most of the repressed transcripts from genes in carbon utilization. Another example is siphovirus p2-infected *Lactococcus lactis subsp. Cremoris MG1363* [104]; infected cells produced a set of proteins unique to infected cells, with >200 bacterial proteins differentially produced after phage infection.
3. Altered sensitivity to substances such as antibiotics [59].

Although phage therapy had been shunned by most of the global community following the discovery of antibiotics, it is now re-emerging as an attractive therapy option due to the global antibiotic resistance crisis [57]. Notably, lytic phage had been shown to successfully treat infection with resistant bacteria (such as carbapenem-resistant *Enterobacterales* [55], and diarrhoea in animals [116]), while its ability to modulate the gastrointestinal community was shown useful to improve disease profiles in at least one gastrointestinal disease [83]. At least one study [55] demonstrated the efficacy of combined phage therapy and antibiotic therapy to treat a complicated case of infection of a bomb victim with an infection persisting for longer than 700 days. In this case, a combination of antibiotic therapy and phage vB_KpnM_M1 (a myovirus originally isolated from sewage) was used to target pan-drug-resistant *K. pneumoniae*, one of multiple microbes cultured from the bomb victim.

The successful use of phage hydrolyses targeting the bacterial cell (endolysins) is another avenue being explored for phage-based antibacterial therapy [139]. Unfortunately, bacteriophages are known to disseminate bacterial genes between bacteria during the repeated cycles of infection, lysogeny, prophage induction and exit of current host by lysis. Newly packaged phage DNA, after excision from the bacterial chromosome, had been shown to contain bacterial chromosomal material. Furthermore, the use of phage is aimed at treating antibiotic-resistant infections or infections characterized by biofilm formation [58]. Both these scenarios represent the risk of horizontal transfer between microbes, or uptake of bacterial DNA released during lysis of phage-infected cells. This risk is enhanced in a polymicrobial environment. Typically, a biofilm might consist of a polymicrobial ecosystem, consisting of microbes from different phyla, genera, species, or even strains. It is thus feasible that microbial DNA might be shared by microbes living in proximity [115, 123]. HGT between organisms living in biofilms in aquatic environments is hypothesized to play a major role for dissemination of drug resistance determinants [2]. At least one study provides evidence of HGT between marine organisms living in an algal biofilm [170]. These could be via plasmids and phage released from bacteria into the surrounding biofilm environment.

The impact of phage on microbial sensitivity to antimicrobials had been reported by various studies. Notably, one study noted that exposure of antibiotic-sensitive *A. baumannii* to extra-cellular DNA of another *A. baumannii* strain known to carry

AMR genes (*armA*, (aminoglycoside resistance), *bla*TEM-1 (lactam resistance), and *tet(B)* (tetracycline resistance) leads to acquisition of resistance by the formerly sensitive strain. The presence of phage-like particles was observed and shown to be the vectors spreading the AMR phenotype [189].

7.6 Conclusion

AMR is a multifaceted issue that arises from the spread of antimicrobial-resistant strains and mobile genetic elements. Although AMR development is a natural phenomenon which existed for millennia, anthropogenic activities have accelerated occurrence of resistant isolates detected. Globally, strains of multidrug-resistant organisms have increased exponentially and we now sit with a problem where few therapeutic options remain available to treat infections caused by AMR organisms. This has resulted in negative patient outcomes, such as longer hospital stays, and higher morbidity and mortality. What may be less apparent, and not as extensively investigated, are the intricate challenges involved in the transmission of AMR between wild animals and their natural habitats, as well as the emergence of genes encoding novel AMR. Unlike the selective pressure exerted by anthropogenic factors, this phenomenon arises from uncontrolled and complex processes within wildlife populations. To further compound this issue, climate change and alterations in wildlife behaviours (such as residing near urban areas and sedentary tendencies observed in some migratory birds) are likely to exert significant influences on the future evolution of antimicrobial resistance. It is therefore reasonable to argue that appropriate preventative measures and risk management strategies in the fight against AMR should include knowledge of all contributing factors, combined with surveillance of environments for the presence of pre-existing AMR determinants. The complex interplay of environmental reservoirs, human activities, genetic exchange mechanisms, and cutting-edge technologies underscores the need for a multidisciplinary approach to tackle AMR. By gaining a deeper understanding of these factors and their implications, we can better address the growing threat of antimicrobial resistance and safeguard public health.

References

1. Abbassi MS, Badi S, Lengliz S, Mansouri R, Salah H, Hynds P. Hiding in plain sight—wildlife as a neglected reservoir and pathway for the spread of antimicrobial resistance: a narrative review. FEMS Microbiol Ecol. 2022;98:fiac045.
2. Abe K, Nomura N, Suzuki S. Biofilms: hot spots of horizontal gene transfer (HGT) in aquatic environments, with a focus on a new HGT mechanism. FEMS Microbiol Ecol. 2020;96:fiaa031.

3. Ahmad A, Ghosh A, Schal C, Zurek L. Insects in confined swine operations carry a large antibiotic resistant and potentially virulent enterococcal community. BMC Microbiol. 2011;11:23.
4. Alderton I, Palmer BR, Heinemann JA, Pattis I, Weaver L, Gutiérrez-Ginés MJ, Horswell J, Tremblay LA. The role of emerging organic contaminants in the development of antimicrobial resistance. Emerg Contamin. 2021;7:160–71.
5. Algarni S, Gudeta DD, Han J, Nayak R, Foley SL. Genotypic analyses of IncHI2 plasmids from enteric bacteria. Sci Rep. 2024;14:9802.
6. Anacarso I, Iseppi R, Sabia C, Messi P, Condò C, Bondi M, De Niederhäusern S. Conjugation-mediated transfer of antibiotic-resistance plasmids between Enterobacteriaceae in the digestive tract of Blaberus craniifer (Blattodea: Blaberidae). J Med Entomol. 2016;53:591–7.
7. Andrade FF, Silva D, Rodrigues A, Pina-Vaz C. Colistin update on its mechanism of action and resistance, present and future challenges. Microorganisms. 2020;8:1716.
8. Andremont A, Cervesi J, Bandinelli P-A, Vitry F, De Gunzburg J. Spare and repair the gut microbiota from antibiotic-induced dysbiosis: state-of-the-art. Drug Discov Today. 2021;26:2159–63.
9. Atashgahi S, Sánchez-Andrea I, Heipieper HJ, Van Der Meer JR, Stams AJ, Smidt H. Prospects for harnessing biocide resistance for bioremediation and detoxification. Science. 2018;360:743–6.
10. Atterby C, Ramey AM, Hall GG, Järhult J, Börjesson S, Bonnedahl J. Increased prevalence of antibiotic-resistant E. coli in gulls sampled in Southcentral Alaska is associated with urban environments. Infect Ecol Epidemiol. 2016;6:32334.
11. Bailey JK, Pinyon JL, Anantham S, Hall RM. Commensal Escherichia Coli of healthy humans: a reservoir for antibiotic-resistance determinants. J Med Microbiol. 2010;59:1331.
12. Banerjee A, Batabyal K, Singh AD, Joardar SN, Dey S, Isore DP, Sar TK, Dutta TK, Bandyopadhyay S, Samanta I. Multi-drug resistant, biofilm-producing high-risk clonal lineage of Klebsiella in companion and household animals. Lett Appl Microbiol. 2020;71:580–7.
13. Baquero F, Martínez JL, Lanza VF, Rodríguez-Beltrán J, Galán JC, San Millán A, Cantón R, Coque TM. Evolutionary pathways and trajectories in antibiotic resistance. Clin Microbiol Rev. 2021;34:e0005019.
14. Barancheshme F, Munir M. Strategies to combat antibiotic resistance in the wastewater treatment plants. Front Microbiol. 2017;8:2603.
15. Barka EA, Vatsa P, Sanchez L, Gaveau-Vaillant N, Jacquard C, Meier-Kolthoff JP, Klenk HP, Clément C, Ouhdouch Y, Van Wezel GP. Taxonomy, physiology, and natural products of Actinobacteria. Microbiol Mol Biol Rev. 2016;80:1–43.
16. Bengtsson-Palme J, Kristiansson E, Larsson DGJ. Environmental factors influencing the development and spread of antibiotic resistance. FEMS Microbiol Rev. 2018;42:fux053.
17. Benjamin L, Atwill ER, Jay-Russell M, Cooley M, Carychao D, Gorski L, Mandrell RE. Occurrence of generic Escherichia coli, E. coli O157 and Salmonella spp. in water and sediment from leafy green produce farms and streams on the Central California coast. Int J Food Microbiol. 2013;165:65–76.
18. Bennett PM. Plasmid encoded antibiotic resistance: acquisition and transfer of antibiotic resistance genes in bacteria. Br J Pharmacol. 2008;153:S347–57.
19. Blasi MF, Migliore L, Mattei D, Rotini A, Thaller MC, Alduina R. Antibiotic resistance of gram-negative bacteria from wild captured loggerhead sea turtles. Antibiotics. 2020;9:162.
20. Boiocchi F, Davies MP, Hilton AC. An examination of flying insects in seven hospitals in the United Kingdom and carriage of bacteria by true flies (Diptera: Calliphoridae, Dolichopodidae, Fanniidae, Muscidae, Phoridae, Psychodidae, Sphaeroceridae). J Med Entomol. 2019;56:1684–97.
21. Bojarski B, Kot B, Witeska M. Antibacterials in aquatic environment and their toxicity to fish. Pharmaceuticals. 2020;13:189.
22. Bonnedahl J, Järhult JD. Antibiotic resistance in wild birds. Ups J Med Sci. 2014;119:113.

23. Bornbusch SL, Harris RL, Grebe NM, Roche K, Dimac-Stohl K, Drea CM. Antibiotics and fecal transfaunation differentially affect microbiota recovery, associations, and antibiotic resistance in lemur guts. Animal Microbiome. 2021;3:1–15.
24. Boyanova L, Kolarov R, Mitov I. Recent evolution of antibiotic resistance in the anaerobes as compared to previous decades. Anaerobe. 2015;31:4–10.
25. Brahmi S, Touati A, Dunyach-Remy C, Sotto A, Pantel A, Lavigne JP. High prevalence of extended-Spectrum β-Lactamase-producing Enterobacteriaceae in wild fish from the Mediterranean Sea in Algeria. Microb Drug Resist. 2018;24:290–8.
26. Brockhurst MA, Harrison E, Hall JPJ, Richards T, Mcnally A, Maclean C. The ecology and evolution of Pangenomes. Curr Biol. 2019;29:R1094–103.
27. Buffet-Bataillon S, Tattevin P, Maillard J-Y, Bonnaure-Mallet M, Jolivet-Gougeon A. Efflux pump induction by quaternary ammonium compounds and fluoroquinolone resistance in bacteria. Future Microbiol. 2016;11:81–92.
28. Cabello F, Tomova A, Ivanova L, Godfrey H. Aquaculture and mcr colistin resistance determinants. MBio. 2017;8:e01229–17.
29. Cabello FC, Godfrey HP. Aquaculture, exaptation, and the origin of mcr-positive Colistin resistance. Antimicrob Agents Chemother. 2018;62:e01903.
30. Cacace D, Fatta-Kassinos D, Manaia CM, Cytryn E, Kreuzinger N, Rizzo L, Karaolia P, Schwartz T, Alexander J, Merlin C, Garelick H, Schmitt H, De Vries D, Schwermer CU, Meric S, Ozkal CB, Pons M-N, Kneis D, Berendonk TU. Antibiotic resistance genes in treated wastewater and in the receiving water bodies: a pan-European survey of urban settings. Water Res. 2019;162:320–30.
31. Caneschi A, Bardhi A, Barbarossa A, Zaghini A. The use of antibiotics and antimicrobial resistance in veterinary medicine, a complex phenomenon: a narrative review. Antibiotics. 2023;12:487.
32. Cao TP, Yi H, Dhanasingh I, Ghosh S, Choi JM, Lee KH, Ryu S, Kim HS, Lee SH. Non-catalytic-region mutations conferring transition of class A β-Lactamases into ESBLs. Front Mol Biosci. 2020;7:598998.
33. Chen Q, Li D, Beiersmann C, Neuhann F, Moazen B, Lu G, Müller O. Risk factors for antibiotic resistance development in healthcare settings in China: a systematic review. Epidemiol Infect. 2021;149:e141.
34. Chong C, Alkatheeri A, Ali N, Tay Z, Lee Y, Paramasivam S, Jeevaratnam K, Low W, Lim S. Association of antimicrobial resistance and gut microbiota composition in human and non-human primates at an urban ecotourism site. Gut Pathogens. 2020;12:1–12.
35. Chung DM, Ferree ED, Simon DM, Yeh PJ. Patterns of Bird–Bacteria Associations. EcoHealth. 2018;15(3):627.
36. Clarke BO, Smith SR. Review of 'emerging' organic contaminants in biosolids and assessment of international research priorities for the agricultural use of biosolids. Environ Int. 2011;37:226–47.
37. Cohen SN, Chang AC, Hsu L. Nonchromosomal antibiotic resistance in bacteria: genetic transformation of Escherichia coli by R-factor DNA. Proc Natl Acad Sci USA. 1972;69:2110–4.
38. Cycoń M, Mrozik A, Piotrowska-Seget Z. Antibiotics in the soil environment—degradation and their impact on microbial activity and diversity. Front Microbiol. 2019;10:338.
39. D'costa VM, Mcgrann KM, Hughes DW, Wright GD. Sampling the antibiotic resistome. Science. 2006;311:374–7.
40. Dai K, He L, Chang YF, Cao S, Zhao Q, Huang X, Wu R, Huang Y, Yan Q, Han X, Ma X, Wen X, Wen Y. Basic characterization of natural transformation in a highly transformable haemophilus parasuis Strain SC1401. Front Cell Infect Microbiol. 2018;8:32.
41. Darby EM, Trampari E, Siasat P, Gaya MS, Alav I, Webber MA, Blair JMA. Molecular mechanisms of antibiotic resistance revisited. Nat Rev Microbiol. 2023;21:280–95.
42. Darwish RM, Matar SG, Snaineh AAA, Alsharif MR, Yahia AB, Mustafa HN, Hasabo EA. Impact of antimicrobial stewardship on antibiogram, consumption and incidence of multi drug resistance. BMC Infect Dis. 2022;22:916.

43. Davies J, Davies D. Origins and evolution of antibiotic resistance. Microbiol Mol Biol Rev. 2010;74:417–33.
44. Davis BM, Kimsey HH, Chang W, Waldor MK. The Vibrio cholerae O139 Calcutta bacteriophage CTXphi is infectious and encodes a novel repressor. J Bacteriol. 1999;181:6779–87.
45. Davis CE, Anandan J. The evolution of r factor. A study of a "preantibiotic" community in Borneo. N Engl J Med. 1970;282:117–22.
46. De Briyne N, Atkinson J, Borriello SP, Pokludová L. Antibiotics used most commonly to treat animals in Europe. Vet Rec. 2014;175:325.
47. De Freitas MC, Resende JA, Ferreira-Machado AB, Saji GD, De Vasconcelos AT, Da Silva VL, Nicolás MF, Diniz CG. Exploratory investigation of Bacteroides fragilis transcriptional response during in vitro exposure to subinhibitory concentration of Metronidazole. Front Microbiol. 2016;7:1465.
48. De Vries LE, Vallès Y, Agersø Y, Vaishampayan PA, García-Montaner A, Kuehl JV, Christensen H, Barlow M, Francino MP. The gut as reservoir of antibiotic resistance: microbial diversity of tetracycline resistance in mother and infant. PLoS One. 2011;6:e21644.
49. Deng Y, Bao X, Ji L, Chen L, Liu J, Miao J, Chen D, Bian H, Li Y, Yu G. Resistance integrons: class 1, 2 and 3 integrons. Ann Clin Microbiol Antimicrob. 2015;14:45.
50. Dethlefsen L, Huse S, Sogin ML, Relman DA. The pervasive effects of an antibiotic on the human gut microbiota, as revealed by deep 16S rRNA sequencing. PLoS Biol. 2008;6:e280.
51. Diwan V, Tamhankar AJ, Khandal RK, Sen S, Aggarwal M, Marothi Y, Iyer RV, Sundblad-Tonderski K, Stålsby-Lundborg C. Antibiotics and antibiotic-resistant bacteria in waters associated with a hospital in Ujjain, India. BMC Public Health. 2010;10:1–8.
52. ECDC&WHO. Antimicrobial resistance surveillance in Europe 2023–2021 data. Stockholm: European Centre for Disease Prevention and Control and World Health Organization; 2023.
53. Ekwanzala MD, Dewar JB, Kamika I, Momba MNB. Systematic review in South Africa reveals antibiotic resistance genes shared between clinical and environmental settings. Infect Drug Resist. 2018;11:1907–20.
54. Elhadi N. Prevalence of extended-spectrum-β-lactamase-producing Escherichia coli in imported frozen freshwater fish in Eastern Province of Saudi Arabia. Saudi J Med Med Sci. 2016;4:19.
55. Eskenazi A, Lood C, Wubbolts J, Hites M, Balarjishvili N, Leshkasheli L, Askilashvili L, Kvachadze L, Van Noort V, Wagemans J, Jayankura M, Chanishvili N, De Boer M, Nibbering P, Kutateladze M, Lavigne R, Merabishvili M, Pirnay J-P. Combination of pre-adapted bacteriophage therapy and antibiotics for treatment of fracture-related infection due to pandrug-resistant Klebsiella pneumoniae. Nat Commun. 2022;13:302.
56. Evans R, Beckerman AP, Wright RCT, Mcqueen-Mason S, Bruce NC, Brockhurst MA. Eco-evolutionary dynamics set the tempo and trajectory of metabolic evolution in multispecies communities. Curr Biol. 2020;30:4984–4988.e4.
57. Federici S, Nobs SP, Elinav E. Phages and their potential to modulate the microbiome and immunity. Cell Mol Immunol. 2021;18:889–904.
58. Fernández L, Gutiérrez D, García P, Rodríguez A. The perfect bacteriophage for therapeutic applications-a quick guide. Antibiotics (Basel). 2019;8:126.
59. Fernández L, Rodríguez A, García P. Phage or foe: an insight into the impact of viral predation on microbial communities. ISME J. 2018;12:1171–9.
60. Fischer J, Schmoger S, Jahn S, Helmuth R, Guerra B. Ndm-1 carbapenemase-producing Salmonella enterica subsp. enterica serovar Corvallis isolated from a wild bird in Germany. J Antimicrob Chemother. 2013;68:2954–6.
61. Ford C, Yusim K, Ioerger T, Feng S, Chase M, Greene M, Korber B, Fortune S. Mycobacterium tuberculosis – Heterogeneity revealed through whole genome sequencing. Tuberculosis. 2012;92:194–201.
62. Founou LL, Founou RC, Essack SY. Antibiotic resistance in the food chain: a developing country-perspective. Front Microbiol. 2016;7:1881.

63. Fruciano DE, Bourne S. Phage as an antimicrobial agent: d'Herelle's heretical theories and their role in the decline of phage prophylaxis in the West. Can J Infect Dis Med Microbiol. 2007;18:19–26.
64. Fukuda A, Usui M, Okamura M, Dong-Liang H, Tamura Y. Role of flies in the maintenance of antimicrobial resistance in farm environments. Microb Drug Resist. 2019;25:127–32.
65. Gal M, Brazier JS. Metronidazole resistance in Bacteroides spp. carrying nim genes and the selection of slow-growing metronidazole-resistant mutants. J Antimicrob Chemother. 2004;54:109–16.
66. García J, García-Galán MJ, Day JW, Boopathy R, White JR, Wallace S, Hunter RG. A review of emerging organic contaminants (EOCs), antibiotic resistant bacteria (ARB), and antibiotic resistance genes (ARGs) in the environment: increasing removal with wetlands and reducing environmental impacts. Bioresour Technol. 2020;307:123228.
67. Granato ET, Meiller-Legrand TA, Foster KR. The evolution and ecology of bacterial warfare. Curr Biol. 2019;29:R521–37.
68. Gueimonde M, Salminen S, Isolauri E. Presence of specific antibiotic (tet) resistance genes in infant faecal microbiota. Fems Immunol Med Microbiol. 2006;48:21–5.
69. Gwenzi W, Chaukura N, Muisa-Zikali N, Teta C, Musvuugwa T, Rzymski P, Abia ALK. Insects, rodents, and pets as reservoirs, vectors, and sentinels of antimicrobial resistance. Antibiotics (Basel). 2021;10:68.
70. Haggoud A, Reysset G, Azeddoug H, Sebald M. Nucleotide sequence analysis of two 5-nitroimidazole resistance determinants from Bacteroides strains and of a new insertion sequence upstream of the two genes. Antimicrob Agents Chemother. 1994;38:1047–51.
71. Hall WH, Spink WW. In vitro sensitivity of Brucella to Streptomycin:development of resistance during Streptomycin treatment.∗. Proc Soc Exp Biol Med. 1947;64:403–6.
72. Hammerum AM, Heuer OE, Emborg HD, Bagger-Skjot L, Jensen VF, Rogues AM, Skov RL, Agerso Y, Brandt CT, Seyfarth AM, Muller A, Hovgaard K, Ajufo J, Bager F, Aarestrup FM, Frimodt-Moller N, Wegener HC, Monnet DL. Danish integrated antimicrobial resistance monitoring and research program. Emerg Infect Dis. 2007;13:1632–9.
73. Hand K. Antibiotic stewardship. Clin Med. 2013;13:499.
74. Harding-Crooks R, Smith D, Fanning S, Fox EM. Dissemination of carbapenemase-producing Enterobacteriaceae and associated resistance determinants through global food systems. Compr Rev Food Sci Food Saf. 2023;22:2706–27.
75. Hartmann A, Locatelli A, Amoureux L, Depret G, Jolivet C, Gueneau E, Neuwirth C. Occurrence of CTX-M producing Escherichia coli in soils, cattle, and farm environment in France (Burgundy Region). Front Microbiol. 2012;3:83.
76. Hasan B, Melhus Å, Sandegren L, Alam M, Olsen B. The gull (Chroicocephalus brunnicephalus) as an environmental bioindicator and reservoir for antibiotic resistance on the coastlines of the Bay of Bengal. Microb Drug Resist. 2014;20:466–71.
77. Hassan J, Eddine RZ, Mann D, Li S, Deng X, Saoud IP, Kassem II. The mobile colistin resistance gene, mcr-1.1, is carried on IncX4 plasmids in multidrug resistant E. coli isolated from rainbow trout aquaculture. Microorganisms. 2020;8:1636.
78. Hassen B, Jouini A, Elbour M, Hamrouni S, Maaroufi A. Detection of extended-spectrum β-lactamases (ESBL) producing enterobacteriaceae from fish trapped in the lagoon area of Bizerte, Tunisia. Biomed Res Int. 2020;2020:7132812.
79. Hawley JS, Murray CK, Jorgensen JH. Development of colistin-dependent Acinetobacter baumannii-Acinetobacter calcoaceticus complex. Antimicrob Agents Chemother. 2007;51:4529–30.
80. Heuer H, Smalla K. Horizontal gene transfer between bacteria. Environ Biosaf Res. 2007;6:3–13.
81. Hidalgo L, Hopkins KL, Gutierrez B, Ovejero CM, Shukla S, Douthwaite S, Prasad KN, Woodford N, Gonzalez-Zorn B. Association of the novel aminoglycoside resistance determinant RmtF with NDM carbapenemase in Enterobacteriaceae isolated in India and the UK. J Antimicrob Chemother. 2013;68:1543–50.

82. Hölzel C, Bauer J. Salmonella spp. in Bavarian liquid pig manure: occurrence and relevance for the distribution of antibiotic resistance. Zoonoses Public Health. 2008;55:133–8.
83. Ichikawa M, Nakamoto N, Kredo-Russo S, Weinstock E, Weiner IN, Khabra E, Ben-Ishai N, Inbar D, Kowalsman N, Mordoch R, Nicenboim J, Golembo M, Zak N, Jablonska J, Sberro-Livnat H, Navok S, Buchshtab N, Suzuki T, Miyamoto K, Teratani T, Fujimori S, Aoto Y, Konda M, Hayashi N, Chu P-S, Taniki N, Morikawa R, Kasuga R, Tabuchi T, Sugimoto S, Mikami Y, Shiota A, Bassan M, Kanai T. Bacteriophage therapy against pathological Klebsiella pneumoniae ameliorates the course of primary sclerosing cholangitis. Nat Commun. 2023;14:3261.
84. Inda-Diaz JS, Lund D, Parras-Molto M, Johnning A, Bengtsson-Palme J, Kristiansson E. Latent antibiotic resistance genes are abundant, diverse, and mobile in human, animal, and environmental microbiomes. Microbiome. 2023;11:44.
85. Jang J, Hur HG, Sadowsky MJ, Byappanahalli MN, Yan T, Ishii S. Environmental *Escherichia coli*: ecology and public health implications - a review. J Appl Microbiol. 2017;123:570.
86. Janik-Karpinska E, Brancaleoni R, Niemcewicz M, Wojtas W, Foco M, Podogrocki M, Bijak M. Healthcare waste-a serious problem for global health. Healthcare (Basel). 2023;11:242.
87. Jutkina J, Marathe NP, Flach CF, Larsson DGJ. Antibiotics and common antibacterial biocides stimulate horizontal transfer of resistance at low concentrations. Sci Total Environ. 2018;616:172–8.
88. Kalli M, Noutsopoulos C, Mamais D. The fate and occurrence of antibiotic-resistant bacteria and antibiotic resistance genes during advanced wastewater treatment and disinfection: a review. Water. 2023;15:2084.
89. Kalová A, Gelbíčová T, Overballe-Petersen S, Litrup E, Karpíšková R. Characterisation of Colistin -resistant Enterobacterales and Acinetobacter strains carrying mcr genes from Asian aquaculture products. Antibiotics (Basel). 2021;10:838.
90. Khan SA, Imtiaz MA, Sayeed MA, Shaikat AH, Hassan MM. Antimicrobial resistance pattern in domestic animal - wildlife - environmental niche via the food chain to humans with a Bangladesh perspective; a systematic review. BMC Vet Res. 2020;16:302.
91. Kim SA, Moon H, Lee K, Rhee MS. Bactericidal effects of triclosan in soap both in vitro and in vivo. J Antimicrob Chemother. 2015;70:3345–52.
92. Kinashi H. Giant linear plasmids in Streptomyces: a treasure trove of antibiotic biosynthetic clusters. J Antibiot. 2011;64:19–25.
93. Kiss C, Connoley D, Connelly K, Horne K, Korman T, Woolley I, Lau JSY. Long-term outcomes in patients on life-long antibiotics: a five-year Cohort study. Antibiotics (Basel). 2022;11:62.
94. Kiu R, Caim S, Alexander S, Pachori P, Hall LJ. Probing genomic aspects of the multi-host pathogen Clostridium perfringens reveals significant Pangenome diversity, and a diverse array of virulence factors. Front Microbiol. 2017;8:2485.
95. Klous G, Huss A, Heederik DJ, Coutinho RA. Human–livestock contacts and their relationship to transmission of zoonotic pathogens, a systematic review of literature. One Health. 2016;2:65–76.
96. Koonin EV, Makarova KS, Wolf YI. Evolution of microbial genomics: conceptual shifts over a quarter century. Trends Microbiol. 2021;29:582–92.
97. Kopotsa K, Osei Sekyere J, Mbelle NM. Plasmid evolution in carbapenemase-producing Enterobacteriaceae: a review. Ann N Y Acad Sci. 2019;1457:61–91.
98. Kümmerer K. Resistance in the environment. J Antimicrob Chemother. 2004;54:311–20.
99. Künzl C, Kaiser S, Meier E, Sachser N. Is a wild mammal kept and reared in captivity still a wild animal? Horm Behav. 2003;43:187–96.
100. Kupc M, Paunkov A, Strasser D, Sóki J, Leitsch D. Initial expression levels of nimA are decisive for protection against metronidazole in Bacteroides fragilis. Anaerobe. 2022;77:102630.
101. Lacey JA, Keyburn AL, Ford ME, Portela RW, Johanesen PA, Lyras D, Moore RJ. Conjugation-mediated horizontal gene transfer of Clostridium perfringens plasmids in the chicken gas-

trointestinal tract results in the formation of new virulent strains. Appl Environ Microbiol. 2017;83:e01814.
102. Lahiri N, Shah RR, Layre E, Young D, Ford C, Murray MB, Fortune SM, Moody DB. Rifampin resistance mutations are associated with broad chemical remodeling of mycobacterium tuberculosis. J Biol Chem. 2016;291:14248–56.
103. Lau JSY, Korman TM, Woolley I. Life-long antimicrobial therapy: where is the evidence? J Antimicrob Chemother. 2018;73:2601–12.
104. Lemay ML, Otto A, Maas S, Plate K, Becher D, Moineau S. Investigating Lactococcus lactis MG1363 response to phage p2 infection at the proteome level. Mol Cell Proteomics. 2019;18:704–14.
105. Levy SB, Marshall B. Antibacterial resistance worldwide: causes, challenges and responses. Nat Med. 2004;10:S122–9.
106. Li LY, Shoemaker NB, Wang GR, Cole SP, Hashimoto MK, Wang J, Salyers AA. The mobilization regions of two integrated Bacteroides elements, NBU1 and NBU2, have only a single mobilization protein and may be on a cassette. J Bacteriol. 1995;177:3940–5.
107. Li S, Zhang C, Li F, Hua T, Zhou Q, Ho S-H. Technologies towards antibiotic resistance genes (ARGs) removal from aquatic environment: a critical review. J Hazard Mater. 2021;411:125148.
108. Li X, Wang W, Jin X, Zhang X, Zou X, Ma Q, Hu Q, Huang H, Tu Y. Emergence of plasmids Co-Harboring Carbapenem resistance genes and tmexCD2-toprJ2 in sequence type 11 Carbapenem resistant Klebsiella pneumoniae strains. Front Cell Infect Microbiol. 2022;12:902774.
109. Lo Scrudato M, Blokesch M. The regulatory network of natural competence and transformation of vibrio cholerae. PLoS Genet. 2012;8:e1002778.
110. Loayza F, Graham JP, Trueba G. Factors obscuring the role of E. coli from domestic animals in the global antimicrobial resistance crisis: an evidence-based review. Int J Environ Res Public Health. 2020;17:3061.
111. Löfmark S, Fang H, Hedberg M, Edlund C. Inducible metronidazole resistance and nim genes in clinical Bacteroides fragilis group isolates. Antimicrob Agents Chemother. 2005;49:1253–6.
112. Löfmark S, Jernberg C, Billström H, Andersson DI, Edlund C. Restored fitness leads to long-term persistence of resistant Bacteroides strains in the human intestine. Anaerobe. 2008;14:157–60.
113. Lundborg CS, Tamhankar AJ. Antibiotic residues in the environment of South East Asia. BMJ. 2017;358:j2440.
114. Macfadyen LP, Ma C, Redfield RJ. A 3′,5′ cyclic AMP (cAMP) phosphodiesterase modulates cAMP levels and optimizes competence in Haemophilus influenzae Rd. J Bacteriol. 1998;180:4401–5.
115. Madsen JS, Burmølle M, Hansen LH, Sørensen SJ. The interconnection between biofilm formation and horizontal gene transfer. FEMS Immunol Med Microbiol. 2012;65:183–95.
116. Mao X, Wu Y, Ma R, Li L, Wang L, Tan Y, Li Z, Liu H, Han K, Cao Y, Li Y, Peng H, Li X, Hu C, Wang X. Oral phage therapy with microencapsulated phage A221 against Escherichia coli infections in weaned piglets. BMC Vet Res. 2023;19:165.
117. Marston HD, Dixon DM, Knisely JM, Palmore TN, Fauci AS. Antimicrobial Resistance. JAMA. 2016;316:1193–204.
118. Mast Y, Stegmann E. Actinomycetes: the antibiotics producers. Antibiotics (Basel). 2019;8:105.
119. Matzaras R, Nikopoulou A, Protonotariou E, Christaki E. Gut microbiota modulation and prevention of Dysbiosis as an alternative approach to antimicrobial resistance: a narrative review. Yale J Biol Med. 2022;95:479–94.
120. Mays TD, Smith CJ, Welch RA, Delfini C, Macrina FL. Novel antibiotic resistance transfer in Bacteroides. Antimicrob Agents Chemother. 1982;21:110–8.

121. Mcdonough KA, Rodriguez A. The myriad roles of cyclic AMP in microbial pathogens: from signal to sword. Nat Rev Microbiol. 2012;10:27–38.
122. Mcewen SA, Collignon PJ. Antimicrobial resistance: a one health perspective. Microbiol Spectr. 2018;6(2):ARBA-0009-2017
123. Michaelis C, Grohmann E. Horizontal gene transfer of antibiotic resistance genes in biofilms. Antibiotics (Basel). 2023;12:328.
124. Mitsuhashi S. The R factors*. J Infect Dis. 1969;119:89–100.
125. Mojardín L, Salas M. Global transcriptional analysis of virus-host interactions between phage ф29 and Bacillus subtilis. J Virol. 2016;90:9293–304.
126. Monecke S, Gavier-Widén D, Hotzel H, Peters M, Guenther S, Lazaris A, Loncaric I, Müller E, Reissig A, Ruppelt-Lorz A, Shore AC, Walter B, Coleman DC, Ehricht R. Diversity of Staphylococcus aureus isolates in European wildlife. PLoS One. 2016;11:e0168433.
127. Muenchhoff M, Healy M, Singh R, Roider J, Groll A, Kindra C, Sibaya T, Moonsamy A, Mcgregor C, Phan MQ, Palma A, Kloverpris H, Leslie A, Bobat R, Larussa P, Ndung'u T, Goulder P, Sobieszczyk ME, Archary M. Malnutrition in HIV-infected children is an indicator of severe disease with an impaired response to antiretroviral therapy. AIDS Res Hum Retrovir. 2018;34:46–55.
128. Narciso-Da-Rocha C, Rocha J, Vaz-Moreira I, Lira F, Tamames J, Henriques I, Martinez JL, Manaia CM. Bacterial lineages putatively associated with the dissemination of antibiotic resistance genes in a full-scale urban wastewater treatment plant. Environ Int. 2018;118:179–88.
129. Niegowska M, Sanseverino I, Navarro A, Lettieri T. Knowledge gaps in the assessment of antimicrobial resistance in surface waters. FEMS Microbiol Ecol. 2021;97:fiab140.
130. Norris MH, Heacock-Kang Y, Zarzycki-Siek J, Bluhm AP, Mcmillan IA, Schweizer HP, Hoang TT. Burkholderia pseudomallei natural competency and DNA catabolism: identification and characterization of relevant genes from a constructed fosmid library. PLoS One. 2017;12:e0189018.
131. Ogawara H. Comparison of antibiotic resistance mechanisms in antibiotic-producing and pathogenic bacteria. Molecules. 2019;24:3430.
132. Okoh AI, Fadare FT. Integrons as emerging contaminants facilitating the widespread of antimicrobial resistance in Enterobacteriaceae. Adv Biomed Health Sci. 2022;1:68.
133. Oravcova V, Janecko N, Ansorge A, Masarikova M, Literak I. First record of vancomycin-resistant E nterococcus faecium in C anadian wildlife. Environ Microbiol Rep. 2014a;6:210–1.
134. Oravcova V, Svec P, Literak I. Vancomycin-resistant enterococci with vanA and vanB genes in Australian gulls. Environ Microbiol Rep. 2017;9:316–8.
135. Oravcova V, Zurek L, Townsend A, Clark AB, Ellis JC, Cizek A, Literak I. A merican crows as carriers of vancomycin-resistant enterococci with vanA gene. Environ Microbiol. 2014b;16:939–49.
136. Pal C, Bengtsson-Palme J, Kristiansson E, Larsson DG. The structure and diversity of human, animal and environmental resistomes. Microbiome. 2016;4:54.
137. Paredes-Amaya CC, Ulloa MT, García-Angulo VA. Fierce poison to others: the phenomenon of bacterial dependence on antibiotics. J Biomed Sci. 2023;30:67.
138. Paunkov A, Hummel K, Strasser D, Sóki J, Leitsch D. Proteomic analysis of metronidazole resistance in the human facultative pathogen Bacteroides fragilis. Front Microbiol. 2023;14:1158086.
139. Petrovic Fabijan A, Iredell J, Danis-Wlodarczyk K, Kebriaei R, Abedon ST. Translating phage therapy into the clinic: recent accomplishments but continuing challenges. PLoS Biol. 2023;21:e3002119.
140. Pilmis B, Le Monnier A, Zahar JR. Gut microbiota, antibiotic therapy and antimicrobial resistance: a narrative review. Microorganisms. 2020;8:269.
141. Pitout JD, Devinney R. Escherichia coli ST131: a multidrug-resistant clone primed for global domination. F1000Res. 2017;6:F1000 Faculty Rev-195.
142. Podglajen I, Breuil J, Rohaut A, Monsempes C, Collatz E. Multiple mobile promoter regions for the rare carbapenem resistance gene of Bacteroides fragilis. J Bacteriol. 2001;183:3531–5.

143. Pournaras S, Ntokou E, Zarkotou O, Ranellou K, Themeli-Digalaki K, Stathopoulos C, Tsakris A. Linezolid dependence in Staphylococcus epidermidis bloodstream isolates. Emerg Infect Dis. 2013;19:129–32.
144. Queen J, Zhang J, Sears CL. Oral antibiotic use and chronic disease: long-term health impact beyond antimicrobial resistance and Clostridioides difficile. Gut Microbes. 2020;11:1092–103.
145. Quesada-Gómez C. Bacteroides mobilizable and conjugative genetic elements: antibiotic resistance among clinical isolates. Rev Esp Quimioter. 2011;24:184–90.
146. Rashid M, Rakib MM, Hasan B. Antimicrobial-resistant and ESBL-producing Escherichia coli in different ecological niches in Bangladesh. Infect Ecol Epidemiol. 2015;5:26712.
147. Redfield RJ, Cameron AD, Qian Q, Hinds J, Ali TR, Kroll JS, Langford PR. A novel CRP-dependent regulon controls expression of competence genes in Haemophilus influenzae. J Mol Biol. 2005;347:735–47.
148. Rhodes A, Evans LE, Alhazzani W, Levy MM, Antonelli M, Ferrer R, Kumar A, Sevransky JE, Sprung CL, Nunnally ME, Rochwerg B, Rubenfeld GD, Angus DC, Annane D, Beale RJ, Bellinghan GJ, Bernard GR, Chiche JD, Coopersmith C, De Backer DP, French CJ, Fujishima S, Gerlach H, Hidalgo JL, Hollenberg SM, Jones AE, Karnad DR, Kleinpell RM, Koh Y, Lisboa TC, Machado FR, Marini JJ, Marshall JC, Mazuski JE, Mcintyre LA, Mclean AS, Mehta S, Moreno RP, Myburgh J, Navalesi P, Nishida O, Osborn TM, Perner A, Plunkett CM, Ranieri M, Schorr CA, Seckel MA, Seymour CW, Shieh L, Shukri KA, Simpson SQ, Singer M, Thompson BT, Townsend SR, Van Der Poll T, Vincent JL, Wiersinga WJ, Zimmerman JL, Dellinger RP. Surviving sepsis campaign: International Guidelines for Management of Sepsis and Septic Shock: 2016. Intensive Care Med. 2017;43:304–77.
149. Rizzo L, Gernjak W, Krzeminski P, Malato S, Mcardell CS, Perez JAS, Schaar H, Fatta-Kassinos D. Best available technologies and treatment trains to address current challenges in urban wastewater reuse for irrigation of crops in Eu countries. Sci Total Environ. 2020;710:136312.
150. Rizzo L, Manaia CM, Merlin C, Schwartz T, Dagot C, Ploy M-C, Michael I, Fatta-Kassinos D. Urban wastewater treatment plants as hotspots for antibiotic resistant bacteria and genes spread into the environment: a review. Sci Total Environ. 2013;447:345.
151. Roberts MC, No DB, Marzluff JM, Delap JH, Turner R. Vancomycin resistant Enterococcus spp. from crows and their environment in metropolitan Washington State, USA: is there a correlation between VRE positive crows and the environment? Vet Microbiol. 2016;194:48–54.
152. Rodgers K, Mclellan I, Peshkur T, Williams R, Tonner R, Hursthouse AS, Knapp CW, Henriquez FL. Can the legacy of industrial pollution influence antimicrobial resistance in estuarine sediments? Environ Chem Lett. 2019;17:595–607.
153. Rousham EK, Unicomb L, Islam MA. Human, animal and environmental contributors to antibiotic resistance in low-resource settings: integrating behavioural, epidemiological and One Health approaches. Proc Biol Sci. 2018;285:20180332.
154. Rozwandowicz M, Brouwer MSM, Fischer J, Wagenaar JA, Gonzalez-Zorn B, Guerra B, Mevius DJ, Hordijk J. Plasmids carrying antimicrobial resistance genes in Enterobacteriaceae. J Antimicrob Chemother. 2018;73:1121–37.
155. Salyers AA, Gupta A, Wang Y. Human intestinal bacteria as reservoirs for antibiotic resistance genes. Trends Microbiol. 2004;12:412–6.
156. Salyers AA, Shoemaker NB. Resistance gene transfer in anaerobes: new insights, new problems. Clin Infect Dis. 1996;23(Suppl 1):S36–43.
157. Samreen AI, Malak HA, Abulreesh HH. Environmental antimicrobial resistance and its drivers: a potential threat to public health. J Global Antimicrob Resist. 2021;27:101–11.
158. Santos ACDM, Santos FD, Silva RM, Gomes TAT. Diversity of hybrid- and hetero-pathogenic *Escherichia coli* and their potential implication in more severe diseases. Front Cell Infect Microbiol. 2020;10:339.
159. Sapugahawatte DN, Li C, Zhu C, Dharmaratne P, Wong KT, Lo N, Ip M. Prevalence and characteristics of extended-spectrum-β-lactamase-producing and carbapenemase-producing

Enterobacteriaceae from freshwater fish and pork in wet markets of Hong Kong. Msphere. 2020;5 https://doi.org/10.1128/msphere.00107-20.
160. Sarowska J, Futoma-Kołoch B, Jama-Kmiecik A, Frej-Mądrzak M, Książczyk M, Bugla-Płoskońska G, Choroszy-Król I. Virulence factors, prevalence and potential transmission of Extraintestinal pathogenic *Escherichia coli* isolated from different sources: recent reports. Gut Pathogens. 2019;11:10.
161. Schaumann R, Petzold S, Fille M, Rodloff AC. Inducible metronidazole resistance in nim-positive and nim-negative bacteroides fragilis group strains after several passages metronidazole containing columbia agar plates. Infection. 2005;33:368–72.
162. Sevillano G, Paz YMA, Solís MB, Vaca JP, Zurita-Salinas C, Zurita J. First report of antibiotic resistance markers cfiA and nim among Bacteroides fragilis group strains in Ecuadorian patients. Microb Drug Resist. 2023;29(11):533.
163. Seville LA, Patterson AJ, Scott KP, Mullany P, Quail MA, Parkhill J, Ready D, Wilson M, Spratt D, Roberts AP. Distribution of tetracycline and erythromycin resistance genes among human oral and fecal metagenomic DNA. Microb Drug Resist. 2009;15:159–66.
164. Sharma VK, Johnson N, Cizmas L, Mcdonald TJ, Kim H. A review of the influence of treatment strategies on antibiotic resistant bacteria and antibiotic resistance genes. Chemosphere. 2016;150:702–14.
165. Siguier P, Gourbeyre E, Chandler M. Bacterial insertion sequences: their genomic impact and diversity. FEMS Microbiol Rev. 2014;38:865–91.
166. Silva V, Igrejas G, Carvalho I, Peixoto F, Cardoso L, Pereira JE, Del Campo R, Poeta P. Genetic characterization of vanA-Enterococcus faecium isolates from wild red-legged partridges in Portugal. Microb Drug Resist. 2018;24:89–94.
167. Sirot D, Recule C, Chaibi EB, Bret L, Croize J, Chanal-Claris C, Labia R, Sirot J. A complex mutant of TEM-1 beta-lactamase with mutations encountered in both IRT-4 and extended-spectrum TEM-15, produced by an Escherichia coli clinical isolate. Antimicrob Agents Chemother. 1997;41:1322–5.
168. Sizemore RK, Colwell RR. Plasmids carried by antibiotic-resistant marine bacteria. Antimicrob Agents Chemother. 1977;12:373–82.
169. Sóki J, Hedberg M, Patrick S, Bálint B, Herczeg R, Nagy I, Hecht DW, Nagy E, Urbán E. Emergence and evolution of an international cluster of MDR Bacteroides fragilis isolates. J Antimicrob Chemother. 2016;71:2441–8.
170. Song W, Wemheuer B, Steinberg PD, Marzinelli EM, Thomas T. Contribution of horizontal gene transfer to the functionality of microbial biofilm on a macroalgae. ISME J. 2021;15:807–17.
171. Sousa M, Silva N, Borges V, Gomes JP, Vieira L, Caniça M, Torres C, Igrejas G, Poeta P. MRSA CC398 recovered from wild boar harboring new SCCmec type IV J3 variant. Sci Total Environ. 2020a;722:137845.
172. Sousa M, Silva V, Silva A, Silva N, Ribeiro J, Tejedor-Junco MT, Capita R, Chenouf NS, Alonso-Calleja C, Rodrigues TM. Staphylococci among wild European rabbits from the Azores: a potential zoonotic issue? J Food Prot. 2020b;83:1110–4.
173. Stalder T, Press MO, Sullivan S, Liachko I, Top EM. Linking the resistome and plasmidome to the microbiome. ISME J. 2019;13:2437–46.
174. Stanton IC, Bethel A, Leonard AFC, Gaze WH, Garside R. Existing evidence on antibiotic resistance exposure and transmission to humans from the environment: a systematic map. Environ Evid. 2022;11:8.
175. Sun D. Pull in and push out: mechanisms of horizontal gene transfer in bacteria. Front Microbiol. 2018;9:2154.
176. Teshome BF, Vouri SM, Hampton N, Kollef MH, Micek ST. Duration of exposure to antipseudomonal β-Lactam antibiotics in the critically ill and development of new resistance. Pharmacotherapy. 2019;39:261–70.
177. Tettelin H, Masignani V, Cieslewicz MJ, Donati C, Medini D, Ward NL, Angiuoli SV, Crabtree J, Jones AL, Durkin AS, Deboy RT, Davidsen TM, Mora M, Scarselli M, Margarit Y,

Ros I, Peterson JD, Hauser CR, Sundaram JP, Nelson WC, Madupu R, Brinkac LM, Dodson RJ, Rosovitz MJ, Sullivan SA, Daugherty SC, Haft DH, Selengut J, Gwinn ML, Zhou L, Zafar N, Khouri H, Radune D, Dimitrov G, Watkins K, O'connor KJ, Smith S, Utterback TR, White O, Rubens CE, Grandi G, Madoff LC, Kasper DL, Telford JL, Wessels MR, Rappuoli R, Fraser CM. Genome analysis of multiple pathogenic isolates of Streptococcus agalactiae: implications for the microbial "pan-genome". Proc Natl Acad Sci USA. 2005;102:13950–5.

178. Thomson M, Liu Y, Nunta K, Cheyne A, Fernandes N, Williams R, Garza-Garcia A, Larrouy-Maumus G. Expression of a novel mycobacterial phosphodiesterase successfully lowers cAMP levels resulting in reduced tolerance to cell wall-targeting antimicrobials. J Biol Chem. 2022;298:102151.
179. Tian B, Fadhil NH, Powell JE, Kwong WK, Moran NA. Long-term exposure to antibiotics has caused accumulation of resistance determinants in the gut microbiota of honeybees. MBio. 2012;3:e00377.
180. Trinh S, Haggoud A, Reysset G, Sebald M. Plasmids pIP419 and pIP421 from Bacteroides: 5-nitroimidazole resistance genes and their upstream insertion sequence elements. Microbiology (Reading). 1995;141(Pt 4):927–35.
181. Udaondo Z, Abram KZ, Kothari A, Jun SR. Insertion sequences and other mobile elements associated with antibiotic resistance genes in enterococcus isolates from an inpatient with prolonged bacteraemia. Microb Genom. 2022;8:mgen000855.
182. Union E. Regulation (EU) No 528/2012 of the European Parliament and of the Council of 22 May 2012 concerning the making available on the market and use of biocidal products. Off J Eur Union L. 2012;167:1–123.
183. Van Boeckel TP, Brower C, Gilbert M, Grenfell BT, Levin SA, Robinson TP, Teillant A, Laxminarayan R. Global trends in antimicrobial use in food animals. Proc Natl Acad Sci. 2015;112:5649–54.
184. Varela AR, Manageiro V, Ferreira E, Guimarães MA, Da Costa PM, Caniça M, Manaia CM. Molecular evidence of the close relatedness of clinical, gull and wastewater isolates of quinolone-resistant Escherichia coli. J Glob Antimicrob Resist. 2015;3:286–9.
185. Ventola CL. The antibiotic resistance crisis: part 1: causes and threats. P T. 2015;40:277–83.
186. Vicetti Miguel CP, Mejias A, Leber A, Sanchez PJ. A decade of antimicrobial resistance in Staphylococcus aureus: a single center experience. PLoS One. 2019;14:e0212029.
187. Vittecoq M, Laurens C, Brazier L, Durand P, Elguero E, Arnal A, Thomas F, Aberkane S, Renaud N, Prugnolle F, Solassol J, Jean-Pierre H, Godreuil S, Renaud F. VIM-1 carbapenemase-producing Escherichia coli in gulls from southern France. Ecol Evol. 2017;7:1224–32.
188. Vlaanderen EJ, Ghaly TM, Moore LR, Focardi A, Paulsen IT, Tetu SG. Plastic leachate exposure drives antibiotic resistance and virulence in marine bacterial communities. Environ Pollut. 2023;327:121558.
189. Wachino JI, Jin W, Kimura K, Arakawa Y. Intercellular transfer of chromosomal antimicrobial resistance genes between Acinetobacter baumannii strains mediated by prophages. Antimicrob Agents Chemother. 2019;63:e00334.
190. Waldor MK, Mekalanos JJ. Lysogenic conversion by a filamentous phage encoding cholera toxin. Science. 1996;272:1910–4.
191. Wang J, Ma Z-B, Zeng Z-L, Yang X-W, Huang Y, Liu J-H. The role of wildlife (wild birds) in the global transmission of antimicrobial resistance genes. Zool Res. 2017;38:55.
192. Wieck S, Olsson O, Kümmerer K. Possible underestimations of risks for the environment due to unregulated emissions of biocides from households to wastewater. Environ Int. 2016;94:695–705.
193. Wise EM Jr, Alexander SP, Powers M. Adenosine 3′:5′-cyclic monophosphate as a regulator of bacterial transformation. Proc Natl Acad Sci USA. 1973;70:471–4.
194. Woolhouse M, Ward M, Van Bunnik B, Farrar J. Antimicrobial resistance in humans, livestock and the wider environment. Philos Trans R Soc Lond Ser B Biol Sci. 2015;370:20140083.

195. Woolhouse ME, Ward MJ. Microbiology. Sources of antimicrobial resistance. Science. 2013;341:1460–1.
196. World Health, O. Joint FAO/OIE/WHO Expert Workshop on non-human antimicrobial usage and antimicrobial resistance: scientific assessment: Geneva, December 1–5, 2003. Geneva: World Health Organization; 2004.
197. Wu R, Zhao M, Li J, Gao H, Kan B, Liang W. Direct regulation of the natural competence regulator gene tfoX by cyclic AMP (cAMP) and cAMP receptor protein (CRP) in Vibrios. Sci Rep. 2015;5:14921.
198. Yang Y, Song W, Lin H, Wang W, Du L, Xing W. Antibiotics and antibiotic resistance genes in global lakes: a review and meta-analysis. Environ Int. 2018;116:60–73.
199. Zhang H, He H, Chen S, Huang T, Lu K, Zhang Z, Wang R, Zhang X, Li H. Abundance of antibiotic resistance genes and their association with bacterial communities in activated sludge of wastewater treatment plants: geographical distribution and network analysis. J Environ Sci. 2019;82:24–38.
200. Zheng XF, Liu XQ, Peng SY, Zhou Q, Xu B, Yuan H, Tang GL. Characterization of the Rifamycin-degrading monooxygenase from Rifamycin producers implicating its involvement in Saliniketal biosynthesis. Front Microbiol. 2020;11:971.
201. Zhu YG, Johnson TA, Su JQ, Qiao M, Guo GX, Stedtfeld RD, Hashsham SA, Tiedje JM. Diverse and abundant antibiotic resistance genes in Chinese swine farms. Proc Natl Acad Sci USA. 2013;110:3435–40.

Chapter 8
Multi-Omics Approaches to Resolve Antimicrobial Resistance

Dung Thuy Tran and Amber Dahlin

8.1 Introduction

In recent years, AMR has emerged as a global health concern that greatly impacts the effective prevention and treatment of infectious diseases [1]. The emergence of bacteria that are resistant to one or more antimicrobial drugs has contributed to a corresponding rise in infections that are resistant to treatment [1]. For example, pneumonia, which was treatable after the introduction of penicillin, now requires additional antibiotics for treatment [1]. According to the 2019 Antibiotic Resistance Threats report, more than 30% of cases of pneumococcal disease fail to respond to treatment with one or more antibiotics [2]. The World Health Organization (WHO) placed AMR among the top 10 global public health threats [3]. Within the USA, more than 2.8 million antimicrobial-resistant infections occur each year, with at least 35,000 dying as a result [2]. The U.S. Center for Disease Control and Prevention estimated the cost of AMR as $55 billion per year [1]. In Europe, the overall economic burden of AMR was estimated to be at least 1.5 billion euros [1]. Thus, AMR represents a large-scale and growing concern for public health worldwide.

Antimicrobial resistance occurs when bacteria exposed to antibiotic drugs can survive and multiply due to selective pressures [1]. AMR is on the rise in part due to the misuse and overuse of antibiotics, particularly in regions where antibiotics can be obtained without a prescription from a doctor [1]. In 2022, the COVID-19 U.S Impact on Antimicrobial Resistance report concluded that AMR is worsening because of the frequent use of antibiotics. Although antibiotics are not used to treat COVID-19, they were frequently prescribed due to uncertainty in the diagnosis of respiratory symptoms and concern for comorbid infections [4]. During the first year

D. T. Tran (✉) · A. Dahlin
Channing Division of Network Medicine, Department of Medicine, Brigham and Women's Hospital and Harvard Medical School, Boston, MA, USA
e-mail: dung.t.tran@channing.harvard.edu; amber.dahlin@channing.harvard.edu

V. Soni, A. S. Akhade (eds.), *Antimicrobial Resistance: Factors to Findings*,
https://doi.org/10.1007/978-3-031-65986-7_8

of the pandemic, over 29,400 deaths related to AMR infections occurred, and nearly 40% of patients acquired infections while they were hospitalized [5]. In 2020, AMR in the USA increased 15% compared to 2019, with 78% resulting from carbapenem-resistant *Acinetobacter* infection, and 32% in multidrug-resistant *Pseudomonas aeruginosa* infections [5]. The environment also plays a role in the spread of AMR. For instance, several antimicrobials are derived from soil microorganisms (such as *Actinomycete* genus Streptomyces) that are naturally resistant to antibiotics [6], and they are spread by contamination of water or improperly disposed medical waste [1]. Therefore, it is necessary to develop innovative strategies to reduce the spread of AMR and accelerate the process of discovering novel antibiotics and alternative treatments for infectious diseases [7].

Systems biology represents an innovative approach to address AMR that incorporates levels of molecular information provided by multiple "omics" data, generated through genomics, transcriptomics, proteomics, and metabolomics techniques [8]. There are two main approaches for using systems biology to integrate multi-omics data in biomedical studies [9]. The first approach focuses on understanding disease pathways and mechanisms by integrating information from knowledge bases, intending to obtain insights into the disease and identify key molecular players [9]. The second approach involves analyzing multi-omics datasets from the same patient samples and identifying correlations and patterns to understand dysregulated mechanisms [9]. Multi-omics integration could be applied in AMR research to reveal molecular mechanisms [10, 11], classify the risk level of AMR [12], identify new target drugs [13], and predict patient responses to antibiotics [14]. However, there are several hurdles in implementing this approach for AMR, including choosing the best omics combination within budget constraints, selecting computational methods for data integration, and accessing data and sample repositories. In this chapter, we provide an overview of the methods of integration for multi-omics workflows and current tools in AMR research. Additionally, we highlight specific examples of the application of integrative multi-omics data in AMR and also discuss current challenges and future directions in this growing area of research. Figure 8.1 illustrates the general application of multi-omics in the context of AMR.

8.2 Computational Approaches for Integrating Multi-Omics Data in AMR

Through profiling the genome (DNA), transcriptome (mRNA), proteome (proteins), and metabolome (metabolites), large-scale "omics" data are generated that can provide valuable insights into the molecular mechanisms underlying AMR. These high-throughput approaches can be applied to interrogate "omics" data from bacterial strains or species and to detect potential molecular targets for antibiotic response or resistance. In contrast to profiling individual species, metagenomics (defined as the study of the genetic material derived from a bulk sample of multiple

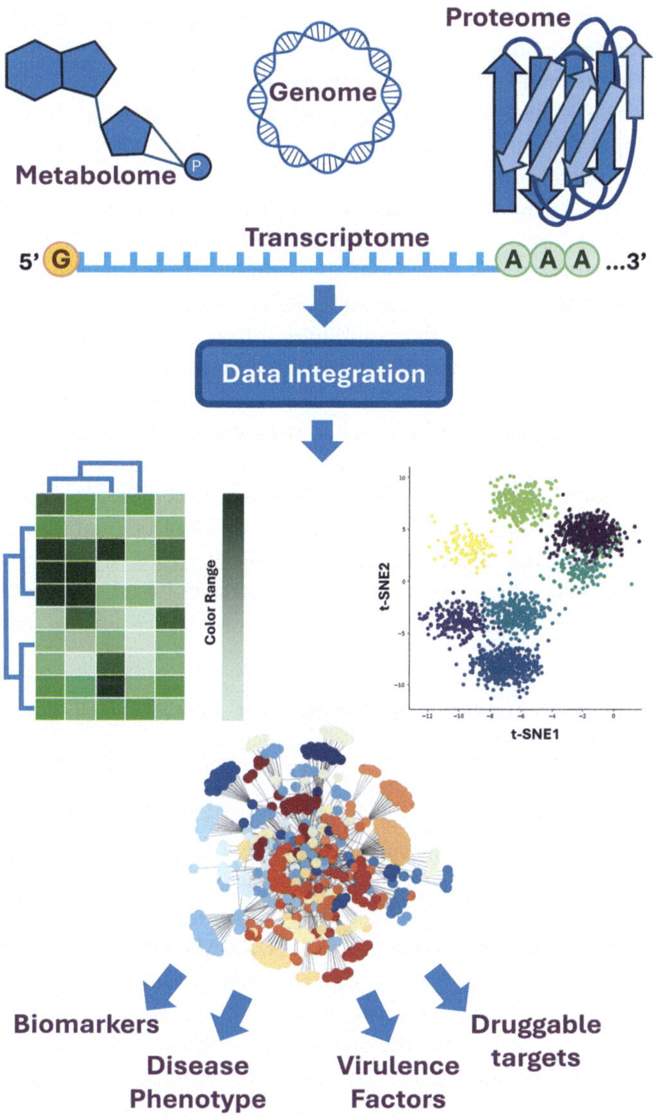

Fig. 8.1 Multi-omics data integration and application. (Source: Ball et al. [15])

microorganisms) allows for the exploration of resistance genes across entire microbial communities, facilitating the understanding of resistance-related mechanisms, such as horizontal gene transfer [16]. Transcriptomics reveals changes in gene expression patterns occurring in a single organism in response to antibiotic drug exposure, yielding information about the mechanism of antimicrobial action and potential drug targets [17]. Through advances in applications and technologies utilizing mass spectrometry, proteomics and metabolomics are increasingly relevant to investigations of AMR. These approaches are powerful as they can be used to profile all metabolites and proteins within a sample or an organism (untargeted approach) or focus on a smaller set of metabolites or proteins (e.g., within specific molecular pathways) that are selected a priori (targeted approach). It is important to note that a single omics layer captures only one specific aspect of biological systems, while biological processes often involve interactions between multiple omics [18]. Additionally, as the fluctuations in biological processes are dynamic and can occur rapidly in response to exposure, a single omics layer is unable to capture these temporal changes effectively [18]. AMR is a complex phenomenon that is affected by various molecular and environmental factors, with a capacity for change and evolution over time [1]. Therefore, combining different sources of knowledge generated through multi-omics approaches enables a more comprehensive understanding of AMR.

The workflow of multi-omics integration in biological research typically follows a structured process, beginning with the collection of various "omics" data sets. Based on available resources and study objectives, two or more omics layers are integrated using different statistical approaches to identify relationships, correlations, and patterns across different layers. To date, there are two established means of integration of omics data: knowledge-based and data-based approaches [19].

8.2.1 Knowledge-Based Approaches for Integrating Multi-Omics Data

A knowledge-based strategy for integrating multi-omics data utilizes a priori biological information derived from various databases and scientific studies, via specialized software packages and algorithms designed for this purpose. The generation of large-scale multi-omics networks can help clarify interactions across omics layers [14]. Genomic-scale metabolomic models (GSMMs) combine genome sequencing with gene transcript profiles, enzymatic reactions, and metabolite databases, which are broadly used for investigating metabolic changes in response to external exposures such as antibiotic treatment or dietary changes [20]. Since 1993, more than 6000 GSMMs have been constructed covering more than 183 organisms, functioning as a repository for the biochemical reactions of organisms [21]. GSMMs use Flux Balance Analysis (FBA) and Flux Variability Analysis (FVA) to provide genes, transcripts, metabolites, and metabolic pathways associated with simulated environments to predict the growth rates of organisms [22]. The beginning step of GSMM is the reconstruction of metabolic networks of the organisms of interest, beginning with genome annotation and generation of organism-specific Gene-Protein-Reaction associations, followed by validation of resulting models using canonical databases such as KEGG and PubMed [20, 22]. After completing the draft reconstruction, the network is converted into a mathematical representation of the entire metabolic

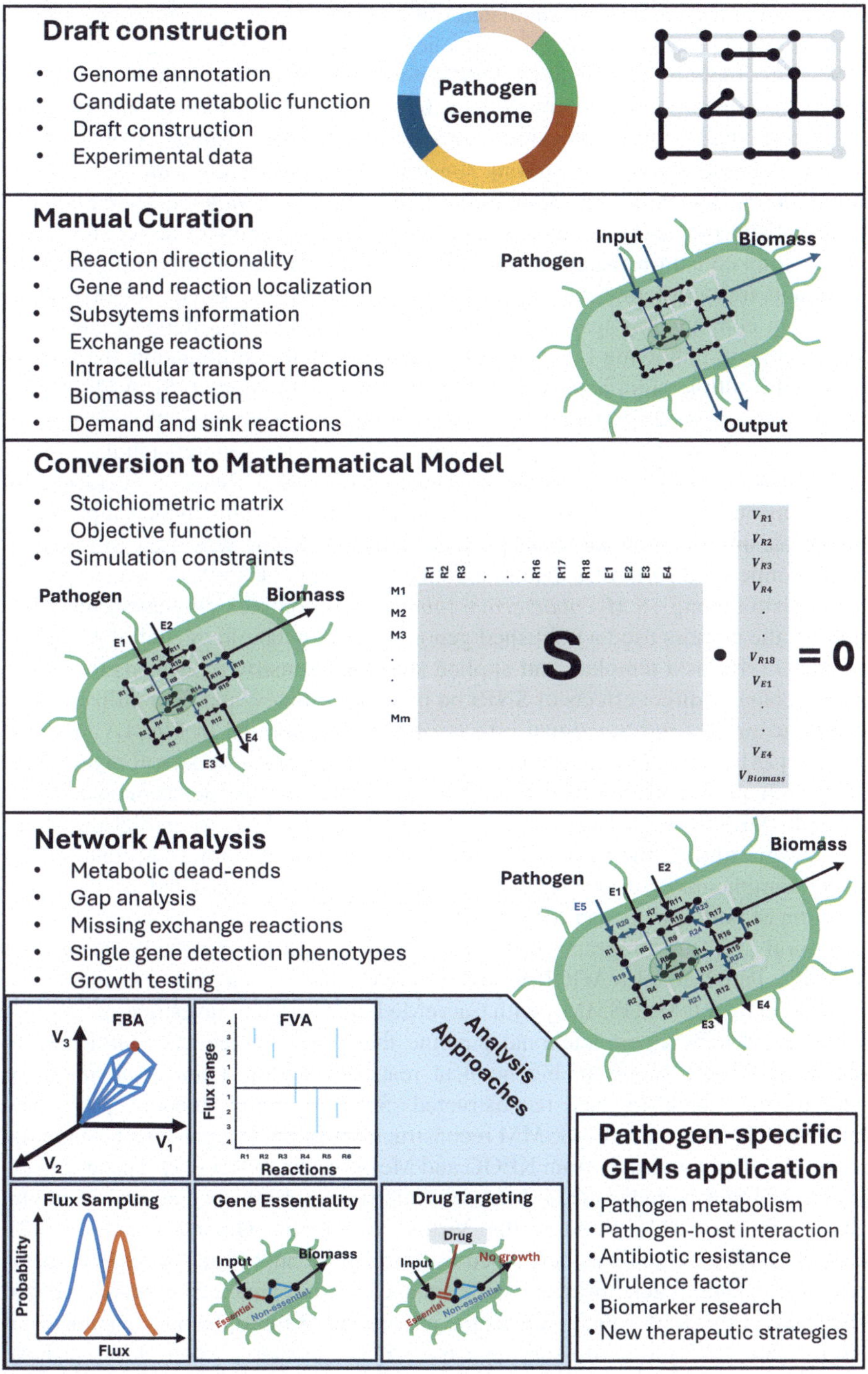

Fig. 8.2 Reconstruction process of pathogen-specific GEMs. (Source: Sertbas and Ulgen [23])

network and these models are simulated using FBA/FVA [20]. The workflow of this reconstruction process for bacterial pathogens is presented in Fig. 8.2.

With the rapid development of GSMMs and the FBA/FVA methods, numerous studies have incorporated various omics data such as transcriptomic and genomic data to test whether the pathogen develops resistance under treatment with antibiotics. For example, through integrating constraint-based modeling with gene expression data, the metabolic reprogramming that occurs in *Acinetobacter baumannii* under colistin treatment revealed a set of genes that are primary targets for novel drugs within colistin-resistant strains [24]. Another example of integrating metabolomic and transcriptomic data is shown by the synergistic effect of polymyxin-rifampicin combination against *Pseudomonas aeruginosa* at different time points post-treatment. Following 1 and 24 h of treatment with the combination, the authors identified multiple significantly altered genes and metabolites [25]. Additionally, Jenior, Dickenson [26] integrated transcriptomic meta-analyses with a previously published GSMM of *Klebsiella pneumoniae* (iYL1228) to generate unique isolate-type specific models of metabolism using a tool called Reaction Inclusion by Parsimony and Transcript Distribution to create context-specific models representative of parsimonious metabolism [27]. Moreover, GSMM was used to integrate metabolomics and genomics data to classify the effects of SNPs on strain-specific metabolism among 18 *Mycobacterium tuberculosis* complex clinical strains [28]. In brief, the authors used a published genome-scale metabolic model of *M. tuberculosis* G37Rv as a template and applied structural sensitivity analysis to predict the direct and indirect effects of SNPs on metabolic fluxes, and then utilized these models to predict the functional effects of SNPs (presented in Fig. 1A in Øyås, Borrell [28]). The flux distribution and metabolic shift of streptomycin-and-chloramphenicol-resistant *Chromobacterium violaceum* were determined using its GSMM (iDB858) with metabolomics data [29]. Using FVA to predict the metabolic programming, the authors identified enhanced production of acetate related to chloramphenicol resistance, while streptomycin resistance led to increased secretion of both acetate and formate [29].

Several software programs have been developed to automatically reconstruct GSMMs. Path2Models, AGORA, and CarveMe have been used extensively to reconstruct over 6000 GSMMs, with CarveMe largely used in investigations of bacteria. CarveMe is a computational pipeline that relies on sequence similarity to assign confidence scores to biochemical reactions within a universal model of metabolism. CarveMe has reconstructed 5587 bacterial models [30]. The Path2Models represents the GSMM reconstructions of 2630 organisms through the extraction of pathway data from KEGG and MetaCyc databases [31]. The latest version of AGORA includes the GSMMs of 7302 strains, 1738 species, and 25 phyla, which are accessible through the Virtual Metabolic Human database [32]. AutoKEGGRec, which was developed by Karlsen, Schulz [33], uses KEGG databases to create draft genome-scale models that can be implemented in MATLAB. An advantage of this tool is that it allows the processing of multiple microorganisms in one run but it is limited to using only the KEGG database, which does not have biomass, transport, and exchange reactions. Another tool is AuReme (Automatic

Table 8.1 List of genome-scale reconstruction tools

Tool	Description	Citation
MetaDraft	Python-based software using internal databases	[35]
RAVEN version 2	Run in MATLAB and use KEGG, MetaCyc	[36]
ModelSEED version 2	Web tool and automate every step, taking ~48 h to reconstruct a metabolic model from an assembled genome sequence	[37]
Pathway Tools	Integrate information about genes, proteins, metabolic networks, and regulatory networks to develop organism-specific databases	[38]
Merlin	Java application based on the KEGG database	[39]
Kbase	Open-source software includes annotation, reconstruction, curation, and modeling	[40]
CoReCO	Simultaneous reconstruction of multiple related species based on KEGG stoichiometry data	[41]
MetExplore	Web-based application includes interactive tools for metabolic network curation, exploration, and omics data analysis	[42]
METABOLIC	Profile metabolic and biogeochemical traits and functional networks by integrating annotation of proteins using KEGG, TIGRfam, Pfam, custom hidden Markow model databases	[43]

Reconstruction of Metabolic Models), which applies information from different databases such as MetaCyc and offers a user-friendly experience with personalized pipelines [34]. An overview of these tools is presented in Table 8.1.

8.2.2 Data-Based Approaches for Integrating Multi-Omics Data

Thanks to advancements in machine learning, the methods for integration of multi-omics data leverage statistical models as well as deep learning to uncover relationships between omics data layers. In contrast to the knowledge-based approach, existing information such as biological relationships or annotation is not required. Machine learning models are classified into two types: supervised and unsupervised learning. In supervised learning, the model first learns from a training dataset and makes predictions for unseen observations. Common models in supervised learning include the support vector machine (which segregates data observations into two classes by maximizing the distance between data points), random forests (an ensemble of decision trees with a randomly selected subset of features), Decision Trees (a tree-like model where nodes represent attributes and branches denote attribute values), K-nearest neighbors (KNN: an instance-based method comparing new observations with trained examples), regression algorithms (which estimate the functional relationship between input features), and Neural Networks (comprising interconnected units called neurons that receive weighted inputs and perform a weighted

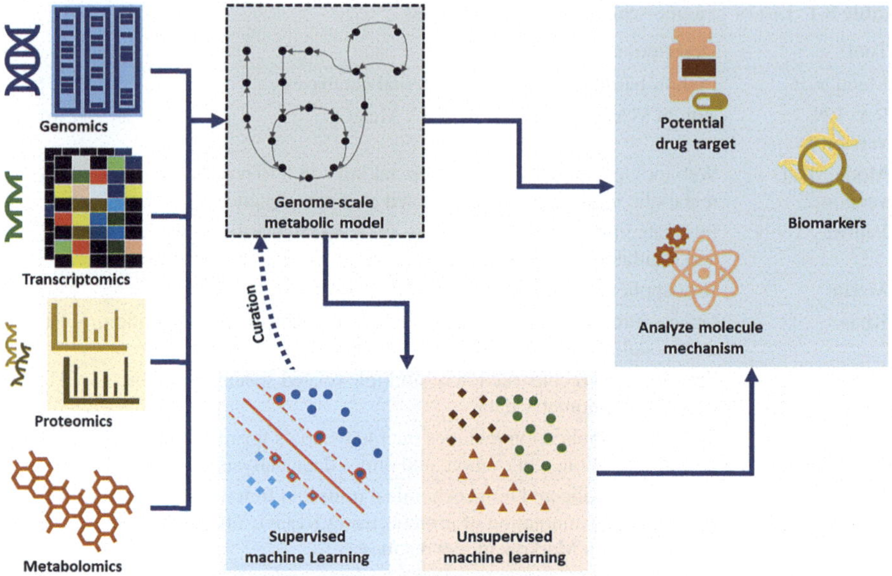

Fig. 8.3 Integration of machine learning and genome-scale metabolic modeling and applications. (Adapted from Passi et al. [22])

sum of those inputs). In contrast, unsupervised models identify the structure and pattern of unlabeled data including clustering and classification methods.

Machine learning has been integrated with GSMM to reveal more accurate flux predictions (Fig. 8.3). For example, the GSMM-based machine learning classifier, the Metabolic Association Classifier (MAC) was used to classify 1595 antibiotic-resistant and susceptible *M. tuberculosis* strains based on their genetic variation and biochemical pathway differences [44]. In brief, the MAC uses genome sequences of particular *M. tuberculosis* strains and classifies them as either resistant or susceptible using an allele-parameterized form of FBA, which takes antibiotic-specific objective functions and steady-state space and upper/lower bounds for each reaction flux as constraints [44]. GSMM and machine learning have been applied to identify the interplay between the genetic determinants of AMR and metabolic adaptations in *E. coli.* Pearcy, Hu [45] employed a machine learning KNN and SNP-informed approach to identify the genomic features of *E. coli* and then used Gradient Boosting Classifiers to rank these features based on their strength of correlation with AMR and phenotypes of interest. Finally, GSMMs were utilized to assess functional genetic determinants by applying FBA for the prediction of metabolic effects [45]. Additionally, Condition Specific Antibiotic Regimen Assessment using Mechanistic Learning (CARAMeL) was developed to understand the impact of different drugs and changes in drug efficacy depending on growth conditions. This method used Random Forests and GSMMs of *E.coli* and *M. tuberculosis* to predict combination therapy outcomes, as well as understand how these pathogens adapted to the first drug and how this adaptation affected sensitivity to the second drug [46].

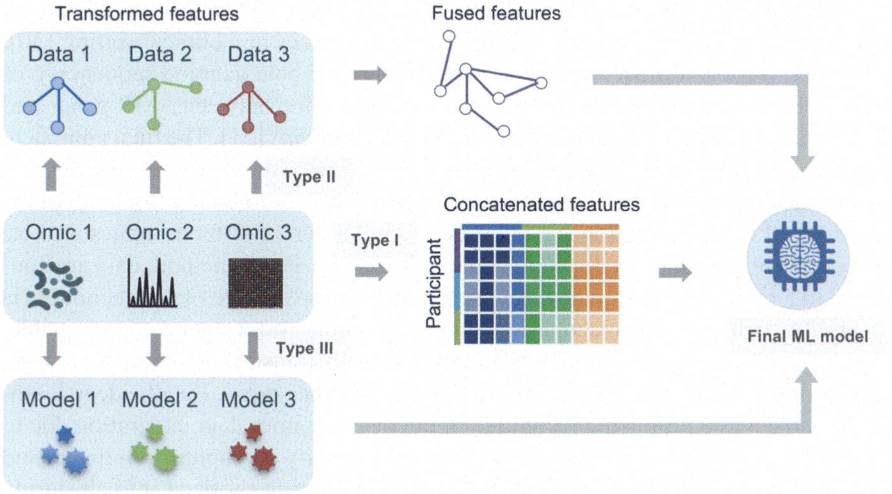

Fig. 8.4 Integration approach of multi-omics data using machine learning. (Source: Li et al. [57])

Beyond the integration of GSMMs, machine learning has been used to integrate different omics data (Fig. 8.4). There are three methods that can be used for omics integration: concatenation, model-based, and transformed-based methods [47]. The concatenation method starts with combining all omics data types into one comprehensive matrix of multi-omics data [47]. Then, this matrix is analyzed using unsupervised or supervised methods. In an investigation of 4141 drug-resistant clinical *Pseudomonas aeruginosa* isolates, genomic data and transcriptomics data were integrated to determine antimicrobial susceptibility and resistance to four antibiotics including ceftazidime, meropenem, ciprofloxacin, and tobramycin, using machine learning classifiers. Khaledi, Weimann [48] concatenated SNPs, gene presence/absence, and expression features into one full dataset and then split them into training and test sets, wherein the training set comprised 80% of the initial data, while 20% of the remaining data was treated as the test set. Here, a support vector machine (SVM) with a linear kernel was used to predict susceptibility and resistance to four antibiotics. The combination of all these three features resulted in high sensitivity and predictive values (AUROC > 0.8) [48]. The second approach, known as model-based integration, involves the creation of multiple models using various types of data as training sets [47]. The models are then combined to create a final integrated model from a machine learning algorithm such as a Bayesian network. The technique involves first generating single omics data and phenotype data, developing single models for each omics dataset, and later integrating them into a joint final model to identify the interaction between different levels of omics data associated with phenotypes of interest [47]. Ruiz-Perez, Lugo-Martinez [49] developed a computational pipeline for the analysis of longitudinal multi-omics data, which used dynamic Bayesian networks to integrate genomic sequence and metabolomics data from microbiome samples to identify the interaction between taxa, their genes, and metabolites. Additive Bayesian networks were used to evaluate antibiotic cross-resistance in five bacterial species from different sources in a clinical setting [50]. Moreover, Bayesian modeling was used to integrate whole genome sequencing of 3901 isolates from 880 studies to discover the association between AMR genes and the persistence of methicillin resistance in *S. aureus strains* [51]. The final approach, transformation-based integration, involves the conversion of each omics dataset into an intermediate format such as a graph or kernel matrix before merging [47]. In the graph-based method, relationships between samples across different omics are formally represented and transformed, while in the kernel-based method, data are converted to a higher dimensional feature space [47]. An advantage of this technique is its ability to merge diverse omics data when unique information is accessible. However, one of the drawbacks is that each dataset is transformed separately, making it more challenging to detect interactions among data types. Finally, deep learning has arisen as a promising technique in the field of omics data integration due to its exceptional predictive performance and its capacity to capture nonlinear and hierarchical features. Deep learning-based methods are categorized into clustering for feature selection/reduction, outcome prediction, survival analysis, and subtype discovery [52]. The typical workflow includes data preprocessing, feature selection/reduction, and concatenation of omics data types into one unified dataset, followed by performing data integration using different deep learning models [47]. Deep learning software tools for integration have been developed such as DeepOmix [53], DeepProg [54], and DeepDRK [55]. The recurrent neural network and other machine learning models were applied to build an integrative model for predicting molecular states, which integrates different omics layers including genomics, proteomics, metabolomics, and phenotypic data to study bacterial growth dynamics [56].

8.3 Applications of Multi-Omics in AMR Research

In this section, we discuss how multi-omics methods have been applied to investigate the evolution and emergence of AMR. We also discuss recently published research utilizing multi-omics approaches for AMR and its role in drug development for antimicrobials. Finally, we present some of the key limitations for incorporating multi-omics data in AMR investigations.

Multiple omics data have greatly assisted efforts toward taxonomic classification of bacterial genomes in addition to characterizing numerous aspects of comparative microbial genomics. For example, using a pan-genome and phylogenomic analysis of fully sequenced *Clostridium* isolates, Bhardwaj and Somvanshi [58] clarified diverse host symptomatic responses and correlated these with pathogenic effects. *Clostridium botulinum* is among the most dangerous infectious pathogens affecting human and animal health, and its ability to evade immune responses has prompted detailed genetic investigations to characterize its evolutionary patterns of resistance. Utilizing publicly available genome sequences, the whole genomes for 13

Clostridium lineages were aligned and a phylogenetic tree was generated and mapped to gene transcripts; evolutionary relationships and patterns were then inferred using various bioinformatics tools [58]. A very high degree of sequence similarity among *Clostridium* lineages revealed that they also share a large percentage of genomic segments, including those involved in pathogenesis [58]. Numerous ARGs were also shared across lineages, and their gene products were inferred; a number of these directed resistance to chloramphenicol, vancomycin, B-lactams, and other antimicrobials [58]. The high level of overlap of these genes across *Clostridium* lineages suggests the presence of universal resistance to vancomycin and chloramphenicol [58]. The authors also characterized the virulence potential for these lineages and identified multiple shared toxin and antitoxin families that could potentially serve as novel therapeutic targets [58].

8.3.1 Recent Applications of Multi-Omics Approaches in AMR

The advantage of including multi-omics data in AMR studies is that multiple layers of molecular interaction can be measured in a single study, and often in a single sample. However, individual omics have generated well-defined signatures for AMR. For example, by genomic sequencing of *Acinetobacter baumannii* mutants, candidate genes associated with extreme resistance to tobramycin, an aminoglycoside, were identified [59]. Notably, these resistance-conferring genes governed three major core regulatory functions including bacterial envelope homeostasis, regulation of phosphate, and phospholipid biosynthesis [59]. As *A. baumannii* strains are among the most resistant bacterial species to aminoglycosides, identifying core functions involved in extreme resistance is critical for developing alternative treatment strategies. Using a targeted metabolomics approach, metabolites in two isogenic *Staphylococcus aureus* strains with differential susceptibility to methicillin were measured [60]; alterations in purine, pyrimidine, and amino acid pathways were observed after 6 h of co-incubation with antibiotics [60]. Although individual omics studies have identified important molecular drivers of AMR recently, multi-omics perspectives have accelerated the understanding of mechanisms, predictors, and biomarkers of AMR, guiding new treatment strategies for complex and infectious diseases.

Citations in the NCBI PubMed database that report using multi-omics data or approaches in AMR span a decade. The rapid increase in the number of citations over the past several years reflects the emergence of multi-omics as an important tool to address the problem of AMR in human health and disease. These studies have revealed novel biochemical relationships and the presence of crosstalk between the gut microbiome, organ systems, pathogens, and disease states. In a recent study, to investigate the mechanisms underlying heterogeneity in patient responsiveness to preoperative neoadjuvant chemoradiotherapy for the treatment of rectal tumors, Teng, Wang [61] profiled and integrated multi-omics data from sera and fecal specimens from patients and determined that nucleotide biosynthesis mediated by

Bacteroides vulgatus was associated with poor responsiveness and treatment resistance in tumors. Furthermore, nonresponsive tumors also demonstrated an increased expression in genes involved in DNA repair and nucleoside transport; cancer cells exposed to either nucleosides or *B. vulgatus* were protected against treatment, indicating the presence of crosstalk between tumor cells and the gut microbiome [61]. Recent research efforts applying multi-omics data have also focused on combating *E. coli*, the top pathogen contributing to deaths related to AMR [62]. Rihacek, Kosaristanova [63] performed a phenotypic and multi-omics analysis to investigate the effect of treatment with sub-lethal concentrations of zinc, a common additive to animal feed and fertilizer that may increase drug resistance in *E. coli* [63]. Exposure to even sub-lethal concentrations of zinc resulted in the development of multi-drug resistance as well as important phenotypic changes related to changes in the expression of genes and proteins involved in AMR [63].

The damage caused by environmentally acquired infections with protozoal species, leading to dysbiosis, is not just limited to the gut but affects multiple organs that interact (e.g., gut-liver axis) by altering the host microbiome. Populations of healthy bacteria are also reduced in individuals infected by protozoa species, while pathogenic populations are comparatively increased [64, 65]. In a mouse model of giardiasis, elevated metabolite biomarkers for oxidative stress were observed in the liver, small intestine, and serum as a result of the corresponding depletion of *Akkermansiaceae* spp. in the gut [64]. Relationships between the host, parasite, and microbiome are complex and reflect significant alterations in the expression of genes, proteins, metabolites, and transcripts for these entities. Recent studies have employed multi-omics efforts to understand the host-pathogen-microbiome interactions [64, 65]. Recently, Karpe, Beale [64] generated and integrated metabolomic, genomic, and metaproteomic data using LC/GC-MS-based and bioinformatic approaches and determined key microbiome species that are affected by giardiasis as well as the most significant pathways involved in pathogenic stress within the gut-liver axis.

With the rise in next-generation sequencing, genome annotation tools, and advances in protein and metabolite profiling methodologies, the number of new bioinformatics resources has increased rapidly, necessitating improved workflows and guidelines for investigators to utilize these tools effectively. While entire and partial genomes of numerous bacterial species have been sequenced and deposited into public repositories, complete (and therefore useful in research efforts) sequenced genomes exist for only 1882 different bacterial genera representing 3732 distinct species [66]. Currently, there is a focus on using genomics to identify microbial antibiotic resistance genes (ARGs) that can predict AMR more rapidly, thereby enabling clinicians to select alternative treatments or reduce the risk of treatment failure. In addition, rapid profiling of virulence factors and bacterial toxins related to resistance is critical for understanding the mechanisms and evolution of AMR. To this end, ARG and virulence factor/toxin prediction tools that utilize simple algorithms to more complex machine learning tools have been developed to facilitate profiling or integrating multi-omics data [66]. In a recent review article, Samantray,

Table 8.2 Bioinformatics databases of antibiotic resistance genes

Database	Description	Citation	Link
AMRFinderPlus	A tool for the identification of AMR genes, resistance-associated point mutation	[67]	https://github.com/ncbi/amr
ARG-ANNOT	Identification of existing and new AMR genes in bacterial genomes	[68]	https://www.mediterranee-infection.com/acces-ressources/base-de-donnees/arg-annot-2
ARGMiner	A web tool to curate and retrieve all information available from several ARD resources	[69]	http://bench.cs.vt.edu/argminer
ARGs-OAP	The database of AMR gene and analysis tools	[70]	https://galaxyproject.org/use/args-oap/
CARD	Curated collection of characterized determinants and associated abiotic and tools for the analysis of molecular sequences	[71]	https://card.mcmaster.ca
Mustard	Antibiotic resistance determinants database	[72]	http://mgps.eu/Mustard/
Resfams	A curated database of protein families and used Hidden Markov Models to classify AMR-related protein sequences	[73]	https://github.com/dantaslab/resfams
ResFinder 4.0	Online resources for the identification of AMR genes using NGS	[74]	https://cge.cbs.dtu.dk/services/ResFinder/
PATRIC	Web-based analysis and visualization tools with over 250,000 annotated and publicly bacterial genomes	[75]	https://www.bv-brc.org/
PointFinder	A web tool for chromosomal point mutations associated with AMR using WGS	[76]	https://bitbucket.org/genomicepidemiology/pointfinder_db/src

Table 8.3 Computational resources available for toxins or virulence factor prediction

Resource	Description	Reference	Link
DBETH	A database of bacterial exotoxins for humans	[77]	http://www.hpppi.iicb.res.in/btox
HyperVR	A hybrid deep ensemble learning approach for predicting AR genes	[78]	https://github.com/jiboyalab/HyperVR
T1TAdb	A database of Type I Toxin-Antitoxin systems	[79]	https://d-lab.arna.cnrs.fr/t1tadb
TASmania	A large database of bacterial genomes for predicting toxins and antitoxins	[80]	https://shiny.bioinformatics.unibe.ch/apps/tasmania
VFDB 2022	Comprehensive knowledge of bacterial virulence factors	[81]	http://www.mgc.ac.cn/VFs/

Adapted from Samantray et al. [66]

Tanwar [66] provided a comprehensive summary of these multi-omics-focused computational tools and databases (included in Tables 8.2 and 8.3).

8.3.2 Use of Multi-Omics Data in Antimicrobial Drug Development

The development of bioinformatics workflows to integrate and analyze multi-omics data is critical for escalating novel antimicrobial drug discovery, in addition to preventing AMR. The use of multi-omics methodologies is also anticipated to improve specificity in rational drug design and development of novel antimicrobials. However, antimicrobial drug development programs to prevent AMR have largely been focused on improving the properties of existing drugs rather than characterizing novel compounds. Multi-omics studies of treatment responses to antimicrobials have identified perturbations in cellular pathways that could be exploited as potential novel drug targets. However, the activity of these molecules may differ across bacterial species, and under different environmental conditions.

Multi-omics approaches have been applied to identify new potential compounds and intervention strategies targeting the most virulent and opportunistic pathogens: *Klebsiella pneumonia*, a major cause of hospital- and community-acquired infections, and methicillin-resistant *Staphylococcus aureus* (MRSA). *K. pneumonia* strains that have developed resistance to beta-lactams via upregulation of carbapenemase are a major public health threat worldwide and have prompted alternative treatment with polymyxins; despite this, resistance to polymyxins is also on the rise. As a result, the search for alternative therapies is increasing and the application of multi-omics data in these efforts bears promise. To address this, Ramos et al. developed an integrative multi-omics (incorporating genomics, transcriptomics, structuromics, and metabolomics) framework to identify candidate proteins for susceptibility to these antimicrobials [11]. The authors performed in silico screening and prioritized 29 "druggable" proteins with features that made them promising targets for antimicrobial development [11]. These proteins are involved in multiple bacterial homeostasis mechanisms. Notably, the authors also identified a subset of proteins representing both high- and intermediate-value targets for drug development, representing novel candidate targets that are involved in phospholipid synthesis [11]. These represent valuable potential candidates for antimicrobial targeting of *K. pneumoniae* and await further characterization.

A recent approach to address AMR conferred by *S. aureus* profiled multi-omics data from clinical isogenic strain isolates to evaluate the effects of oxacillin on methicillin- susceptible/-resistant cultures [82]. The isogenic model was introduced to limit background effects and facilitate the evaluation of only genetic elements related to AMR and differed only by the presence of a mobile genetic element called staphylococcal cassette chromosome mec (SCCmec) type IV [82]. The strains were then treated with oxacillin, followed by quantitative profiling of metabolome and proteome [82]. Multiple proteins involved in peptidoglycan biosynthesis, two-component systems, oxidative stress, and stringent response, were differentially regulated by oxacillin treatment, while metabolites related to peptidoglycan biosynthesis were decreased in the oxacillin-treated MRSA, indicating that treatment response variation to oxacillin is a result of SCCmec type IV. This integrative multi-omics approach provides a new understanding of antibiotic resistance in MRSA and yields insight into the development of novel antimicrobials.

8.3.3 Limitations of Multi-Omics to Address AMR

While the use of multi-omics bears promises for future treatment strategies to overcome AMR, it has several limitations. A notable limitation is that although an increasing number of potential biomarkers and microbial targets in the form of genes and proteins have been identified to date, their utility in clinical practice is not well understood, and few prospective studies to establish their effectiveness and reproducibility have been conducted. In many cases, the biological functions of the putative targets are not clear. Furthermore, many multi-omics studies to date represent proof-of-concept studies and require validation in large-scale cohorts in appropriate patient populations. Access to technological resources and analytical or sequencing platforms for high-throughput profiling of multiple omics signatures from different samples, tissues, or biofluids also presents a challenge. Workflows for integrating these data may be nascent, or not straightforward to implement. While there is a rapid rise in the development of computational tools, algorithms, and databases to integrate and analyze multiple omics data, these resources vary in their complexity and utility. Finally, multi-omics approaches generate an unprecedented amount of information, and the computational resources and data storage needs required for untargeted profiling of all genes, metabolites, proteins, and transcripts within large populations (and often at multiple time points or conditions) may be prohibitive for some researchers. As multi-omics data also generate multiple forms of personal health data, the development of practical and accessible downstream data-sharing practices that preserve the privacy of patients is also an important consideration.

8.4 Future Directions

In this chapter, we have described how the threat of AMR to public health demands innovative strategies for effective intervention and presented multiple examples and applications as well as computational methods delineating how multi-omics approaches have emerged as a promising avenue to understand AMR. Here, we elaborate on the key emerging trends, opportunities, and challenges in the application of multi-omics methodologies to address the growing crisis of AMR.

By integrating multiple layers of molecular data using robust bioinformatics and computational workflows, researchers can develop a holistic understanding of the molecular mechanisms of AMR. Through leveraging machine learning algorithms for both data integration as well as statistical analysis, meaningful insights and novel potential drug targets can be obtained. Furthermore, predictive models can also aid in identifying and prioritizing potential antimicrobial drug targets revealed through multi-omics approaches, as well as inform therapeutic strategies. The ability to profile multiple layers of omics through analysis of individual cells or isolates, in addition to profiling metagenomic or entire microbiome communities, provides further insights into understanding the heterogeneity of microbial populations and host cells, clarifying host-pathogen relationships, and empowers the identification of rare sub-populations driving resistance. The integration of multi-omics reveals several opportunities for advancement in thwarting AMR through precision medicine approaches that can leverage resistance mechanisms specific to individuals or populations or require early intervention. This information can also be applied to assist in drug development or repurposing of existing antimicrobials.

Despite these promising advantages, the use of multi-omics data in AMR research is challenged by methodological, analytical, and statistical issues; difficulty with reproducibility and validation; and the need for safeguarding personal health data. Due to the large amount of data generated in these studies, integrating omics layers remains a technical challenge, and robust methods for integrating and interpreting these data are required for interpreting biological meaning. As many multi-omics workflows and platforms are still developing, reproducibility of results and the development of best practices can be problematic. A number of potential targets await clinical validation, and putative drug targets require screening and functional validation to determine their suitability and relevance. Finally, balancing the need for investigators to share results and personal health information of patients from whom the multi-omics data originated, with ethical and privacy guidelines, is paramount.

In conclusion, multi-omics approaches represent a promising future for combating AMR. While recent studies have revealed numerous potential realizations for applying precision medicine and drug discovery in AMR, significant challenges remain that must be addressed. Overcoming these limitations will contribute to developing innovative solutions in the battle against AMR.

References

1. Prestinaci F, Pezzotti P, Pantosti A. Antimicrobial resistance: a global multifaceted phenomenon. Pathog Glob Health. 2015;109(7):309–18.
2. Centers for Disease Control and Prevention. Antibiotic resistance threats in the United States 2019. CDC; 2019.
3. EclinicalMedicine. Antimicrobial resistance: a top ten global public health threat. eClinicalMedicine. 2021;41:101221.
4. Langford BJ, et al. Antibiotic resistance associated with the COVID-19 pandemic: a systematic review and meta-analysis. Clin Microbiol Infect. 2023;29(3):302–9.
5. CDC. COVID 19: US impact on antimicrobial resistance, special report 2022. Atlanta: US Department of Health and Human Services, CDC; 2022.
6. Forsberg KJ, et al. The shared antibiotic resistome of soil bacteria and human pathogens. Science. 2012;337(6098):1107–11.
7. Willemsen A, Reid S, Assefa Y. A review of national action plans on antimicrobial resistance: strengths and weaknesses. Antimicrob Resist Infect Control. 2022;11(1):90.
8. Francine P. Systems biology: new insight into antibiotic resistance. Microorganisms. 2022;10(12):2362.
9. Athieniti E, Spyrou GM. A guide to multi-omics data collection and integration for translational medicine. Comput Struct Biotechnol J. 2023;21:134–49.
10. Spänig S, et al. A multi-omics study on quantifying antimicrobial resistance in European freshwater lakes. Environ Int. 2021;157:106821.
11. Ramos PIP, et al. An integrative, multi-omics approach towards the prioritization of Klebsiella pneumoniae drug targets. Sci Rep. 2018;8(1):10755.
12. Zhang A-N, et al. An omics-based framework for assessing the health risk of antimicrobial resistance genes. Nat Commun. 2021;12(1):4765.
13. Paananen J, Fortino V. An omics perspective on drug target discovery platforms. Brief Bioinform. 2020;21(6):1937–53.
14. Yow HY, et al. Optimizing antimicrobial therapy by integrating multi-omics with pharmacokinetic/pharmacodynamic models and precision dosing. Front Pharmacol. 2022;13:915355.
15. Ball B, Langille M, Geddes-McAlister J. Fun(gi)omics: advanced and diverse technologies to explore emerging fungal pathogens and define mechanisms of antifungal resistance. mBio. 2020;11(5):e01020-20.
16. de Abreu VAC, Perdigão J, Almeida S. Metagenomic approaches to analyze antimicrobial resistance: an overview. Front Genet. 2020;11:575592.
17. Howden BP, et al. Analysis of the small RNA transcriptional response in multidrug-resistant Staphylococcus aureus after antimicrobial exposure. Antimicrob Agents Chemother. 2013;57(8):3864–74.
18. Sun YV, Hu YJ. Integrative analysis of multi-omics data for discovery and functional studies of complex human diseases. Adv Genet. 2016;93:147–90.
19. Wörheide MA, et al. Multi-omics integration in biomedical research – a metabolomics-centric review. Anal Chim Acta. 2021;1141:144–62.
20. Hao T, et al. The genome-scale integrated networks in microorganisms. Front Microbiol. 2018;9:296.
21. Zhu Y, Zhao J, Li J. Genome-scale metabolic modeling in antimicrobial pharmacology. Eng Microbiol. 2022;2(2):100021.

22. Passi A, et al. Genome-scale metabolic modeling enables in-depth understanding of big data. Metabolites. 2021;12(1):14.
23. Sertbas M, Ulgen KO. Genome-scale metabolic modeling for unraveling molecular mechanisms of high threat pathogens. Front Cell Dev Biol. 2020;8:566702.
24. Presta L, et al. Constraint-based modeling identifies new putative targets to fight colistin-resistant A. baumannii infections. Sci Rep. 2017;7(1):3706.
25. Mahamad Maifiah MH, et al. Integrated metabolomic and transcriptomic analyses of the synergistic effect of polymyxin–rifampicin combination against Pseudomonas aeruginosa. J Biomed Sci. 2022;29(1):89.
26. Jenior ML, Dickenson ME, Papin JA. Genome-scale metabolic modeling reveals increased reliance on valine catabolism in clinical isolates of Klebsiella pneumoniae. NPJ Syst Biol Appl. 2022;8(1):41.
27. Jenior ML, et al. Transcriptome-guided parsimonious flux analysis improves predictions with metabolic networks in complex environments. PLoS Comput Biol. 2020;16(4):e1007099.
28. Øyås O, et al. Model-based integration of genomics and metabolomics reveals SNP functionality in Mycobacterium tuberculosis. Proc Natl Acad Sci. 2020;117(15):8494–502.
29. Banerjee D, Raghunathan A. Constraints-based analysis identifies NAD+ recycling through metabolic reprogramming in antibiotic resistant Chromobacterium violaceum. PLoS One. 2019;14(1):e0210008.
30. Machado D, et al. Fast automated reconstruction of genome-scale metabolic models for microbial species and communities. Nucleic Acids Res. 2018;46(15):7542–53.
31. Büchel F, et al. Path2Models: large-scale generation of computational models from biochemical pathway maps. BMC Syst Biol. 2013;7(1):116.
32. Heinken A, et al. Genome-scale metabolic reconstruction of 7,302 human microorganisms for personalized medicine. Nat Biotechnol. 2023;41(9):1320–31.
33. Karlsen E, Schulz C, Almaas E. Automated generation of genome-scale metabolic draft reconstructions based on KEGG. BMC Bioinformatics. 2018;19(1):467.
34. Aite M, et al. Traceability, reproducibility and wiki-exploration for "à-la-carte" reconstructions of genome-scale metabolic models. PLoS Comput Biol. 2018;14(5):e1006146.
35. Hanemaaijer M, et al. Model-based quantification of metabolic interactions from dynamic microbial-community data. PLoS One. 2017;12(3):e0173183.
36. Wang H, et al. RAVEN 2.0: a versatile toolbox for metabolic network reconstruction and a case study on Streptomyces coelicolor. PLoS Comput Biol. 2018;14(10):e1006541.
37. Henry CS, et al. High-throughput generation, optimization and analysis of genome-scale metabolic models. Nat Biotechnol. 2010;28(9):977–82.
38. Karp PD, et al. Pathway Tools version 19.0 update: software for pathway/genome informatics and systems biology. Brief Bioinform. 2016;17(5):877–90.
39. Dias O, et al. Reconstructing genome-scale metabolic models with merlin. Nucleic Acids Res. 2015;43(8):3899–910.
40. Arkin AP, et al. KBase: the United States Department of Energy Systems Biology Knowledgebase. Nat Biotechnol. 2018;36(7):566–9.
41. Pitkänen E, et al. Comparative genome-scale reconstruction of gapless metabolic networks for present and ancestral species. PLoS Comput Biol. 2014;10(2):e1003465.
42. Cottret L, et al. MetExplore: collaborative edition and exploration of metabolic networks. Nucleic Acids Res. 2018;46(W1):W495–502.
43. Zhou Z, et al. METABOLIC: high-throughput profiling of microbial genomes for functional traits, metabolism, biogeochemistry, and community-scale functional networks. Microbiome. 2022;10(1):33.
44. Kavvas ES, et al. A biochemically-interpretable machine learning classifier for microbial GWAS. Nat Commun. 2020;11(1):2580.
45. Pearcy N, et al. Genome-scale metabolic models and machine learning reveal genetic determinants of antibiotic resistance in Escherichia coli and unravel the underlying metabolic adaptation mechanisms. mSystems. 2021;6(4):e0091320.

46. Chung CH, Chandrasekaran S. A flux-based machine learning model to simulate the impact of pathogen metabolic heterogeneity on drug interactions. PNAS Nexus. 2022;1(3):pgac132.
47. Chen C, et al. Applications of multi-omics analysis in human diseases. MedComm (2020). 2023;4(4):e315.
48. Khaledi A, et al. Predicting antimicrobial resistance in Pseudomonas aeruginosa with machine learning-enabled molecular diagnostics. EMBO Mol Med. 2020;12(3):e10264.
49. Ruiz-Perez D, et al. Dynamic Bayesian networks for integrating multi-omics time series microbiome data. mSystems. 2021;6(2):e01105-20. https://doi.org/10.1128/msystems.01105-20.
50. Cherny SS, Chowers M, Obolski U. Bayesian network modeling of patterns of antibiotic cross-resistance by bacterial sample source. Commun Med. 2023;3(1):61.
51. Ojala F, et al. Bayesian modeling of the impact of antibiotic resistance on the efficiency of MRSA decolonization. PLoS Comput Biol. 2023;19(10):e1010898.
52. Kang M, Ko E, Mersha TB. A roadmap for multi-omics data integration using deep learning. Brief Bioinform. 2022;23(1):bbab454.
53. Zhao L, et al. DeepOmix: a scalable and interpretable multi-omics deep learning framework and application in cancer survival analysis. Comput Struct Biotechnol J. 2021;19:2719–25.
54. Poirion OB, et al. DeepProg: an ensemble of deep-learning and machine-learning models for prognosis prediction using multi-omics data. Genome Med. 2021;13(1):112.
55. Wang Y, et al. DeepDRK: a deep learning framework for drug repurposing through kernel-based multi-omics integration. Brief Bioinform. 2021;22(5):bbab048.
56. Kim M, et al. Multi-omics integration accurately predicts cellular state in unexplored conditions for Escherichia coli. Nat Commun. 2016;7:13090.
57. Li P, et al. Machine learning for data integration in human gut microbiome. Microb Cell Factories. 2022;21(1):241.
58. Bhardwaj T, Somvanshi P. Pan-genome analysis of Clostridium botulinum reveals unique targets for drug development. Gene. 2017;623:48–62.
59. Gallagher LA, Lee SA, Manoil C. Importance of core genome functions for an extreme antibiotic resistance trait. mBio. 2017;8(6):e01655-17.
60. Schelli K, Zhong F, Zhu J. Comparative metabolomics revealing Staphylococcus aureus metabolic response to different antibiotics. Microb Biotechnol. 2017;10(6):1764–74.
61. Teng H, et al. Gut microbiota-mediated nucleotide synthesis attenuates the response to neoadjuvant chemoradiotherapy in rectal cancer. Cancer Cell. 2023;41(1):124–138.e6.
62. Murray CJL, et al. Global burden of bacterial antimicrobial resistance in 2019: a systematic analysis. Lancet. 2022;399(10325):629–55.
63. Rihacek M, et al. Zinc effects on bacteria: insights from Escherichia coli by multi-omics approach. mSystems. 2023;8:e0073323.
64. Karpe AV, Beale DJ, Tran CD. Intelligent biological networks: improving anti-microbial resistance resilience through nutritional interventions to understand protozoal gut infections. Microorganisms. 2023;11(7):1800.
65. McKenney EA, et al. Down for the count: Cryptosporidium infection depletes the gut microbiome in Coquerel's sifakas. Microb Ecol Health Dis. 2017;28(1):1335165.
66. Samantray D, et al. A comprehensive bioinformatics resource guide for genome-based antimicrobial resistance studies. Omics. 2023;27(10):445–60.
67. Feldgarden M, et al. AMRFinderPlus and the Reference Gene Catalog facilitate examination of the genomic links among antimicrobial resistance, stress response, and virulence. Sci Rep. 2021;11(1):12728.
68. Gupta SK, et al. ARG-ANNOT, a new bioinformatic tool to discover antibiotic resistance genes in bacterial genomes. Antimicrob Agents Chemother. 2014;58(1):212–20.
69. Arango-Argoty GA, et al. ARGminer: a web platform for the crowdsourcing-based curation of antibiotic resistance genes. Bioinformatics. 2020;36(9):2966–73.
70. Yang Y, et al. ARGs-OAP: online analysis pipeline for antibiotic resistance genes detection from metagenomic data using an integrated structured ARG-database. Bioinformatics. 2016;32(15):2346–51.

71. Alcock BP, et al. CARD 2020: antibiotic resistome surveillance with the comprehensive antibiotic resistance database. Nucleic Acids Res. 2020;48(D1):D517–d525.
72. Ruppé E, et al. Prediction of the intestinal resistome by a three-dimensional structure-based method. Nat Microbiol. 2019;4(1):112–23.
73. Gibson MK, Forsberg KJ, Dantas G. Improved annotation of antibiotic resistance determinants reveals microbial resistomes cluster by ecology. ISME J. 2015;9(1):207–16.
74. Zankari E, et al. Identification of acquired antimicrobial resistance genes. J Antimicrob Chemother. 2012;67(11):2640–4.
75. Antonopoulos DA, et al. PATRIC as a unique resource for studying antimicrobial resistance. Brief Bioinform. 2017;20(4):1094–102.
76. Zankari E, et al. PointFinder: a novel web tool for WGS-based detection of antimicrobial resistance associated with chromosomal point mutations in bacterial pathogens. J Antimicrob Chemother. 2017;72(10):2764–8.
77. Chakraborty A, et al. DBETH: a database of bacterial exotoxins for human. Nucleic Acids Res. 2012;40(Database issue):D615–20.
78. Ji B, et al. HyperVR: a hybrid deep ensemble learning approach for simultaneously predicting virulence factors and antibiotic resistance genes. NAR Genom Bioinform. 2023;5(1):1–17.
79. Tourasse NJ, Darfeuille F. T1TAdb: the database of type I toxin-antitoxin systems. RNA. 2021;27(12):1471–81.
80. Akarsu H, et al. TASmania: a bacterial Toxin-Antitoxin Systems database. PLoS Comput Biol. 2019;15(4):e1006946.
81. Liu B, et al. VFDB 2022: a general classification scheme for bacterial virulence factors. Nucleic Acids Res. 2022;50(D1):D912–7.
82. Jo SH, et al. Multi-omics based characterization of antibiotic response in clinical isogenic isolates of methicillin-susceptible/-resistant Staphylococcus aureus. RSC Adv. 2020;10(46):27864–73.

Chapter 9
Systems Biology Approaches to Study Antimicrobial Resistance

Sukriti Pal and Nagasuma Chandra

Abbreviations

ABM	Agent-based model
ALE	Adaptive laboratory evolution
AMR	Antimicrobial resistance
ATP	Adenosine triphosphate
DFE	Distribution of fitness effects
FBA	Flux balance analysis
GSEA	Gene set enrichment analysis
GWAS	Genome-wide association studies
MDR	Multidrug resistance
ODE	Ordinary differential equations
PAMP	Pathogen-associated molecular patterns
PBP	Penicillin-binding proteins
PMF	Proton motive force
PRR	Pattern recognition receptor
ROS	Reactive oxygen species
SNP	Single nucleotide polymorphism
TCA	Tricarboxylic acid

S. Pal
Department of Biochemistry, Indian Institute of Science, Bangalore, Karnataka, India

N. Chandra (✉)
Department of Biochemistry, Indian Institute of Science, Bangalore, Karnataka, India

Department of Bioengineering, Indian Institute of Science, Bangalore, Karnataka, India
e-mail: nchandra@iisc.ac.in

V. Soni, A. S. Akhade (eds.), *Antimicrobial Resistance: Factors to Findings*,
https://doi.org/10.1007/978-3-031-65986-7_9

9.1 Introduction

The discovery of penicillin by Alexander Fleming led to the "golden age" of antibiotics, during which different classes of antibiotics, both natural and synthetic, were introduced into the clinic. However, within a short span (~10 years), bacterial strains that were resistant to the action of antibiotics started emerging, with penicillin-resistant *Staphylococcus aureus* being one of the first superbugs [1]. Owing to their high cell division rate, bacterial cells are capable of rapidly accumulating mutations to adapt to various environmental stresses within a small timescale. This leads to the threat of antimicrobial resistance (AMR) during which most pathogenic bacteria evolve various mechanisms that make them insensitive to the action of multiple drugs. Treating infectious diseases is becoming increasingly challenging due to the emergence and global spread of antibiotic-resistant pathogens. Antimicrobial resistance (AMR) is rising to alarmingly high levels across the world. According to CDC's 2019 AR threats report [2], more than 2.8 million antibiotic-resistant infections occur in the USA each year, and more than 35,000 people die as a result. Multidrug-resistant (MDR) ESKAPE (*Enterococcus faecium, Staphylococcus aureus, Klebsiella pneumoniae, Acinetobacter baumannii, Pseudomonas aeruginosa*, and *Enterobacter* species) pathogens present the biggest challenges in a clinical setting. Without effective treatment options for drug-resistant infections, the number of deaths due to treatment failures will increase in the near future, which is very concerning.

Early research on antimicrobial resistance involved identification of resistance genes and antibiotic targets using genetic screens and biochemical approaches (like pulldowns) [3, 4]. At the time, a newly encountered resistant pathogen would usually be dealt with, by releasing a new antibiotic [1]. The golden age of antibiotics (1950–1970), during which a number of new antimicrobials were discovered, was shortly followed by a "discovery void," due to an exhaustion of the natural reserves of antibiotics and the sharp rise in multidrug-resistant pathogens. This called for newer methods to tackle the problem. Several attempts have since been made to overcome the problem of resistance, although it is far from being solved. Current research is trying to address this from different perspectives. The first is from a microbiology perspective, to find novel targeting strategies that are less prone to evolution and novel antimicrobials that can target the resistance-causing machinery. The second is from a clinical perspective, where different treatment regimens are being used, which include antibiotic cycling, combination therapy, and adaptive treatments with respect to their efficacy in clearing resistant infections in the host and also in minimizing the selection of resistant pathogens. The third is from a public health perspective, where improvements in AMR detection methods are being sought for more precise and faster clinical surveillance of drug resistance, since culture-based methods can be time-consuming for slow-growing bacterial species. Despite these efforts, the burden of AMR is on the rise, warranting a deeper understanding of the emergence of resistance.

Overall, the problem of addressing AMR can be dissected into four broad parts: (i) detecting the resistance, (ii) identifying the underlying AMR mechanism in that strain, (iii) developing an effective tackling strategy, and (iv) implementing that strategy. To start with, better methods to enable faster and more efficient detection of drug resistance are much needed for progress on all these fronts. Through decades of evolution, microbial communities have accumulated numerous mechanisms that confer resistance to multiple antibiotics. Although several of these mechanisms are well elucidated in literature, there remains a gap in our understanding of the evolutionary process that drives such adaptations. A deeper understanding of the cues that are sensed by bacterial populations and the precise response that drives such adaptations is necessary for tackling AMR at its source.

9.1.1 The Molecular Basis of Antimicrobial Resistance and the Need for a Systems Approach

Over the years, research in the field of AMR has revealed ways in which bacteria become insensitive to antibiotics and these mechanisms of resistance have been studied in different microbial pathogens. The primary mechanisms that are known to lead to AMR are (a) modification of the drug, (b) efflux of the drug, (c) acquisition of mutations in the drug target, (d) overexpression of the target, and (e) bypassing the target (e.g., cell wall modifications) [5]. Bacterial cells can acquire resistance endogenously through vertical transfer or exogenously through horizontal transfer [5]. In the case of vertical evolution, some bacterial cells within a population acquire spontaneous mutations in their genome that offer some survival advantage over the rest of the population [6]. These mutations can gradually get selected in the presence of the selection pressure caused by the presence of antibiotics in the environment [5]. Such mutations are often present in the drug target and cause resistance by affecting the binding affinity of the antibiotic to its cellular target [5]. Resistance acquisition can also be through the attainment of exogenous genetic material by conjugation, transformation, or transduction, from other bacterial species in the population [5]. This leads to the dissemination of resistance genes across species and within a population. Such genes, for example, the *mec* and *van* operons, usually code for a bypass mechanism around the primary cellular target [7, 8]. These bypass routes are not inhibited by the antibiotic in question, thereby leading to resistance. A well-known resistance mechanism is the acquisition of drug-inactivation enzymes such as beta-lactamases that inactivate the penicillin class of antibiotics and the macrolide esterases inactivating macrolides such as erythromycin [9]. Apart from these, modulation of the expression of existing genes can also contribute to resistance, for example, overexpression of efflux pumps that expel drug molecules from within the bacterial cell [5].

Despite the identification of several AMR mechanisms, the problem is still far from being solved due to the inevitability of the emergence of resistance, making it necessary to gain a deeper understanding of the various molecular mechanisms that lead to the resistant phenotype. At a mechanistic level, although resistance-causing genes have been well elucidated, associated with these effectors is an entire molecular network that might not directly be responsible for the phenotype, but can act as sensors and regulators that mediate it. For example, expression of the enzyme beta-lactamase (responsible for the degradation of beta-lactam antibiotics) is dependent on regulators that are induced in the presence of beta-lactams in the environment [10]. Similarly, overexpression of antibiotic efflux pumps is regulated by specific regulators in response to various cues, like oxidative stress [11]. It is useful to study these molecular networks in order to understand the information flow that leads to the expression of these resistance-conferring modules. Such insights can be translated into the discovery of novel cellular targets. Furthermore, resistance-causing genes and mutations might be accompanied by cellular alterations which can give rise to key trade-offs that arise in conjunction with resistance [12–14]. Genome-wide changes can also be directly linked with primary resistance mechanisms. For instance, expression of alternative pathways to drug targets and metabolic readjustments can contribute directly to resistance to antibiotics, as will be discussed further in the coming sections. The interaction of the drug with the target can induce several downstream events that arise due to the inhibition of essential cellular processes. The response of a bacterial cell to antibiotics in the environment can determine the treatment outcome. Timely activation of stress defenses and the necessary metabolic modulations can lead to cell survival during antibiotic treatments followed by adaptations that cause resistance [15, 16]. Studying such evolutionary aspects of resistance can reveal targets that can curb the rapid emergence of resistance. In this regard, exploration of epistatic interactions among multiple genetic loci has helped reveal evolutionary constraints that can be exploited to tackle the development of resistance.

Systems biology promises to provide a global view of molecular patterns associated with resistance and therefore, identify factors that can be targeted to curb the emergence of antimicrobial resistance. Systems biology is based on the idea that the system's properties are greater than the sum of its parts and any behavior of the system cannot solely be explained by just a single part alone [17]. Biological systems are complex with various levels of organization (cells, tissues, and organs) and multiple dynamic states. This makes it important to study the complex interplay among its components for a better understanding of the underlying physiology. With the advent of high-throughput techniques, started a new era of antimicrobial research. The increasing availability of molecular-level data has opened new ways of addressing gaps in the field. Increasing numbers of resistance determinants have since been identified through genome sequencing approaches. In contrast to the traditional methods of studying one gene at a time, we can now profile multiple molecules at the same time. This allows us to measure global changes associated with any given condition which can further be probed to gain a more complete understanding of the biological state. By combining experimental high-throughput molecular data with computational models, systems biology approaches allow us to address the complex interplay among multiple cellular processes that guide the development of antibiotic resistance.

In this chapter, we will discuss different types of systems biology approaches and their applications in understanding the mechanisms of antibiotic resistance evolution in bacterial pathogens. This chapter begins with an introduction of different high-throughput experimental data and computational models, and explores how these methods can be used to gain holistic insights into a biological system. Specifically in the context of AMR, we discuss the gaps in this field, how they have been addressed using systems biology approaches, and how such insights can aid in the discovery of more effective targeting strategies. Finally, we discuss the applications of dynamic modeling in the evolution and spread of AMR in the population and the guidance of appropriate treatment strategies.

9.2 The Systems Approach

Biological systems are complex with multiple levels of organization. Genes synthesize proteins, which interact among themselves and other biomolecules to perform various cellular functions. Cells are further organized into tissues, and tissues into organs. There are two broad methods in systems biology to address this hierarchical structure: top-down and bottom-up (Fig. 9.1). The top-down systems models aim to provide a holistic understanding of the AMR dynamics. They typically consider the complex interplay of biological, ecological, and sociological factors to simulate the spread of antimicrobial-resistant pathogens within populations. Evolutionary models apply game theory concepts to microbial populations, and model sensitivity and resistance to antibiotics as cooperation and defection strategies [18]. Evolutionary dynamics under different antibiotic selection pressures are analyzed to understand the stability of resistance and the emergence of multidrug-resistant strains.

The spread of resistant strains and its impact on the public health can be studied through different modeling approaches, which take into account pathogen transmission dynamics and influence of factors such as human contact patterns and healthcare settings. The widely used among them are the epidemiological ordinary differential equations (ODE)-based susceptible-infected-recovered (SIR) type of compartmental models, which analyze the impact of different intervention strategies such as vaccination or antibiotic stewardship on AMR prevalence in a community [19, 20]. Agent-based models (ABMs) have also been used to simulate individual agents and their interactions within a defined environment, model various aspects such as the movement and interactions of patients, healthcare workers, and pathogens within a hospital to study how different infection control measures affect the spread of resistant bacteria and predict population-level outcomes [21].

The genetic relatedness of bacterial strains with resistant phenotypes has been studied through comparative genomics and phylogenetic analysis [22]. By studying genomic sequences of these microbes and specific alterations in them as they acquire drug resistance, common genetic elements associated with antibiotic

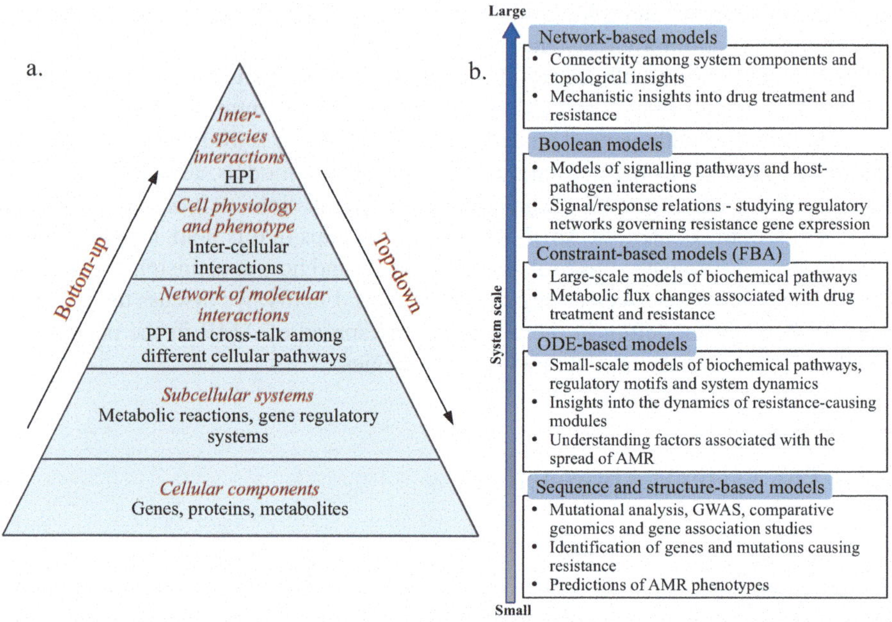

Fig. 9.1 Systems biology approaches and modeling techniques. (**a**) The hierarchical organization of biological systems. The two main approaches in systems biology are top-down and bottom-up. In a bottom-up approach, we put together small functional components to build toward the entire system, and in a top-down approach, we start with a global view of the system and go deeper to study smaller subsystems. (**b**) Different types of models used in systems biology and insights gained from them. The models have been ordered based on their scale, indicating the number of components in the model. FBA Flux balance analysis, ODE Ordinary differential equations, GWAS Genome-wide association studies, HPI Host-pathogen interactions, and PPI Protein-protein interactions

resistance are identified and used for tracking the spread of resistance genes and mutations. Genome-wide omics data provide a global profile of the system that can be used to identify its overall characteristics and obtain data-driven inferences about the molecular interactions. Although such data allow us to correlate molecular perturbations with organismal phenotypes, the complexity of the system often makes it difficult to make causal inferences [23]. In a bottom-up approach, individual molecules and information about their interactions are put together to gain insights into a larger system (Fig. 9.1). This approach can provide mechanistic insights into the interplay among different units that work together to form a system. However, in this case, as the system gets bigger, the details of the model often make it less interpretable and it may become difficult to replicate the overall capabilities of the system [23]. One can also use a middle-out approach, where a subset of the system's components that are known to be relevant to the phenotype are studied [23]. Below, we discuss the types of data and models that are frequently used in the realm of molecular systems biology.

9.2.1 Data

In order to study a system as a whole, we need information about all its components. Advances in high-throughput technologies such as genomics, transcriptomics, proteomics, and metabolomics, often collectively known as the "omics" technologies, have paved the way for the measurement of multiple molecular units at the same time and enabled the identification of genes at a genome-scale and high-throughput quantification of abundances of mRNAs, proteins, and metabolites in cells under a given condition. Using this vast amount of data, we can estimate changes in cellular components associated with a particular physiological state.

Genomics provides us with a large number of genome sequences using which we can investigate single nucleotide polymorphisms (SNPs), copy number variations, insertion and deletions (indels), and DNA methylation sites (epigenetics), affecting a particular phenotype. These data are often used to identify genetic variants associated with a disease, often at a genome-scale in studies termed as genome-wide association studies (GWAS) [24]. Such studies provide comprehensive and precise descriptions of the genotypes with a potential to serve as starting points to rationalize and possibly predict the phenotypes through a systems-level mapping and analysis. In bacterial comparative genomic studies, whole genome sequences have been used to study drug-resistant strains, find genetic determinants of a phenotype, and track phylogenetic relationships.

Apart from variations in the genetic code, gene expression changes are either directly orchestrated by genetic variations or indirectly by some other downstream agent, and can also provide information about the molecular mechanisms underlying a given physiological state. Transcriptomics, which refers to the genome-wide quantification of gene transcripts in a given sample, reveals genes with a statistically significant difference in expression between a pair of conditions. This information can be used to answer a variety of questions such as unraveling underlying mechanisms, classifying physiological states based on gene expression patterns, or identifying characteristic biomarkers of a disease. Differentially expressed genes as obtained from the analysis are typically followed by functional enrichment of these genes to identify specific pathways that may be perturbed in that disease. This is accomplished through specific statistical methods like Gene Set Enrichment Analysis [25] (GSEA), which checks for gene sets (or processes) that are significantly represented by the genes of interest obtained from the sample. These methods help us extract biological meaning from the high-dimensional data.

Similarly, proteins and metabolites are also studied primarily by using mass-spectrometry techniques. Proteomics and metabolomics are widely applied to profile changes in protein and metabolite levels under a specific condition, to identify functionally active proteins through post-translational modifications, or to study protein interactions [23, 26]. The use of statistical and computational models is necessary to analyze and interpret such high-dimensional data. Low-throughput, high-fidelity experimental methods have also contributed significantly to the vast knowledge base that is used to establish ground truth in systems-level models.

Several assays like affinity chromatography, co-immunoprecipitation, yeast-two-hybrid assays, and ChIP-Seq provide molecular interaction data that are put together to construct molecular pathways and networks [27]. Integration of these data with models can provide a deeper understanding of the system's physiology by revealing molecular players and interactions that orchestrate cellular signal-response relations.

9.2.2 Models

Proteins interact in complex cellular pathways in order to orchestrate their function. Interactions can be in the form of signaling pathways, metabolic reactions, or physical binding of proteins to bring about a specific response. Antibiotics interact with their cellular targets, eliciting a response in the bacterial system which involves extensive cellular rewiring of multiple processes. The information flows in the form of multiple protein interactions involving bacterial two-component systems and regulatory networks that govern the expression of stress-defense and resistance genes. It is important to study such molecular networks to understand the response of bacteria under antibiotic stress and point out the specific network components that are involved in drug resistance. In these networks, typically nodes represent proteins and the edges represent interactions between them, which could be physical binding, transcriptional regulation, signaling, or metabolic interactions. Apart from protein interactions, nodes in a network can also represent metabolites as in the case of biochemical pathways. In this case, metabolites would be connected by an edge if they participate in the same enzyme-catalyzed reaction. Such molecular interactions are typically identified by classical biochemical experiments supplemented with high-throughput experimental methods and are cataloged in comprehensive databases (Table 9.1), the most common being KEGG [32] and STRING [36]. Furthermore, the functional annotation of proteins has allowed us to classify them with respect to the cellular processes they are involved in. For example,

Table 9.1 Commonly used data and their repositories

Data	Resources
Sequence and annotations	NCBI [28], UniProt [29], Ensembl [30], BV-BRC [31]
Biochemical reactions and pathways	KEGG [32], Reactome [33], BioCyc [34], WikiPathways [35]
Protein interactions	STRING [36], BioGRID [37], IntAct [38], MINT [39]
High-throughput omics data	NCBI GEO [40], ArrayExpress [41], PRIDE [42], PeptideAtlas [43], MetaboLights [44]
Kinetic parameters of biochemical reactions	BRENDA [45], SABIO-RK [46]
AMR phenotypes and mechanisms	
Resistance phenotypes of bacterial strains	BV-BRC [31]
Resistance genes and mutations	CARD [47], TBDReaM [48], ResFinder [49], MegaRes [50]

enzymes that participate in glycolysis would be grouped together under the glycolytic metabolic process. Such functional linkages along with knowledge-based physical interactions among proteins are also documented in some of these databases. This extensive biological knowledge is also organized into gene ontology terms where specific genes or proteins are associated with their corresponding biological processes, molecular function, and cellular components. Several initiatives like the Gene Ontology [51, 52] (GO) and databases like Kyoto Encyclopedia of Genes and Genomes [32] (KEGG), Reactome [33], Wikipathways [35], and Molecular Signatures database [53] (MSigDB) contain information about annotated proteins and the pathways they participate in (Table 9.1). Although the interactome is not as well characterized for some bacteria as it is for humans, enough information is available for the commonly occurring MDR pathogens, enabling us to carry out a system-wide network analysis for drug-resistant strains. Put together, these resources are frequently queried for the reconstruction of knowledge-based biological protein interaction networks. Graph theoretical approaches can then be applied to study the topology and information flow through these networks [54]. Dissecting the network topology can allow for the identification of important hubs and regulators that are key to mediating the resistant phenotype.

While a network provides an overview of the functional organization and interactions, it does not capture the perturbations of interactions under different conditions. The integration of transcriptomic data with knowledge-based networks generates a condition-specific network which can then be interpreted meaningfully. Methods to integrate data into networks can be based on node weights [55] or edge weights [56]. Although node-based scores capture all the perturbed genes and their interactions, it does not prioritize the edges that are specifically active in the given condition [57]. Edge-based scores are particularly useful in this aspect as it has the potential to capture the nodes as well as the interactions that are active under the specific condition [56, 58]. Several data-driven approaches are also available that do not rely on a prior knowledge-based network to obtain contextualized networks. WGCNA [59] and ARACNE [60] are methods that construct context-specific networks based solely on statistically significant associations obtained from expression data. As expected, in order to obtain meaningful conclusions from such methods, we need a large amount of data to gain the required statistical power. It is also important to emphasize that the interactions obtained from these methods do not necessarily represent "real" biological links, but coupled expression patterns indicative of co-expressed and co-regulated genes.

Networks can further be dissected into subnetworks to study components having high interconnectivity or motifs that represent functional modules of the system. Using networks, we can also predict the essentiality of a gene, by simulating node deletions and assessing the resulting disconnectivity in the network. They can also be bipartite, which contain two different types of nodes, like metabolites and enzymes in biochemical pathways, or drug-target interaction networks. Furthermore, the wide availability of a number of software packages for carrying out these network analyses has made it increasingly accessible to the research community [61, 62].

Biological systems are dynamic in the sense that their states can change over time. For example, metabolic reactions and other protein interactions all occur at different rates giving rise to a dynamic property that needs to be taken into account while studying systems. Networks are often composed of motifs that correspond to small subnetworks having a specific architecture [63]. Such motifs are often associated with specific types of dynamic regulatory patterns, for example, feedback, feedforward, or an oscillatory pattern of expression. Although graph-based models contain dynamic elements like the flow of signal from a receptor to the downstream effectors, they do not capture the spatiotemporal progression of the system. The dynamics of signal propagation through specific subnetworks of interest can be studied by converting the components into a set of ordinary differential equations (ODEs) [23] (Fig. 9.1). This has widely been used to model biochemical reactions, where a set of metabolic reactions is converted into a system of differential equations using knowledge of their rate kinetics. Simulations with ODE-based models allow us to study the time evolution of the model, thereby improving predictions about the system's behavior. A big limiting step in this method is the availability of kinetic data that relies on experiments.

System dynamics can also be studied using Boolean models where nodes take a value of either "1" or "0" representing an "ON" or "OFF" state, respectively [64] (Fig. 9.1). The state of each node changes over the simulation period depending on logical rules governing its activity. Boolean models represent a discrete dynamic modeling framework where the system progresses over discrete timesteps. These models do not rely on kinetic parameters, making them more suitable just for qualitative analyses. Constraint-based modeling methods like flux balance analysis (FBA) are also one of the commonly used modeling approaches in systems biology [65] (Fig. 9.1). In this case, biochemical reactions are modeled as a system of linear equations with an $m \times n$ stoichiometry matrix, where there are m reactions and n metabolites. Constraints on this model, in terms of mass balance at steady state and optimization of the selected objective function, produce a solution to the system of equations. This represents calculated fluxes through each reaction under steady state.

In the context of AMR, apart from the expression dynamics of resistance operons, the emergence and spread of resistance can also be studied using dynamic models. The evolutionary dynamics and the dissemination of resistance in the community can be studied using models that consider antibiotic selection pressures, different antibiotic treatment regimens, heterogeneity in the bacterial population, pressure due to the host immune system, contacts, and pathogen transmission dynamics. While large-scale network models studied through graph-based methods provide an idea of the global organization of the system, mathematical models of smaller network components shed light on the dynamics of such systems. In order to fully understand biological systems and to make accurate predictions about their behavior, an integration of both static and dynamic models is most useful.

9.3 Systems-Level Understanding of AMR

Associated with the primary resistance machinery are global alterations within the cell that are imperative in generating or sustaining the resistance phenotype. Furthermore, the inherent tendency of biological systems to adapt to challenging environments makes it necessary to comprehensively understand the mechanisms by which bacteria respond to the drug and activate necessary processes that ultimately give rise to the adaptation. A lack of comprehensive understanding of how resistance emerges in bacteria upon drug treatment limits our ability to develop newer more effective strategies.

While genome sequencing and genetic screens have allowed us to identify resistance genes and mutations in drug targets that mediate the resistance phenotype, we can now profile changes in molecular expression patterns with the help of multi-omics technologies. Using this extensive amount of data and systems-level models, we can address some of the gaps in the understanding of AMR discussed so far. The following sections provide an overview of the global perturbations that are associated with drug resistance, drug response, and genetic interactions that have improved our understanding of resistance development. We also discuss how such findings provide new opportunities for drug discovery to treat drug-resistant infections.

9.3.1 Molecular Reprogramming Often Accompanies Antibiotic Exposure and Drug Resistance

Bactericidal antibiotics target several essential processes in bacteria to exert their effect. The most important among them are the cell wall-inhibiting drugs such as beta-lactams and glycopeptides which bind to cell wall synthesizing enzymes (PBPs) and peptidoglycan building blocks; ribosome-targeting drugs such as aminoglycosides and tetracyclines; nucleic acid synthesis inhibitors such as sulfonamides; RNA transcription inhibitors like rifamycins; and DNA gyrase inhibitors such as quinolones (Fig. 9.2). While these are the primary targets of antibiotics, the effect of an antibiotic can have far-reaching consequences for the bacterial cell. For example, treatment with quinolones not only affects DNA replication by targeting DNA gyrase, but also introduces DNA double-strand breaks that require the recruitment of DNA repair mechanisms, which in turn may alter cellular nucleotide metabolism [66]. Similarly, several drug-target interactions have been known to alter cellular reactive oxygen species (ROS) levels that might lead to the activation of redox response and the SOS response pathways which can promote emergence of drug-resistance [67] (Fig. 9.2). Extensive metabolic modulation and activation of oxidative stress responses have been widely associated with antibiotic treatment and drug resistance. Furthermore, certain modules of stringent stress, acid stress, and heat shock responses have also been linked to antimicrobial resistance [11]. Several genome-wide studies that have explored this extensive cellular rewiring are discussed in the following sections.

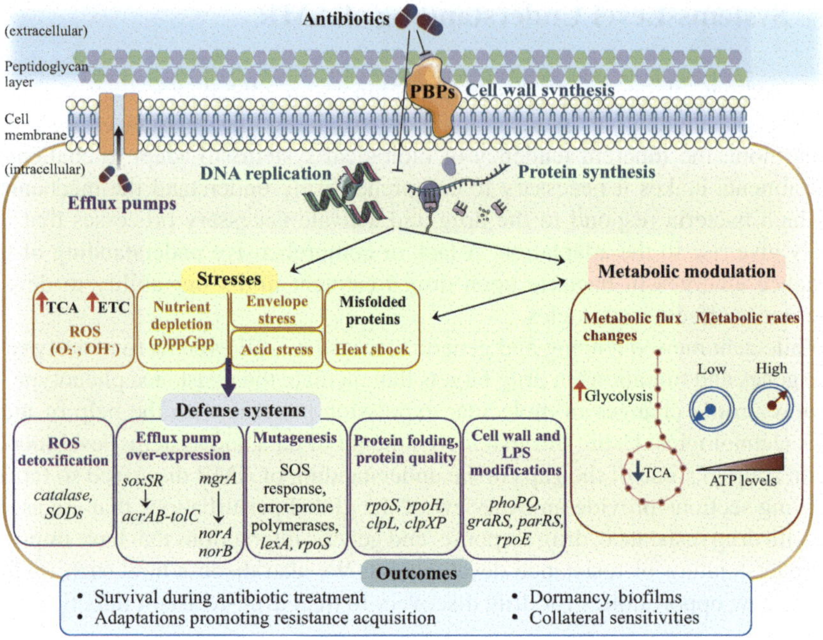

Fig. 9.2 System-wide changes associated with antibiotic treatment response and resistance. Drug-target interactions can cause perturbations in different cellular pathways. Various cellular stress defense systems and altered cellular metabolism can be associated with antibiotic treatment failures. Components of the stress defense machinery might induce necessary downstream alterations that promote survival and give rise to adaptations. Metabolic alterations are frequently associated with biofilms, swarming cells, and dormant conditions like persisters. These system-wide changes can also uncover collateral sensitivities to other antibiotics. PBPs Penicillin-binding proteins, TCA Tricarboxylic acid cycle, ETC Electron transport chain, ROS Reactive oxygen species, SOD Superoxide dismutase, ATP Adenosine triphosphate, and LPS Lipopolysaccharide. Protein-23, membrane-2d-yellowblue, translation, and DNA-nucleotides-forked icons by Server https://smart.servier.com/ are licensed under CC-BY 3.0 Unported https://creativecommons.org/licenses/by/3.0/

9.3.1.1 Modulations in Cellular Metabolism Affect Treatment Efficacy

Metabolic reprogramming is often associated with cell wall modulation, in response to DNA damage and ROS production, or biofilm production during which cells switch to an anaerobic state. Inhibition of cellular targets by bactericidal antibiotics stimulates the generation of hydroxyl radicals via the Fenton reaction which involves the TCA cycle and the electron transport chain [66]. Since the efficacy of the antibiotics is dependent on the activity of these metabolic networks, alterations in metabolism are expected to influence antibiotic susceptibility. A study has shown that increased oxygen levels can enhance antibiotic efficacy against mycobacteria by elevating the extent of oxidative damage caused by metabolic byproducts of antibiotic treatment [68]. Treatment with bedaquiline, an F_1F_0 ATP synthase inhibitor, has been shown to induce extensive metabolic rewiring involving activation of dormancy and ATP-generating pathways in mycobacteria, that attempt to compensate for the ATP crisis [69]. These examples illustrate that interactions between the antibiotic and the cellular target can cause changes in the metabolic state of the bacteria.

Independent studies have revealed that increasing the basal respiration of *E. coli* and *S. aureus* improves antibiotic efficacy [70, 71]. Heightened metabolic states during antibiotic treatment can contribute to lethality by increasing the build-up of toxic byproducts. Energy-consuming futile cycles of synthesis and degradation that exhaust cellular resources, lead to an increase in the metabolic rate and ultimately to death of the bacteria. This has been shown in the work of Cho et al. [72], where the beta-lactam mecillinam causes lethality by inducing a futile cycle of cell wall synthesis and degradation in *E. coli*. The repression of metabolism in tolerant and persisting bacterial populations leads to survival under treatment with bactericidal antibiotics. Likewise, bacteriostatic antibiotics reduce the efficacy of bactericidal antibiotics by slowing down the metabolic rate of the bacterial cells [70]. Therefore, drug-resistant bacteria often exhibit alterations in their metabolic states such that the antibiotic treatment is rendered ineffective. For example, central carbon metabolites have been shown to be downregulated in resistant organisms, which corresponds with a decreased level of ROS, thereby causing less oxidative damage during drug treatment [73]. Yet another type of modification is seen with polymyxin-resistance in gram-negative bacteria, where cell wall modifications that cause extensive rewiring of the fatty acid biosynthetic pathways are observed. Metabolomics and metabolic flux analysis demonstrated that alterations in lipid biosynthesis in polymyxin-resistant bacteria are linked with changes in energy metabolism, namely, increased flux through glycolysis and reduced flux through TCA cycle, suggesting a switch to glucose fermentation for energy production [74–76].

Despite the role of metabolism in antibiotic action, mutations in metabolism-related genes are less commonly mentioned in the context of antibiotic resistance. Mutations associated with drug resistance are frequently found in target proteins, drug-inactivating enzymes, and efflux pumps that directly interact with the drug to counter its effect. More recently, mutations in metabolic genes have been identified that modulate the metabolic state of the cell, such that it allows growth and survival of cells in the presence of antibiotics. By performing adaptive laboratory evolution experiments on *E. coli*, with gradually increasing antibiotic concentrations as well as temperature (forcing the cells to a highly active metabolic state), researchers have identified mutations in metabolism-related genes (central carbon and energy metabolism) that can help evade the metabolic pressure induced by the action of antibiotics [77]. Some of these mutations were also found in a large number of clinical drug-resistant isolates, thus suggesting that metabolism-related mutations are implicated in the resistance phenotype [77]. Studying metabolic adaptations in response to antibiotic treatment can therefore open up novel ways of targeting resistant pathogens and allow for the identification of synergistic drug combinations that can help increase the efficacy of existing treatments.

9.3.1.2 The Role of Stress Defense Systems in Modulating Antibiotic Resistance

Several studies have identified redox response to play a very important role in resistance acquisition and survival under antibiotic stress [11, 78–82]. It was found that the rate of mutagenesis increased with ROS levels during sublethal doses of antibiotic treatment [80]. This study further demonstrated that these effects in conjunction with SOS-induced mutagenesis led to the formation of mutants that were cross-resistant to different antibiotics. Apart from directly or indirectly causing mutagenesis, oxidative stresses also control the expression of a lot of defense systems like efflux pumps, porins, biofilm-related genes, and toxins which are mediated by redox-responsive regulators [11].

In an independent study, trimethoprim resistance in *E. coli* was found to be associated with cross talk among acid stress response, SOS response, and folate metabolism [83]. A critical dependence on *glyA* was observed in this particular case, which was found to be important for the synthesis of metabolic intermediates required for nucleotide biosynthesis and biofilm formation [83]. A recent study offered a genome-wide view of cellular processes that decrease antibiotic susceptibility and induce tolerance to specific antibiotics in *S. pneumoniae* [84]. Using Tn-seq in *S. pneumoniae* exposed to different antibiotics, the authors identified fitness changes associated with processes that were seemingly distant in function from the primary drug targets. Apart from known patterns, such as the involvement of DNA repair genes in fluoroquinolone treatment, cell wall and peptidoglycan synthesis genes during treatment with beta-lactams and glycopeptides, several unknown effects were also identified. Modulation of processes such as glycolysis, purine metabolism, (p)ppGpp and ATP synthesis, and transport systems, among others, were found to alter fitness during antibiotic selection [84]. Importantly, inhibition of a given pathway could simultaneously be beneficial or detrimental for the pathogen depending on the specific antibiotic in the environment. The alarmone (p)ppGpp forms a part of the bacterial stringent response and coordinates several responses like the downregulation of ribosomal RNA synthesis and the upregulation of stress-related genes and amino acid biosynthetic pathways [85]. Apart from modulating cellular metabolism in response to nutrient limitation, (p)ppGpp has also been implicated in antibiotic resistance, for example, in mediating vancomycin tolerance in *E. faecalis* [86] and antibiotic persistence in several species. Aminoglycosides and beta-lactams have been shown to induce heat shock response in several bacteria [11]. It is hypothesized that the heat shock components, through the elimination of misfolded proteins, might reduce toxicity due to the accumulation of aberrantly folded polypeptides during antibiotic treatment [87]. In a separate study, an extensive metabolic shift has been reported in response to nitrogen limitation, giving rise to swarmer cells. Swarming has been shown to reduce antibiotic susceptibility, which can be attributed to the cell envelope changes (thickening of the envelope and surface charge affect antibiotic permeability) associated with metabolic alterations [88]. Put together, these system-wide alterations in multiple metabolic and stress defense pathways point toward deeper implications of antibiotic treatment on bacterial cells (Fig. 9.2).

It is becoming increasingly evident that resistance mechanisms are not limited to single molecules interacting directly with the antibiotic. The cellular response to an antibiotic can influence the treatment outcome and promote adaptations that pave the way for resistance acquisition. By combining high-throughput omics data with systems-level models, we can probe alteration patterns in multiple cellular pathways in response to different antibiotic classes, which can in turn guide the development of newer antimicrobials and discovery of novel cellular targets.

9.3.2 Understanding the Evolution of AMR

Apart from characterizing the resistome and elucidating the mechanisms of resistance, it is important to understand the evolutionary principles that drive the process of resistance acquisition, so as to both predict and prevent resistance evolution. System-wide profiling of bacterial clones obtained from different points along the evolutionary trajectory helps determine mutations and molecular alterations responsible for the gradual, stepwise adaptations. The emergence and fixation of a resistance mutation in the population depends on factors like the mutation rate, the strength of the selection pressure in terms of the concentration of the antibiotic being used and the duration of treatment, and the fitness differences associated with the resistance-conferring mutations [89]. Several studies have addressed this by performing highly parallel laboratory evolution experiments with tightly controlled selection pressures and population sizes [90, 91] In vitro laboratory evolution experiments (adaptive laboratory evolutions or ALEs) are frequently used as tools to replay the process of resistance evolution in a controlled laboratory setup and to identify factors that drive the process [90, 91]. In this regard, studying lab-evolved strains is more useful as compared to clinical resistant isolates for which we do not have any information about their evolutionary history. To add context to the insights obtained from such lab-evolved strains, the findings can further be tested on clinical isolates. Integrating in vitro evolutions with omics data and systems-level approaches expands our understanding of the evolution of AMR [90, 92] and offers opportunities for developing novel therapeutic strategies, predicting potential future resistance patterns and for finding ways to prolong the effectiveness of existing therapeutics in clinical settings.

9.3.2.1 Predicting Evolutionary Trajectories

A major limitation in the prediction of resistance evolution is posed by the intrinsic stochasticity of the evolutionary process. Therefore, independently evolving populations may not follow the same genetic trajectory [92]. Heterogeneity in drug response and genetic background leading to complex intergenic epistasis can alter the outcomes. By tracking numerous lineages using high-throughput techniques, several groups have tried to understand the complex interplay among various factors underlying the dynamics of resistance evolution.

Fitness or adaptive landscapes of antibiotic resistance evolution comprise a high-dimensional genotype space with corresponding fitness evaluations. This is an example of the top-down approach wherein high-throughput data from multiple ALE experiments are used to explore multiple routes to high fitness and make predictions about the evolutionary trajectories. During bacterial evolution under a selection pressure, mutations in different genetic loci can originate spontaneously and alter the fitness of the cell. These mutants then compete toward fixation in the population. Taking this into account, in vitro laboratory evolution experiments are performed with the goal of replicating bacterial evolution as it happens in the natural environment.

Owing to the inherent randomness of the process, multiple laboratory evolution experiments give rise to diverging genotypes that limit replicability of evolutionary trajectories [92]. Therefore, the high dimensionality of the genotype space makes it difficult to draw predictions based on genotypic data. However, different mutations are often associated with redundant effects, for example, loss of function of a gene can be the result of mutations in distinct regions of the locus. These redundancies imply that, as opposed to genotypes, functional effects of the mutations are often repeatable and therefore, we do not need to know the exact path taken by the population in the genotype space if we are to predict functional changes [92]. Several studies have therefore suggested the use of phenotypic profiles, as opposed to genotypic data, to construct adaptive landscapes that hold higher predictive power [92]. In a recent study, multiple phenotypic measurements that constituted the phenotypic space were used to construct an adaptive landscape for the evolution of antibiotic resistance in *E. coli* [93]. By performing highly parallel laboratory evolutions with eight different antibiotics starting from different points in the phenotype space, the authors sampled enough points to construct a phenotype-fitness landscape. Fitness evaluated as IC50 values against the eight different antibiotics and trajectories projected on this eight-dimensional resistance space were used to infer the antibiotic resistance landscape. The directions of evolution predicted by this landscape corresponded well with the observed trajectories and the fitness peaks represented distinct antibiotic resistance mechanisms. Although the inclusion of more antibiotics or a different set of antibiotics can change the predictions from the landscape, this method showed how phenotypic profiles could be used to infer fitness landscapes and can allow better predictability of evolutionary trajectories [93].

Over the years, several groups have elucidated mutational landscapes of specific genes by studying fitness changes associated with large-scale mutations in the gene of interest, followed by the approximation of the distribution of fitness effects (DFEs) [94–97]. The shape of this distribution reveals important aspects of the evolutionary dynamics, for example, a DFE with a higher variance implies a higher probability for the emergence of beneficial mutations and therefore greater evolvability [98]. Several studies have explored the DFEs of mutations in specific resistance genes [94–97, 99]. Particularly, mutations in TEM-1, a beta-lactamase gene, have been studied extensively for their potential to alter fitness in the presence of beta-lactams. Few mutations were found to confer a much greater change in fitness, revealing constraints in the mutational landscape, and thereby implying that evolution of resistance genes might be more predictable than expected [94, 97]. In another study, it was found that genetic interactions can also increase the number of feasible pathways to adaptive peaks, observed during the evolution of trimethoprim resistance in *E. coli* [100]. In this study, the authors inferred a multi-peaked adaptive landscape by reconstituting all resistance-conferring mutations in DHFR (dihydrofolate reductase; trimethoprim target) and estimating their corresponding fitness effects. Epistatic interactions among the mutations were found to increase the accessibility of adaptive peaks by generating paths that involved multiple gains and losses of mutations. While this expansion of the number of alternate paths to a specific fitness peak made it difficult to draw predictions from the landscape, it indicated the complexity of the genetic interactions that underlie the evolutionary process of resistance acquisition.

9.3.2.2 Epistatic Interactions in the Evolution of Resistance

Recent studies have identified that certain bacterial strains evolve resistance at a higher frequency [101–107], meaning that there are specific genetic backgrounds (presence, absence, or mutations in some genetic loci) that may open new routes to easily accessible fitness peaks. In this regard, epistatic interactions have been deemed important in constraining evolutionary trajectories by altering the fitness costs associated with specific evolutionary routes. Such interactions are difficult to study intuitively because they might involve seemingly unrelated genes and can be orchestrated through multiple levels of cross talk. Several studies have used ALEs and omics approaches to identify high-evolvability lineages and the genes that influence the rate of adaptation to antibiotics in these lineages [101, 105, 106]. The effect of the interactions between a specific genetic background and the resistance-causing modules on the evolutionary outcome can be studied using systems-level models followed by experimental validation.

In a long-term evolution experiment of *E. coli*, reported by the Lenski group, certain clones were identified that showed a higher probability of evolving a citrate utilization phenotype, as compared to the wild-type parent population [102]. Such populations of cells that were phenotypically similar to the wild-type population but had a higher potency to evolve in the near future were thought to be potential intermediates in the evolutionary process and this phenomenon was referred to as "potentiation" [102]. Similar observations were also made in the context of antimicrobial resistance, where the evolution of resistance was thought to be contingent on the particular history of the lineage [101]. Specific mutations or genes could open up different pathways to resistance giving access to previously inaccessible fitness peaks.

Further investigation from multiple groups revealed "potentiator" genes and mutations, and molecular patterns in specific strains, were at a higher risk of evolving resistance to antibiotics. Specifically, efflux pumps, transcriptional regulators, and drug-deactivating enzymes were found to play a role [103, 105–107]. In *P. aeruginosa*, parallel in vitro evolutions coupled to genome sequencing identified evolutionary constraints faced by the bacteria during colistin resistance evolution [103]. In this case, strong epistatic interactions among multiple loci were found to determine the fate of a population in terms of resistance level to colistin. Mutations in transcriptional regulators *pmrB* and *phoQ* were associated with increased evolvability by modulation of the expression of *arnB* operon which is required for colistin resistance [103]. Another study compared eight different strains of *P. aeruginosa* with respect to their ability to evolve resistance to the antibiotic ceftazidime. The differences in evolvability were attributed to the presence of a transcription factor *ampR*, which modulates the expression of *ampC* (beta-lactamase) in response to the accumulation of peptidoglycan catabolites during beta-lactam treatment [105]. Through complex intergenic epistatic interactions, potentiator genes can alter the fitness effects of subsequent mutations thereby directing the cells along a constrained evolutionary trajectory. In *S. aureus*, the efflux pump *norA* was associated with increasing the fitness benefit conferred by DNA topoisomerase mutations under ciprofloxacin treatment [106]. High-throughput evolution experiments have been performed to screen gene deletion strains of *E. coli*, for the identification of genes that affect resistance evolvability [108]. In this study, evolution experiments, tightly controlled for selection pressure, were performed with 98 gene deletions representing diverse cellular processes that had only minor fitness defects. Some of these deletions accelerated the mutation rate thus increasing evolvability of resistance to tetracycline. While several deletions showed a pattern of diminishing-returns epistasis, deletions of efflux pump genes significantly lowered evolvability, with *tolC* (part of several efflux pumps) deletion completely blocking resistance evolution. Additionally, genes related to protein folding and lipopolysaccharide biosynthesis also contributed to the evolvability of resistance.

One of the mechanisms by which potentiator genes increase evolvability is by favoring the survival of bacteria in the presence of antibiotics, thereby increasing the probability of acquiring mutations that can be beneficial. Transcriptional regulators control the expression of a wide range of targets, therefore mutations or alterations in these can open up multiple possibilities of resistance acquisition. Genes and genetic variations that have been linked to increasing the evolvability of strains can act as targets for preventing the rapid emergence of resistance. Identifying such evolutionary patterns could also mean that evolution might be repeatable in some cases and therefore more predictable than previously thought. Put together, these findings improve our understanding of resistance evolution which can guide the identification of more effective inhibition points.

9.3.3 Exploiting Emergent Vulnerabilities in Drug-Resistant Strains to Identify Novel Targets and Antibiotic Combinations

The alterations in different cellular processes and the complex interactions that underlie the resistance phenotype have led us to several strategies to tackle resistant pathogens. So far, we have established that diverse mechanisms exist for causing resistance and sometimes these systems can be distant from the drug targets. This indicates that complex cellular networks exist that coordinate the information flow to the appropriate components of the resistance-causing effectors. Gene expression data of drug response also indicates alterations in different cellular modules that are seemingly remote from the primary resistance-causing machinery. Using a network of protein-protein interactions, Raman et al. [109] discovered possible pathways that might be responsible for giving rise to drug resistance. The study highlights controlling hubs in the network that mediate resistance and suggest their use as "co-targets." Co-targets have been referred to as genes that are activated in response to drug treatment and need to be inhibited along with the primary targets in order to keep the emergence of resistance in check. The study reported probable routes to resistance to specific drugs. Furthermore, it also suggested that different targets might have different propensities for developing resistance, implying that bacteria might have a higher chance of developing resistance to certain drugs as compared to others. Further work in this direction will throw light on whether these are viable targeting strategies and can help control resistance evolution.

A systems perspective of the molecular networks and pathways involved in generating resistance confers opportunities for understanding emergent properties of that system, specifically identification of conditional essentiality of any specific protein, pathway, or process leading to new vulnerabilities. Cellular modulations and molecular reprogramming in resistant pathogens can lead to trade-offs and vulnerabilities that can give rise to collateral sensitivities. Collateral sensitivity occurs when resistance acquisition against one antibiotic is accompanied by increased susceptibility to a different antibiotic [12, 13]. Although more effort is required to probe these systematically, there are some well-known instances of it. As we have seen earlier, resistance mechanisms are often associated with genome-wide alterations and such alterations can incur a fitness cost on the cell which can be exploited for finding collateral sensitivities. For example, resistance to aminoglycosides in gram-negative bacteria is mediated through mutations that alter the cell membrane potential, thereby affecting the uptake of the antibiotics [12]. Such mutations cause changes in oxidative phosphorylation, or proton/potassium ion symport, leading to a decrease in the proton motive force (PMF) [12]. However, this affects the functioning of PMF-dependent efflux pumps which are responsible for the removal of other antibiotics from the cell. The AcrAB efflux pump system in *E. coli*, a major contributor of multidrug resistance is hugely affected, thereby conferring collateral sensitivity to various antimicrobials [110].

Resistance due to mutations in molecular targets is often accompanied by susceptibility to other antimicrobials. For instance, mutation in the gene *fusA* (elongation factor G) which causes fusidic acid resistance in gram-positive bacteria, results in slow growth rate, reduced heme and (p)ppGpp levels, and heightened damage due to oxidative stress and DNA damage [12]. Consequently, such mutants exhibit collateral sensitivity to several antibiotic classes like beta-lactams, fluoroquinolones, rifampicin, aminoglycosides, and chloramphenicol [111]. System-wide changes in resistant pathogens can also give rise to patterns of collateral sensitivities. In the case of fluoroquinolones, resistance is mediated by mutations in its target, *gyrA* (DNA gyrase). A specific *gyrA* mutant was found to be susceptible to doxycycline [112] and it is hypothesized that the resulting genome-wide modulation of transcription, most likely affected the expression of stress-related genes like *rpoE, rpoS,* and *recA* [12]. In a separate study, collateral sensitivities in isoniazid-resistant mycobacteria were discovered using contextualized transcriptome-integrated protein interaction networks [82]. Mutation in the catalase gene (*katG*) along with alterations in oxidative stress response pathways, led to a diminished potential of the resistant bacteria to counter redox stress. Consequently, clinical isolates of isoniazid-resistant mycobacteria were found to be susceptible to oxidative stress-inducing antibiotics like vancomycin [82].

Collateral sensitivities are also observed in the case of horizontal transfer of antibiotic resistance genes which is often energetically expensive for cells and causes extensive metabolic rewiring [14]. In one study, the fitness costs associated with resistance gene-carrying plasmids were explored to look for effective antibiotics in resistant *E. coli* [13]. Most of these plasmids also conferred cross-resistance to other antibiotics apart from beta-lactams. While a lot of collateral sensitivity patterns were observed, of note, was the plasmid pOXA-48 which confers resistance to carbapenem antibiotics. This plasmid resulted in conserved collateral sensitivity to azithromycin, colistin or both across phylogenetically diverse strains of *E. coli*. Although the mechanisms were not discussed, the study provides an overview of plasmid-mediated collateral sensitivity patterns that can be explored further. Collectively, these studies reveal that evolutionary trade-offs often arise as a consequence of antibiotic resistance. Identification of such collateral sensitivities can help curb the rapid emergence of resistance by guiding antibiotic cycling and antibiotic combination treatment regimes.

9.4 Modeling System Dynamics

Studying system-wide alterations in molecular patterns associated with antimicrobial resistance allows us to better understand the mechanistic aspects of resistance and its emergence. The topics discussed so far mostly offer a static picture of cellular modulations associated with resistance. However, biological systems are highly dynamic, in the sense that several time-dependent factors play a role in the expression of a phenotype. For example, drug response itself is orchestrated through the timely activation of subcellular regulatory networks that depend on the kinetics of the interactions. As will be discussed in the next section, the architecture of such regulatory motifs governs the time-dependent and coordinated expression of the response modules which also affects the antibiotic treatment outcome. At a population level, the dynamics of the interactions among bacterial cells under antibiotic selection pressure and the host immune system determine the emergence and spread of resistant bacterial populations in the community. The following sections focus on the use of mathematical models to study the temporal aspects of cellular networks that regulate the expression of resistance determinants and the guidance of effective treatment strategies based on the observation of infection dynamics with hosts.

9.4.1 Dynamics of Gene Regulatory Networks

Bacteria respond to their environmental cues via multiple molecular interactions. Sensor molecules present on the cell membrane can bind to drug molecules in the extracellular milieu and transmit this signal through the cell to modulate the expression of downstream effector genes that bring about the response. Complex biological networks are often dissected by identifying subnetworks with specific architectures referred to as network motifs that influence specific outcomes [63]. The type of connections in the motif defines its dynamic behavior and is specific to the type of response needed, for example, positive or negative feedback. Such regulatory network motifs are often involved in the modulation of resistance gene expression. Studying the architecture and the resulting dynamics of such regulatory motifs can shed light on the fine-tuning and the timely expression of resistance genes during antibiotic treatment, thereby, expanding our understanding of drug response and in turn, offering new approaches for tackling AMR.

The usefulness of studying the dynamics of a gene regulatory network is illustrated by Schultz et al. [113], who quantitatively studied the dynamics of the *tet* operon in *E. coli*. Operons of resistance genes and their regulators are often transferred across bacterial species via horizontal gene transfer, examples being the *mec*, *tet*, and *van* operons. These regulatory systems form small network motifs that mostly function as independent units exempt from cellular control. The *tet* operon is comprised of *tetA*, an efflux pump for tetracycline, and *tetR*, a repressor for *tetA*. During antibiotic exposure both genes are transcribed, thereby keeping the cellular burden due to the overexpression of the efflux pump under check. Combining mathematical modeling with experimental data, the authors identified that timely expression of *tetR* is the key to cell survival under drug treatment [113]. The rate of *tetR* expression resulted in three distinct cell fates—survival due to a high rate of *tetR* expression, death due to excess *tetA* (suboptimal *tetR* expression), and death from the drug due to a low rate of expression.

Several other resistance gene operons share a similar architecture, such as the *mec* operon in *S. aureus* that regulates β-lactam resistance. Similar to the *tet* motif, the *mec* module also expresses an inhibitor along with the resistance gene during antibiotic treatment. Studying the fitness trade-offs similarly as Schultz et al. [113] may shed light on the workings of these resistance motifs under environmental stress. The dynamics of another efflux pump, the AcrAB-TolC efflux pump system, in gram-negative bacteria has also been studied in detail [114]. The AcrAB-TolC and AcrEF-TolC efflux pump systems are major contributors to MDR in bacteria. The gene regulatory network governing their expression under antibiotic treatment is complex and its understanding is important for designing inhibitors against these systems. Youlden et al. [114] built quantitative ODE-based mathematical models to assess the efflux pump expression dynamics in a wild-type and a resistant mutant strain. Steady-state analysis of these models revealed targets that, if inhibited, can suppress the expression of both efflux pumps during antibiotic stress. Furthermore, the study also reported the temporal progression of events leading to overexpression of the efflux pumps and identified targets that can act at shorter timescales. Therefore, dissecting the constraints of these resistance modules expands our understanding beyond molecular mechanisms to the quantitative dynamics of antibiotic response that can uncover alternative targeting strategies.

9.4.2 Infection Dynamics in the Host and the Guidance of Antibiotic Treatment Strategies

An important aspect of controlling the emergence of resistance is to identify how evolutionary principles can guide effective treatment strategies. The spread of resistance in the population depends on the selection and fixation of resistant pathogens in the population. This fixation of resistance is determined by factors like the mutation rate of bacteria, fitness costs of the resistance-causing alterations, the heterogeneity of the bacterial population, and the strength of the selection pressure [89]. Within-host mathematical models have been used to describe factors that determine the competition among strains in a heterogeneous bacterial population and the interplay of host immune responses with antibiotic treatment [115–119]. The competition among sensitive and resistant strains was studied using models that identified factors like antibiotic concentration and strength of the selection pressure to be important in determining the survival, extinction, or co-existence of resistant and sensitive strains [120]. The constant competition between resistant and sensitive bacterial strains in the environment has given rise to mutants that have low fitness costs associated with the resistance mechanism. The evolution of compensatory mutations allows such strains to employ the resistance machinery only in the presence of antibiotics. Therefore, lowering the selection pressure imposed by antibiotics often does not work in selecting against the resistant bacterial population. Treatment strategies involving antibiotic combinations and antibiotic cycling have been suggested to be effective in controlling resistance evolution [121]. Specific treatment regimens like fixed-dose-fixed-duration, and adaptive treatments based on the infection progression have also been explored [115, 116]. These strategies have been evaluated extensively using mathematical models, to find out the best treatment protocol that can effectively clear infections and curb the spread of resistance.

Apart from drug-target interactions in the pathogen, the interaction of the pathogen with the host immune system also determines the outcome of the infection in terms of bacterial clearance or persistence in the host. Infection with a bacterial pathogen induces several components of the immune system that try to tackle the infection by clearing the pathogen from the system. Apart from some interactions being specific to a pathogen, there are more general immune responses that are activated during bacterial infections. A typical host response to a bacterial infection starts with the recognition of bacterial signatures (pattern-associated molecular patterns or PAMPs) by pattern recognition receptors (PRRs) that are present on the surfaces of immune cells like macrophages, and dendritic cells to name a few [122]. This interaction leads to the recognition and ingestion of the pathogen (called phagocytosis) by macrophages followed by bacterial lysis mediated by various cellular mechanisms like acidic pH, oxidative stress, and lytic enzymes [123]. The presentation of bacterial fragments by dendritic cells can further trigger helper T cells that secrete inflammatory molecules like cytokines which can activate cell-mediated or humoral immunity [122].

Within-host mathematical models have been used to study the effect of different antibiotic treatment strategies and their interplay with the host immune system in determining infection outcomes such as clearance time, emergence, and selection of resistant bacteria. Gjini and Brito et al. [115] explored the differences in treatment efficacy between classical (fixed-duration and fixed-time) and adaptive (follows infection outcomes and patient symptoms) treatment strategies in relation to the host immune response. While classical treatment strategies were found to be more effective in immunocompromised hosts, adaptive treatments performed better in controlling the selection of resistance by utilizing lower antibiotic doses and in promoting host immunization [115]. In both cases, treatment timing and the host immune state were found to be important parameters for a successful outcome. In another study, models of host infection dynamics have been used to study the evolution of resistance-conferring mutations under the interplay of host immunity and antibiotic dosage regimes [116]. Results from this study suggested that the probability of high-level resistance evolving declines with increasing dosage of the drug used for treatment, which is supported by experimental data. On the other hand, lower antibiotic doses can delay bacterial clearance due to a combination of low bactericidal activity and reduced activation of the host immune response. The authors suggest that the emergence and spread of resistance mutations depend on a complex cross talk between the pharmacokinetics, pharmacodynamics, and the host immune system.

Mathematical models have also been used to understand the factors governing the spread of resistance among individuals in hospitals and community settings [124–126]. Single or multiple treatment regimens have been evaluated using compartmental models where uninfected hosts and patients infected with sensitive or resistant bacteria are considered distinct compartments [127]. In the case where two drugs are considered, additional compartments are included, representing resistance to either of the two or both drugs [127]. In the work of Bergstrom et al. [125], antibiotic cycling and antibiotic mixing (combination treatment) have been evaluated for their importance in preventing the formation of new resistant strains and in limiting the spread of resistant strains present in the population. Antibiotic mixing was found to be more effective than antibiotic cycling in controlling resistance, which is in line with the poor success observed for clinical trials involving antibiotic cycling [125]. The authors explain that mixing imposes greater variation than cycling, thereby slowing down the selection and spread of resistance.

Put together, simulations with mathematical models have allowed the exploration of several factors and treatment strategies for controlling the spread of AMR in the population. However, some of these models need to be validated experimentally using in vivo data. Instead of following generalized treatment regimes, individualized treatments that combine appropriate dosing strategies and shorter treatment durations might prove to be more effective at clearing infections and preventing the emergence of resistance.

9.5 Conclusion

An extensive body of research on AMR has revealed different mechanisms of resistance and approaches to find novel targeting strategies and antibiotics. Yet, the problem of antibiotic resistance is still far from being solved because of the inevitability of antibiotic resistance. More recently, research in AMR has mostly focused on the identification of factors that contribute to the evolution of resistance. Predictions of antibiotic fitness landscapes rely on a large amount of ALE data and are still difficult to generalize for different bacteria and different antibiotics. The inherent heterogeneity of bacterial populations in responding to antibiotics also adds to the complexity of the evolutionary process.

It is evident that antibiotic resistance is the result of a complex interplay among various components of the system and therefore it requires the use of holistic models to address the gaps in this field (Fig. 9.3). The advancements in high-throughput omics technologies and the development of systems-level models have contributed greatly to gaining insights into the mechanisms of resistance and the factors that play a role in its emergence. An extensive molecular network senses and responds to antibiotics in the environment by activating necessary components of the system that detoxify and counter the stress due to antibiotics. Therefore, perturbations in molecular components that are seemingly disconnected from the primary target have important implications for drug resistance. We now know that several mechanisms exist that can be located away from the primary drug target. Several of these response systems also pave the way for resistance acquisition by various mechanisms like inducing mutagenesis, or by promoting survival under antibiotic treatment. With regard to the evolution of resistance, we have discussed how genetic interactions among multiple loci can influence the evolutionary trajectory of antibiotic resistance by balancing the fitness costs and benefits of mutations. This has led

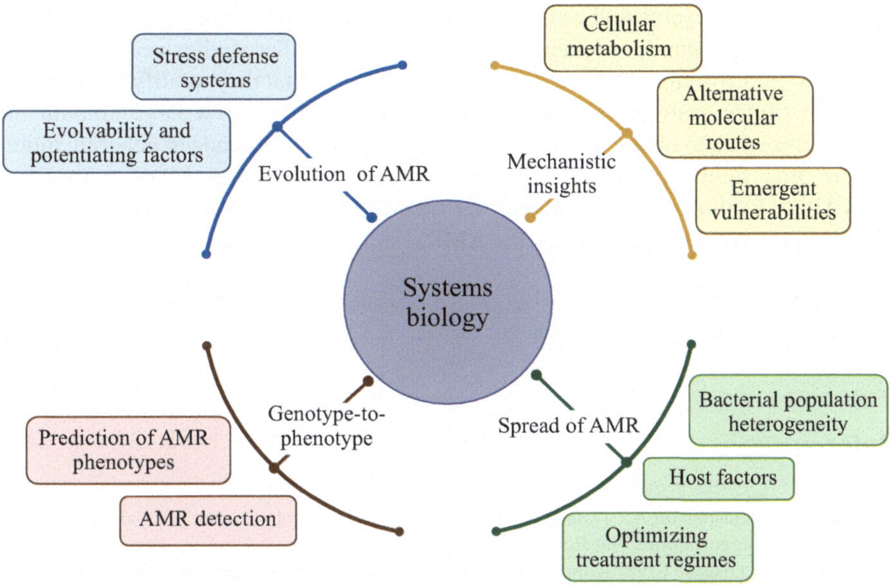

Fig. 9.3 Gaps in AMR that can be addressed by systems biology. Systems biology approaches can expand our understanding of AMR. Novel mechanistic insights can reveal alternate routes, metabolic adjustments, and trade-offs that can be exploited for the discovery of novel targets. Systems biology approaches can also be used in the identification of factors promoting the evolution and spread of resistance through the use of dynamic models. Furthermore, the availability of a large amount of genomic data can aid in the prediction of AMR phenotypes and help in faster AMR detection. AMR Antimicrobial resistance

to the discovery of specific genetic backgrounds with higher evolvability of resistance than some others, thereby revealing critical inhibition points for controlling the emergence of resistance. Furthermore, resistance mechanisms can be accompanied by emergent vulnerabilities and such trade-offs can be exploited for discovering collateral sensitivities to different antibiotics (Fig. 9.3). This allows for the development of newer and possibly more effective treatment strategies like antibiotic combinations and antibiotic cycling regimes. Using dynamic models, the efficacy of different antibiotic treatment regimens has also been studied with respect to their ability in controlling the spread of resistance in the host. Such efforts collectively aim to curb the rapid rise in the emergence and spread of resistance.

It is also desirable to obtain a genotype-to-phenotype map so as to enable faster and more efficient detection of AMR. Although culture-based antimicrobial susceptibility tests are accurate for the determination of drug susceptibility of clinical bacterial isolates, they are limited to the testing of metabolically active and fast-growing species. An alternative to this could be the use of known AMR genes as markers to detect the presence of resistant isolates. The availability of a large number of genome sequences of infection-causing pathogens, along with information about their resistance profiles, has opened up opportunities to make genotype-to-phenotype predictions. Using sequence-based comparison approaches and machine learning models, several studies have identified novel AMR loci and predicted AMR phenotypes from genome sequence data [128–130]. Further studies need to be carried out to improve the accuracy of such predictions so that they qualify the clinical standards. Since systems biology approaches rely heavily on a large amount of experimental data to obtain useful insights, efforts need to be made to generate more data in this field, specifically for the commonly occurring MDR pathogens (ESKAPE). Combining such efforts, we can aim to find novel targets and more effective treatment strategies to tackle AMR.

Acknowledgments Adithya Chedere, Shrisruti Sriraman, and Naren Chandran Sakthivel are acknowledged for reading and reviewing this chapter.

References

1. Crofts TS, Gasparrini AJ, Dantas G. Next-generation approaches to understand and combat the antibiotic resistome. Nat Rev Microbiol. 2017;15:422–34.
2. Centers for Disease Control and Prevention (U.S.). Antibiotic resistance threats in the United States, 2019. Atlanta: US Department of Health and Human Services; 2019. https://stacks.cdc.gov/view/cdc/82532, https://doi.org/10.15620/cdc:82532
3. da Cunha BR, Fonseca LP, Calado CRC. Antibiotic discovery: where have we come from, where do we go? Antibiotics. 2019;8:45.
4. Liu M, Van Voorhis WC, Quinn RJ. Development of a target identification approach using native mass spectrometry. Sci Rep. 2021;11:2387.
5. Laws M, Shaaban A, Rahman KM. Antibiotic resistance breakers: current approaches and future directions. FEMS Microbiol Rev. 2019;43:490–516.
6. Francine P. Systems biology: new insight into antibiotic resistance. Microorganisms. 2022;10:2362.
7. Périchon B, Courvalin P. VanA-type vancomycin-resistant Staphylococcus aureus. Antimicrob Agents Chemother. 2009;53:4580–7.
8. Li G, Walker MJ, De Oliveira DMP. Vancomycin resistance in enterococcus and Staphylococcus aureus. Microorganisms. 2022;11:24.
9. Egorov AM, Ulyashova MM, Rubtsova MY. Bacterial enzymes and antibiotic resistance. Acta Nat. 2018;10:33–48.
10. Zeng X, Lin J. Beta-lactamase induction and cell wall metabolism in Gram-negative bacteria. Front Microbiol. 2013;4:128.
11. Poole K. Bacterial stress responses as determinants of antimicrobial resistance. J Antimicrob Chemother. 2012;67:2069–89.
12. Pál C, Papp B, Lázár V. Collateral sensitivity of antibiotic-resistant microbes. Trends Microbiol. 2015;23:401–7.
13. Herencias C, et al. Collateral sensitivity associated with antibiotic resistance plasmids. eLife. 2021;10:e65130.
14. San Millan A, et al. Integrative analysis of fitness and metabolic effects of plasmids in Pseudomonas aeruginosa PAO1. ISME J. 2018;12:3014–24.
15. Levin-Reisman I, et al. Antibiotic tolerance facilitates the evolution of resistance. Science. 2017;355:826–30.
16. Liu J, Gefen O, Ronin I, Bar-Meir M, Balaban NQ. Effect of tolerance on the evolution of antibiotic resistance under drug combinations. Science. 2020;367:200–4.
17. Raman K, Chandra N. Systems biology. Resonance. 2010;15:131–53.

18. Niehus R, Oliveira NM, Li A, Fletcher AG, Foster KR. The evolution of strategy in bacterial warfare via the regulation of bacteriocins and antibiotics. eLife. 2021;10:e69756.
19. Durazzi F, et al. Modelling antimicrobial resistance transmission to guide personalized antimicrobial stewardship interventions and infection control policies in healthcare setting: a pilot study. Sci Rep. 2023;13:15803.
20. Changruenngam S, Modchang C, Bicout DJ. Modelling of the transmission dynamics of carbapenem-resistant Klebsiella pneumoniae in hospitals and design of control strategies. Sci Rep. 2022;12:3805.
21. Almagor J, et al. The impact of antibiotic use on transmission of resistant bacteria in hospitals: insights from an agent-based model. PLoS One. 2018;13:e0197111.
22. Zhang R, et al. Epidemiological and phylogenetic analysis reveals Flavobacteriaceae as potential ancestral source of tigecycline resistance gene tet(X). Nat Commun. 2020;11:4648.
23. Tavassoly I, Goldfarb J, Iyengar R. Systems biology primer: the basic methods and approaches. Essays Biochem. 2018;62:487–500.
24. Uffelmann E, et al. Genome-wide association studies. Nat Rev Methods Primer. 2021;1:59.
25. Subramanian A, et al. Gene set enrichment analysis: a knowledge-based approach for interpreting genome-wide expression profiles. Proc Natl Acad Sci. 2005;102:15545–50.
26. Richards AL, Eckhardt M, Krogan NJ. Mass spectrometry-based protein–protein interaction networks for the study of human diseases. Mol Syst Biol. 2021;17:e8792.
27. Zhou M, Li Q, Wang R. Current experimental methods for characterizing protein–protein interactions. ChemMedChem. 2016;11:738–56.
28. NCBI Resource Coordinators. Database resources of the National Center for biotechnology information. Nucleic Acids Res. 2017;45:D12–7.
29. The UniProt Consortium, et al. UniProt: the universal protein knowledgebase in 2023. Nucleic Acids Res. 2023;51:D523–31.
30. Martin FJ, et al. Ensembl 2023. Nucleic Acids Res. 2023;51:D933–41.
31. Olson RD, et al. Introducing the Bacterial and Viral Bioinformatics Resource Center (BV-BRC): a resource combining PATRIC, IRD and ViPR. Nucleic Acids Res. 2023;51:D678–89.
32. Kanehisa M. KEGG: Kyoto encyclopedia of genes and genomes. Nucleic Acids Res. 2000;28:27–30.
33. Gillespie M, et al. The reactome pathway knowledgebase 2022. Nucleic Acids Res. 2022;50:D687–92.
34. Karp PD, et al. The BioCyc collection of microbial genomes and metabolic pathways. Brief Bioinform. 2019;20:1085–93.
35. Martens M, et al. WikiPathways: connecting communities. Nucleic Acids Res. 2021;49:D613–21.
36. Szklarczyk D, et al. The STRING database in 2023: protein–protein association networks and functional enrichment analyses for any sequenced genome of interest. Nucleic Acids Res. 2023;51:D638–46.
37. Stark C, et al. BioGRID: a general repository for interaction datasets. Nucleic Acids Res. 2006;34:D535–9.
38. del Toro N, et al. The IntAct database: efficient access to fine-grained molecular interaction data. Nucleic Acids Res. 2022;50:D648–53.
39. Chatr-aryamontri A, et al. MINT: the molecular INTeraction database. Nucleic Acids Res. 2007;35:D572–4.
40. Barrett T, et al. NCBI GEO: archive for functional genomics data sets—update. Nucleic Acids Res. 2012;41:D991–5.
41. Parkinson H, et al. ArrayExpress–a public database of microarray experiments and gene expression profiles. Nucleic Acids Res. 2007;35:D747–50.
42. Perez-Riverol Y, et al. The PRIDE database and related tools and resources in 2019: improving support for quantification data. Nucleic Acids Res. 2019;47:D442–50.
43. Desiere F. The PeptideAtlas project. Nucleic Acids Res. 2006;34:D655–8.

44. Haug K, et al. MetaboLights: a resource evolving in response to the needs of its scientific community. Nucleic Acids Res. 2019;48:gkz1019. https://doi.org/10.1093/nar/gkz1019.
45. Chang A, et al. BRENDA, the ELIXIR core data resource in 2021: new developments and updates. Nucleic Acids Res. 2021;49:D498–508.
46. Wittig U, Rey M, Weidemann A, Kania R, Müller W. SABIO-RK: an updated resource for manually curated biochemical reaction kinetics. Nucleic Acids Res. 2018;46:D656–60.
47. Alcock BP, et al. CARD 2023: expanded curation, support for machine learning, and resistome prediction at the comprehensive antibiotic resistance database. Nucleic Acids Res. 2023;51:D690–9.
48. Sandgren A, et al. Tuberculosis drug resistance mutation database. PLoS Med. 2009;6:e1000002.
49. Zankari E, et al. Identification of acquired antimicrobial resistance genes. J Antimicrob Chemother. 2012;67:2640–4.
50. Lakin SM, et al. MEGARes: an antimicrobial resistance database for high throughput sequencing. Nucleic Acids Res. 2017;45:D574–80.
51. Ashburner M, et al. Gene ontology: tool for the unification of biology. Nat Genet. 2000;25:25–9.
52. Gene Ontology Consortium, et al. The gene ontology knowledgebase in 2023. Genetics. 2023;224:iyad031.
53. Liberzon A, et al. The molecular signatures database hallmark gene set collection. Cell Syst. 2015;1:417–25.
54. Albert R. Network inference, analysis, and Modeling in systems biology. Plant Cell. 2007;19:3327–38.
55. Ideker T, Ozier O, Schwikowski B, Siegel AF. Discovering regulatory and signalling circuits in molecular interaction networks. Bioinformatics. 2002;18:S233–40.
56. Sambarey A, Prashanthi K, Chandra N. Mining large-scale response networks reveals 'topmost activities' in Mycobacterium tuberculosis infection. Sci Rep. 2013;3:2302.
57. Ravichandran S, Chandra N. Interrogation of genome-wide networks in biology: comparison of knowledge-based and statistical methods. Int J Adv Eng Sci Appl Math. 2019;11:119–37.
58. Banerjee U, et al. PathTracer comprehensively identifies hypoxia-induced dormancy adaptations in mycobacterium tuberculosis. J Chem Inf Model. 2023;63:6156–67.
59. Zhang B, Horvath S. A general framework for weighted gene co-expression network analysis. Stat Appl Genet Mol Biol. 2005;4:Article17.
60. Margolin AA, et al. ARACNE: an algorithm for the reconstruction of gene regulatory networks in a mammalian cellular context. BMC Bioinf. 2006;7:S7.
61. Shannon P, et al. Cytoscape: a software environment for integrated models of biomolecular interaction networks. Genome Res. 2003;13:2498–504.
62. Csardi G, Nepusz T. The igraph software package for complex network research. Int J Complex Syst. 2006;1695:1–9.
63. Liu J, Gefen O, Balaban NQ. Tackling antibiotic resistance with systems-level perspective. Cell Syst. 2017;5:546–8.
64. Schwab JD, Kühlwein SD, Ikonomi N, Kühl M, Kestler HA. Concepts in Boolean network modeling: what do they all mean? Comput Struct Biotechnol J. 2020;18:571–82.
65. Orth JD, Thiele I, Palsson BØ. What is flux balance analysis? Nat Biotechnol. 2010;28:245–8.
66. Kohanski MA, Dwyer DJ, Collins JJ. How antibiotics kill bacteria: from targets to networks. Nat Rev Microbiol. 2010;8:423–35.
67. Shin B, Park C, Park W. Stress responses linked to antimicrobial resistance in Acinetobacter species. Appl Microbiol Biotechnol. 2020;104:1423–35.
68. Grant SS, Kaufmann BB, Chand NS, Haseley N, Hung DT. Eradication of bacterial persisters with antibiotic-generated hydroxyl radicals. Proc Natl Acad Sci. 2012;109:12147–52.
69. Koul A, et al. Delayed bactericidal response of mycobacterium tuberculosis to bedaquiline involves remodelling of bacterial metabolism. Nat Commun. 2014;5:3369.

70. Lobritz MA, et al. Antibiotic efficacy is linked to bacterial cellular respiration. Proc Natl Acad Sci. 2015;112:8173–80.
71. Stokes JM, Lopatkin AJ, Lobritz MA, Collins JJ. Bacterial metabolism and antibiotic efficacy. Cell Metab. 2019;30:251–9.
72. Cho H, Uehara T, Bernhardt TG. Beta-lactam antibiotics induce a lethal malfunctioning of the bacterial Cell Wall synthesis machinery. Cell. 2014;159:1300–11.
73. Zhang S, Yang M, Peng B, Peng X, Li H. Reduced ROS -mediated antibiotic resistance and its reverting by glucose in Vibrio alginolyticus. Environ Microbiol. 2020;22:4367–80.
74. Aye SM, et al. Lipid A profiling and metabolomics analysis of paired polymyxin-susceptible and -resistant MDR Klebsiella pneumoniae clinical isolates from the same patients before and after colistin treatment. J Antimicrob Chemother. 2020;75:2852–63.
75. Mahamad Maifiah MH, et al. Global metabolic analyses identify key differences in metabolite levels between polymyxin-susceptible and polymyxin-resistant Acinetobacter baumannii. Sci Rep. 2016;6:22287.
76. Koen N, Van Breda SV, Loots DT. Elucidating the antimicrobial mechanisms of colistin sulfate on Mycobacterium tuberculosis using metabolomics. Tuberculosis. 2018;111:14–9.
77. Lopatkin AJ, et al. Clinically relevant mutations in core metabolic genes confer antibiotic resistance. Science. 2021;371:eaba0862.
78. Dwyer DJ, et al. Antibiotics induce redox-related physiological alterations as part of their lethality. Proc Natl Acad Sci. 2014;111:E2100-9.
79. Dwyer DJ, Kohanski MA, Collins JJ. Role of reactive oxygen species in antibiotic action and resistance. Curr Opin Microbiol. 2009;12:482–9.
80. Kohanski MA, DePristo MA, Collins JJ. Sublethal antibiotic treatment leads to multidrug resistance via radical-induced mutagenesis. Mol Cell. 2010;37:311–20.
81. Mishra S, et al. Efficacy of β-lactam/β-lactamase inhibitor combination is linked to WhiB4-mediated changes in redox physiology of Mycobacterium tuberculosis. elife. 2017;6:e25624.
82. Padiadpu J, et al. Identifying and tackling emergent vulnerability in drug-resistant mycobacteria. ACS Infect Dis. 2016;2:592–607.
83. Bhosle A, et al. A strategic target rescues trimethoprim sensitivity in Escherichia coli. iScience. 2020;23:100986.
84. Leshchiner D, et al. A genome-wide atlas of antibiotic susceptibility targets and pathways to tolerance. Nat Commun. 2022;13:3165.
85. Jain V, Kumar M, Chatterji D. ppGpp: stringent response and survival. J Microbiol. 2006;44:1–10.
86. Abranches J, et al. The molecular Alarmone (p)ppGpp mediates stress responses, vancomycin tolerance, and virulence in enterococcus faecalis. J Bacteriol. 2009;191:2248–56.
87. Davis BD, Chen LL, Tai PC. Misread protein creates membrane channels: an essential step in the bactericidal action of aminoglycosides. Proc Natl Acad Sci. 1986;83:6164–8.
88. Overhage J, Bains M, Brazas MD, Hancock REW. Swarming of Pseudomonas aeruginosa is a complex adaptation leading to increased production of virulence factors and antibiotic resistance. J Bacteriol. 2008;190:2671–9.
89. Hughes D, Andersson DI. Evolutionary trajectories to antibiotic resistance. Ann Rev Microbiol. 2017;71:579–96.
90. Conrad TM, Lewis NE, Palsson BØ. Microbial laboratory evolution in the era of genome-scale science. Mol Syst Biol. 2011;7:509.
91. Furusawa C, Horinouchi T, Maeda T. Toward prediction and control of antibiotic-resistance evolution. Curr Opin Biotechnol. 2018;54:45–9.
92. Lässig M, Mustonen V, Walczak AM. Predicting evolution. Nat Ecol Evol. 2017;1:0077.
93. Iwasawa J, et al. Analysis of the evolution of resistance to multiple antibiotics enables prediction of the Escherichia coli phenotype-based fitness landscape. PLoS Biol. 2022;20:e3001920.
94. Jacquier H, et al. Capturing the mutational landscape of the beta-lactamase TEM-1. Proc Natl Acad Sci. 2013;110:13067–72.

95. Firnberg E, Labonte JW, Gray JJ, Ostermeier M. A comprehensive, high-resolution map of a gene's fitness landscape. Mol Biol Evol. 2014;31:1581–92.
96. Weinreich DM, Delaney NF, DePristo MA, Hartl DL. Darwinian evolution can follow only very few mutational paths to fitter proteins. Science. 2006;312:111–4.
97. Schenk MF, Szendro IG, Krug J, De Visser JAGM. Quantifying the adaptive potential of an antibiotic resistance enzyme. PLoS Genet. 2012;8:e1002783.
98. Lukačišinová M, Bollenbach T. Toward a quantitative understanding of antibiotic resistance evolution. Curr Opin Biotechnol. 2017;46:90–7.
99. Chevereau G, et al. Quantifying the determinants of evolutionary dynamics leading to drug resistance. PLoS Biol. 2015;13:e1002299.
100. Palmer AC, et al. Delayed commitment to evolutionary fate in antibiotic resistance fitness landscapes. Nat Commun. 2015;6:7385.
101. Card KJ, LaBar T, Gomez JB, Lenski RE. Historical contingency in the evolution of antibiotic resistance after decades of relaxed selection. PLoS Biol. 2019;17:e3000397.
102. Blount ZD, Borland CZ, Lenski RE. Historical contingency and the evolution of a key innovation in an experimental population of Escherichia coli. Proc Natl Acad Sci. 2008;105:7899–906.
103. Jochumsen N, et al. The evolution of antimicrobial peptide resistance in Pseudomonas aeruginosa is shaped by strong epistatic interactions. Nat Commun. 2016;7:13002.
104. San Millan A, Escudero JA, Gifford DR, Mazel D, MacLean RC. Multicopy plasmids potentiate the evolution of antibiotic resistance in bacteria. Nat Ecol Evol. 2016;1:0010.
105. Gifford DR, et al. Identifying and exploiting genes that potentiate the evolution of antibiotic resistance. Nat Ecol Evol. 2018;2:1033–9.
106. Papkou A, Hedge J, Kapel N, Young B, MacLean RC. Efflux pump activity potentiates the evolution of antibiotic resistance across S. aureus isolates. Nat Commun. 2020;11:3970.
107. Gifford DR, et al. Mutators can drive the evolution of multi-resistance to antibiotics. PLoS Genet. 2023;19:e1010791.
108. Lukačišinová M, Fernando B, Bollenbach T. Highly parallel lab evolution reveals that epistasis can curb the evolution of antibiotic resistance. Nat Commun. 2020;11:3105.
109. Raman K, Chandra N. Mycobacterium tuberculosis interactome analysis unravels potential pathways to drug resistance. BMC Microbiol. 2008;8:234.
110. Lázár V, et al. Bacterial evolution of antibiotic hypersensitivity. Mol Syst Biol. 2013;9:700.
111. Macvanin M, Hughes D. Hyper-susceptibility of a fusidic acid-resistant mutant of Salmonella to different classes of antibiotics. FEMS Microbiol Lett. 2005;247:215–20.
112. Lázár V, et al. Genome-wide analysis captures the determinants of the antibiotic cross-resistance interaction network. Nat Commun. 2014;5:4352.
113. Schultz D, Palmer AC, Kishony R. Regulatory dynamics determine cell fate following abrupt antibiotic exposure. Cell Syst. 2017;5:509–517.e3.
114. Youlden GH, et al. Time dependent asymptotic analysis of the gene regulatory network of the AcrAB-TolC efflux pump system in gram-negative bacteria. J Math Biol. 2021;82:31.
115. Gjini E, Brito PH. Integrating antimicrobial therapy with host immunity to fight drug-resistant infections: classical vs. adaptive treatment. PLoS Comput Biol. 2016;12:e1004857.
116. Ankomah P, Levin BR. Exploring the collaboration between antibiotics and the immune response in the treatment of acute, self-limiting infections. Proc Natl Acad Sci. 2014;111:8331–8.
117. Day T, Read AF. Does high-dose antimicrobial chemotherapy prevent the evolution of resistance? PLoS Comput Biol. 2016;12:e1004689.
118. Paterson IK, Hoyle A, Ochoa G, Baker-Austin C, Taylor NGH. Optimising antibiotic usage to treat bacterial infections. Sci Rep. 2016;6:37853.
119. Geli P, Laxminarayan R, Dunne M, Smith DL. "One-size-fits-all"? Optimizing treatment duration for bacterial infections. PLoS One. 2012;7:e29838.
120. Garber AM. Antibiotic exposure and resistance in mixed bacterial populations. Theor Popul Biol. 1987;32:326–46.

121. Tetteh JNA, Matthäus F, Hernandez-Vargas EA. A survey of within-host and between-hosts modelling for antibiotic resistance. Biosystems. 2020;196:104182.
122. Medzhitov R. Recognition of microorganisms and activation of the immune response. Nature. 2007;449:819–26.
123. Weiss G, Schaible UE. Macrophage defense mechanisms against intracellular bacteria. Immunol Rev. 2015;264:182–203.
124. Lipsitch M, Bergstrom CT, Levin BR. The epidemiology of antibiotic resistance in hospitals: paradoxes and prescriptions. Proc Natl Acad Sci. 2000;97:1938–43.
125. Bergstrom CT, Lo M, Lipsitch M. Ecological theory suggests that antimicrobial cycling will not reduce antimicrobial resistance in hospitals. Proc Natl Acad Sci. 2004;101:13285–90.
126. Haber M, Levin BR, Kramarz P. Antibiotic control of antibiotic resistance in hospitals: a simulation study. BMC Infect Dis. 2010;10:254.
127. Bonhoeffer S, Lipsitch M, Levin BR. Evaluating treatment protocols to prevent antibiotic resistance. Proc Natl Acad Sci. 1997;94:12106–11.
128. Yang Y, et al. Machine learning for classifying tuberculosis drug-resistance from DNA sequencing data. Bioinformatics. 2018;34:1666–71.
129. Arango-Argoty G, et al. DeepARG: a deep learning approach for predicting antibiotic resistance genes from metagenomic data. Microbiome. 2018;6:23.
130. Stokes JM, et al. A deep learning approach to antibiotic discovery. Cell. 2020;180:688–702.e13.

Chapter 10
Host-Directed Omics Approaches to Tackle Antimicrobial Resistance

Shivangi Rastogi and Pallavi Chandra

Abbreviations

AI	Artificial Intelligence
AMR	Antimicrobial Resistance
ARG	Antimicrobial Resistance Genes
AST	Antibiotic Susceptibility Testing
DNA	Deoxyribonucleic acid
ESKAPE-E	Acronym for *Enterococcus faecium, Staphylococcus aureus, Klebsiella pneumoniae, Acinetobacter baumanii, Pseudomonas aeruginosa, Enterobacter* species, and *Escherichia coli*
FACS	Fluorescence-Activated Cell Sorting
GC-MS	Gas Chromatography-Mass Spectrometry
GWAS	Genome-wide association studies
HDAC	Histone deacetylase
HDT	Host-directed Therapy
LC-MS	Liquid Chromatography-Mass Spectrometry
lncRNA	Long non-coding RNA
MALDI-MS	Matrix-assisted laser desorption/ionization-Mass Spectrometry
MDR	Multi-drug resistant
ML	Machine learning
mRNA	Messenger RNA
MS	Mass Spectrometry
Mtb	*Mycobacterium tuberculosis*
MTBC	*Mycobacterium tuberculosis* complex

S. Rastogi (✉)
Department of Cell Biology and Molecular Genetics, University of Maryland,
College Park, MD, USA
e-mail: srastog1@umd.edu

P. Chandra (✉)
Department of Medicine, Division of Infectious Diseases, Washington University in St. Louis,
St. Louis, MO, USA
e-mail: chandra.pallavi@wustl.edu

V. Soni, A. S. Akhade (eds.), *Antimicrobial Resistance: Factors to Findings*,
https://doi.org/10.1007/978-3-031-65986-7_10

NGS	Next-generation DNA sequencing
NMR	Nuclear Magnetic Resonance
PAMP	Pathogen-associated Molecular Pattern
PCR	Polymerase Chain Reaction
PDIM	Phthiocerol dimycocerosates
PK/PD	Pharmacokinetics/Pharmacodynamics
RNA	Ribonucleic acid
RNA-seq	RNA Sequencing
TB	Tuberculosis
TGS	Third-generation DNA sequencing
WGS	Whole Genome Sequencing
WMS	Whole Metagenome Sequencing

10.1 Introduction to HDTs

Antimicrobial resistance and the lack of broad-spectrum pathogen-targeting medications remain significant challenges despite the recent surge in the discovery of antivirals and antibiotics, and more alternative ways to treat infectious diseases are immediately required. Host-directed therapy (HDT) is a complementary approach that modifies the host response against pathogen infection by using a variety of therapeutic agents, including repurposed drugs, small molecules, synthetic nucleic acids, cytokines, cellular therapies, and recombinant proteins. In addition to modifying host factors that exacerbate inflammation and unbalance immune responses at the pathological site, HDT can target pathways that are disrupted by a pathogen and lead to hyper-inflammation, interfere with host mechanisms that a pathogen needs for productive replication or persistence, and stimulate host defense mechanisms against the pathogen to boost the protective immune response [1]. Over time, bacterial pathogens have developed numerous strategies to evade the host immune system [2]. *Mycobacterium tuberculosis* (Mtb) is the causative agent of tuberculosis (TB). This fatal bacterial infection is difficult to treat since mycobacteria have a high antibiotic resistance profile and can alter intracellular signaling pathways to evade host-defense mechanisms in human cells [3–6]. The primary goals of HDT are to circumvent this immune evasion and maintain the host cell pathways, thereby eliminating infections. HDT, by manipulating the host immune response and the metabolic state of bacteria, enhances host cell function, which has numerous significant advantages over existing antibiotics and, therefore, is a viable adjunct therapy to antibiotic therapy for treating intracellular bacterial infections. HDT counters pathogen-induced immune modulation. A revolutionary new approach is using HDT to treat infectious diseases, particularly tuberculosis, the most common cause of infection-related death (Table 10.1 and Fig. 10.1).

Table 10.1 Host-directed therapies against tuberculosis infection

Drug/Compound	Host Target	Mode of Action	Effect	References
Verapamil	L-type calcium channel	Blocks calcium ion channel activity	Decreased permeability of calcium channel; reduced efflux of calcium from cytosol; vasodilation; potentiate anti-TB drugs in the clinic	[7, 8]
Aspirin and ibuprofen	COX-1 and COX-2	Eicosanoid pathway/ Arachidonic acid catabolism	Regulation of TNF alpha levels via LXA4/LTB4 balance; controlled immunopathology; improved survival; activation of vitamin D antimicrobial activities	[9, 10]
Imatinib	BCR-ABL oncogenic tyrosine kinase	Induces V-ATPase pump subunit expression and recruitment	Promotes phagosomal acidification	[11]
Imiquimod	Autophagy	Stimulates TLR7	Activates autophagy; enhances NO production	[12]
Valproic acid, suberoylanilide hydroxamic acid, and lithium prochlorperazine	Histone deacetylases, autophagy	Decrease phosphorylation of mTOR and PI3- kinase; histone modification	Histone acetylation; increase gene expression; inhibition of TBK-1; xenophagy; reduce TNF alpha and tissue damage	[13]
Clofazimine	Pro-caspase-3	Caspase-3 and PARP-mediated apoptosis	Induction of apoptosis; release of apoptotic vesicles containing immunomodulatory molecules and antigens	[14]
Diazoxide	Potassium ion channel	Caspase-1 dependent cell death (pyroptosis)	Potassium efflux; NLRP3 inflammasome activation; induction of pyroptosis; activation of T cells	[15]

(continued)

Table 10.1 (continued)

Drug/Compound	Host Target	Mode of Action	Effect	References
Metformin	Autophagy;HIF1α, CD39, CD73, AMPK-DACHi-CXCL1	Reduced phosphorylation of mTOR and p70S6k	Induction of mitochondrial ROS; promotes phagolysosome fusion; induction of apoptosis	[16]
Vitamin D3	Autophagy via cathelicidin	Stimulates vitamin D receptors to induce cathelicidin expression; upregulates the expression of Atg5 and Beclin-1	Immunomodulation and direct antimicrobial activity	[10]
Carbamazepine	Autophagy	AMPK activation	Promotes autophagy	[17]
Trimetazidine	Fatty acid oxidation inhibitor	Generates mitochondria-derived ROS	Promotes autophagy; reduce pathogen burden in infected mice	[18]
AZD0530	Src/Abl inhibitor	Blocks the ATP binding site of Src kinases	Reduces pathogen burden and reduced necrotic granulomas	[19]

10.2 Advantages of HDTs

In contrast to conventional antibiotics, HDTs offer several significant advantages. First, HDT can be effective against mycobacteria that are extensively drug-resistant (XDR) and multidrug-resistant (MDR), which are resistant to traditional antibiotics. Second, host-targeting medications have a lower risk of developing drug resistance since mycobacteria are not under selection pressure. Third, by targeting different host pathways, HDT may enable the shortening of the duration of TB antibiotic treatment regimens, a reduction in the disease's lung damage, and a decreased chance of relapse or reinfection. Fourth, when treating latent tuberculosis infection (LTBI), host-targeting drugs may be able to kill metabolically dormant, non-replicating bacteria that are tolerant to or resistant to standard antibiotic treatments. Fifth, HDT might make it possible to cut the dose of conventional antibiotics, reducing toxicity without compromising efficacy. Finally, combinatorial regimens might promote possible synergistic treatment effects because HDT and mycobacterium-targeting drugs (i.e., antibiotics) work through various mechanisms [19–21].

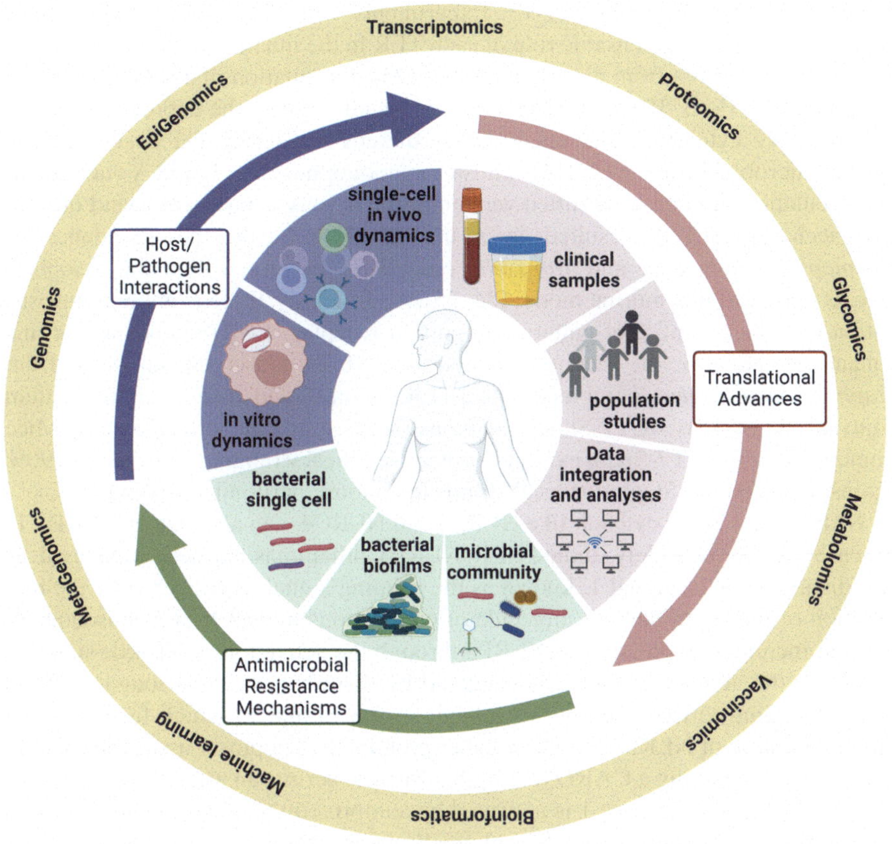

Fig. 10.1 Combined omics approaches (shown in yellow circle) can provide multi-layered insights into disease pathogenesis. Omics technologies can identify how microbes acquire AMR (shown in green), and influence host/pathogen interactions (shown in purple) to shape infection outcomes. Omics can be used to study infections at a population level (in pink) to understand AMR transmission and heterogeneity in infection disease dynamics leading to the rational development of therapies, vaccines, biomarkers, and diagnostics

10.3 Novel Strategies for HDTs: Bacterial Infections

10.3.1 Modulating Pathways of Pattern Recognition Receptor (PRR) Signaling

Various pattern recognition receptors (PRRs), including C-type lectin receptors, Toll-like receptors (TLRs), AIM2-like receptors, NOD-like receptors (NLRs), RIG-I-like receptors, and intracellular DNA sensors like cGAS, identify pathogen-associated molecular patterns (PAMPs), which are consistent signatures of pathogens [22]. In humans, the TLR family encompasses TLR1-10, while in mice, it includes TLR1-9 and TLR11-13. Experiments with TLR-deficient mice have demonstrated the indispensable role of each TLR in the innate immune response, as they identify PAMPs from diverse microbes [23]. Stimulation of TLRs triggers the activation of NF-κB, IRFs, or MAP kinases, which regulate the expression of pro-inflammatory cytokines, and chemokines, thereby bolstering the host's defense against microbial invasion [23]. Administering monophosphoryl lipid A via intranasal inoculation in mice, a modified version of the lipid A component found in lipopolysaccharide (LPS), resulted in a notable decrease in the abundance of *Haemophilus influenzae* and *Moraxella catarrhalis* bacteria within the nasopharynx, as observed in a murine model [24, 25]. Additionally, there is ongoing investigation into the potential of pattern recognition receptor (PRR) ligands as vaccine enhancers. Notably, a vaccine incorporating five conserved antigens from *Staphylococcus aureus*, combined with a TLR7 agonist, and absorbed onto an alum adjuvant (known as 4CT7-Staph), demonstrated significant effectiveness. Mice immunized with 4CT7-Staph exhibit a protective efficacy ranging from 80% to 90% against four distinct Staphylococcal strains in a mouse peritonitis model [25].

NLRs found in the cytosol, like NLRP3 and NLRC4, can also detect PAMPs. In the case of the most widely studied NLR, NLRP3, inflammasome activation involves two distinct signals: signal 1, known as the priming signal, is triggered by the recognition of pathogen-derived molecule or endogenous molecule and consequently leads to increased expression of NLRP3 through the activation of NF-κB; signal 2, known as the activation signal, is triggered by disturbances in various ion fluxes thereby leading to the assembly of inflammasome complex, which facilitates the oligomerization of NLRP3 with an adapter protein, the apoptosis-associated speck-like protein containing a CARD (ASC) that further recruits an effector protein, pro-caspase-1 [26]. Pro-caspase 1 is activated by autoproteolytic cleavage and converts pro-interleukin-1 beta (pro-IL-1β) and pro-IL-18 into their active forms, as well as cleaves gasdermin D, resulting in an inflammatory form of cell death known as pyroptosis [26]. Several bacterial infections, such as those caused by *Salmonella* Typhimurium *(Stm)*, *S. flexneri*, *P. aeruginosa*, and *B. pseudomallei*, can be treated by targeting the NLRC4 inflammasome. *B. pseudomallei* induces NLRC4-dependent pyroptosis, which restricts the growth of bacteria inside cells, to manage intracellular pathogen infections. NLRP3 inflammasome activation results in an increase in IL-1β release, which facilitates *B. pseudomallei* replication, draws an excessive number of neutrophils to the lungs and damages tissue [27]. Intraperitoneal (i.p.) administration of glyburide, a small chemical that prevents IL-1β production, followed by Intranasal infection with *B. pseudomallei* significantly lessens bacterial dissemination into the liver and spleen in the mouse model of *B. pseudomallei* infection [28]. The association between glyburide-mediated anti-inflammatory responses and the course of the disease is still debatable; hence, it is unclear whether glyburide is therapeutically beneficial for individuals infected with *B. pseudomallei*. The NLRP3 inflammasome is inhibited by the small molecule MCC950 [29]. It is necessary to further investigate the function of MCC950 in preventing bacterial infection. Clemastine, an FDA-approved medicine that is well-tolerated and widely available, alters macrophage calcium flux through potentiation of the purinergic receptor P2RX7 that further enhances inflammasome activation and hence serves as a novel HDT strategy for improved control of mycobacterial infections [30]. Additionally, Mtb inhibits the NLRP3 inflammasome activation by limiting cell surface ligand exposure through its hydrolase Hip1, inhibiting the host cell's cathepsin G protease through Rv3364c, a Mtb secreted effector, and lastly by limiting intracellular signal two triggers (K+ and Cl- efflux and cytosolic ROS production) through its serine/threonine kinase PknF [5, 31]. Identifying small compounds that selectively block IL-1β secretion in response to NLRP3 inflammasome activation might be an ideal therapeutic strategy for controlling bacterial infections.

10.3.2 Autophagy Induction/Targeting Autophagy

A cellular mechanism called autophagy promotes the breakdown of harmful protein aggregates and malfunctioning lysosome organelles. Lysosomes are used by cells during the autophagic process to break down and recycle the contents of their cytoplasm. Inextricably connected to cellular metabolism, autophagy is a conserved mechanism for preserving homeostasis. Furthermore, it contributes to the cellular defense mechanisms against various intracellular microorganisms [32]. The broad term "autophagy" can be further broken down into the specialized but overlapping processes known as macro-autophagy, mitophagy, xenophagy, LC3-associated phagocytosis, and chaperone-mediated autophagy [33].

Several significant primary pathogens such as Mtb, Salmonella, and Legionella have developed defensive tactics to evade different stages of the autophagy pathway. These strategies include inhibiting the signal for autophagy induction, evading recognition by the autophagy machinery, disrupting components of the autophagy process, or impeding the fusion of autophagosomes with lysosomes [34]. Recent research has demonstrated that small molecules capable of altering diverse autophagic pathways responsible for processing cytoplasmic contents can stimulate autophagy [35], as observed in the context of viruses and the intracellular pathogen Mtb [36]. Alternatively, another strategy entails selectively addressing host proteins that pathogens engage with during the manipulation of autophagy.

Pathogens such as *Stm*, *Escherichia coli*, *B. pseudomallei* and *Mycobacterium marinum*, Mtb induce LC3-associated phagocytosis (LAP) [37]. Through the induction of LAP using the mTOR inhibitor sirolimus, *B. pseudomallei* enhances its colocalization with LC3 in phagosomes, facilitating phagosome maturation and phagocytosis. This process results in a simultaneous decrease in *B. pseudomallei* intracellular survival during sirolimus therapy [26, 38]. Conversely, inhibiting autophagy activation significantly enhances intracellular *B. pseudomallei* survival [39]. Therapeutically facilitating bacterial clearance involves enhancing autophagy activity by blocking negative regulators of the pathway such as p38 and cathepsin B [40]. Treatment of macrophages with AMG548, a p38 inhibitor, promotes Mtb clearance by activating autophagy. Gefitinib, an inhibitor of the epidermal growth factor receptor, diminishes Mtb-induced p38 phosphorylation, thus enhancing autophagy by elevating LC3 expression and suppressing Mtb growth [41]. Additionally, both autophagy and lysosomal biogenesis are negatively regulated by lysosomal cathepsin B [42]. Cathepsin-B-deficient macrophages increase lysosome biogenesis and expression of autophagy-related genes, resulting in a 40% increase in the survival rate of mice infected with *Francisella novicida* compared to wild-type mice. Developing improved cathepsin B inhibitors could offer new possibilities for host-directed antimicrobial therapy. Vitamin D exhibits protective effects against Mtb infections by regulating Mtb growth intracellularly through induction of the AMP LL-37. LL-37 has direct anti-mycobacterial activity in vitro, with a hypothesis suggesting its mediation of autophagy activation [43]. Vitamin D stimulates the expression of autophagy-related genes Atg5 and Beclin-1 in human macrophages [44]. Recent studies demonstrated that vitamin D can normalize lysosomal acidity in a Helicobacter pylori model by activating the PDIA3 receptor through overexpression of Ca++ channels. A meta-analysis of 1850 participants in eight interventional trials found that vitamin D3 enhances sputum culture conversion in patients with MDR Mtb isolates but not in patients with susceptible Mtb isolates [45]. Numerous clinical trials of vitamin D supplementation as adjunctive therapy for pulmonary TB have been conducted due to its potential effects on autophagy and inhibition of Mtb growth in human macrophages. The anti-parasitic drug nitazoxanide, the antidepressant fluoxetine, and the TB drugs isoniazid and pyrazinamide have all been shown to induce autophagy. Additionally, carbamazepine and valproic acid, two anti-epileptic medications, stimulate autophagy and inhibit Mtb replication in human macrophages.

Shigella flexneri (*S. flexneri*), a clinically significant enteric bacterium living in the cytosol by evading membrane-bound compartments, was more easily killed when Atg16l1 was depleted in myeloid cells. The study used TMT-based multiplexed proteomic analysis, a method that allows for almost complete characterization of the global proteome, and this method finds numerous unique post-translational modifications (PTMs) in the fundamental autophagy machinery, innate sensing, and inflammatory cytokine signaling components that develop because of *S. flexneri* infection [46]. Quantification of PTMs on thousands of different proteins is possible when isobaric multiplexing approaches are combined with enrichment [47]. Quantitative multiplexed proteomics revealed that ATG16L1 deletion significantly boosted proteins involved in the glutathione-mediated antioxidant response to counteract increased oxidative stress, simultaneously promoting clearance of *S. flexneri* [48]. Hence, autophagy may be a desirable target for HDT as several lines of evidence point to its significance in the pathophysiology of infectious diseases.

10.3.3 Enhancement of Phagosome Maturation

The intricate relationship between Mtb and the phagosome raises the possibility of manipulating several pathways to limit Mtb intracellular survival and proliferation. Phagosomal maturation and acidification offer potential host-directed targets for intervention. Imatinib can chemically inhibit Abelson tyrosine kinase (Abl), which is involved in bacterial uptake by altering cytoskeletal dynamics in host cells, and imatinib therapy controls phagosomal acidification and decreases Mtb internalization by human macrophages [11]. AZD0530, a Src inhibitor that reduced disease burden in Mtb-infected guinea pigs by promoting phagosomal acidification, further highlights the possibility of blocking host tyrosine kinases to impede intracellular mycobacterial survival [19]. The Mtb-secreted protein 1tuberculosinyladenosine-antacid (1TbAd) also prevents phagolysosome formation and subsequent acidification. The buildup of 1TbAd in acidic intracellular compartments caused phagosomes to expand and eventually burst, allowing mycobacterial egress into the cytosol [49].

Many human infections hinder phagocytosis or prevent phagosome remodeling to circumvent the innate immune system's microbicidal reaction. *Salmonella* and *Shigella* are two genera of bacteria that aggressively try to invade mammalian cells. Once inside, they persist by stealing the host cell's signaling and trafficking machinery. Beta-oxidation, gluconeogenesis, and the glyoxylate shunt are vital for the survival of Mtb within the phagosomes of macrophages during persistence, a state characterized by glucose deficiency and fatty acid abundance. Isocitrate lyase (ICL), a pivotal enzyme in the glyoxylate shunt, initiates the conversion of isocitrate to succinate and glyoxylate. This enzyme is crucial for carbon anaplerosis in the TCA cycle when utilizing C2 substrates like fatty acids. While the glyoxylate shunt is absent in vertebrates, it is prevalent in prokaryotes, lower eukaryotes, and plants. MtbICL is considered a promising drug target for combating persistent infections and has been shown to be inhibited by quercetin, salicylanilide, benzanilide, pthalazinyl derivatives, fluoroquinolone derivatives, and pyruvate–isoniazid analogs [50, 51]. Work-intensive microscopic procedures are utilized in current experimental paradigms to research phagocytosis, phagosome maturation, and bacterial invasion. The number of experimental conditions that can be investigated is constrained by the time-consuming requirement for individual observation and scoring of each particle or organelle. Such methods cannot be used to effectively screen chemical libraries for new medications or small interference RNA libraries for crucial molecules involved in host-pathogen interactions. To overcome this limitation, there is an improved and automated system for investigating phagocytosis, phagosome maturation, and bacterial invasion. Using this technology, researchers can quickly analyze particle engulfment and/or phagosome remodeling in many samples based on the recent development of robotic, high-throughput fluorescence microscopes [52, 53]. A potent method to examine changes in the phagosomal proteome is the latex bead phagocytosis model system coupled with mass spectrometry. Based on current research, phagosomes serve as signaling platforms that integrate several intraphagosomal, intracellular, and extracellular signals. According to proteomic investigations, numerous proteins, including receptors and kinases, which are known to be involved in signaling, are found in phagosomes, and the phosphorylation status of these proteins is crucial and influences phagosome maturation. Future studies on the effects of immunological signals, particularly when a pathogen disrupts phagosome maturation to circumvent host defenses, could facilitate the development of novel HDT treatments that improve pathogen elimination and adaptive immunity.

10.3.4 Inhibition of Lipid Body Biogenesis

Mtb subverts the host macrophage's cellular machinery and promotes lipid droplet accumulation, forming a foamy cell phenotype and building the cellular environment essential for pathogen intracellular survival during TB infection. Mtb degrades host-derived fatty acids from foamy macrophages in intracytoplasmic lipid inclusions as triacylglycerols, which are principally used as carbon and energy sources during its persistence in the diseased host. The Mtb genome codes for 15 triacylglycerol synthases, 24 lipase/esterases, and seven cutinase-like proteins, which the bacilli most likely use to obtain fatty acids to meet their energy needs during dormancy and reactivation. Moreover, many of these enzymes are used by Mtb in cell wall remodeling during the active replication phase of infection [54–56]. Recent studies indicate that Mtb induces lipid body formation by activating GPR109A and an antilipolytic pathway, which results in an accumulation of lipid bodies that may have a protective impact on Mtb infection [57]. Treatment with Mepenzolate bromide (MPN), an inhibitor of GPR109A, increases the death of Mtb in THP1 cells, human peripheral blood monocyte-derived macrophages, and mouse studies [57].

Nuclear receptors that sense lipids, such as testicular receptor 4 (TR4), liver X receptors alpha and beta (LXR α, β), and peroxisome proliferator-activated receptor gamma (PPARγ), control the development and metabolism of macrophage lipid bodies in Mtb pathogenesis [58]. PPARγ acts as a promising HDT target since it controls the secretion of cytokines, the formation of lipid bodies, and the replication of Mtb in macrophages. Treatment of human and mouse macrophages with siRNA or the GW9662 antagonist blocking the PPARγ decreased Mtb and BCG intracellular survival [59]. Human THP-1 macrophage's ability to kill mycobacteria was improved by suppressing TR4 using RNA interference (RNAi) technology. The preventive and therapeutic administration of an LXR agonist (TO91317 or GW3965) to WT mice increased Mtb clearance. These findings imply that lipid-sensing nuclear receptors are likely candidates for therapeutic targeting in the pathogenesis of TB.

10.3.5 Cytokine-Mediated Activation

TNFα, IFNγ, and IL-1β are three major cytokines that stimulate macrophage antimicrobial activity. TNFα enhances responsiveness to IFNγ and increases killing of the pathogen through ROS production, IFNγ triggers phagosome maturation and autophagy, IL-1β increases AMP synthesis, upregulates the expression of the TNF receptor, and prevents the production of type 1 IFN. IFNγ or TNFα modulators are the two main cytokine types used in current HDT methods. For instance, it has been demonstrated that administering aerosolized IFNγ to patients with TB shortens the time until sputum conversion without consistently improving the outcome of the chest radiological examination [60]. IFNγ adjunct therapy promoted lung healing in small studies conducted in cohorts of individuals with drug-resistant TB and enhanced bacterial killing [60]. However, cytokine-based HDT approaches that control TNFα as an adjuvant to traditional antibiotic therapy appear more promising because they significantly shorten the length of antibiotic treatment, improve disease pathology, and prevent TB relapse in animal models. Combination therapy with zileuton and PGE2 increases the production of host protective cytokine, IL-1β, thereby reducing type I IFN production and inhibiting the growth of Mtb [61].

10.4 Using Multi-omics Techniques to Identify Host Targets That Are Involved in Bacterial Infection

10.4.1 Utilization of Host Kinases During Bacterial Infection

A recent study investigated how kinase inhibitors might be utilized as host-directed therapeutics to treat infectious illnesses caused by intracellular bacteria, particularly *Stm* and Mtb. This study utilizes the human cell lines and primary macrophages as intracellular infection models for *Stm* and Mtb to screen 827 ATP-competitive kinase inhibitors with established target profiles from two published kinase inhibitor sets (PKIS1 and PKIS2). A total of 14 hit compounds for *Stm* and 19 hit compounds for Mtb were found in the kinase inhibitor screen to be effective against intracellular bacteria and non-toxic to host cells. Morpholine-imidazole/triazole-pyrimidinones, which mainly target the kinases PIK3CB and PIK3CD, as well as 2-aminobenzimidazoles, which target BLK, ABL1, and TRKA, were effective against intracellular Mtb infection [62].

AKT1 signaling plays a crucial role in regulating the host cell's response to various bacterial infections, rendering it a promising target for host-directed antibacterial therapies. A kinome screening employing small interfering RNA (siRNA) unveiled that *Stm* secretes an effector protein named SopB, which interferes with the activation of host AKT1, resulting in impaired phagosome-lysosome fusion [26, 63]. Screening of a phosphatase-specific siRNA library in the presence of *Stm*- and Mtb-infected host cells showed that approximately half of the phosphatases involved in AKT-related pathways were affected. During *Pa* infection, phosphatidylinositol 3-kinase (PI3K)-driven synthesis of phosphatidylinositol 2,4,5-triphosphate recruits AKT1, thereby promoting bacterial transcytosis [64]. *Klebsiella pneumoniae* triggers an anti-inflammatory reaction by engaging a signaling pathway that includes AKT1, resulting in the suppression of the NF-κB pathway.

Inhibitors of the PI3K/AKT pathway have shown promise in curing bacterial infections due to the critical role of the AKT1 kinase network in controlling the growth of intracellular bacteria. The celecoxib derivative AR-12, devoid of cyclooxygenase 2 (COX-2) inhibitory effects, suppresses the phosphorylation of PDK-1, thereby inhibiting the phosphorylation and activation of AKT1, potentially blocking the PI3K/AKT signaling pathway [65]. AR-12 has been proposed as an inducer of autophagy as shown in THP-1 macrophages, AR-12 induces autophagy and effectively reduces the intracellular growth of *Francisella novicida* and *Francisella tularensis* [66, 68]. Oral administration of AR-12 to Stm-infected mice significantly reduced bacterial burden in the spleen and liver, delayed initiation of illness, and increased survival rates [67]. Cells may activate the defensive mechanism of inducing autophagy to eradicate Mtb infection. Considering that the enzyme mTOR kinase negatively regulates autophagy, mTOR kinase inhibitors may serve as effective host-directed therapy options for eradicating Mtb infection [71].

Reverse phase protein microarray (RPMA) and mass spectrometry (MS) offer promising tools for understanding host-pathogen interactions and identifying novel therapeutic targets. RPMA detected phosphorylated forms of AMPK-1, Src, and glycogen synthase kinase 3 (GSK-3) during Burkholderia infection, suggesting the involvement of metabolic pathways and innate immune pathway control [70]. Bacterial internalization involves the recruitment of nonreceptor tyrosine kinases and their signaling cascade. A study targeting conserved genes involved in actin cytoskeleton regulation revealed that *Pa* uptake requires Abl tyrosine kinase, the adapter protein Crk, small GTPases Rac1 and Cdc42, and p21-activated kinase [26, 69]. Protein phosphorylation, which activates or deactivates receptors, signaling intermediates, and transcription factors, is a critical regulatory mechanism during pathogen infections. Therapeutic interventions could potentially benefit from modifying the phosphorylation states of signaling molecules.

10.4.2 Pharmacological Screening to Find Host-Directed Small Compounds That Control Bacterial Infection

Researchers investigated 640 FDA-approved medications to identify pharmaceuticals capable of impeding the intracellular replication of *Legionella pneumophila*, *Rickettsia conorii*, *C. burnetii*, and *Brucella abortus* within a monocytic cell line derived from humans. They discovered that compounds targeting G-protein-coupled receptors (GPCRs) and intracellular calcium fluxes effectively inhibited the replication of these pathogens. Furthermore, compounds influencing membrane cholesterol distribution were found to reduce the replication of *C. burnetii* and *L. pneumophila* [72]. Additionally, a study introduced a comprehensive screening platform aimed at uncovering a novel anti-virulence drug targeting *Pseudomonas aeruginosa* (Pa) and Stm by quantifying pathogen-induced host-cell death. By screening more than 10,000 small, diverse molecules, the study finds a few compounds with intriguing effects on Pa and Stm pathogenicity in vitro infection models. A unique chemical compound with an inhibitory effect on the function of the type 3 secretion system (T3SS) of Stm was also found, which is significant since it potently inhibits bacteria-induced death of host cells [73]. The most widely utilized in vitro Mtb infection models include RAW 264.7 and J774A.1 murine macrophage cell lines. A library of 56,984 synthetic chemicals was screened in RAW 264.7 cells infected with Mtb (H37Rv-GFP) and used a high throughput assay. Mtb

H37Rv-GFP-infected cells were distributed into separate wells of screening plates preincubated with compounds. After 5 days, plates were scanned inside BSL3 utilizing an entirely automated platform (EVOscreen-MarkIII). In total, 135 active compounds with strong intracellular anti-mycobacterial activity and no host cell toxicity were found in the screen [74]. Dinitrobenzamide derivatives (DNB) found in this screen displayed effectiveness against the XDR strains as well as having an impact on the intracellular growth of Mtb [74]. Hung et al. conducted a chemical screen utilizing a high-content microscopic test to find small compounds that inhibit mycobacterial growth in macrophages by targeting host cell signaling pathways. The identified host-targeted inhibitors that restricted bacterial growth mostly fall into five categories: GPCR modulators, ion channel inhibitors, membrane transport proteins, anti-inflammatories, and kinase modulators [41]. Moreover, the study discovered that when administered to infected macrophages, fluoxetine, a selective serotonin reuptake inhibitor, increases the release of the pro-inflammatory cytokine TNF-α and triggers autophagy. Gefitinib, an inhibitor of the EGFR, promotes autophagy and inhibits Mtb growth. Drug repurposing as a first step towards HDT is attractive since it might accelerate drug discovery and establish a pipeline for new therapeutic modalities. A recent study using a human cell-based intracellular infection model evaluated a library of chemicals with specified autophagy-inducing or -inhibitory activities and discovered two antipsychotic medications of the diphenylbutylpiperidine-class, Fluspirilene, and Pimozide, as interesting hits among several potential candidates [75]. Two licensed antipsychotic medications, Pimozide and Fluspirilene, are promising and quickly adaptable options for HDT against Mtb and Stm infections and affect the autophagic/lysosomal response through various mechanisms.

10.5 Interplay Between Antibiotic Resistance and Virulence Factors

10.5.1 PathoFact: A System for Predicting Virulence Factors and Genes Associated with Antibiotic Resistance in Metagenomic Data

Virulence is described as the capacity of a pathogen to infect the host and cause a disease. Virulence factors, also called pathogenicity factors, are the cellular structures, molecules, and regulatory systems that enable bacteria to colonize the host and play an essential role in immunoevasion and immunosuppression, ultimately leading to inflammation and tissue damage. Antimicrobial resistance genes (ARGs) are the genes that make pathogens resistant to certain antibiotics. Identifying virulence factors and antimicrobial resistance genes in situ is crucial to understanding causal links between microbiome compositions, functioning, and diseases. There are numerous tools available for predicting ARGs themselves, including DeepARG

[76], RGI [77], ResFinder [78], and ARGsOAP [79]; however, there are only a few tools available for predicting virulence factors, including MP3 [80] and VirulentPred [81]. There is a lack of recent computational approaches that can concurrently predict virulence factors in metagenomic samples. Laura de Nies et al. have created PathoFact, a bioinformatics tool to identify simultaneously virulence factors, bacterial toxins, and ARGs in metagenomic data with high sensitivity, accuracy, and specificity [82]. PathoFact is a simple-to-use, modular, and reproducible pipeline. PathoFact also integrates the detection of mobile genetic elements with predicting these pathogenicity factors. PathoFact is a flexible and adaptable tool since its modules for virulence factors, toxins, and ARGs can each be used individually. PathoFact is an open source, and access to its datasets and models is free at https://pathofact.lcsb.uni.lu. Toxins, ARGs, and virulence factors are all identified with the aid of PathoFact, which is crucial for comprehending the pathogenic status of microbiomes.

10.5.2 HyperVR: A Hybrid Deep Ensemble Learning Technique for Concurrently Predicting Virulence Factors and ARGs

Virulence factors (VFs) and ARGs have characteristics that are essential for pathogenic bacteria to adapt to and thrive in a hostile microbial environment, although they have followed separate evolutionary trajectories. VFs and ARGs typically spread between bacteria by horizontal gene transfer, and both use comparable mechanisms (such as two-component systems, efflux pumps, cell wall modifications, and porins) to stimulate or suppress the expression of numerous genes. Therefore, identifying VFs and ARGs is necessary to comprehend the causal link between microbiome structures, functions, and diseases. Most bioinformatic tools can be divided into computational technique-based tools and tools based on the "best hit" method. By potentially learning the standard best hit scores and statistical gene sequence patterns of VFs and ARGs at the same time, a recent study suggested a hybrid computational deep ensemble learning approach termed HyperVR [83]. HyperVR outperforms the most advanced prediction tools currently available. Only gene sequence information is used by HyperVR, and there are no hard cutoff limits, making prediction simple and accurate. For training and prediction, HyperVR integrates several important genetic features, such as the one-hot encoding feature, the bit score-based similarity feature, the physicochemical property-based feature, and the evolutionary information-based feature. Compared to the presently available prediction tools, HyperVR is more efficient and vigorous and tackles the limitations of conventional alignment-based solutions. Based on HyperVR's prediction scores, researchers can minimize the scope of conducting biological studies and significantly reduce their time and effort by filtering ARGs and VFs in natural pathogenic strains [83]. Improved models of the evolution of antibiotic-resistance genes may aid in the development of more long-term therapeutic approaches. Furthermore, focusing on these microbial compounds through the assessment of their immunogenicity and application in HDT may prove to be a successful anti-infective strategy.

10.6 Rise of AMR and Future Antimicrobial Resistance Threats

10.6.1 The Importance of Identifying Antibiotic Resistance Genes That Are High-Risk

Bacterial ARGs are common. Nevertheless, not all ARGs represent significant risks to public health, emphasizing the significance of identifying those that do. It can be challenging to determine which ARGs endanger human health. Numerous biological functions, such as efflux systems and cell-cell signaling, are carried out by genes widely distributed throughout bacteria that are thought to impart antibiotic resistance, including those predicted by sequence homology. A very small percentage of these genes are dangerous to human health. Finding "high-risk" resistance genes among the tens of thousands of speculative ARGs is crucial in combating this global issue with low-cost methods. Zhang et al. created an 'omics-based' methodology to assess ARG risk considering (1) enrichment in human-associated environments, (2) gene mobility, and (3) presence/absence in ESKAPE pathogens (host pathogenicity). Four risk categories result from this straightforward paradigm. ARGs that do not achieve the first requirement are placed in Rank IV (lowest risk), those that do so but not the second are placed in Rank III, those that do so but not the third are placed in Rank II, and those that attain all three criteria are placed in Rank I (the highest risk), which were further distinguished as "current threats" (Rank I; 3%)—already present among pathogens—and "future threats" (Rank II; 0.6%)—novel resistance emerging from non-pathogens. Seventy-three "current threat" ARG families were discovered using an omics-based approach. Thirty-five of these belonged to the list of the 37 high-risk ARGs that the World Health Organization and other sources had suggested, and the remaining 38 were markedly enriched in hospital plasmids. Studies verified that ARGs that recently moved into pathogens were highly enriched in Rank II ('future risks') based on an analysis of all pathogen genomes published since framework development. Moreover, this approach was applied to the gut microbiome genomes of donors for fecal microbiota transplants. Only 8.9% of the genomes had high-risk ARGs, even though ARGs were present in 73% of the genomes [84]. This paradigm offers a simple method for identifying present and emerging antimicrobial resistance threats, with potential therapeutic implications such as lowering the risk of microbiome-based therapies.

10.7 Approaches to Tackle AMR: From "Omics" Eyes

Microbes can develop antibiotic tolerance in various ways, such as genetic mutations and horizontal gene transfers, efflux pump-mediated drug exclusion, and modifying the antimicrobial target to decrease the affinity or uptake for the drug. Microbes can also induce global metabolic changes to support resistance

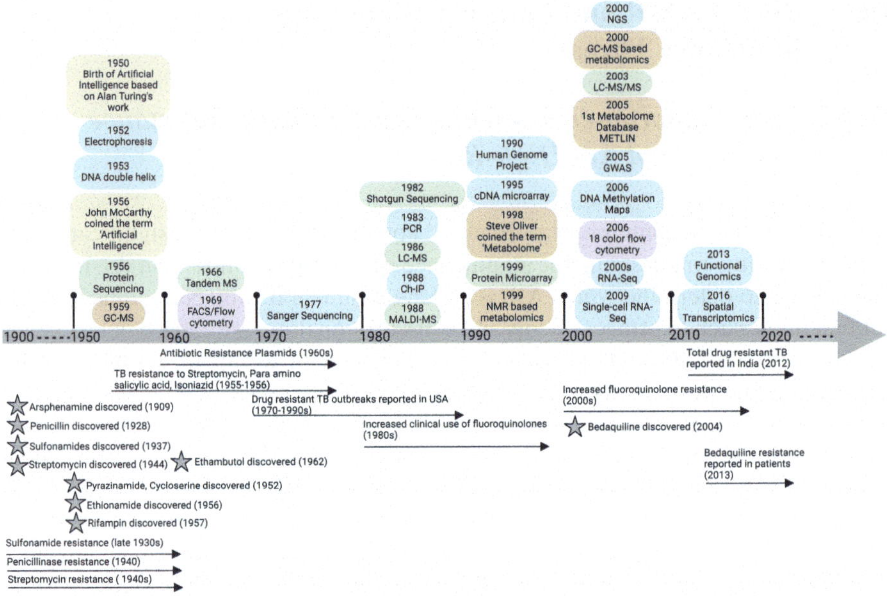

Fig. 10.2 Historical timeline paralleling prominent antibiotics discoveries, the rise of AMR, and development of omics technologies. Omics approaches are color-coded based on their application such as genomics/transcriptomics/epigenomics are shown in blue, proteomics in green, and metabolomics in orange. Computational advances are in yellow, and techniques that enable single-cell analyses are shown in pink

mechanisms. Resistance adaptations in pathogens can orchestrate differential host responses, which shape infection outcomes. A detailed understanding of these complex connections is vital to combinatorial antimicrobial drug discovery and vaccine development. Here, we discuss how high-throughput omics approaches can be leveraged to identify unique patterns in drug-resistant infections at genomic, transcriptional, proteomic, and metabolomics levels. In Fig. 10.2, we show how technological advancement goes hand-in-hand with drug discovery to tackle AMR since the 1950s.

10.7.1 Genomic and Transcriptomic Approaches to Decode AMR Mechanisms

The conventional method to determine microbial drug resistance is antibiotic susceptibility testing (AST), where the bacterial growth is monitored by different antibiotics. AST requires prior knowledge of growth conditions and is largely limited to culturable bacteria. AMR is usually genetically encoded since it can arise from point mutations, overexpression or gene duplication events, and horizontal gene transfer. Therefore, sequence-based methodologies present robust alternatives for

AMR discovery. Genomics decodes the sequence of an organism's DNA and gives a powerful window to infer microbial identity and phylogeny. The development of Sanger sequencing in 1977 initiated decoding the genetic code using DNA polymerase enzyme and dideoxynucleotides [85]. New technologies such as NGS, TGS, and DNA microarrays have been developed to provide relatively faster, inexpensive, and accurate options [86–88]. Whole genome sequencing and whole metagenome sequencing can be utilized to detect AMR emergence and microbial resistomes and develop novel drugs and diagnostics [89]. A helpful tool called total genotyping solution for TB (TGS-TB) uses NGS data to discriminate between Mtb isolates and assess outbreak and surveillance (https://gph.niid.go.jp/tgs-tb/). 454 pyrosequencing was used to identify and sequence the Mtb F_0 subunit of ATP synthase across several isolates, which enabled the development of bedaquiline, a novel class of antibiotics for TB in 40 years [90, 91]. WGS identified that α and β', the subunits of RNA polymerase, allowed the dissemination of drug-resistant MTBC [92]. WGS supported clinical decision-making by identifying genomic versus plasmid-mediated linezolid-resistant *Enterococcus faecalis* [93]. Recently, a novel tetracycline resistance mechanism was discovered in soil-dwelling bacteria using functional metagenomics, demonstrating that ARGs can be horizontally transferred to human pathogens [94].

Microbial adaptation to antibiotics and intracellular stressors requires a global genome rewiring. Transcriptomics techniques such as RNA-seq, high-throughput real-time PCR analyses, and microarrays provide a dynamic snapshot of differentially expressed genes [95], novel virulence factors, predicting AMR [96, 97], and even identify non-coding RNAs [98]. The intracellular milieu presents a significant challenge to Mtb, which was studied using in vivo pulmonary transcriptomics [99] and dual RNA-Seq [100]. Such techniques are invaluable in uncovering underlying host-pathogen interactions that may inform novel therapeutics and vaccines.

Epigenetic modifications have emerged as an unexpected realm for AMR development. Epigenomics enables the identification of DNA and histone modifications, ribosomal RNA methylation, and non-coding RNA expression, which are increasingly appreciated as the basis of drug resistance. The mycobacterial *tlyA* gene targets the second-line antibiotics capreomycin and viomycin. However, Mtb develops resistance to these antibiotics by decreasing TlyA-mediated rRNA methylations [101]. Infection can also induce epigenetic modifications in the host. Exposure to PAMPs can induce epigenetic remodeling in hematopoietic stem cells, favoring the formation of innate immunological memory or trained immunity [102]. *Helicobacter pylori* infection induces DNA hypermethylation in the promoter regions of several tumor suppressor genes in the epithelial cells of the human gastric mucosa, which can initiate oncolytic transformation [103, 104]. A study comparing MDR-TB patients with healthy controls identified a large set of differentially expressed lncRNAs and mRNA exclusive to infection. Epigenomics can help understand pathogenesis mechanisms deeper to guide the development of novel therapeutics, such as HDAC inhibitors that selectively interfere with epigenetic changes. Tubastatin, an HDAC6 inhibitor, was shown to clear infection in human macrophages by enhancing mitochondrial reactive oxygen species production [105].

10.7.2 Understanding Protein Dynamics During Infection Using Proteomics Approaches

The protein landscape undergoes significant changes during infection. Pathogens release siderophores and virulence factors and can directly interact with and modify host proteins. The host responds to infection by altering its proteome globally and contributing to antimicrobial mechanisms mediated by cytokines, antibodies, and other molecular regulators. Proteomics can be grouped into expression proteomics, which looks into qualitative or quantitative protein expression (two-dimensional electrophoresis and mass spectrometry), functional proteomics to study protein-protein interactions and signaling (yeast-two-hybrid combined with mass spectrometry); and structural proteomics which decodes the three-dimensional protein structure (NMR and X-ray crystallography) [106, 107]. Large-scale protein analyses have been made possible by mass spectrometry-based techniques, which provide a snapshot of the protein-protein interactions and dynamic proteome changes during infection or underlying antimicrobial resistance. A top-down proteomics approach studies intact proteins with a mass >200 kDa, which gives the advantage of analyzing post-translational modifications. However, this technique can be limited by poor solubility of proteins and interference by solvents [106]. In the bottom-up or 'shotgun' approach, proteins are first separated using two-dimensional electrophoresis followed by enzymatic digestion into more easily fragmented peptides [106]. Developing a protein-protein interaction (PPI) network involving host proteins and bacterial virulence factors presents an alternate method for thoroughly understanding host-pathogen interactions during infection. Host proteins from humans or mice interacting with various *B. mallei* virulence factors have been detected using a yeast two-hybrid (Y2H) library containing both human and murine proteins [108]. Three previously unrecognized *B. mallei* virulence factors, strongly linked with host proteins, were incorporated into the resultant host-pathogen PPI network, highlighting their functional significance in *B. mallei* virulence, and validating the interactions depicted by the PPI network [108]. In a parallel manner, the Y2H investigation revealed pathways linked to focal adhesion, cytoskeleton modulation, leukocyte transendoepithelial migration, as well as Toll-like receptor (TLR) and mitogen-activated protein kinase (MAPK) signaling pathways subsequent to *Yersinia pestis* infection [26, 109]. Research employing comprehensive mass spectrometry alongside chemical protein cross-linking unveiled networks connecting host structural proteins responsible for regulating host cell-to-cell adhesion and *Acinetobacter baumannii* virulence factors during infection [110]. Additionally, examination of mucosal tissues from chinchillas afflicted with acute otitis media induced by non-typeable *Haemophilus influenzae* (NTHi) revealed changes in metabolites and proteins using targeted quantitative metabolomics and high-throughput, unbiased proteomics. Treatment of human epithelial cells with an inhibitor targeting the actin-related protein Arp2/3 resulted in decreased NTHi infection [26, 111]. Comprehensive mechanistic investigations of these host proteins within the infection context can offer a foundation for identifying novel therapeutic strategies by interfering with host-pathogen interactions.

Proteomics studies using drug-resistant pathogens have shown differences in proteomes compared to their drug-sensitive counterparts, which can contribute to biomarker discovery and, most importantly, provide meaningful insights into drug-resistance mechanisms and the environmental pressures experienced by the pathogens that favor the development of AMR [112–114]. Proteomics studies have shown that pathogens mainly adapt to antibiotic pressure by inducing efflux pumps and changing cell membrane composition. However, antibiotic resistance mechanisms are not restricted to intrabacterial modifications; combining alternate strategies to tackle AMR is essential. For instance, quorum sensing primarily controls pathogen virulence, and targeting biofilms is important to ensure better drug penetration [115].

10.7.3 Capturing the Role of Metabolism, Metabolites, and Co-metabolites in AMR

Metabolomics captures changes in the metabolite levels of a biological system in response to various stimuli. Foundational work using radioisotope tracers has mapped metabolic pathways in great detail. Analytical methods such as NMR spectroscopy and mass spectrometry-based techniques can provide a high-throughput approach to identify metabolic perturbations on a global level [116]. Such approaches can be targeted to identify changes in a pre-selected list of analytes or provide an untargeted view across multiple pathways. The metabolite pool is closely linked to organismal phenotype. It can provide a snapshot of the bacterial response to antibiotic pressure, the adaptations required to sustain AMR, and host adaptations to infections. Metabolomic approaches are essential to identify links between microbial metabolism and AMR mechanisms. Many reports indicate that a decreased antibiotic sensitivity can be attributed to reduce bacterial growth rate owing to host pressures such as oxidative stress, low pH, and hypoxia [117]. During infection, Mtb preferentially utilizes carbon sources such as triacylglycerols and cholesterol, which are crucial for the biosynthesis of virulence lipids such as PDIM and sulfolipids [118]. Lipid utilization results in the accumulation of toxic intermediates, reducing the growth rate of Mtb and contributing to antibiotic tolerance [119]. Tolerance to antibiotic pressure modifies a pathogen's metabolism to facilitate modifications primarily in its cell wall and lipidome. Unbiased mass spectrometry approaches enabled the identification of a variety of metabolites that were differentially expressed in drug-resistant Mtb spanning lipids and glycerophospholipids, nicotinate and nicotinamide metabolism, and purine and pyrimidine metabolism [120–125]. Metabolomics was also used to identify synergizing antibiotics against drug-resistant *Acinetobacter baumanii* [126].

Infection can also alter host immune responses and metabolism. Previous studies have shown the potential of host metabolomics to predict the disease status and development of biomarkers [127, 128]. Drug-resistant pathogens were shown to alter host immune activation and modulate metabolic reprogramming [129]. Infection typically involves multiple immune cell types that cooperate to resolve pathogen spread and inflammation. The effect of drug-resistant infections on host metabolism is poorly understood and requires large-scale experimentation combined with in-depth bioinformatics analyses. As our understanding of pathogenesis grows, we must employ powerful tools to complement mass spectrometry-based analyses and achieve a broader comprehension of the process. Recent tools such as Met-flow and microfluidics can be utilized to capture the metabolic status at a single-cell level during infection with drug-sensitive and resistant bacteria [130, 131].

10.8 Emerging Omics Avenues to Tackle AMR

In addition to perturbations in the genome, proteome, and metabolites, drug-resistant pathogens can have multifaceted effects on host-pathogen interactions. Applying omics to identify such unconventional connections will help broaden our understanding and develop promising avenues to tackle AMR. An emerging field of study is post-translational modifications that can alter protein function and contribute to AMR. Succinylation of different lysine residues of the Mtb isocitrate lyase enzymes is associated with resistance to different antibiotics [132]. Another dominant modification is glycosylation, yet limited information exists on how these are impacted in drug-resistant infections. Estimated to modify about 90% of proteins, sugar residues can influence immune function, antibody efficacy, pathogen virulence, and signaling, underscoring glycomics' importance. Glycolipids are crucial components of the mycobacterial cell envelope and significantly confer drug resistance [133]. Carbanepem-resistance in *Acinetobacter baumannii* is associated with differences in capsule composition [134]. Technological advances have enabled parsing out adaptations at a single-cell level and identifying the drivers of resistance in a given bacterial population. Recently, BacDrop—a droplet-based methodology for bacterial single-cell RNA-seq—showed resistance-associated changes in individual bacteria [135]. Drug-resistant Mtb can also alter host immune responses and techniques like single-cell genomics can effectively chart out cell type-specific responses to infection [120, 129].

A key weapon against AMR is vaccines, which bolster the immune response and develop immunological memory. A recent paradigm in vaccine development is vaccinomics, which explores the inter-individual heterogeneity resulting in a range of immune protection [136, 137]. A TB vaccine has existed for over a century but its efficacy has diminished. Recent studies show how Mtb undermines vaccine efficacy and how the effectiveness of this vaccine can be improved [138, 139]. Thus, vaccinomics can accurately correlate dosage and response to vaccines in personalized medicine.

Pathogens can experience selective environmental pressures, such as biocides, environmental pollutants, industrial chemicals, and even livestock constantly exposed to antibiotics [140]. Pathogens can even acquire resistance genes from interacting with bacteria that do not usually cause human diseases [141]. AMR surveillance is increasingly recognized as a core component in infectious disease management, especially in a post-COVID era [142] (https://www.who.int/publications/i/item/9789240021402). Microbial genomics, metagenomics, and phylogenomics have immense potential as tools for surveillance and understanding resistance dynamics [140, 143, 144].

The application of artificial intelligence presents an excellent futuristic opportunity to integrate information from data generated in laboratory or clinical settings, and predict drugs and drug targets in AMR. Bioinformatics tools available online to detect AMR, such as ARG-ANNOT, CARD, STST2, MEGARes, Genefinder, ARIBA, kmer Resistance, AMRFinder, and ResFinder can be utilized to integrate DNA and amino acid sequences. In 2019, AI was trained to screen an online database of nearly 1.5 billion chemical compounds (ZINC15) and predict a novel antibiotic against a broad spectrum of bacteria, including MDRs. The machine learning algorithm identified an anti-diabetic medication called Halicin, which showed remarkable efficacy against ESKAPE and MDR pathogens [145]. Kavvas et al. combined GWAS with metabolic flux adaptations to develop the ML classifier called metabolic allele classifier to predict drug-sensitive and resistant Mtb isolates [146] successfully. Algorithms are also being developed to predict higher-order combinatorial therapies in TB, which will aid in designing synergistic drug combinations to enhance the current therapeutic regimen and shorten the treatment period against MDR isolates [147]. The lack of diverse training datasets currently limits the application of AI to predict HDT, and this is where omics techniques can bridge the knowledge gap by generating biologically meaningful data [148].

10.9 Integr-Omics: Strengthening the Fight Against AMR by Integrating-Omics and Bioinformatics Tools

Technological and computational advances combined with a better understanding of pathogenesis have enabled us to explore beyond a singular level. Infection involves perturbations at multiple levels, and integration of omics platforms is important to get a cohesive understanding of pathogenesis [149]. For example, combining sequencing, microarray analyses, and proteomics techniques revealed the epigenetic mechanisms underlying drug resistance in Mtb [150]. Combining multiple omics techniques was used to predict individual disease risk and showed that insulin biosynthetic pathway spiked during viral infections [151].

Multi-omics is a meaningful approach that is important in optimizing antimicrobial therapy. Continuous monitoring and dose modifications are necessary for certain antibiotics to work best in individuals with infectious illnesses. The patient, antibiotic, and pathogen determine whether an antibiotic therapy is successful. Due to genotype and disease features, treatment outcomes for some medications vary significantly between individuals. Therefore, with these medications, individualized doses are necessary for effective therapy. When using antibiotics, patients may be in danger of therapeutic failure, which could result in death if the dosage is inaccurate. On the other hand, toxicities could result from excessive doses. Therefore, precision dosing, which tailors doses to specific patients, is essential for antibiotics—especially those with a limited therapeutic index.

The application of omics technologies may significantly aid in the development of small molecules as well as the identification of new targets. Besides exploring the genome information of the microbes and the response of the antimicrobial agents, multi-omics also addresses the inter-individual variability through exploration of the genomic data from the host, which is crucial in optimizing antimicrobial therapy. Dosing regimens have been further optimized using mathematical modeling. These include population PK models, quantitative systems pharmacology (QSP), systems-based, and mechanism-based PK/PD models. Precision antimicrobial chemotherapy in the clinic has considerable potential for quantitative systems pharmacology. Model-informed precision dosing (MIPD) has been used in population PK models. Data from multi-omics are incorporated into metabolic modeling to enhance model predictions and offer thorough insights into cellular networks at the system level to reveal the complexity of the microbial system [152]. Using a genome-scale metabolic model (GSMM) and multi-omics data, Zhu et al. examined how *P. aeruginosa* responded to polymyxins [153]. A system-based model was used by Abdul Rahim et al. to examine the interaction between polymyxin B and chloramphenicol [154]. Wozniak et al. used a computational method and a combination of metabolomics and proteomics study of *S. aureus* bacteremia as biomarkers to estimate mortality risk and suggest individualized treatment [155]. The relationship between genes and proteins and the interaction of biological networks in a system-based model can be uncovered by incorporating high-throughput multi-omics data. The European Union also supports using multi-omics technology as a foundation for creating a personalized treatment plan based on omics to combat the evolution of antibiotic resistance [156]. To create a database that bioinformatics tools can mine to find significant correlations between various variables, this tailored-treatment approach integrates personalized patient data such as identification of genetic predisposition to specific infections (host genomics), host-immune response to infections (proteomics and transcriptomics), microbiome analysis (transcriptomics), and characterization of pathogen resistance profile (microbial genomics). This will eventually aid in the development of treatment algorithms that will execute individualized treatment with the best treatment plan for each patient [156]. Moreover, anti-tuberculosis drug-induced hepatitis has been linked to genetic polymorphisms of drug-metabolizing enzymes, including NAT2*4 and CYP2E1*4 [157]. These pharmacogenomic data are crucial for anticipating the clinical response to antibiotics and customizing antibiotic therapy. Hence, pharmacogenomics data integration into the medication PK/PD model may offer a method for improving antibiotic dosages while reducing antimicrobial agent toxicity and addressing the challenges in managing infectious diseases.

10.10 Conclusions

Microbes will continue to impact human health globally. Throughout history, we have seen how microbes acquire AMR shortly after introducing novel antibiotics. Industrial and farming practices and climate change increases have fueled the ongoing AMR crisis. The rise of antibiotic-resistant superbugs is one of humanity's most imminent threats, underscoring the need to develop novel therapeutics, vaccines, and diagnostics. HDTs provide a promising avenue for designing antimicrobials and immunomodulators independent of pathogen strain diversity. A rational approach to combinatorial therapy is crucial for HDT development since it depends on an in-depth understanding of pathogenesis. Undoubtedly, omics combined with AI-based algorithms can catapult our position in the antibiotics arms race.

References

1. Kaufmann SHE, Dorhoi A, Hotchkiss RS, Bartenschlager R. Host-directed therapies for bacterial and viral infections. Nat Rev Drug Discov. 2018; https://doi.org/10.1038/nrd.2017.162.
2. Finlay BB, McFadden G. Anti-immunology: evasion of the host immune system by bacterial and viral pathogens. Cell. 2006;124(4):767–82.
3. Chandra PGS, Philips JA. Immune evasion and provocation by Mycobacterium tuberculosis. Nat Rev Microbiol. 2022; https://doi.org/10.1038/s41579-022-00763-4.
4. Rastogi S, Ellinwood S, Augenstreich J, Mayer-Barber KD, Briken V. Mycobacterium tuberculosis inhibits the NLRP3 inflammasome activation via its phosphokinase PknF. PLoS Pathog. 2021;17(7):e1009712.
5. Rastogi S, Briken V. Interaction of mycobacteria with host cell inflammasomes. Front Immunol. 2022;13:791136.
6. Rastogi S, Evavold CL, Briken V. Putting the p(hosphor) in pyroptosis. Cell Host Microbe. 2022;30(12):1650–2.
7. Chen C, Gardete S, Jansen RS, Shetty A, Dick T, Rhee KY, et al. Verapamil targets membrane energetics in Mycobacterium tuberculosis. Antimicrob Agents Chemother. 2018;62(5):e02107.
8. Zumla A, Rao M, Parida SK, Keshavjee S, Cassell G, Wallis R, et al. Inflammation and tuberculosis: host-directed therapies. J Intern Med. 2015;277(4):373–87.
9. Tobin DM, Roca FJ, Oh SF, McFarland R, Vickery TW, Ray JP, et al. Host genotype-specific therapies can optimize the inflammatory response to mycobacterial infections. Cell. 2012;148(3):434–46.
10. Byrne ST, Denkin SM, Zhang Y. Aspirin and ibuprofen enhance pyrazinamide treatment of murine tuberculosis. J Antimicrob Chemother. 2007;59(2):313–6.
11. Bruns H, Stegelmann F, Fabri M, Döhner K, van Zandbergen G, Wagner M, et al. Abelson tyrosine kinase controls phagosomal acidification required for killing of Mycobacterium tuberculosis in human macrophages. J Immunol. 2012;189(8):4069–78.
12. Lee HJ, Kang SJ, Woo Y, Hahn TW, Ko HJ, Jung YJ. TLR7 stimulation with Imiquimod induces selective autophagy and controls. Front Microbiol. 2020;11:1684.
13. Rao M, Valentini D, Zumla A, Maeurer M. Evaluation of the efficacy of valproic acid and suberoylanilide hydroxamic acid (vorinostat) in enhancing the effects of first-line tuberculosis drugs against intracellular Mycobacterium tuberculosis. Int J Infect Dis. 2018;69:78–84.
14. Gopal M, Padayatchi N, Metcalfe JZ, O'Donnell MR. Systematic review of clofazimine for the treatment of drug-resistant tuberculosis. Int J Tuberc Lung Dis. 2013;17(8):1001–7.

15. Chen X, Feng L, Yao H, Yang L, Qin Y. Efficacy and safety of diazoxide for treating hyperinsulinemic hypoglycemia: a systematic review and meta-analysis. PLoS One. 2021;16(2):e0246463.
16. Singhal A, Jie L, Kumar P, Hong GS, Leow MK, Paleja B, et al. Metformin as adjunct antituberculosis therapy. Sci Transl Med. 2014;6(263):263ra159.
17. Schiebler M, Brown K, Hegyi K, Newton SM, Renna M, Hepburn L, et al. Functional drug screening reveals anticonvulsants as enhancers of mTOR-independent autophagic killing of Mycobacterium tuberculosis through inositol depletion. EMBO Mol Med. 2015;7(2):127–39.
18. Chandra P, He L, Zimmerman M, Yang G, Köster S, Ouimet M, et al. Inhibition of fatty acid oxidation promotes macrophage control of Mycobacterium tuberculosis. mBio. 2020;11(4):e01139-20.
19. Chandra P, Rajmani RS, Verma G, Bhavesh NS, Kumar D. 2016. Targeting drug-sensitive and -resistant strains of Mycobacterium tuberculosis by inhibition of Src family kinases lowers disease burden and pathology. mSphere 1(2):e00043-15. https://doi.org/10.1128/mSphere.00043-15
20. Kilinç G, Saris A, Ottenhoff THM, Haks MC. Host-directed therapy to combat mycobacterial infections. Immunol Rev. 2021;301(1):62–83.
21. Napier RJ, Rafi W, Cheruvu M, Powell KR, Zaunbrecher MA, Bornmann W, et al. Imatinib-sensitive tyrosine kinases regulate mycobacterial pathogenesis and represent therapeutic targets against tuberculosis. Cell Host Microbe. 2011;10(6):635.
22. Amarante-Mendes GP, Adjemian S, Branco LM, Zanetti LC, Weinlich R, Bortoluci KR. Pattern recognition receptors and the host cell death molecular machinery. Front Immunol. 2018;9:2379.
23. Akira S, Uematsu S, Takeuchi O. Pathogen recognition and innate immunity. Cell. 2006;124(4):783–801.
24. Hirano T, Kodama S, Kawano T, Maeda K, Suzuki M. Monophosphoryl lipid a induced innate immune responses via TLR4 to enhance clearance of nontypeable haemophilus influenzae and Moraxella catarrhalis from the nasopharynx in mice. FEMS Immunol Med Microbiol. 2011;63(3):407–17.
25. Bagnoli F, Fontana MR, Soldaini E, Mishra RP, Fiaschi L, Cartocci E, et al. Vaccine composition formulated with a novel TLR7-dependent adjuvant induces high and broad protection against Staphylococcus aureus. Proc Natl Acad Sci USA. 2015;112(12):3680–5.
26. Chiang CY, Uzoma I, Moore RT, Gilbert M, Duplantier AJ, Panchal RG. Mitigating the impact of antibacterial drug resistance through host-directed therapies: current progress, outlook, and challenges. MBio. 2018;9(1):e01932–17.
27. Sahoo M, Del Barrio L, Miller MA, Re F. Neutrophil elastase causes tissue damage that decreases host tolerance to lung infection with burkholderia species. PLoS Pathog. 2014;10(8):e1004327.
28. Koh GC, Weehuizen TA, Breitbach K, Krause K, de Jong HK, Kager LM, et al. Glyburide reduces bacterial dissemination in a mouse model of melioidosis. PLoS Negl Trop Dis. 2013;7(10):e2500.
29. Coll RC, Hill JR, Day CJ, Zamoshnikova A, Boucher D, Massey NL, et al. MCC950 directly targets the NLRP3 ATP-hydrolysis motif for inflammasome inhibition. Nat Chem Biol. 2019;15(6):556–9.
30. Matty MA, Knudsen DR, Walton EM, Beerman RW, Cronan MR, Pyle CJ, et al. Potentiation of P2RX7 as a host-directed strategy for control of mycobacterial infection. elife. 2019;8:e39123.
31. Rastogi S, Ganesh A, Briken V. Mycobacterium tuberculosis Utilizes Serine/Threonine Kinase PknF to Evade NLRP3 Inflammasome-driven Caspase-1 and RIPK3/Caspase-8 Activation in Murine Dendritic Cells. J Immunol 2024; ji2300753. https://doi.org/10.4049/jimmunol.2300753.
32. Glick D, Barth S, Macleod KF. Autophagy: cellular and molecular mechanisms. J Pathol. 2010;221(1):3–12.

33. Gómez-Virgilio L, Silva-Lucero MD, Flores-Morelos DS, Gallardo-Nieto J, Lopez-Toledo G, Abarca-Fernandez AM, et al. Autophagy: a key regulator of homeostasis and disease: an overview of molecular mechanisms and modulators. Cells. 2022;11(15):2262.
34. Siqueira MDS, Ribeiro RM, Travassos LH. Autophagy and its interaction with intracellular bacterial pathogens. Front Immunol. 2018;9:935.
35. Vakifahmetoglu-Norberg H, Xia HG, Yuan J. Pharmacologic agents targeting autophagy. J Clin Invest. 2015;125(1):5–13.
36. Kim YS, Silwal P, Kim SY, Yoshimori T, Jo EK. Autophagy-activating strategies to promote innate defense against mycobacteria. Exp Mol Med. 2019;51(12):1–10.
37. Yuan J, Zhang Q, Chen S, Yan M, Yue L. LC3-associated phagocytosis in bacterial infection. Pathogens. 2022;11(8):863.
38. Cullinane M, Gong L, Li X, Lazar-Adler N, Tra T, Wolvetang E, et al. Stimulation of autophagy suppresses the intracellular survival of Burkholderia pseudomallei in mammalian cell lines. Autophagy. 2008;4(6):744–53.
39. Rinchai D, Riyapa D, Buddhisa S, Utispan K, Titball RW, Stevens MP, et al. Macroautophagy is essential for killing of intracellular Burkholderia pseudomallei in human neutrophils. Autophagy. 2015;11(5):748–55.
40. Webber JL, Tooze SA. Coordinated regulation of autophagy by p38alpha MAPK through mAtg9 and p38IP. EMBO J. 2010;29(1):27–40.
41. Stanley SA, Barczak AK, Silvis MR, Luo SS, Sogi K, Vokes M, et al. Identification of host-targeted small molecules that restrict intracellular Mycobacterium tuberculosis growth. PLoS Pathog. 2014;10(2):e1003946.
42. Qi X, Man SM, Malireddi RK, Karki R, Lupfer C, Gurung P, et al. Cathepsin B modulates lysosomal biogenesis and host defense against Francisella novicida infection. J Exp Med. 2016;213(10):2081–97.
43. Mily A, Rekha RS, Kamal SM, Arifuzzaman AS, Rahim Z, Khan L, et al. Significant effects of Oral Phenylbutyrate and vitamin D3 adjunctive therapy in pulmonary tuberculosis: a randomized controlled trial. PLoS One. 2015;10(9):e0138340.
44. Chung C, Silwal P, Kim I, Modlin RL, Jo EK. Vitamin D-cathelicidin axis: at the crossroads between protective immunity and pathological inflammation during infection. Immune Netw. 2020;20(2):e12.
45. Jolliffe DA, Ganmaa D, Wejse C, Raqib R, Haq MA, Salahuddin N, et al. Adjunctive vitamin D in tuberculosis treatment: meta-analysis of individual participant data. Eur Respir J. 2019;53(3):e01139-20.
46. Lapek JD, Greninger P, Morris R, Amzallag A, Pruteanu-Malinici I, Benes CH, et al. Detection of dysregulated protein-association networks by high-throughput proteomics predicts cancer vulnerabilities. Nat Biotechnol. 2017;35(10):983–9.
47. Rose CM, Isasa M, Ordureau A, Prado MA, Beausoleil SA, Jedrychowski MP, et al. Highly multiplexed quantitative mass spectrometry analysis of Ubiquitylomes. Cell Syst. 2016;3(4):395–403.e4.
48. Maculins T, Verschueren E, Hinkle T, Choi M, Chang P, Chalouni C, et al. Multiplexed proteomics of autophagy-deficient murine macrophages reveals enhanced antimicrobial immunity via the oxidative stress response. elife. 2021;10:e62320.
49. Buter J, Cheng TY, Ghanem M, Grootemaat AE, Raman S, Feng X, et al. Mycobacterium tuberculosis releases an antacid that remodels phagosomes. Nat Chem Biol. 2019;15(9):889–99.
50. (2015) Insight into the structural flexibility and function of Mycobacterium tuberculosis isocitrate lyase Biochimie 11073-80 https://doi.org/10.1016/j.biochi.2014.12.016.
51. Harish, Shukla Vikash, Kumar Amit Kumar, Singh Shivangi, Rastogi Shaheb Raj, Khan Mohammad Imran, Siddiqi Manju Yasoda, Krishnan Md. Sohail, Akhtar (2015) Isocitrate lyase of Mycobacterium tuberculosis is inhibited by quercetin through binding at N-terminus International Journal of Biological Macromolecules 78137-141 https://doi.org/10.1016/j.ijbiomac.2015.04.005.

52. Steinberg BE, Grinstein S. Analysis of macrophage phagocytosis: quantitative assays of phagosome formation and maturation using high-throughput fluorescence microscopy. Methods Mol Biol. 2009;531:45–56.
53. Steinberg BE, Scott CC, Grinstein S. High-throughput assays of phagocytosis, phagosome maturation, and bacterial invasion. Am J Physiol Cell Physiol. 2007;292(2):C945–52.
54. Rastogi S, Agarwal P, Krishnan MY. Use of an adipocyte model to study the transcriptional adaptation of Mycobacterium tuberculosis to store and degrade host fat. Int J Mycobacteriol. 2016;5(1):92–8.
55. Rastogi S, Singh AK, Pant G, Mitra K, Sashidhara KV, Krishnan MY. Down-regulation of PE11, a cell wall associated esterase, enhances the biofilm growth of mycobacterium tuberculosis and reduces cell wall virulence lipid levels. Microbiology (Reading). 2017;163(1):52–61.
56. Rastogi S, Singh AK, Chandra G, Kushwaha P, Pant G, Singh K, et al. The diacylglycerol acyltransferase Rv3371 of Mycobacterium tuberculosis is required for growth arrest and involved in stress-induced cell wall alterations. Tuberculosis (Edinb). 2017;104:8–19.
57. Singh V, Jamwal S, Jain R, Verma P, Gokhale R, Rao KV. Mycobacterium tuberculosis-driven targeted recalibration of macrophage lipid homeostasis promotes the foamy phenotype. Cell Host Microbe. 2012;12(5):669–81.
58. Leopold Wager CM, Arnett E, Schlesinger LS. Mycobacterium tuberculosis and macrophage nuclear receptors: what we do and don't know. Tuberculosis (Edinb). 2019;116S:S98–S106.
59. de Almeida PE, Pereira de Sousa NM, Rampinelli PG, Silva RVS, Correa JR, D'Avila H. Lipid droplets as multifunctional organelles related to the mechanism of evasion during mycobacterial infection. Front Cell Infect Microbiol. 2023;13:1102643.
60. Chin KL, Anis FZ, Sarmiento ME, Norazmi MN, Acosta A. Role of interferons in the development of diagnostics, vaccines, and therapy for tuberculosis. J Immunol Res. 2017;2017:5212910.
61. Mayer-Barber KD, Andrade BB, Oland SD, Amaral EP, Barber DL, Gonzales J, et al. Host-directed therapy of tuberculosis based on interleukin-1 and type I interferon crosstalk. Nature. 2014;511(7507):99–103.
62. van den Biggelaar RHGA, van Doorn CLR, Eugenia M, de Ries AS, et al. Identification of kinase inhibitors as potential host-directed therapies for intracellular bacteria. bioRxiv. 2023:08.28.555045. https://doi.org/10.1101/2023.08.28.555045.
63. Albers HM, Kuijl C, Bakker J, Hendrickx L, Wekker S, Farhou N, et al. Integrating chemical and genetic silencing strategies to identify host kinase-phosphatase inhibitor networks that control bacterial infection. ACS Chem Biol. 2014;9(2):414–22.
64. Kierbel A, Gassama-Diagne A, Mostov K, Engel JN. The phosphoinositol-3-kinase-protein kinase B/Akt pathway is critical for Pseudomonas aeruginosa strain PAK internalization. Mol Biol Cell. 2005;16(5):2577–85.
65. Koselny K, Green J, DiDone L, Halterman JP, Fothergill AW, Wiederhold NP, et al. The celecoxib derivative AR-12 has broad-spectrum antifungal activity in vitro and improves the activity of fluconazole in a murine model of cryptococcosis. Antimicrob Agents Chemother. 2016;60(12):7115–27.
66. Abdulrahman BA, Abdelaziz D, Thapa S, Lu L, Jain S, Gilch S, et al. The celecoxib derivatives AR-12 and AR-14 induce autophagy and clear prion-infected cells from prions. Sci Rep. 2017;7(1):17565.
67. Hoang KV, Borteh HM, Rajaram MV, Peine KJ, Curry H, Collier MA, et al. Acetalated dextran encapsulated AR-12 as a host-directed therapy to control Salmonella infection. Int J Pharm. 2014;477(1–2):334–43.
68. Chiu HC, Soni S, Kulp SK, Curry H, Wang D, Gunn JS, et al. Eradication of intracellular Francisella tularensis in THP-1 human macrophages with a novel autophagy inducing agent. J Biomed Sci. 2009;16(1):110.
69. Pielage JF, Powell KR, Kalman D, Engel JN. RNAi screen reveals an Abl kinase-dependent host cell pathway involved in Pseudomonas aeruginosa internalization. PLoS Pathog. 2008;4(3):e1000031.

70. Chiang CY, Uzoma I, Lane DJ, Memišević V, Alem F, Yao K, et al. A reverse-phase protein microarray-based screen identifies host signaling dynamics upon Burkholderia spp. infection. Front Microbiol. 2015;6:683.
71. Singh P, Subbian S. Harnessing the mTOR pathway for tuberculosis treatment. Front Microbiol. 2018;9:70.
72. Czyż DM, Potluri LP, Jain-Gupta N, Riley SP, Martinez JJ, Steck TL, et al. Host-directed antimicrobial drugs with broad-spectrum efficacy against intracellular bacterial pathogens. MBio. 2014;5(4):e01534–14.
73. von Ambüren J, Schreiber F, Fischer J, Winter S, van Gumpel E, Simonis A, et al. Comprehensive host cell-based screening assays for identification of anti-virulence drugs targeting. Microorganisms. 2020;8(8):1096.
74. Subhash N, Sundaramurthy V. Advances in host-based screening for compounds with intracellular anti-mycobacterial activity. Cell Microbiol. 2021;23(7):e13337.
75. Heemskerk MT, Korbee CJ, Esselink JJ, Dos Santos CC, van Veen S, Gordijn IF, et al. Repurposing diphenylbutylpiperidine-class antipsychotic drugs for host-directed therapy of Mycobacterium tuberculosis and Salmonella enterica infections. Sci Rep. 2021;11(1):19634.
76. Arango-Argoty G, Garner E, Pruden A, Heath LS, Vikesland P, Zhang L. DeepARG: a deep learning approach for predicting antibiotic resistance genes from metagenomic data. Microbiome. 2018;6(1):23.
77. McArthur AG, Waglechner N, Nizam F, Yan A, Azad MA, Baylay AJ, et al. The comprehensive antibiotic resistance database. Antimicrob Agents Chemother. 2013;57(7):3348–57.
78. Florensa AF, Kaas RS, Clausen PTLC, Aytan-Aktug D, Aarestrup FM. ResFinder – an open online resource for identification of antimicrobial resistance genes in next-generation sequencing data and prediction of phenotypes from genotypes. Microb Genom. 2022;8(1):000748.
79. Yang Y, Jiang X, Chai B, Ma L, Li B, Zhang A, et al. ARGs-OAP: online analysis pipeline for antibiotic resistance genes detection from metagenomic data using an integrated structured ARG-database. Bioinformatics. 2016;32(15):2346–51.
80. Gupta A, Kapil R, Dhakan DB, Sharma VK. MP3: a software tool for the prediction of pathogenic proteins in genomic and metagenomic data. PLoS One. 2014;9(4):e93907.
81. Garg A, Gupta D. VirulentPred: a SVM based prediction method for virulent proteins in bacterial pathogens. BMC Bioinf. 2008;9:62.
82. de Nies L, Lopes S, Busi SB, Galata V, Heintz-Buschart A, Laczny CC, et al. PathoFact: a pipeline for the prediction of virulence factors and antimicrobial resistance genes in metagenomic data. Microbiome. 2021;9(1):49.
83. Ji B, Pi W, Liu W, Liu Y, Cui Y, Zhang X, et al. HyperVR: a hybrid deep ensemble learning approach for simultaneously predicting virulence factors and antibiotic resistance genes. NAR Genomics Bioinf. 2023;5(1):lqad012.
84. Zhang AN, Gaston JM, Dai CL, Zhao S, Poyet M, Groussin M, et al. An omics-based framework for assessing the health risk of antimicrobial resistance genes. Nat Commun. 2021;12(1):4765.
85. Sanger F. Determination of nucleotide sequences in DNA. Biosci Rep. 2004;24(4–5):237–53.
86. Quail MA, Smith M, Coupland P, Otto TD, Harris SR, Connor TR, et al. A tale of three next generation sequencing platforms: comparison of ion torrent, Pacific Biosciences and Illumina MiSeq sequencers. BMC Genomics. 2012;13:341.
87. Jain M, Fiddes IT, Miga KH, Olsen HE, Paten B, Akeson M. Improved data analysis for the MinION nanopore sequencer. Nat Methods. 2015;12(4):351–6.
88. Schena M, Shalon D, Davis RW, Brown PO. Quantitative monitoring of gene expression patterns with a complementary DNA microarray. Science. 1995;270(5235):467–70.
89. D'Costa VM, McGrann KM, Hughes DW, Wright GD. Sampling the antibiotic resistome. Science. 2006;311(5759):374–7.
90. Andries K, Verhasselt P, Guillemont J, Göhlmann HW, Neefs JM, Winkler H, et al. A diarylquinoline drug active on the ATP synthase of Mycobacterium tuberculosis. Science. 2005;307(5707):223–7.

91. Rothberg JM, Leamon JH. The development and impact of 454 sequencing. Nat Biotechnol. 2008;26(10):1117–24.
92. Comas I, Borrell S, Roetzer A, Rose G, Malla B, Kato-Maeda M, et al. Whole-genome sequencing of rifampicin-resistant Mycobacterium tuberculosis strains identifies compensatory mutations in RNA polymerase genes. Nat Genet. 2011;44(1):106–10.
93. Diaz L, Kiratisin P, Mendes RE, Panesso D, Singh KV, Arias CA. Transferable plasmid-mediated resistance to linezolid due to cfr in a human clinical isolate of Enterococcus faecalis. Antimicrob Agents Chemother. 2012;56(7):3917–22.
94. Forsberg KJ, Patel S, Wencewicz TA, Dantas G. The tetracycline destructases: a novel family of tetracycline-inactivating enzymes. Chem Biol. 2015;22(7):888–97.
95. Karim AF, Chandra P, Chopra A, Siddiqui Z, Bhaskar A, Singh A, et al. Express path analysis identifies a tyrosine kinase Src-centric network regulating divergent host responses to mycobacterium tuberculosis infection. J Biol Chem. 2011;286(46):40307–19.
96. Khaledi A, Schniederjans M, Pohl S, Rainer R, Bodenhofer U, Xia B, et al. Transcriptome profiling of antimicrobial resistance in Pseudomonas aeruginosa. Antimicrob Agents Chemother. 2016;60(8):4722–33.
97. Roe C, Williamson CHD, Vazquez AJ, Kyger K, Valentine M, Bowers JR, et al. Bacterial genome wide association studies (bGWAS) and Transcriptomics identifies cryptic antimicrobial resistance mechanisms in. Front Public Health. 2020;8:451.
98. Dar D, Sorek R. Regulation of antibiotic-resistance by non-coding RNAs in bacteria. Curr Opin Microbiol. 2017;36:111–7.
99. Coppola M, Lai RP, Wilkinson RJ, Ottenhoff THM. The in vivo transcriptomic blueprint of Mycobacterium tuberculosis in the lung. Front Immunol. 2021;12:763364.
100. Pisu D, Huang L, Grenier JK, Russell DG. Dual RNA-Seq of Mtb-infected macrophages in vivo reveals ontologically distinct host-pathogen interactions. Cell Rep. 2020;30(2):335–50.e4.
101. Johansen SK, Maus CE, Plikaytis BB, Douthwaite S. Capreomycin binds across the ribosomal subunit interface using tlyA-encoded 2'-O-methylations in 16S and 23S rRNAs. Mol Cell. 2006;23(2):173–82.
102. Netea MG, Joosten LA, Latz E, Mills KH, Natoli G, Stunnenberg HG, et al. Trained immunity: a program of innate immune memory in health and disease. Science. 2016;352(6284):aaf1098.
103. Chan AO, Lam SK, Wong BC, Wong WM, Yuen MF, Yeung YH, et al. Promoter methylation of E-cadherin gene in gastric mucosa associated with Helicobacter pylori infection and in gastric cancer. Gut. 2003;52(4):502–6.
104. Ando T, Yoshida T, Enomoto S, Asada K, Tatematsu M, Ichinose M, et al. DNA methylation of microRNA genes in gastric mucosae of gastric cancer patients: its possible involvement in the formation of epigenetic field defect. Int J Cancer. 2009;124(10):2367–74.
105. Ariffin JK, das Gupta K, Kapetanovic R, Iyer A, Reid RC, Fairlie DP, et al. Histone deacetylase inhibitors promote mitochondrial reactive oxygen species production and bacterial clearance by human macrophages. Antimicrob Agents Chemother. 2015;60(3):1521–9.
106. Al-Amrani S, Al-Jabri Z, Al-Zaabi A, Alshekaili J, Al-Khabori M. Proteomics: concepts and applications in human medicine. World J Biol Chem. 2021;12(5):57–69.
107. Torres-Sangiao E, Giddey AD, Leal Rodriguez C, Tang Z, Liu X, Soares NC. Proteomic approaches to unravel mechanisms of antibiotic resistance and immune evasion of bacterial pathogens. Front Med (Lausanne). 2022;9:850374.
108. Memisević V, Zavaljevski N, Pieper R, Rajagopala SV, Kwon K, Townsend K, et al. Novel Burkholderia mallei virulence factors linked to specific host-pathogen protein interactions. Mol Cell Proteomics. 2013;12(11):3036–51.
109. Yang H, Ke Y, Wang J, Tan Y, Myeni SK, Li D, et al. Insight into bacterial virulence mechanisms against host immune response via the Yersinia pestis-human protein-protein interaction network. Infect Immun. 2011;79(11):4413–24.
110. Schweppe DK, Harding C, Chavez JD, Wu X, Ramage E, Singh PK, et al. Host-microbe protein interactions during bacterial infection. Chem Biol. 2015;22(11):1521–30.

111. Harrison A, Dubois LG, St John-Williams L, Moseley MA, Hardison RL, Heimlich DR, et al. Comprehensive proteomic and Metabolomic signatures of Nontypeable haemophilus influenzae-induced acute otitis media reveal bacterial aerobic respiration in an immunosuppressed environment. Mol Cell Proteomics. 2016;15(3):1117–38.
112. Maus A, Bisha B, Fagerquist C, Basile F. Detection and identification of a protein biomarker in antibiotic-resistant Escherichia coli using intact protein LC offline MALDI-MS and MS/MS. J Appl Microbiol. 2020;128(3):697–709.
113. Josten M, Dischinger J, Szekat C, Reif M, Al-Sabti N, Sahl HG, et al. Identification of agr-positive methicillin-resistant Staphylococcus aureus harbouring the class A mec complex by MALDI-TOF mass spectrometry. Int J Med Microbiol. 2014;304(8):1018–23.
114. Flores-Treviño S, Garza-González E, Mendoza-Olazarán S, Morfín-Otero R, Camacho-Ortiz A, Rodríguez-Noriega E, et al. Screening of biomarkers of drug resistance or virulence in ESCAPE pathogens by MALDI-TOF mass spectrometry. Sci Rep. 2019;9(1):18945.
115. Savijoki K, Myllymäki H, Luukinen H, Paulamäki L, Vanha-Aho LM, Svorjova A, et al. Surface-shaving proteomics of Mycobacterium marinum identifies biofilm subtype-specific changes affecting virulence, tolerance, and persistence. mSystems. 2021;6(3):e0050021.
116. Kok M, Maton L, van der Peet M, Hankemeier T, van Hasselt JGC. Unraveling antimicrobial resistance using metabolomics. Drug Discov Today. 2022;27(6):1774–83.
117. Samuels AN, Wang ER, Harrison GA, Valenta JC, Stallings CL. Understanding the contribution of metabolism to. Front Cell Infect Microbiol. 2022;12:958555.
118. Peyron P, Vaubourgeix J, Poquet Y, Levillain F, Botanch C, Bardou F, et al. Foamy macrophages from tuberculous patients' granulomas constitute a nutrient-rich reservoir for M. tuberculosis persistence. PLoS Pathog. 2008;4(11):e1000204.
119. Wilburn KM, Fieweger RA, VanderVen BC. Cholesterol and fatty acids grease the wheels of Mycobacterium tuberculosis pathogenesis. Pathog Dis. 2018;76(2):fty021. https://doi.org/10.1093/femspd/fty021. PMID: 29718271; PMCID: PMC6251666
120. Bobba S, Khader SA. Rifampicin drug resistance and host immunity in tuberculosis: more than meets the eye. Trends Immunol. 2023;44(9):712–23.
121. Wang L, Ying R, Liu Y, Sun Q, Sha W. Metabolic profiles of clinical isolates of drug-susceptible and multidrug-resistant. Infect Drug Resist. 2023;16:2667–80.
122. Pal R, Hameed S, Kumar P, Singh S, Fatima Z. Comparative lipidomics of drug sensitive and resistant. 3 Biotech. 2017;7(5):325.
123. Lahiri N, Shah RR, Layre E, Young D, Ford C, Murray MB, et al. Rifampin resistance mutations are associated with broad chemical remodeling of Mycobacterium tuberculosis. J Biol Chem. 2016;291(27):14248–56.
124. Loots DT. New insights into the survival mechanisms of rifampicin-resistant Mycobacterium tuberculosis. J Antimicrob Chemother. 2016;71(3):655–60.
125. Chaiyachat P, Kaewseekhao B, Chaiprasert A, Kamolwat P, Nonghanphithak D, Phetcharaburanin J, et al. Metabolomic analysis of Mycobacterium tuberculosis reveals metabolic profiles for identification of drug-resistant tuberculosis. Sci Rep. 2023;13(1):8655.
126. Maifiah MH, Creek DJ, Nation RL, Forrest A, Tsuji BT, Velkov T, et al. Untargeted metabolomics analysis reveals key pathways responsible for the synergistic killing of colistin and doripenem combination against Acinetobacter baumannii. Sci Rep. 2017;7:45527.
127. Collins JM, Siddiqa A, Jones DP, Liu K, Kempker RR, Nizam A, et al. Tryptophan catabolism reflects disease activity in human tuberculosis. JCI Insight. 2020;5(10):e137131.
128. Chandra P, Coullon H, Agarwal M, Goss CW, Philips JA. Macrophage global metabolomics identifies cholestenone as host/pathogen cometabolite present in human Mycobacterium tuberculosis infection. J Clin Invest. 2022;132(3):e152509.
129. Howard NC, Marin ND, Ahmed M, Rosa BA, Martin J, Bambouskova M, et al. Mycobacterium tuberculosis carrying a rifampicin drug resistance mutation reprograms macrophage metabolism through cell wall lipid changes. Nat Microbiol. 2018;3(10):1099–108.

130. Ahl PJ, Hopkins RA, Xiang WW, Au B, Kaliaperumal N, Fairhurst AM, et al. Met-flow, a strategy for single-cell metabolic analysis highlights dynamic changes in immune subpopulations. Commun Biol. 2020;3(1):305.
131. Kraly JR, Holcomb RE, Guan Q, Henry CS. Review: microfluidic applications in metabolomics and metabolic profiling. Anal Chim Acta. 2009;653(1):23–35.
132. Zhou M, Xie L, Yang Z, Zhou J, Xie J. Lysine succinylation of Mycobacterium tuberculosis isocitrate lyase (ICL) fine-tunes the microbial resistance to antibiotics. J Biomol Struct Dyn. 2017;35(5):1030–41.
133. Garcia-Vilanova A, Chan J, Torrelles JB. Underestimated manipulative roles of mycobacterium tuberculosis cell envelope glycolipids during infection. Front Immunol. 2019;10:2909.
134. Yakovlieva L, Fülleborn JA, Walvoort MTC. Opportunities and challenges of bacterial glycosylation for the development of novel antibacterial strategies. Front Microbiol. 2021;12:745702.
135. Ma P, Amemiya HM, He LL, Gandhi SJ, Nicol R, Bhattacharyya RP, et al. Bacterial droplet-based single-cell RNA-seq reveals antibiotic-associated heterogeneous cellular states. Cell. 2023;186(4):877–91.e14.
136. Omersel J, Karas KN. Vaccinomics and adversomics in the era of precision medicine: a review based on HBV, MMR, HPV, and COVID-19 vaccines. J Clin Med. 2020;9(11):3561.
137. Rawat BS, Kumar D, Soni V, Rosenn EH. Therapeutic potentials of immunometabolomic modulations induced by tuberculosis vaccination. Vaccines (Basel). 2022;10(12):2127.
138. Khan N, Downey J, Sanz J, Kaufmann E, Blankenhaus B, Pacis A, et al. M. Tuberculosis reprograms hematopoietic stem cells to limit Myelopoiesis and impair trained immunity. Cell. 2020;183(3):752–70.e22.
139. Darrah PA, Zeppa JJ, Maiello P, Hackney JA, Wadsworth MH, Hughes TK, et al. Prevention of tuberculosis in macaques after intravenous BCG immunization. Nature. 2020;577(7788):95–102.
140. Djordjevic SP, Jarocki VM, Seemann T, Cummins ML, Watt AE, Drigo B, et al. Genomic surveillance for antimicrobial resistance – a one health perspective. Nat Rev Genet. 2023; https://doi.org/10.1038/s41576-023-00649-y.
141. Salinas L, Cárdenas P, Johnson TJ, Vasco K, Graham J, Trueba G. Diverse commensal Escherichia coli clones and plasmids disseminate antimicrobial resistance genes in domestic animals and children in a semirural community in Ecuador. Msphere. 2019;4(3):e00316-19.
142. Watsa M, Group WDSF. Rigorous wildlife disease surveillance. Science. 2020;369(6500):145–7.
143. Spänig S, Eick L, Nuy JK, Beisser D, Ip M, Heider D, et al. A multi-omics study on quantifying antimicrobial resistance in European freshwater lakes. Environ Int. 2021;157:106821.
144. Soni VAA, Bhattacharya CY, Bukhari Z, Gupta SL, Basu S, Jain S, Soni Y. Chapter 4: Genomic surveillance of bacterial pathogens: expanding horizons. In: Pandey R, editor. Genomic surveillance and pandemic preparedness. Academic; 2023. p. 71–117.
145. Stokes JM, Yang K, Swanson K, Jin W, Cubillos-Ruiz A, Donghia NM, et al. A deep learning approach to antibiotic discovery. Cell. 2020;180(4):688–702.e13.
146. Kavvas ES, Yang L, Monk JM, Heckmann D, Palsson BO. A biochemically-interpretable machine learning classifier for microbial GWAS. Nat Commun. 2020;11(1):2580.
147. Larkins-Ford J, Degefu YN, Van N, Sokolov A, Aldridge BB. Design principles to assemble drug combinations for effective tuberculosis therapy using interpretable pairwise drug response measurements. Cell Rep Med. 2022;3(9):100737.
148. Anahtar MN, Yang JH, Kanjilal S. Applications of machine learning to the problem of antimicrobial resistance: an emerging model for translational research. J Clin Microbiol. 2021;59(7):e0126020.
149. Payne DJ, Gwynn MN, Holmes DJ, Pompliano DL. Drugs for bad bugs: confronting the challenges of antibacterial discovery. Nat Rev Drug Discov. 2007;6(1):29–40.

150. Chen L, Li H, Chen T, Yu L, Guo H, Chen Y, et al. Genome-wide DNA methylation and transcriptome changes in Mycobacterium tuberculosis with rifampicin and isoniazid resistance. Int J Clin Exp Pathol. 2018;11(6):3036–45.
151. Chen R, Mias GI, Li-Pook-Than J, Jiang L, Lam HY, Miriami E, et al. Personal omics profiling reveals dynamic molecular and medical phenotypes. Cell. 2012;148(6):1293–307.
152. Fondi M, Liò P. Multiomics and metabolic modelling pipelines: challenges and tools for systems microbiology. Microbiol Res. 2015;171:52–64.
153. Zhu Y, Czauderna T, Zhao J, Klapperstueck M, Maifiah MHM, Han ML, et al. Genome-scale metabolic modeling of responses to polymyxins in Pseudomonas aeruginosa. Gigascience. 2018;7(4):giy021.
154. Abdul Rahim N, Cheah SE, Johnson MD, Zhu Y, Yu HH, Sidjabat HE, et al. Transcriptomic responses of a New Delhi metallo-β-lactamase-producing Klebsiella pneumoniae isolate to the combination of polymyxin B and chloramphenicol. Int J Antimicrob Agents. 2020;56(2):106061.
155. Wozniak JM, Mills RH, Olson J, Caldera JR, Sepich-Poore GD, Carrillo-Terrazas M, et al. Mortality risk profiling of Staphylococcus aureus Bacteremia by multi-omic serum analysis reveals early predictive and pathogenic signatures. Cell. 2020;182(5):1311–27.e14.
156. Cohen A, Bont L, Engelhard D, Moore E, Fernández D, Kreisberg-Greenblatt R, et al. A multifaceted 'omics' approach for addressing the challenge of antimicrobial resistance. Future Microbiol. 2015;10(3):365–76.
157. Kim SH, Bahn JW, Kim YK, Chang YS, Shin ES, Kim YS, et al. Genetic polymorphisms of drug-metabolizing enzymes and anti-TB drug-induced hepatitis. Pharmacogenomics. 2009;10(11):1767–79.

Chapter 11
Applications of Artificial Intelligence and Machine Learning in Antimicrobial Resistance Study

Ayush Praveen, Nicholas Bartelo, and Vijay Soni

Abbreviations

AI	Artificial Intelligence
AMR	Antimicrobial Resistance
ARG	Antimicrobial Resistance Genes
AST	Antimicrobial Susceptibility Test
CDSS	Clinical Decision Support Systems
MALDI-TOF	Matrix-Assisted Laser Desorption and Ionization-Time-of-Flight
ML	Machine Learning
SHAP	Shapley Additive Explanations
SVM	Support Vector Machines
WGS	Whole-Genome Sequencing
WHO	World Health Organization
XAI	Explainable AI

A. Praveen
Elucidata Data Consulting Pvt Ltd., New Delhi, India

N. Bartelo
Department of Physiology and Biophysics, Weill Cornell Medicine, Institute for Computational Biomedicine, Englander Institute for Precision Medicine, New York, NY, USA

V. Soni (✉)
Division of Infectious Diseases, Weill Department of Medicine, Weill Cornell Medicine, New York, NY, USA
e-mail: vis2032@med.cornell.edu

V. Soni, A. S. Akhade (eds.), *Antimicrobial Resistance: Factors to Findings*,
https://doi.org/10.1007/978-3-031-65986-7_11

11.1 Introduction

Healthcare data such as electronic health and medical records, wearable data, omics data, medical images, and scientific literature data carries a person's biological footprint that contains important hidden information and insights that can be used for different applications [1, 2]. The healthcare data usually have several different types of measurements using numerous features, thus, as a perfect example of big data, but lies behind the "curse of dimensionality" [3]. To move pass the curse of dimensionality, it is required that proper data science, machine learning, and artificial intelligence approaches are followed to deal with noise and extract meaningful insights from the data. In the particular context of antimicrobial resistance (AMR), following artificial intelligence methods to analyze the data can help in AMR diagnosis & prediction, developing new antibiotics, prescribing antibiotics, and assisting Clinical Decision Support Systems (CDSS) [4].

11.1.1 Diagnosis and Prediction of AMR

Currently, AMR diagnosis is done using the Antimicrobial Susceptibility Test (AST), which requires screening of different drug regimens on the derived culture for days and weeks before getting some actionable insights [5]. In contrast, whole-genome sequencing (WGS)-based approaches can provide quick results. However, the diagnosis from WGS is usually following rule-based approaches of checking for specific gene alterations. Machine learning in this case can bridge the gap by deriving the patterns from WGS data itself and extracting signatures associated with AMR which then can be quickly checked within the collected samples, providing fast results. Multiple studies have shown the efficacy of such approaches for the diagnosis of AMR at least a lab scale. Rishishwar, Lavanya, et al. [6] used a logistic regression-based approach to distinguish between vancomycin-intermediate *S. aureus* (VISA) and vancomycin-susceptible *S. aureus* (VSSA) with high accuracy. Santerre et al. [7] used a random forest model trained k-mer representation of bacterial genomes of *A. baumanii*, *M. tuberculosis*, and *S. pneumoniae* for AMR prediction against common antimicrobial drugs achieving as high accuracy as 92% in certain cases. It is not just the NGS-based methods that have the potential to identify, diagnose, and predict AMR. MALDI Biotyper CA (MBT-CA) System (Bruker Daltonics Inc., Billerica, MA) was approved by the FDA as an ML-based MALDI algorithm. It has been successful in the prediction of pathogens with high accuracy of 99.8% (2258/2263) at genus level and 98.2% (2222/2263) at species level [8].

11.1.2 Developing New Antibiotics

The initial discovery of majority of antibiotics was based on screening of secondary metabolites from soil-dwelling bacteria for their activity against pathogenic bacteria [9]. However, further antibiotics cannot be discovered using the conventional methods considering the pace of antimicrobial resistance development. Many antibiotic development programs have turned to screening large chemical libraries. However, these chemical libraries often suffer from two problems. First, these chemical libraries usually have millions of compounds which makes the screening difficult, and second, the compounds often have limited chemical diversity which potentially reduces the chances of finding something novel with good efficacy [10]. Machine learning and deep learning approaches in this case can aid in screening potential antibiotic compounds from the large pool of chemical libraries as well as identify novel compounds based on the chemical structure of existing compounds [11]. Stokes, Jonathan M., et al. [12] used a neural network-based deep learning approach with training data of 2335 compounds on growth inhibition of *E. coli*. Post getting an optimal model, they used the model to screen a library of greater than 107 million compounds. They found Halicin as a potential antibiotic with activity not just against *E. coli* but against multiple drug-resistant strains of bacteria. It had bactericidal activity against *M. tuberculosis* and showed strong growth inhibitions against carbapenem-resistant Enterobacteriaceae and *A. baumannii* isolates, thereby acting as Broad-Spectrum Bactericidal Antibiotic. Additionally, eight compounds were also identified and validated by them from ZINC15 database for their activity against *E. coli*. Antimicrobial peptides are being explored with hope as a new class of compounds with strong activity against bacteria. These are naturally synthesized by different organisms from microbes to humans with the ability to either directly kill the microbe or indirectly modulate the host immune response against microbes [13]. There are different studies that have tried to identify these potential antimicrobial peptides using machine learning approaches. Fields et al. for instance [14], used machine learning in conjunction with biophysical properties of minimal bacteriocin domains to design a pipeline for bacteriocin peptide design. In the first phase, they designed 20 amino acid long peptides using the approach and arrived at 28,895 total 20-mer candidate peptides. Further, postscreening of these peptides using the biophysical filters, they narrowed down to 16 sequences. Upon synthesis and testing, they showed strong antimicrobial activity against *E. coli* and *Pseudomonas aeruginosa*.

11.1.3 Prescribing Antibiotics

The efficacy of an antibiotic treatment is dependent on the correct matching of antibiotic with the pathogen to treat. Prescribing antibiotics without consideration of the patient's infection and antibiotic regimen history can have repercussions of ineffective treatment and further development of AMR [15]. Medical records in this regard provide data for previous infections, diseases, and antibiotic use and machine learning approaches can identify patterns from the data to prescribe the right antibiotic regime for the current infection based on the past data. Yelin et al. [16], used machine learning approach over a 10-year longitudinal dataset. The data had a record of 700,000 community-acquired urinary tract infections with over 5,000,000 mapped records of antibiotic purchases. Logistic regression and gradient-boosting decision trees were trained over the training data to identify patterns of current drug regimen match or mismatch based on the previous regimens. Together the models could predict cases of resistance development, with demographics, sample history, and drug purchase history being the strong factors for predicting such cases. Based on the data, the study claims that the chances of mismatch treatment which could potentially lead to the development of resistance could have been reduced by machine learning algorithms. Although patient medical history provides substantial data to predict correct antibiotic regimen, but it is not comprehensive. Especially with the cases of development of resistance of a different strain of bacteria as opposed to one treated for infection. Stracy et al. [17] paired whole-genome analysis of 1113 pre- and posttreatment bacterial isolates from microbiome samples and machine learning analysis of medical health records of 140,349 UTI and 7365 wound infection cases. They used multivariate logistic regression as a machine learning approach to provide personalized drug prescriptions to minimize antibiotic resistance at an individual level.

11.1.4 Assisting Clinical Decision Support Systems (CDSS)

Clinical decision support systems (CDSS) are used by clinicians to access the updated clinical practice guidelines based on patient clinical case and characteristics. These are computerized systems which help in quick decision-making [18]. Current CDSS are expert and knowledge-driven systems. Nathan et al. in their review found 60 machine learning-assisted clinical decision-making systems. About 20 of them were designed for diagnosis of infection, 18 for early detection, 13 for prediction of treatment response, 4 for prediction of AMR, 3 for assisting choice of antibiotic regimen, and two for assisting the choice of combinatorial antiretroviral therapy. These ML-CDSS can be used for different cases such as prediction of sepsis, diagnosis of infection, antibiotic prescription, and others; however, the current generation of ML-CDSSs have been trained with limited quantity of data and often lacking required data quality. These are usually trained on open-access databases or datasets with limited variables for patient clinical information. To develop comprehensive ML-CDSSs, it is required to meet the lacking criteria for achieving their practical usage in different healthcare settings [4, 19].

11.2 Artificial Intelligence (AI) and Machine Learning (ML) Fundamentals for OMICs Studies

While exploring individual genes, metabolites, proteins, and their alterations, reductionist approaches have supported the understanding of complex mechanisms of AMR [20]. Viewing AMR emergence through a systems biology lens can provide deep mechanisms of interaction between different components of microbes and host [21] that individual components of the system cannot provide. By not restricting the viewpoint to one biomolecule at a time, information at the level of genome, transcriptome, proteome, and metabolome can be studied and their multifaceted interactions can be discovered to figure out mechanisms that are not accessible via traditional methods.

11.2.1 Systems Biology and Omics for Exploring AMR

Systems biology approaches can provide overall insights about the changes in gene expression due to the mutation, which can be correlated with changes in proteome and metabolome, leading to variations in the cellular functions and processes [22]. Different omics methodologies explore separate levels of the biological systems, and their explorations can provide insights about cellular states and the alterations of which can direct the evolution of AMR. Genomics can help in understanding new Antimicrobial Resistance Genes (ARGs) and mutations in already discovered ARGs. Epigenomics can help us in studying how changes in environmental factors can drive epigenetic modification such as methylation over specific ARGs. Transcriptomics can provide a snapshot of gene expression changes due to the development of mutations in existing genes, acquisition of new ARGs or changes in the environment. Proteomics provides an understanding of protein level changes in the system. Metabolomics gives a viewpoint of end effector metabolic changes in the system. Further metagenomics goes one step further by using genomic sequencing to understand bacterial communities and their interactions. Altogether different omics methods can be used for different applications in AMR research. We will be exploring some of these applications in the next section.

11.2.2 Omics for AMR Research and Clinical Applications

WGS has revolutionized different areas of AMR research. It has granted us the ability to explore genome-wide regions with the ultra-resolution of a nucleotide. It has been helpful in identification of AMR, classification of the pathogen, detecting virulence of the microbes, and also for tracing ancestry [23]. Use of traditional culture-based methods for identification and diagnosis of AMR has been restrictive due to their time-intensive nature. WGS can provide identification and discovery of ARGs and AMR strains from the clinical samples without the need of isolating and culturing [24, 25]. This ability has been extended by metagenomic sequencing which not only eliminates the need to culture bacteria but can identify strains from the environmental samples and also help in understanding how complex interactions and sharing of ARGs happen within the community [26].

Transcriptomics analysis provides a glimpse of gene expression changes due to environmental factors or in response to the administration of an antibiotic which can provide cues about the mechanism of antibiotic resistance [27]. Multiple studies have used transcriptomics to understand the mechanism of resistance to specific antibiotics, for instance Li et al. [28] elucidated antibiotic Polymyxin resistance in *Salmonella* Typhimurium ATCC13311 (AT). The study found that *S.* Typhimurium acquires resistance by increasing expression of *phoP* and *phoQ* genes. This results in changes in lipid membrane of the bacterium by decreasing the negative charge on the membrane which makes it less adsorbent for positively charged polymixin. Although transcriptomics can predict the overall changes in the biological system, it is proteomics that provides an accurate estimation of functional changes. Singh et al. [29] found an increase in expression of 27 proteins in clinical isolates of multidrug-resistant (MDR) strains of *M. tuberculosis*. These differentially expressed proteins were enriched for pathways related to lipid metabolism, intermediary metabolism, respiration, cell wall, and cell processes.

Metabolomics can be used for studying the changes in metabolism favoring resistance due to administration of antibiotics [30]. Metabolomics reveals the phenotypic expression for the organism because of global view of all the biomolecules that change in response to factors such as environmental changes, mutations, and bacterial growth. Schelli et al. [31] found changes in the metabolism of amino acids, pyrimidines, and purines between isogenic methicillin-susceptible and -resistant *Staphylococcus aureus* strains against antibiotics from different classes (β-lactams, aminoglycosides, and quinolones).

Studying the bacterial system using specific omics method provides an understanding of the regulation of mechanism of resistance at that level (genome, epigenome, transcriptome, proteome or metabolome); however, approaching the system from a multiomics standpoint provides the global view of the system. Wherever possible, multiomics methodologies should be used to understand the impact of different factors such as mutations, environment, antibiotic stress, and others in a holistic manner. Mu, Klare, Baines, Pang, Guérillot, and Harbison-Price et al. [32] checked multiple clinical isolates of four sepsis-causing pathogens *E. coli*, *Klebsiella pneumoniae* species complex, *S. aureus*, and *Streptococcus pyogenes* using multiomics approach and found conserved and divergent changes across the pathogens. Altogether genomics, transcriptomics, proteomics, and metabolomics revealed changes in fatty acid and lipid biosynthesis, other metabolic pathways required for cell envelope remodeling and nutrient adaptation for Osmo protection. They also found the acquisition of cholesterol as a common mechanism across species.

11.2.3 Artificial Intelligence and Machine Learning for Omics Analysis

Omics approaches are leading the forefront of research and clinical applications in the field of AMR. However, as we have discussed in the previous chapters, like any biomedical data, omics data suffers from curse of dimensionality [3], perhaps more than any other biomedical data. Even for a simple experiment, every single sample requires measurement of tens of thousands of features (genes, proteins, metabolites, SNPs, etc.). Identifying patterns from the omics data using data science, machine learning, and artificial intelligence can have compounding effects in terms of speed, depth of discovery, and applications.

Vilne et al. [33] have reviewed the use of machine learning-based tools from the stance of speed and accuracy of detection of WGS-based knowledge in case of AMR and foodborne disease outbreaks. The group found, through different studies, that ML-based tools exist for all the four steps namely, de novo genome assembly, genome characterization, comparative genomics, and inference of phylogeny or phylogenomics of WGS analysis.

11.2.3.1 Omics Data and Dimensionality Reduction

Omics data often has a much higher number of features (in this case, genes, transcripts, metabolites, and proteins) measured than the number of samples measured. This is the classical problem of "Curse of Dimensionality." A large number of features in the dimensional space are redundant and can lead to inaccurate predictions by ML methods. Thus, it is often required to pick the right set of features for training any model [3, 34]. Dimensionality reduction methods help in dealing with redundant features and can be classified into two categories: feature selection and feature engineering. Feature selection approaches can be classified into filter, wrapper, and embedded methods and they deal with selecting the right sets of features important for training. Feature engineering methods reduce dimensions by engineering a set of representative features from the original set of features. This leads to decrease in dimensions in the order of few tens of thousands to tens of engineered features. Examples of feature engineering methods are Principal Component Analysis (PCA), Multidimensional Scaling (MDS), Linear Discriminant Analysis (LDA), t-distributed Stochastic Neighbor Embedding (t-SNE), Uniform Manifold Approximation and Projection (UMAP), and many others [34].

11.2.3.2 AI Methods for Clustering, Regression, and Classification

Based on the learning approach and the outcome categorization, computational methods can be used to group data points (Clustering) created on the values of input features or calculate an outcome that could be continuous (Regression) or discrete

(Classification) in nature. Reel et al. [35] reviewed commonly used ML and DL methods and grouped them based on the following categories: probability or Bayesian, tree, error, similarity, support vectors, and neural network.

Probability-based or Bayesian methods [36], like Naive Bayes or Bayesian networks, rely on probability to arrive at an outcome. Bayesian methods are often fast and less complex to implement; however, the built models are difficult to interpret and explain.

Decision trees [37] are used for classification where decision-making is based on branches of trees with several subdecisions. They are fast and relatively easy to interpret. Random forest [38] is an ensemble method that utilizes multiple decision trees to improve accuracy and reliability. The improvement in accuracy however comes at the cost of speed and explainability.

Error-based methods are used for regression [39]. Based on the nature of the relationship between independent variable and outcome, there are linear regression, logistic regression, and partial linear regression methods. The regression-based approaches are simple with high speed and low compute complexity but are often not as accurate as other methods.

Similarity-based methods such as k-nearest neighbors [40] are often used for clustering data points based on shared characteristics. K-nearest neighbors is a routinely used clustering method which is computationally simple, fast, and easy to implement. It tries to divide the samples or features into k distinct clusters. If k is however not chosen wisely, it can lead to over or under fitting.

Support vector-based [41] methods such as Linear or Nonlinear Support Vector Machines (SVM) are supervised learning methods used for classification or regression tasks. They are particularly good with high-dimensional data, and it is very robust to outliers (one of the applications of SVM is in outlier detection). SVM optimizes to find a hyperplane that can divide different samples into different classes by trying to maximize the margin between the closest points of different classes. Linear SVMs are used when the hyperplane can be defined by a linear boundary between different classes. If a hyperplane cannot be defined using a linear boundary, then Nonlinear SVMs are used where Sigmoid, Polynomial or Gaussian functions are used.

Neural networks have been highly useful for modeling complex relationships between features in high-dimensional data. They are organized in the form of three (one input, at least one hidden, and one output) or more layers of artificial neurons or nodes. Neural networks are the underlying methods in deep learning [42]. Each layer in a neural network performs operations over the features for classification exercises from complex and seemingly unrelated data points. Neural Networks have pushed the boundaries of accuracy of prediction, however due to their inherent layered structure, they take longer training time, computationally intensive, and have very low explainability.

11.2.3.3 Explainable AI (XAI) to Decode "Black Box" Models

AI provides an unprecedented advantage to predict decisions with high confidence from the high-dimensional and complex omics data. These end decisions could be susceptibility to antibiotics or development of resistance in a community. However, AI methods whether machine learning or deep learning based are considered "black boxes" because of their inherent inability to explain the basis of prediction. Explainability of models is crucial for biomedical applications to know not just the end decision but to also understand the mechanism or biological phenomenon behind the decision. Explainable AI (XAI) solves the problem of explainability of models which can offer applications of its own for extracting knowledge of the biological system through its omics data. These are the set of methods that can be used on the top of ML or DL models to interpret the reasoning for decisions. In the case of biological applications, the reasoning is very important to deep dive into the mechanism behind the biological phenomenon.

Based on different working principles, there are many XAI methods. SHAP [43] and LIME [44] have recently gathered high usability for different applications around biological data. SHAP works on game theory, where considering a cooperative game, payoffs for players participating are decided in a fair manner. For machine learning models, both local for a class and global importance (payoff) for each feature are assigned using Shapely values. LIME which is Local Interpretable Model-Agnostic Explanations, doesn't need to peek into the model to predict importance but instead perturbs the input around its neighborhood to understand how the model behaves to assign importance to features. It does so in an instance-specific manner where an instance is taken (for example, whether subject 10 out of 200 subjects has diabetes or not) and feature importance is calculated.

Both SHAP and LIME have been used for different applications around OMICs data. Benkwitz-Bedford, Sam, et al. [45] used SHAP to understand the impact of different antimicrobial resistance genes on doubling time and growth yield after 8 h for 1407 strains of *E. coli* for TransPred project treated with subinhibitory antimicrobial concentrations of different antimicrobials. Lemsara et al. [46] created PathME, a pathway-based tool for clustering of patients using multimodal data. They have used DeepSHAP (an implementation of SHAP for deep neural networks) for understanding of cluster-specific pathway scores. In a study by Yagin et al. [47], LIME and SHAP were used to identify biomarkers for COVID-19 from a pool of 15,979 gene expressions of 234 patients (141 COVID-19 negative and 93 COVID-19 positive).

11.2.3.4 Tree-Based Methods and Tabular Omics Data

Although omics data generates a lot of raw data (sequences, annotations, etc.), the final processed data that can be used for analysis is tabular in nature. Tree-based methods are very fast as compared to the deep learning methods due to their simplistic nature. Grinsztajn et al. [48], compared tree-based models and deep learning methods for their performance over the tabular data. The data used by the authors was however nonomics data, thus it is important to consider the learnings within some guardrails. The approach of using simplistic tree-based methods can provide quicker and reasonable insights from the omics data.

11.3 Role of AI/ML-Based Approaches in AMR Diagnosis

Artificial intelligence provides two tools that have been increasingly used for the discovery of biomarkers and therapeutics: machine learning and deep learning. Through the combination of large amounts of omics data and/or additional sample data with learning algorithms, numerous insights into the biological mechanisms by which antimicrobial resistance develops have been discovered, providing increased predictive ability compared with other methods. For example, a comparison between a rules-based and machine learning-based approach with the goal of predicting antimicrobial resistance determined a higher accuracy with novel variants using the machine learning-based approach [49]. The power of AI to discover new biomarkers was demonstrated through the utilization of machine learning techniques and a combination of proteomics and metabolomics by Wozniak et al.[50]. The goal of these researchers was to discover prognostic biomarkers of mortality due to *Staphylococcus aureus* bacteremia using proteomics and metabolomics from 217 patients, including 25 controls, 79 infected patients who died, and 113 infected patients who survived. An ensemble of machine learning algorithms was fit to proteomics and metabolomics separately, then combined, to determine the predictive power of each feature set using the metric area under the receiver operating characteristic (AUC). When separate, proteomics and metabolomics performed similarly with an AUC ~0.75, but when combined, the predictive power increased to an AUC of 0.86, demonstrating the ability for multiple modalities of omics data to enhance the discrimination of different phenotypes. Top protein biomarkers included fetuin B, heparin cofactor II, and carnosine dipeptidase 1, and top metabolite biomarkers encompassed 2-hexadecanoylthio-1-ethylphosphorylcholine, sphingosine-1-phosphate, thyroxine, and decanoyl-carnitine. This work portrays how machine learning algorithms can be employed to find novel markers associated with mortality due to an antimicrobial-resistant strain of bacteria, providing an insight into the change of phenotype due to the pathogen in a patient, allowing for the potential to use these biomarkers to determine and prioritize patients close to death.

Huang et al. used five machine learning algorithms to find protein biomarkers of carbapenem-resistant *K. pneumoniae* by measuring proteomics of 46 resistant and 49 sensitive strains [51]. The analysis purposed that as the top-ranked peaks were added to the model, the accuracy was not optimal when the most peaks were used in the model. This means that the addition of some peaks actually resulted in creating more noise that limited the predictive power of the machine learning algorithms, underscoring the concept that the features input into the model must be carefully considered. However, once a set of optimal biomarkers is chosen, additional validation must be performed with other data to ensure there is no overfitting to the specific dataset. With the optimal number of 80 peaks determined by sensitivity analysis involving running the model with 50–100 peaks at intervals of 10, the random forest model achieved an accuracy of 97% in prediction of resistant or sensitive bacteria samples. While the authors used tenfold cross-validation for the models, it is unknown whether the mean accuracy, or the highest accuracy, is presented. If the stated accuracy is the mean, the authors should have reported the standard error of the tenfold cross-validation accuracies for the reader to understand the variability of the predictive power of the model. If the highest accuracy is stated, the impressive accuracy could be representing the overfitting to a small part of the data which is not generalizable to the rest of the samples. This study shows potential for a great discrimination between carbapenem-resistant and carbapenem-sensitive *K. pneumoniae* using proteomics. Future studies should validate the predictive ability of the 80 proteins which produced the highest accuracy to ensure the biomarkers are generalizable to other datasets. If the proteins are discovered to truly separate the sensitive and resistant bacteria samples, it may be possible to create a diagnostic tool to assess if a patient's *K. pneumoniae* infection is resistant to carbapenems or not, allowing for the prescription of drugs that are optimal for each patient.

One limitation of using learning algorithms is the necessity for a large sample size, which is sometimes difficult due to cost or time needed to develop and measure samples. One of the most impressive studies with respect to sample size for one bacteria species in the field of AMR was conducted by Yu et al. [52]. Proteomics was measured for 20,359 clinical isolates to determine biomarkers for rapid identification of methicillin-resistant *Staphylococcus aureus* using the light gradient boosting machine algorithm. The feature importance was determined by shapley additive explanations (SHAP) values due to the low interpretability of the machine learning algorithm. The model was evaluated using five separate databases, achieving an AUC greater than 0.75 for every population. The main predictor of methicillin resistance was annotated as UPF0337 protein SACOL1680. In addition, molecular docking simulations were run which determine a potential reason for the development of drug resistance is the inability of methicillin to bind with UPF0337 protein SACOL1680. While the identification of this protein as a biomarker is promising, controlled experiments in the laboratory should be conducted to verify the mechanism by which the protein influences the resistance phenotype before using these results in the clinic. Nevertheless, this experiment has extensive impacts, as it demonstrates the ability of machine learning to discover a new biomarker candidate associated with resistance that provides additional knowledge to our understanding of the biological mechanism responsible for the development of drug resistance in bacteria.

Machine learning algorithms can be used to discern and compare the predictive ability of a set of features to discriminate between sensitive and resistant samples for multiple bacteria species. Rees et al. measured metabolites of 60 *K. pneumoniae*, 37 *E. coli*, and 20 *Enterobacter cloacae* samples, each consisting of similar numbers of carbapenem-resistant and carbapenem-sensitive samples [53]. The goals of the researchers were to ascertain the metabolites which provide the greatest discrimination ability between the carbapenem-resistant and -sensitive samples and report the AUC for the algorithms to fit to these features. Three machine learning algorithms were fit to the data: random forest, linear support vector machine, and partial least squares discriminant analysis. After combining all species, the optimal AUC of 0.912 resulted from fitting the linear support vector machine. Top annotated metabolites for discrimination of sensitive and resistant samples were 2-phenylacetate, 2-nonanone, one aldehyde, and two 2-ketones. To determine if species-specific modeling would improve the AUC, the authors fit separate models for each of the three species, revealing an increase in discriminatory ability for *K. pneumoniae* and *E. cloacae*, but a substantial decrease for *E. coli*. This illustrates the specificity of the metabolic changes due to carbapenem resistance in different bacteria species. While the model fit with the combination of all the samples performed well, it is possible that the majority of its correct predictions were from *K. pneumoniae* and *E. cloacae*. By dissecting the model to species-specific groups, the authors have shown that the changes in the measured metabolic profiles due to carbapenem resistance can be determined for some bacteria species, but either other parts of the metabolome must be measured to find good biomarkers of carbapenem resistance, or the metabolome does not provide features that can adequately discriminate the sensitive and resistant samples at all in *E. coli*. This is an important lesson because if the initial classifier had been implemented into a clinical setting for all three bacteria species, many infections of carbapenem-resistant *E. coli* would potentially have been undiagnosed.

Weis et al. applied two machine learning algorithms (logistic regression and gradient-boosted decision trees), a deep neural network classifier, and multilayer perceptron, to methicillin resistant and sensitive strains of *S. aureus*, and ceftriaxone resistant strains of *E. coli* and *K. pneumoniae* [54]. The researchers measured the proteins of the samples in an attempt to discriminate between resistant and sensitive samples, as well as determine biomarkers of the resistance phenotype. For methicillin-resistant versus methicillin-sensitive *S. aureus*, an AUC of 0.80 was achieved. Both ceftriaxone-resistant versus ceftriaxone-sensitive *E. coli* and *K. pneumoniae* resulted in an AUC of 0.74. An additional model was developed to determine if pooling all samples together, with the rationale that an increased sample size may increase each model's discrimination of overall resistance versus sensitive phenotypes, would outperform each species-specific model. The models made with the combined sample resulted in lower performance than any of the three species-specific models, validating the previous work by Rees et al. that best prediction results when the samples come from the same bacteria species. However, the authors demonstrated that although increased samples from different bacteria species did not enhance the model, it was likely that the addition of samples from the same species would increase the performance of each classifier, as the AUC seemed to continue to increase without a plateau in the species-specific models. SHAP values were calculated for each protein, which showed some proteins found in the literature to be associated with antibiotic resistance, as well as new diagnostic biomarkers. A retrospective clinical case study was performed to test the efficacy of the model, resulting in accurate prediction of antibiotic resistance in 51 of these 54 cases. This study reiterates the idea that predictive power, and thus the ability to find biomarkers of antimicrobial resistance, is better when focused on one species of bacteria. Also, the ability for these researchers to test their model in the clinic, and the high accuracy achieved when tested on the clinical samples, illustrates the potential for machine learning to aide in clinical decision-making.

Deep learning uses multiple layers of artificial neurons to create a neural network with the goal of transforming input data to predict a desired output. An advantage of deep learning is that it can take as input numerous forms of unprocessed data, such as image, audio, etc., and create a coherent output. However, deep learning requires a large amount of data to correctly learn its task, which is sometimes difficult due to expense and time limitations. Ahmad et al. designed a pipeline for the rapid identification of bacteria species, strain-level classification, and the presence or absence of AMR, which features a convolutional neural network to predict each identification task separately [55]. The input to the convolutional neural network was an isolated bacteria image, and the predictions for each task were aggregated for each sample. The accuracies using all 63 bacteria images for each task were greater than 90%, portraying the ability for a deep neural network to take one image as input and determine multiple characteristics of the bacteria sample. While these results are promising, one issue with deep learning models using convolutional neural networks is the difficulty of interpreting why the model made its final prediction. Although the authors assessed the confusion matrix for the gram stain prediction, this alone is not sufficient to understand the features within the images that resulted in the decision-making of the model. Therefore, future work should probe the network's layers in an attempt to discover the feature responsible for the impressive accuracies. Once this is understood, the model's utility in the clinic will be much greater. An artificial neural network was created by Hsu et al. to predict the resistance of methicillin-resistant *S. aureus* using nasal cultures for patients from two hospitals, denoted hospital A and hospital B [56]. The data was initially split into a 75% training set and a 25% testing set. Of the 46 risk factors used as input to the model, feature selection was performed which determined only 17 risk factors were necessary for an accuracy above 90% for the prediction of resistance in hospital A, and 20 risk factors necessary for an accuracy above 90% for prediction of resistance in hospital B. When cross-validation was performed with the data, methicillin-resistant Staphylococcus aureus colonization was predicted with an average accuracy of 85.6%, demonstrating a high-predictive ability for resistance development using an artificial neural network.

In addition to deep learning classification models, a tool called DeepARG has been developed by Arango-Argoty et al. which uses deep learning for the prediction of antibiotic resistance genes from metagenomic data [57]. The goal of the researchers was to account for the high rate of false negatives predicted by next-generation sequencing of metagenomic data when searching against databases by using a deep learning approach that makes use of a dissimilarity matrix created based on all known categories of antibiotic resistance genes at the time. Two deep learning models were created, one for short-read sequences, and one for full-gene length sequences (Fig. 11.1) [57]. These models illustrate the common structure of deep

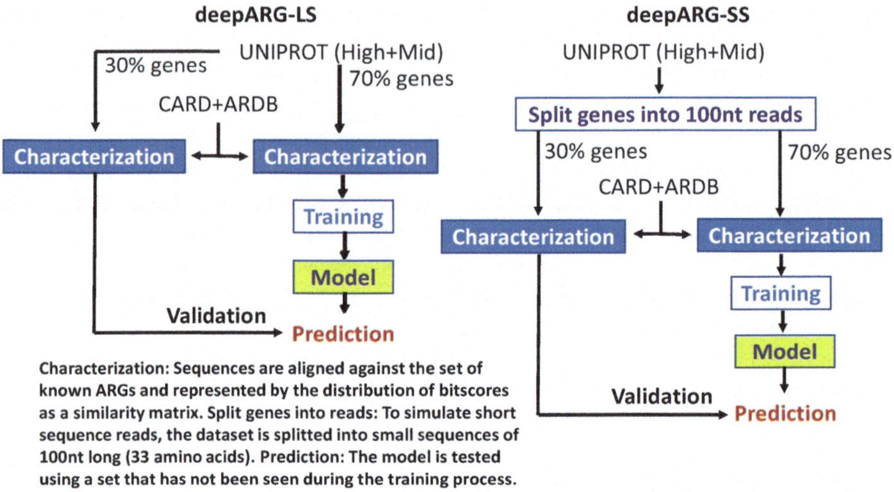

Fig. 11.1 Classification framework. UNIPROT genes were used for validation and training whereas the CARD and ARDB databases were used as features. The distance between genes from UNIPROT to ARGs databases is computed using the sequence alignment bit score. Alignments are done using DIAMOND with permissive cutoffs allowing a high number of hits for each UNIPROT gene. This distribution (showed in panels a and b) is used to train and validate the deep learning models (The panel in the figure provides additional description on the training of the models). (Figure was adapted from citation no 9 [57])

learning models, as they involve a separation of data into training and validation sets. The training data is used to calculate the parameters for the model, and the validation data is used to determine if the model generalizes well to other data, or if it overfits the training data.

Direct comparison with the typical best hit approach revealed better recall and f1-score for both the short-read and full-gene length models. This study shows the high-predictive power of a deep learning model applied to the task of annotation of antibiotic resistance genes, demonstrating the versatility of deep learning tasks. One exciting aspect of the potential of deep learning in the future is its increasing predictive power as more data is fed into the model. Therefore, as new data is generated, fed into existing or new models, and desired outputs are predicted with higher accuracy, it is possible that current deep learning applications will greatly improve. As the cost is high and the time to measure data is long in some biological applications, it is likely that new data will continually increase as new technologies are developed which lower cost and time of measurement, a trend continually seen in many biological applications, such as omics.

11.4 Role of AI/ML-Based Approaches in Therapeutic Development Against AMR

CDSS for antimicrobials has been designed to identify the inappropriate prescription of these drugs. Beaudoin et al. evaluated a knowledge base alone versus coupled with a supervised learning module, which learns whether to give an alert or not of inappropriate antimicrobial prescriptions using patient information including dose, dose frequency, discontinuation of therapy, early switch from intravenous to oral therapy, and redundant antimicrobial spectrum [58]. The researchers used pharmacist reviews of prescriptions as the gold standard for comparing the results of the models. The study revealed a higher positive predictive value for alerts of the inappropriate use of prescriptions compared with the positive predictive value given by the knowledge base alone. This demonstrates the utility of machine learning to aide in clinical settings compared to simpler models. However, the authors also reported that the number of false positives increased in the coupled model compared with the knowledge base alone, meaning that more patients were predicted to have inappropriate prescriptions when the ground truth, i.e., the pharmacist reviews, stated that the prescriptions were not inappropriate. An interesting question in all clinical cases is 'what the outcome is and when the wrong decision is made'. For example, a false positive diagnosis represents a prediction that suggests a patient will be given an inappropriate prescription when the prescription is actually appropriate, while a false negative represents a prediction that suggests a patient will be given an appropriate prescription when the prescription is actually inappropriate. Although both of these scenarios result in errors, it may be the case that one is preferred over the other. Since this model was created to alert clinicians, the clinician may find it more useful to have more false positives than false negatives because this leads to more inappropriate prescriptions being identified that may have been missed. As the system is implemented for longer in the clinic and new data is input into the model, its predictions may become more accurate, leading to less healthcare burden and more appropriate prescriptions for patients.

In addition to engineering an AI model such that the output is the most useful for patients and healthcare providers, the design of the experiment surrounding the biological question must be carefully considered, including the growth of the biological sample, data measurement, data processing, data analysis, and finally biological interpretation. To demonstrate the importance of the study design and the variability of results depending on the drug, dataset, resistance metric, and species, Hicks et al. developed machine learning models using whole-genome sequence data and antibiotic susceptibility phenotypes from seven large *Neisseria gonorrhoeae* datasets, as well as *K. pneumoniae* and *A. baumannii* datasets [59]. Resistance prediction resulted in variable performance for antibiotics, revealing that there are inherent differences in the ability to predict the biological mechanism for resistance development. Therefore, it may be more helpful to create separate models for each drug, focusing on optimizing the parameters of the model for each individual drug, to create optimal AI tools in the clinic. In addition, the researchers demonstrated that bacterial strains with higher genetic diversity had lower resistance phenotype predictability. As a result, the authors concluded there are potential limitations of genotype-to-resistance-phenotype models across different species, emphasizing the repeated finding that bacteria species have different genetic mechanisms influencing the development of resistance. Therefore, models should consist of species-specific data, thereby creating an optimal prediction of antimicrobial resistance for a single bacteria species, as adding other bacteria species could cause a decrease in discrimination. While machine learning is becoming more widely applied, it is necessary to construct the model based on answering a targeted biological question and understand the literature written about the data being used in the model, to have an expectation of what the algorithm is likely to return with respect to most influential features to the model prediction. This allows for a more confident understanding of the model and the ability to communicate the biological impacts of the most important features of the model in the biological system.

One example of the use of a large database, composed partially from omics data results, to formulate novel antimicrobial drugs is through Wang et al.'s deep learning long short-term memory (LSTM) generative model and a bidirectional LSTM classification model to design antimicrobial peptide sequences characteristic of antimicrobial activity against *E. coli* [60]. The authors trained the deep learning model using four databases of antimicrobial peptide sequences to classify peptides as antibacterial or nonantibacterial. By tuning hyperparameters using an 80% split of training data and testing a final model on a 20% split of heldout data, the authors achieved an accuracy greater than 80% for all models created, and an area under the receiver operating characteristic greater than 0.80 for all models, with some models achieving very high results greater than 0.98. The models achieving the highest area under the receiver operating characteristic had a more balanced proportion of antibacterial and nonantibacterial peptide sequences, suggesting it may be possible for the models with worse performance to increase their discriminative ability by adding more nonantibacterial peptides to the training data. The authors further generated 3D structures of peptides that were classified as antimicrobial by all model outputs (Fig. 11.2) [60], discovering that some sequences were not found in other databases, confirming that they are novel sequences that may be important candidates for new therapies. The ability to find potential therapeutic peptides and generate their structure for experts in the field to verify may lead to more collaboration between computational and wet lab scientists, as a pipeline to find molecules of interest and investigate them in living systems can be created.

For additional information on the application of machine learning and deep learning on the discovery and design of antimicrobial peptides, please read the review written by Yan et al. [61]. Through the development of in silico methods to integrate omics data and the curation of large databases from the analysis of omics data, the process of drug development may be shortened and less costly, as optimal targets output by the models can provide focused workflows.

Chung et al. created a machine learning model along with genome-scale metabolic modeling using many data types to study time and order dependent correlations and predicts drug interactions over a variety of growth conditions [62]. This

Fig. 11.2 Secondary structures and surface representations of sequences IWRVWRRW (**a**, **b**), KRWWIRWR (**c**, **d**), APLKQLKW (**e**, **f**), PFKKSIHL (**g**, **h**), and APWKQLKW (**i**, **j**). The residues are labeled according to their one-letter abbreviation. Surface representations show the hydrophilicity (blue) and hydrophobicity (orange) of the peptides. White surfaces represent a hydrophobicity of around 0.0. (Figure was adapted from citation no 12 [60])

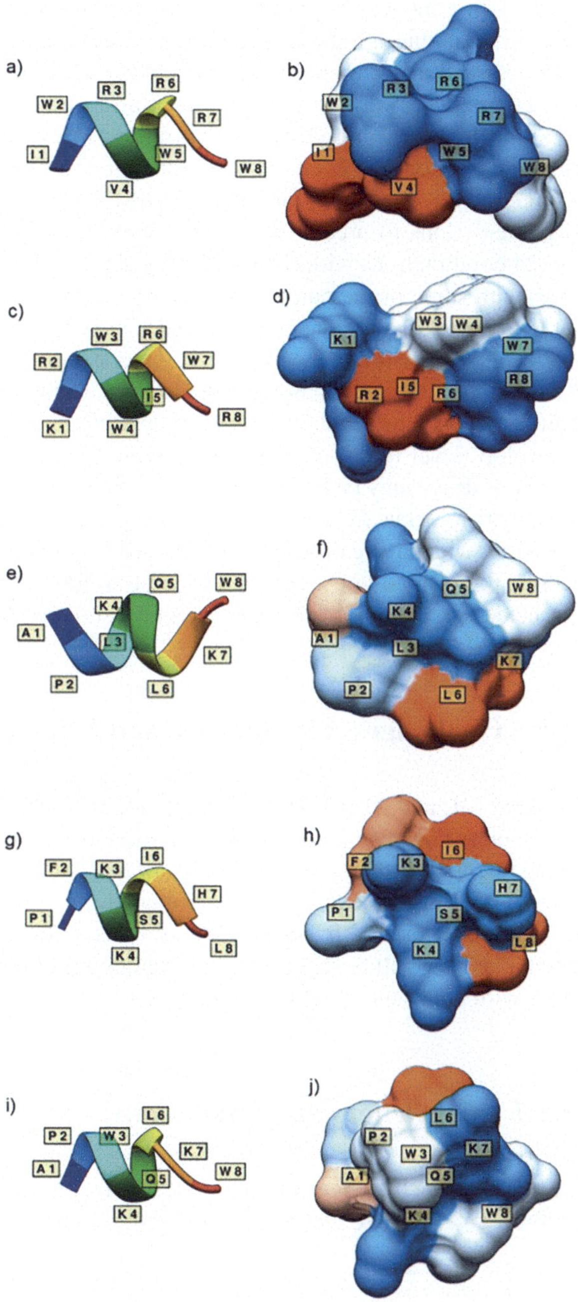

approach first models the flux profiles in response to drug treatment and growth in defined media through the incorporation of omics data and metabolite composition of the extracellular environment, respectively. After the flux is modeled, these profiles, coupled with drug interaction data, are used as input data to train a random forest machine learning classifier that predicts the interaction outcomes for novel drug combinations for *E. coli* and *M. tuberculosis*. The researchers used their model to screen 90,288 pairwise combinations and 2,176,944 three-way combinations of drug interactions to predict if they would be synergistic or not. Results indicated only 24 combinations with robust synergy across growth conditions and interaction cases, with some of the outputs already used clinically, providing some confidence that the other combinations would result in a synergistic effect as well. As bacteria develop new resistance mechanisms, understanding how combinations of drugs effect the vitality of the pathogen could lead to the development of new therapies from repurposed molecules, potentially leading to a reduction in the cost and time to develop novel inhibitors. However, while the synergistic effects resulting from the mix of drugs may be beneficial to the patient, there is also the potential for negative effects to occur from both drug toxicity and unknown alterations of pathways. Therefore, when using drugs that are already approved by the FDA for combination therapy, there should be serious consideration for the regulations and clinical trials conducted for the new combination to make sure patients will be safe.

11.5 Challenges, Limitations, and Ethical Considerations

Applying ML and AI for AMR research and stewardship as discussed in the previous chapters has huge potential and can increase the pace of discovery and disease management. However practical application of the AI models at scale in healthcare systems or drug discovery has many challenges. There are also ethical considerations which comes with the use of AI systems within the healthcare system. Within this chapter, we will be discussing the challenges, limitations, and ethical considerations with the usage of AI for healthcare.

11.5.1 Challenges and Limitations

There are many challenges faced in developing, implementing, and managing AI for healthcare system. We will be discussing these challenges next.

11.5.1.1 Data Availability and Data Quality

Data and model form the basis of any AI system. The accuracy of an AI system is dependent on the data being fed into the model and lack of proper training data can impact the predictions made by the system. Often within the realm of biomedical applications of AI, for instance prediction of a biomarker for a certain infection, a challenge of limited data availability is often faced. To make accurate predictions, a contextually relevant dataset with a high sample size is required to make a reliable AI system. Unfortunately for biomedical data, these two requirements are often unmet, and a lot of expense then goes into generating the data which slows down the impact translation of the systems. Data quality is also an important factor of the data being fed into the system. Unless well-curated data [63] with controlled vocabulary using ontologies or rule-based systems is fed into the model, there can be erroneous predictions. It is routinely observed that biomedical data available in the public domain is often not clean and unfit for feeding directly to AI systems [64].

11.5.1.2 Class Imbalances and Skewed Representation

Apart from the size of the data, diversity within the data is an equally important characteristic. Biomedical data often faces the challenge of not including enough diversity which can lead to biases [65]. Although data subsampling and oversampling can be used within the methods to account for skewed representation within the data, this can often lead to underfitting or overfitting of the data within the method [66, 67]. To deal with the biases, the best solution is to generate data with a good representation of different classes.

11.5.1.3 Evolving Nature of AMR

Concept drift is a constant phenomenon observed in data generated from biological systems. Concept drift is explained as a change in data due to changes in the system from which data is generated. Since biological systems change over time in response to environmental stimuli, such as bacteria in response to antibiotic treatments, the data from a short span of time often faces concept drift. Although the existing methods have been able to deal with the current challenges faced due to concept drift by making accurate future predictions about the evolution of the system, using AI systems built on top of generated long-term data can only lead to a stage of high reliability [68].

11.5.1.4 Biases

Although we have discussed biases in the previous sections, it is important to consider biases in more depth. Biases are differences in predictive ability for different classes or subclasses in an AI system [69]. Biases are often observed for different populations as the training data for one population might not reflect the inherent relationships within the data for another population [70]. Biases often go unnoticed when the models that have been trained are hard to explain or haven't been passed through an explainability framework. There can be different ways to deal with biases. Sampling data to consider the diversity representation within the data, mathematical approaches to learn the diversity within the data to guide balanced predictions, and keeping experts in the loop who can recognize biases can help in avoiding biases. Although these solutions can reduce biases from creeping into the system, it is advocated that rich and diverse data representing different populations when fed into the AI system can mitigate biases to the highest degree.

These challenges often hinder the adoption of AI systems for use in the healthcare system at a practical scale.

11.5.2 Ethical Considerations

The recent developments in the field of artificial intelligence have probed many questions around ethical considerations. Within the field of generative AI, a conversation around the regulation of AI has at least started; however, the regulatory system for scalable use of AI within the healthcare system is still poorly defined. This is all due to several ethical considerations that come with AI systems. In this section, we will be discussing some of those areas.

11.5.2.1 Data Privacy and Ownership

While there are data-sharing policies regulated by different bodies, they are often not comprehensive to protect data privacy. Biomedical data generated often has sensitive patient information and if not protected and regulated can lead to abuse of the information. It is of utmost importance to protect the sensitive information and in no way should present a chance to trace back the origin of the sample [71].

In the realm of AI in healthcare, the question of data ownership and access emerges as a widespread apprehension. Given that the construction of AI algorithms necessitates substantial datasets and continuous collaboration for comprehensive development, there is a keen interest in understanding the prospect of ongoing transfers of personal health information data during the entire lifecycle of AI development. Concerns arise regarding the monitoring mechanisms, potential risks linked to this unrestricted data transfer, and the strategies to preemptively mitigate these risks to prevent adverse effects on the involved patients [72].

11.5.2.2 Appropriate Governance and Oversight

Resolving the concerns of patients and researchers can be done by formalizing regulatory frameworks around the development, usage, and sharing of AI systems by appropriate bodies. This is where the challenges lie as AI within the healthcare and biomedical space is a new and evolving field. The existing regulatory frameworks defined for related fields (such as software development) can be and are being used as a foundation, however, they don't extensively cover all the concerns. Further there are no uniform guidelines within the present frameworks from different bodies [72, 73].

11.6 Recommendations and Future Directions

11.6.1 Recommendations

In the preceding sections, we explored the formidable challenges encountered in the integration of AI into research and healthcare. Addressing these challenges holds the key to enhance the acceptance of AI within the system, offering promising benefits such as timely disease management, deceleration of AMR development, expedited creation of antimicrobials, and advancements in various domains. This section delves into a set of recommendations aimed at fostering increased adoption within the system.

11.6.1.1 Regulations and Stewardship

Highlighted in the previous sections, regulation and stewardship is one of the strong needs for implementing AI for its practical usage in clinics and healthcare. Forming regulatory bodies at institutional, national, and international levels with a shared set of principles and guidelines formed on the basis of ethical use of AI, protection of patient information, data ownership, legal resolution, risk management, transparency of decision and documentation, and other areas [74] can lead to better development, adoption, and use in research and healthcare.

11.6.1.2 Continuous Learning

ML particularly for healthcare is often based on a limited set of data points collected specifically for a certain context to enable decision-making. While this highlights that high-quality data in limited quantity can often make predictions comprehensively and with reasonable accuracy, to avoid biases and increase reproducibility, it is required to train models with diverse datasets with high sample size. Further particularly due to concept drift in AMR, models should be continuously trained with new data to enable better predictions [75].

11.6.1.3 Building Database, Data Collections, and Hosting Public Datasets

As AI is dependent on training data quality and quantity to make reliable predictions, it is an imperative need to build databases and data collections for AMR research. Although public data exists to support research, much more data which is highly contextual in nature (chemical libraries, genomics profile of patients, metagenomics data of microbes, etc.) especially clinical in nature is required in well-curated form to extract knowledge and make predictions. The currently existing data in the public domain can also benefit from data management and curation to make it more usable for AI [76].

11.6.2 *Future Directions*

In our view, the trajectory of AI in the healthcare system hinges significantly on the current level of adoption and the existing knowledge used to train AI. With this in mind, we explore the range of models applicable to clinical scenarios and delve into the use of generative AI applications for accumulating knowledge.

11.6.2.1 Establishing Models for Clinical Applications

Although throughout the chapter, we have explored the use of AI-based methods for AMR research, management, and healthcare, the use of models for clinical purposes is still limited. The first AI-enabled medical device PAPNET testing for automated analysis of conventional (Papanicolaou) cervical smears was approved by the FDA in 1995 [77]. As of October 19, 2023, FDA has approved 692 AI-enabled medical devices with 108 approvals coming in the year 2023 itself. This is a positive sign in the direction of using AI-based approaches within the healthcare system. The category with the highest approvals is radiology which indicates the overall success of imaging-based systems for clinical use. However, models which are approved for clinical use are required to be developed in other areas as well. In particular context to antimicrobial resistance, areas like general hospital models, pathology, microbiology, and clinical decision support systems (CDSS) can have huge applications, particularly for real-time monitoring and decision support [78].

11.6.2.2 LLMs and Knowledge Extraction

Large Language Models like GPT4, LLaMA, and BERT have taken generative AI to new heights with their ability to be trained on large textual data and interact with users in natural languages to provide reasonable responses. As we discussed in the previous section about the FDA-approved AI-enabled medical devices, the use of AI is rising within the clinical system. However, LLMs or even other generative AI methods have been approved for its use in clinical systems. This is partly due to the bias, lack of transparency, and reproducibility in the LLMs which can come from the training data itself. LLMs however have a huge upside for knowledge extraction. Currently, a lot of knowledge remains inside public documents, health records, publications, and other textual knowledge sources which upon extraction and digitization can serve as data for training other ML models. LLMs can serve as the entity to extract knowledge from these sources and further curation in a structured way in the form of organized data points within databases [79, 80]. This can lead to the generation of high-quality data which can be used to solve challenges in areas of R&D, healthcare, and disease management which can benefit from AI.

11.7 Conclusion

Antimicrobial resistance is a global threat that can impact human health significantly and can take us to a preantibiotic era where a simple infection can turn into a life-threatening condition. WHO in their 2023 policy brief on "Global research agenda for antimicrobial resistance in human health" [81] highlighted 40 different research topics under the themes of prevention, diagnosis, treatment and care, cross-cutting, and drug-resistant tuberculosis. Many of the subthemes such as infection prevention and control, immunization, diagnosis and diagnostics development, antibiotic stewardship, antimicrobial use and consumption, antimicrobial development, and others can benefit from a database if not data-driven approach. Further, the solutions to many of the challenges under these research topics can be addressed using AI.

In the different sections of the chapter, we discussed how AI and ML have application for different areas of AMR research, healthcare, clinical management, and decision-making. Starting with the diagnostics, we found how different ML-based approaches have significant advantages over the traditional approach of AST especially in terms of speed of diagnosis. A high-throughput profiling method such as WGS- or MALDI-TOF-based proteomics when combined with ML method can significantly decrease the time to diagnosis and effective decision-making. ML has found a particularly important role in screening through chemical, peptide, and other biomolecular libraries to identify new antibiotics. This approach is both fast and reliably efficacious over any classical approach whether in wet or dry laboratory. Based on the data of chemical structures, composition, and properties of existing and potential antibiotic compounds, peptides or other modalities, new modalities can be discovered. ML has other applications in the areas of prescribing antibiotics and supporting clinical decisions. Based on the patient history from EHR and other data sources, ML can be used to prescribe the right set of antibiotic therapy in case of an infection and for taking other clinical decisions required in case of an emergency due to AMR.

For the scope of the chapter, we next focused on the use of AI and ML for specific applications of navigating omics data to find relevant insights, diagnosis, and therapeutic development. Molecular profiling done at the level of genomics, transcriptomics, proteomics, or metabolomics generates big data that can be used to drive discovery for AMR. System-level profiles can help us in diagnosis, understanding of the biological system and mechanism of resistance, and further identification of new newer therapeutic avenues. ML can drive faster and easier extraction of insights from the data that can be used to understand the biological systems. As we discussed before, different ML methods based on their characteristics of speed, ease of training, accuracy, and explainability can be used over the Omics data to ease discovery and use for several different applications. Explainability is always an issue to address with the ML. Luckily, there are XAI methods being created by computer scientists in different areas to solve this problem. Biomarkers and diagnostics can correctly stratify AMR and patients to increase the chances of a therapy being effective. Sifting through data generated from molecular profiling using ML has a huge potential for biomarker discovery and diagnostics. Antibiotics or other forms of therapy are the key modalities for use in case of an infection. Although we have discussed the use of ML for therapeutic development in the introduction of the chapter, we highlighted it in much more detail in a dedicated section. Other than screening chemical libraries for new biomolecules with potential antimicrobial activity, ML can be used to identify synergistic combinations of these compounds that can work for AMR.

AI and ML can turn the tide of AMR management in our favor. However, adoption of AI faces serious challenges ranging from algorithmic bias, limited data quality and data quantity, and class imbalances. Furthermore, these challenges are further impacted by lack of proper governance frameworks and lack of regulation over data privacy and ownership.

Overall we believe that building knowledge bases through the existing and new research data to guide and regulate ML through governance frameworks, data regulations and stewardship can drive its adoption within the research and clinical setting. We believe investing in these areas can drive many new clinical models and devices based on ML supporting different areas whether for clinical decision-making, diagnosis, and therapeutic administration of AMR.

References

1. Van-Dai T, Chuan-Ming L, Nkabindse GW. Big data stream computing in healthcare real-time analytics. In: 2016 IEEE international conference on cloud computing and big data analysis (ICCCBDA). IEEE; 2016.
2. Cao L. Data science: a comprehensive overview. ACM Comput Surv. 2017;50(3):Article 43.
3. Hassan M, et al. Innovations in genomics and big data analytics for personalized medicine and health care: a review. Int J Mol Sci. 2022;23(9):4645.
4. Peiffer-Smadja N, et al. Machine learning for clinical decision support in infectious diseases: a narrative review of current applications. Clin Microbiol Infect. 2020;26(5):584–95.

5. Pak TR, Kasarskis A. How next-generation sequencing and multiscale data analysis will transform infectious disease management. Clin Infect Dis. 2015;61(11):1695–702.
6. Rishishwar L, et al. Genome sequence-based discriminator for vancomycin-intermediate Staphylococcus aureus. J Bacteriol. 2014;196(5):940–8.
7. Santerre JW, et al. Machine learning for antimicrobial resistance. arXiv:1607.01224. 2016; https://doi.org/10.48550/arXiv.1607.01224.
8. Faron ML, et al. Multicenter evaluation of the Bruker MALDI Biotyper CA system for the identification of clinical aerobic gram-negative bacterial isolates. PLoS One. 2015;10(11):e0141350.
9. Clardy J, Fischbach MA, Walsh CT. New antibiotics from bacterial natural products. Nat Biotechnol. 2006;24(12):1541–50.
10. Tommasi R, et al. ESKAPEing the labyrinth of antibacterial discovery. Nat Rev Drug Discov. 2015;14(8):529–42.
11. Brown DG, et al. Trends and exceptions of physical properties on antibacterial activity for gram-positive and gram-negative pathogens. J Med Chem. 2014;57(23):10144–61.
12. Stokes JM, et al. A deep learning approach to antibiotic discovery. Cell. 2020;181(2):475–83.
13. Mahlapuu M, et al. Antimicrobial peptides: an emerging category of therapeutic agents. Front Cell Infect Microbiol. 2016;6:194.
14. Fields FR, et al. Novel antimicrobial peptide discovery using machine learning and biophysical selection of minimal bacteriocin domains. Drug Dev Res. 2020;81(1):43–51.
15. Kumar A, et al. Initiation of inappropriate antimicrobial therapy results in a fivefold reduction of survival in human septic shock. Chest. 2009;136(5):1237–48.
16. Yelin I, et al. Personal clinical history predicts antibiotic resistance of urinary tract infections. Nat Med. 2019;25(7):1143–52.
17. Stracy M, et al. Minimizing treatment-induced emergence of antibiotic resistance in bacterial infections. Science. 2022;375(6583):889–94.
18. Garg AX, et al. Effects of computerized clinical decision support systems on practitioner performance and patient outcomes: a systematic review. JAMA. 2005;293(10):1223–38.
19. Challen R, et al. Artificial intelligence, bias and clinical safety. BMJ Qual Saf. 2019;28(3):231–7.
20. Martinez JL, Baquero F, Andersson DI. Beyond serial passages: new methods for predicting the emergence of resistance to novel antibiotics. Curr Opin Pharmacol. 2011;11(5):439–45.
21. Gil-Gil T, et al. Antibiotic resistance: time of synthesis in a post-genomic age. Comput Struct Biotechnol J. 2021;19:3110–24.
22. Sandoval-Motta S, Aldana M. Adaptive resistance to antibiotics in bacteria: a systems biology perspective. Wiley Interdiscip Rev Syst Biol Med. 2016;8(3):253–67.
23. NIHR Global Health Research Unit on Genomic Surveillance of AMR. Whole-genome sequencing as part of national and international surveillance programmes for antimicrobial resistance: a roadmap. BMJ Glob Health. 2020;5(11):e002244.
24. Ashton PM, et al. Identification of Salmonella for public health surveillance using whole-genome sequencing. PeerJ. 2016;4:e1752.
25. Argimón S, et al. Integrating whole-genome sequencing within the National Antimicrobial Resistance Surveillance Program in the Philippines. Nat Commun. 2020;11(1):2719.
26. Blake KS, Choi J, Dantas G. Approaches for characterizing and tracking hospital-associated multidrug-resistant bacteria. Cell Mol Life Sci. 2021;78(6):2585–606.
27. Brazas MD, Hancock REW. Using microarray gene signatures to elucidate mechanisms of antibiotic action and resistance. Drug Discov Today. 2005;10(18):1245–52.
28. Li L, et al. Mechanisms of polymyxin resistance induced by Salmonella typhimurium in vitro. Vet Microbiol. 2021;257:109063.
29. Singh A, et al. Comparative proteomic analysis of sequential isolates of Mycobacterium tuberculosis from a patient with pulmonary tuberculosis turning from drug sensitive to multidrug resistant. Indian J Med Res. 2015;141(1):27–45.
30. Kok M, et al. Unraveling antimicrobial resistance using metabolomics. Drug Discov Today. 2022;27(6):1774–83.
31. Schelli K, Zhong F, Zhu J. Comparative metabolomics revealing Staphylococcus aureus metabolic response to different antibiotics. Microb Biotechnol. 2017;10(6):1764–74.

32. Mu A, et al. Integrative omics identifies conserved and pathogen-specific responses of sepsis-causing bacteria. Nat Commun. 2023;14(1):1530.
33. Vilne B, et al. Machine learning approaches for epidemiological investigations of food-borne disease outbreaks. Front Microbiol. 2019;10:1722.
34. Meng C, et al. Dimension reduction techniques for the integrative analysis of multi-omics data. Brief Bioinform. 2016;17(4):628–41.
35. Reel PS, et al. Using machine learning approaches for multi-omics data analysis: a review. Biotechnol Adv. 2021;49:107739.
36. van de Schoot R, et al. Bayesian statistics and modelling. Nat Rev Methods Primers. 2021;1(1):1.
37. Quinlan JR. Induction of decision trees. Mach Learn. 1986;1(1):81–106.
38. Breiman L. Random forests. Mach Learn. 2001;45(1):5–32.
39. Sykes AO. An introduction to regression analysis. Coase-Sandor Institute for Law & Economics Working Paper No. 20. 1993.
40. Cover T, Hart P. Nearest neighbor pattern classification. IEEE Trans Inf Theory. 1967;13(1):21–7.
41. Hearst MA, et al. Support vector machines. IEEE Intell Syst Appl. 1998;13(4):18–28.
42. LeCun Y, Bengio Y, Hinton G. Deep learning. Nature. 2015;521(7553):436–44.
43. Lundberg SM, Lee S-I. A unified approach to interpreting model predictions. Adv Neural Inf Proces Syst. 2017;30:4768–77.
44. Ribeiro MT, Singh S, Guestrin C. "Why should I trust you?": explaining the predictions of any classifier. In: Proceedings of the 22nd ACM SIGKDD international conference on knowledge discovery and data mining. San Francisco: Association for Computing Machinery; 2016. p. 1135–44.
45. Benkwitz-Bedford S, et al. Machine learning prediction of resistance to subinhibitory antimicrobial concentrations from Escherichia coli genomes. mSystems. 2021;6(4):e00346-21. https://doi.org/10.1128/msystems.00346-21.
46. Lemsara A, Ouadfel S, Fröhlich H. PathME: pathway based multi-modal sparse autoencoders for clustering of patient-level multi-omics data. BMC Bioinformatics. 2020;21(1):146.
47. Yagin FH, et al. Explainable artificial intelligence model for identifying COVID-19 gene biomarkers. Comput Biol Med. 2023;154:106619.
48. Grinsztajn L, Oyallon E, Varoquaux G. Why do tree-based models still outperform deep learning on tabular data? arXiv:2207.08815. 2022; https://doi.org/10.48550/arXiv.2207.08815.
49. Boolchandani M, D'Souza AW, Dantas G. Sequencing-based methods and resources to study antimicrobial resistance. Nat Rev Genet. 2019;20(6):356–70.
50. Wozniak JM, et al. Mortality risk profiling of Staphylococcus aureus bacteremia by multi-omic serum analysis reveals early predictive and pathogenic signatures. Cell. 2020;182(5):1311–1327.e14.
51. Huang TS, et al. Detection of carbapenem-resistant Klebsiella pneumoniae on the basis of matrix-assisted laser desorption ionization time-of-flight mass spectrometry by using supervised machine learning approach. PLoS One. 2020;15(2):e0228459.
52. Yu J, et al. Rapid identification of methicillin-resistant Staphylococcus aureus using MALDI-TOF MS and machine learning from over 20,000 clinical isolates. Microbiol Spectr. 2022;10(2):e0048322.
53. Rees CA, et al. Detection of high-risk carbapenem-resistant Klebsiella pneumoniae and Enterobacter cloacae isolates using volatile molecular profiles. Sci Rep. 2018;8(1):13297.
54. Weis C, et al. Direct antimicrobial resistance prediction from clinical MALDI-TOF mass spectra using machine learning. Nat Med. 2022;28(1):164–74.
55. Ahmad A, et al. Highly sensitive quantitative phase microscopy and deep learning aided with whole genome sequencing for rapid detection of infection and antimicrobial resistance. Front Microbiol. 2023;14:1154620.
56. Hsu CC, et al. Validation study of artificial neural network models for prediction of methicillin-resistant Staphylococcus aureus carriage. Infect Control Hosp Epidemiol. 2008;29(7):607–14.
57. Arango-Argoty G, et al. DeepARG: a deep learning approach for predicting antibiotic resistance genes from metagenomic data. Microbiome. 2018;6(1):23.

58. Beaudoin M, et al. Evaluation of a machine learning capability for a clinical decision support system to enhance antimicrobial stewardship programs. Artif Intell Med. 2016;68:29–36.
59. Hicks AL, et al. Evaluation of parameters affecting performance and reliability of machine learning-based antibiotic susceptibility testing from whole-genome sequencing data. PLoS Comput Biol. 2019;15(9):e1007349.
60. Wang C, Garlick S, Zloh M. Deep learning for novel antimicrobial peptide design. Biomolecules. 2021;11(3):471.
61. Yan J, et al. Recent progress in the discovery and design of antimicrobial peptides using traditional machine learning and deep learning. Antibiotics (Basel). 2022;11(10):1451.
62. Chung CH, Chandrasekaran S. A flux-based machine learning model to simulate the impact of pathogen metabolic heterogeneity on drug interactions. PNAS Nexus. 2022;1(3):pgac132.
63. Johnson SG, et al. Application of an ontology for characterizing data quality for a secondary use of EHR data. Appl Clin Inform. 2016;7(1):69–88.
64. Veale M, Zuiderveen Borgesius F. Demystifying the Draft EU Artificial Intelligence Act—analysing the good, the bad, and the unclear elements of the proposed approach. Comput Law Rev Int. 2021;22(4):97–112.
65. Cánovas-Segura B, et al. Improving interpretable prediction models for antimicrobial resistance. In: 2019 IEEE 32nd international symposium on computer-based medical systems (CBMS). IEEE; 2019.
66. Elyan E, Moreno-Garcia CF, Jayne C. CDSMOTE: class decomposition and synthetic minority class oversampling technique for imbalanced-data classification. Neural Comput & Applic. 2021;33(7):2839–51.
67. Bunkhumpornpat C, Sinapiromsaran K, Lursinsap C. DBSMOTE: density-based synthetic minority over-sampling technique. Appl Intell. 2012;36(3):664–84.
68. Puuronen S, Pechenizkiy M, Tsymbal A. Effectiveness of local feature selection in ensemble learning for prediction of antimicrobial resistance. In: 2008 21st IEEE international symposium on computer-based medical systems. IEEE; 2008.
69. Seyyed-Kalantari L, et al. Underdiagnosis bias of artificial intelligence algorithms applied to chest radiographs in under-served patient populations. Nat Med. 2021;27(12):2176–82.
70. Panch T, Mattie H, Atun R. Artificial intelligence and algorithmic bias: implications for health systems. J Glob Health. 2019;9(2):020318.
71. Murdoch B. Privacy and artificial intelligence: challenges for protecting health information in a new era. BMC Med Ethics. 2021;22(1):122.
72. Winter JS. AI in healthcare: data governance challenges. J Hosp Manag Health Policy. 2020;5:8.
73. Reddy S, et al. A governance model for the application of AI in health care. J Am Med Inform Assoc. 2020;27(3):491–7.
74. Reddy S. Navigating the AI revolution: the case for precise regulation in health care. J Med Internet Res. 2023;25:e49989.
75. Hughes JH, et al. Continuous learning in model-informed precision dosing: a case study in pediatric dosing of vancomycin. Clin Pharmacol Ther. 2021;109(1):233–42.
76. Beardmore D, et al. Antimicrobial resistance surveillance: sharing industry data. Technical report, Wellcome Trust/Open Data Institute. 2018. https://wellcome.org/sites.
77. Joshi G, et al. FDA-approved artificial intelligence and machine learning (AI/ML)-enabled medical devices: an updated landscape. Electronics. 2024;13(3):498.
78. U.S. Food and Drug Administration. Artificial intelligence and machine learning (AI/ML)-enabled medical devices. AI/ML-Enabled Medical Devices; 2022.
79. Yoo H, et al. Predicting anti-microbial resistance using large language models. arXiv preprint arXiv:2401.00642. 2024.
80. Schwartz IS, et al. Black box warning: large language models and the future of infectious diseases consultation. Clin Infect Dis. 2023;78:860–6.
81. World Health Organization. Global research agenda for antimicrobial resistance in human health. Geneva: World Health Organization; 2023. https://www.who.int/publications/m/item/global-research-agenda-forantimicrobial-resistance-in-human-health

Chapter 12
Omics-Based Approaches in Developing Treatments Against Antimicrobial Resistance

Zhe Wang, Weile Xie, Dan Luo, Longlong Wang, Yufan Xu, Eric H. Rosenn, and Vijay Soni

Abbreviations

AI	Artificial intelligence
AMP	Antimicrobial peptides
AMR	Antimicrobial resistance
ASO	Antisense oligonucleotides
CDC	Cyclic peptide drug conjugate
CRAB	Carbapenem-resistant *A. baummani*
DAC	Drug antibody conjugate
FDA	Food and Drug Administration
HGT	Horizontal gene transfer
HTS	High-throughput screening
LNP	Lipid nanoparticle
MGE	Mobile genetic elements
MRSA	Methicillin-resistant *S. aureus*
MRSE	Methicillin-resistant *S. epidermis*
PTM	Post-translational modifications
siRNA	Small interfering RNA
SMD	Small molecule drugs
TB	Tuberculosis
WHO	World Health Organization

Z. Wang (✉) · W. Xie · D. Luo · L. Wang · Y. Xu
Shanghai Key Laboratory of Veterinary Biotechnology, School of Agriculture and Biology, Shanghai Jiao Tong University, Shanghai, China

Shanghai Collaborative Innovation Center of Agri-Seeds/School of Agriculture and Biology, Shanghai Jiao Tong University, Shanghai, China
e-mail: wangz@sjtu.edu.cn

E. H. Rosenn · V. Soni
Division of Infectious Diseases, Weill Department of Medicine, Weill Cornell Medicine, New York, NY, USA

V. Soni, A. S. Akhade (eds.), *Antimicrobial Resistance: Factors to Findings*,
https://doi.org/10.1007/978-3-031-65986-7_12

12.1 Introduction

> The greatest triumphs in medicine have been achieved by those who dare to venture beyond the boundaries of convention. – Joseph Lister

Drug development is a long, complex process dependent on numerous factors, often taking years to validate and approve a treatment. An early approach to antibiotic development was the "Waksman platform" [1]. The method was used to find new drugs by isolating actinomycetes from soil and determining the bacteriostatic zone of the culture extracts of different strains. Marked by Waksman's isolation of streptomycin from hundreds of strains in 1944, various classes of antibiotics, including aminoglycosides, chloramphenicols, and tetracyclines were discovered over the next two decades. But after the 1960s, most of these low-hanging fruits had been explored, and this pipeline for new drugs diminished. Around the same time, advances in synthetic chemistry shifted the development of new drugs from screening natural products to synthesizing compounds artificially. This process is clearly demonstrated in the historical discovery of fluoroquinolones. Another case using synthetic design to modify existing antimicrobials is seen in the use of known narrow-spectrum antibiotics against Gram-positive bacteria to derive broad-spectrum antibiotics. An example is ampicillin, which was developed from penicillin, one of the earlier antibiotics famously discovered inadvertently by Fleming [2].

While the ability to yield many compounds with antimicrobial activity has been greatly streamlined via high-throughput screening techniques, no new classes of antibiotics have been discovered since the development of daptomycin in 1986 [3]. As has been discussed many times throughout this text a variety of the most robust antimicrobials had already been developed by the 1970s; previously effective treatments that the latent threat of AMR has made null. Being thus limited to analogs of existing compounds, we for the most part move away from the most effective formulation. Furthermore, AMR can be quickly adapted to a drug that is similar to one for which resistant traits already exist. To combat drug-resistant bacteria, we must then develop completely new classes of antibiotics, new methods of treatment, and novel therapeutic methods.

Systems biology and omics are integrative fields of research that have emerged as a result of a deeper understanding of genetics and biochemistry along with new computational technology that allows us to analyze this information. *Systems biology* aims to inform research endeavors through a better understanding of complex biological interactions by creating functional models and applying "systems thinking".

Omics[1] is the multi-level investigation of all aspects of a biomolecular process using omics techniques [4].

A systems biology approach and the use of multi-omics techniques provide a vast framework for potential therapies and even emerging drug pipelines. In this chapter, we discuss multi-omics approaches being used to conceive novel

---

[1] *ante litteram termini.*

therapeutics, study drug repurposing and repositioning, and inform the rational design of effective drug combinations. We hope this chapter will provide useful background information and confer insights on the MDR problem that will be of interest to researchers and clinicians working in fields such as systems biology, pharmacology, and infectious diseases [5].

Section Highlights

- *Genomics* unveils the genetic mutations and resistance genes that underpin drug resistance. It identifies the blueprint of resistance, offering insights into the origins of resistance mechanisms.
- *Transcriptomics* reveals how genes are dynamically expressed in response to antibiotic exposure. It provides a real-time glimpse into the molecular reactions of microorganisms to the presence of drugs.
- *Proteomics* identifies the proteins actively involved in resistance mechanisms. It unveils the actors executing the resistance strategies within the microbial stage.
- *Metabolomics* profiles the metabolites involved in metabolic adaptations. It offers insights into the metabolic rewiring that enables microorganisms to thrive despite the presence of antibiotics.

12.2 A Brief Commentary on Recent Historic Trends in Drug Development

> I preferred to tell the truth that penicillin started as a chance observation. My only merit is that I did not neglect the observation. – Alexander Fleming

In this and the following sections, we will strive to discuss the discovery and development of novel pharmaceutical compounds, or drugs, as well as various other treatment pathways that may be of particular use in combating AMR. However, without some general knowledge of how therapeutic research and discovery lead to medical application, some concepts may be difficult to understand. For example, the difference between drug discovery and repurposing may seem semantic but in reality, these terms can imply certain industrial regulatory differences. In discussing various therapies, it may be worthwhile to consider pharmaceutical industry practices and regulatory standards as the reader may benefit from understanding hurdles that exist in researching particular medical treatments. There is also much area-specific jargon with terms like "*high-throughput screening*" which might refer to a method of library screening in structural-based design or may refer more generally to identifying the characteristics of any entity like a gene or compound by cross-referencing information with some databases.

First, let us define exactly what a "*class* of drug" is. As we are concerned with drug resistance when discussing *antimicrobials*, we are specifically referencing antibiotics. While a broader class of *antibiotics* includes all compounds with bacteriostatic or bactericidal effects; pharmaceutical treatments can also be divided into classes based on chemical composition or mechanism of action. In terms of pharmaceutical drugs, some major classes include aminoglycosides, macrolides, lactams, tetracyclines, and quinolones. Though a vaccine may have a bactericidal effect, this is a tertiary action via immune activation and not the result of a pharmaceutical formulation acting on some bacterial process. We, therefore, do not call a vaccine a "drug" per se. Other therapeutic methods that fall into gray areas include some genomic therapies, the use of bacteriophages, probiotics, and other immunotherapeutics [6]. These therapeutics are powerful medications and rely heavily on omics techniques in their development. However, we caution the reader not necessarily to equate a bactericidal therapy with an antibiotic compound as varying therapeutics are vastly different in terms of design regulation and practical application.

There are likely many ways to describe drug development, but for the purpose of antimicrobial drugs, we will explain the history of drug discovery that we have classified into three distinct stages, and then we will have a brief overview of the process of a particular drug might go through from conception to development and eventual marketing.

12.2.1 Phenotypic Discovery

The first stage of historic antimicrobial drug development might be coined the era of "phenotypic discovery." As discussed in the case of actinomyces and penicillin, certain compounds were observed to have antimicrobial properties. The compounds were isolated and classified, then applied in medicinal forms. A drug could be altered for more desirable effects, i.e., the effect it would have on the bacteria's phenotype. Initially, organic compounds were thus studied and used to derive synthetic compounds.

This method of drug discovery was most used in the early part of the twentieth century and led to the discovery of many essential modern antimicrobials. This method has been criticized, however, as being a trial-and-error approach; with discoveries being made coincidentally and not as the result of justified avenues of study. It is for this reason that this method became ineffective for discovering compounds with novel modes of action after many "low-hanging fruits" had been revealed. (The idiom "low-hanging fruits" refers to scientific discoveries, here pharmaceutical compounds, that were "easier to discover" either as a result of being likely to observe or simple to deduce.)

12.2.2 Rational Design

With the advent of high-throughput screening methods, the search for antimicrobials turned to the strategy defining the era of *rational design*. Rational or structural-based design is a top-down strategy that starts with a target, such as a key bacterial protein, and then screens the protein against a library of compounds. This gives a number of “hits”—compounds with activity against the protein—which are then further screened to generate lead compounds. *Lead compounds* are extensively studied and optimized to generate pharmaceutical formulations that could be developed into marketable medications. While possibly the most streamlined method for identifying new drugs, and despite growing computational power; discovery via this method has been costly in relation to the discoveries it has yielded.

12.2.3 Current Novel Design Methods

The third era we are now moving into is an incredibly new field that relies heavily on a systems biology understanding of multi-level causes and effects that can only be analyzed and predicted by complex artificial intelligence (*AI*) programs developed just in the last few years [7]. Unlike rational design, which is a top-down process starting with a specific target, this method might be used to predict or generate a particular compound by taking in the whole complexity of a bacterial target’s biology and environment. As opposed to in vivo and in vitro testing, an in silico method tests a phenomenon using completely computer-based methods or simulated models.

The use of AI and such in silico methods is in nascent stages of development now in 2024, however, is already being broadly applied in research and throughout the pharmaceutical industry. The benefit of using these programs is that they save time and money. More advanced programs can handle large amounts of data more quickly, and require less user input, oversight, and maintenance. Automation of labor-intensive processes can enable optimization, producing accurate results more efficiently. However, in research, many such methods have not yet been validated. As such, while new technology presents the potential for exploratory capacity, practically it is being used for exclusionary purposes. For instance, AI programs that determine compounds’ pharmacodynamic properties are being used to exclude those with undesirable characteristics when analyzing drug candidates, saving the effort of examining many un-useful compounds [8].

12.2.4 Drug Discovery to Development

The process of drug discovery starting with basic and translational research that leads to subsequent development into a drug for the pharmaceutical market, or another form of approved medical treatment, can be referred to as a *developmental pipeline*. Once a drug formula has been finalized, it enters the period of *clinical trials*. It is first tested on animals, then human subjects, and lastly on larger groups. Throughout this testing, pharmacokinetic and safety properties are determined. Regulatory agencies like the *FDA* review these studies and decide if a treatment is ready for successive trials or approved to be marketed and used in medical practice.

Here, we see the difference between a drug formulation and a different kind of therapy like a vaccine—there are various rules and regulations that apply, such as pharmacokinetic standards based on the therapy's mechanism of action. For instance, a vaccine and a pharmaceutical compound may have the same bactericidal purpose but comparing a metric like dose-response would be meaningless. Some therapeutics like genomic medicines, compared to small molecule drugs or biologics, can have many off-target effects and thus face heaver regulations and safety concerns; being approved by the FDA at much lower rates. *Cost-benefit* outcomes must also be taken into consideration when allocating resources to develop extremely expensive personalized therapies that may be very effective but impractical for treating at the population levels.

We thus focus much of our effort on repurposing, formulating known drug regimens, and coming up with novel ways to combat AMR with the drugs already at our disposal in the following sections.

12.3 Application of Omics Approaches in Novel AMR Therapeutic Discovery

Multi-omics methodologies interrelate many levels of information and specifically use large amounts of expression pathway data and metabolic mapping to inform research initiatives. Before delving into more practical design strategies, we will discuss omics techniques more generally, and focus on how these methods may contribute to basic research on novel therapeutics to combat AMR. We will investigate how these approaches more broadly inform therapeutic research directions as well as give practical examples of previous applications.

12.3.1 Genomics

Genomic analysis identifies specific genes responsible for antibiotic resistance, known as resistance genes or determinants. When these loci are disrupted, it may render the microorganism re-susceptible to antibiotics [9]. Researchers can thus develop targeted interventions to counteract resistance mechanisms. A notable practical example includes cases of plazomicin use, a member of the aminoglycoside class, strategically designed to alleviate the impact of aminoglycoside-modifying enzymes. Another common application is the utilization of β-lactamase inhibitors, such as in the combination drug avibactam–ceftazidime [10, 11].

Genomics is a collaborative field that plays a crucial role for researchers, clinicians, and public health agencies in fighting AMR. Global initiatives like the global microbial identifier (GMI) and the global initiative on sharing all influenza data (GISAID) are instrumental in swiftly identifying and tracking drug-resistant strains.

This proactive effort enables a rapid response to emerging threats by providing an analysis of genetic fingerprints, facilitating the tracing of pathogen transmissions and outbreaks [12].

Genomics also significantly influences precision medicine in treating infectious diseases, especially those involving resistant pathogens. Unlike a generic approach, clinicians can tailor treatments based on the genomic characteristics of the infecting microorganism. Regarding *Mycobacterium tuberculosis*, analyzing the genomic profile of a particular patient's infecting TB strain enables clinicians to administer precise treatment regimens tailored to the individual's infection resistance profile. Versus the use of standardized drug combinations, this enhances treatment efficacy while simultaneously minimizing antibiotic overuse [13, 14].

Genomics further plays a crucial role in the selection of antibacterial targets, focusing on genes vital for in vitro growth and those expressed exclusively in vivo or within the infected host [15]. Inhibiting the genes found essential in vitro, results in cellular growth arrest or pathogen death, while inhibiting genes involved in host invasion may impact infection progression and the pathogen's survival in the host. A validated target, peptide deformylase, has reaped significant benefits from genomic information, leading to the discovery of multiple inhibitors. Ongoing intensive research is directed toward other targets, encompassing tRNA synthetases, DNA gyrase subunit GyrB, and the fatty acid biosynthetic pathway (FabI) [16]. Furthermore, genomic sequence data play a crucial role in determining numerous protein structures through high-throughput strategies. The analysis of enzyme active sites, virtual screening, and de novo design contributes to the identification of potential lead compounds [17].

Genomic research has also significantly advanced our understanding regarding the role of epigenetics in bacterial drug resistance. Epigenomics considers epigenetic modifications, those that do not alter DNA sequences but profoundly influence gene expression. Bacteria undergo specific changes in DNA methylation under antibiotic pressure that are intricately tied to the emergence of resistance. Bacterial plasmids, known carriers of resistance genes, have also been identified as potential inducers of epigenetic changes during gene transfer [18].

12.3.2 Transcriptomics

Transcriptomics, a powerful tool providing real-time insights into gene expression, plays a crucial role in deciphering how microorganisms respond to antibiotics and other stressors [19]. Beyond elucidating changes in gene expression, transcriptomics facilitates the identification of key regulatory proteins that function as master switches, controlling the activation or suppression of multiple genes in response to specific signals. For instance, regarding methicillin-resistant *Staphylococcus aureus* (MRSA), transcriptomics has shed light on the critical role of regulatory proteins like the mecA gene product in conferring resistance to β-lactam antibiotics. Understanding these regulatory networks provides insights into potential strategies for overcoming or disrupting resistance [20]. Although transcriptomics has more utility in understanding the underlying mechanisms of antimicrobial therapeutics, as opposed to identifying specific drug targets, it is essential for the development of the systems biology perspective bolstering drug development.

12.3.3 Proteomics

Proteins, essential to cellular functions, collectively form the proteome. Proteomics, which directly identifies overexpressed, suppressed, or modified proteins in drug-resistant pathogens, plays a crucial role in combating antimicrobial resistance (AMR) [21]. By comparing the proteomes of drug-resistant and susceptible strains, researchers can pinpoint specific proteins that actively neutralize or bypass the effects of antibiotics. For instance, proteomics identifies efflux pumps in Gram-negative bacteria. Consequently, targeting these efflux pumps with specific inhibitors, identified through proteomic studies, aims to restore the efficacy of existing antibiotics.

Additionally, proteomics unveils post-translational modifications (PTMs), often in response to environmental cues or stressors like antibiotics. Phosphorylation, acetylation, and methylation of specific proteins can modulate their activity, impacting antibiotic resistance. Understanding these modifications enables the development of interventions that disrupt resistance mechanisms mediated by PTMs [22].

The synergy of proteomics with other omics data provides a holistic view of molecular interactions, aiding in identifying key nodes and pathways central to resistance mechanisms. Targeting protein clusters conferring resistance through combination therapies disrupts multiple resistance mechanisms simultaneously, thereby enhancing treatment effectiveness [23].

As AMR research advances, metaproteomics emerges as a promising frontier. *Metaproteomics* involves analyzing the entire protein complement of complex microbial communities, such as those in the human microbiome or environmental ecosystems. This approach offers insights into the resistance dynamics of mixed microbial populations and the potential for horizontal gene transfer of resistance genes. Consequently, understanding how resistance mechanisms evolve and spread within microbial communities is crucial for devising strategies to control the dissemination of resistance [24].

12.3.4 Metabolomics

Metabolomics provides a detailed perspective on a microorganism's metabolism, encompassing essential metabolites such as sugars, lipids, amino acids, and energy-related molecules. Playing a crucial role in the fight against antimicrobial resistance

(AMR), metabolomics reveals how microorganisms respond to antibiotics and unveils metabolic signatures associated with drug resistance [25].

In drug design, compounds are specifically targeted to these essential metabolites or the enzymes involved in their biosynthesis. Metabolomics identifies metabolic changes in drug-resistant microorganisms, showcasing their ability to adapt and bypass antibiotic effects [26]. Disrupting these metabolic pathways can be harnessed to design effective treatments. For example, Wang et al. [27] found that cells tolerating the anti-cancer drug sorafenib upregulate the expression of the glycolytic enzyme HK2. By combining the glycolytic inhibitor 2-DG with sorafenib, they effectively hampered persisters cell growth and reduced tumor growth in mice. This strategy is particularly promising for pathogens with limited treatment options, such as extensively drug-resistant tuberculosis (XDR-TB) [28]. Furthermore, metabolomics studies have revealed certain drug-resistant bacteria upregulate specific amino acids or metabolic intermediates during antibiotic exposure; representing potential targets for therapeutic intervention [29].

By tailoring treatment regimens based on the metabolite profiles of pathogens causing infections, this approach not only improves treatment efficacy but also reduces the risk of further resistance development. Metabolomics also holds promise as a diagnostic tool, i.e. specific metabolite signatures associated with drug resistance can enable healthcare providers to identify resistant strains quickly and make informed treatment decisions [30].

12.3.5 Computational Tools in Omics Analysis

Due to the complexity and volume of the data obtained through omics, specialized software is required to process it. In this section, we introduce some tools which are used in genomic and multi-omics analysis. Whether it is genomic analysis or other, the quality of the validated data needs to be evaluated by tools such as FastQC, Trimmomatic, and Fastp, in order to rule out erroneous data or other contamination. Principal component analysis (PCA), co-inertia analysis (CIA) and canonical correspondence analysis (CCA) are usually used to analyze and show correlations between data. Tools such as PMA, MCIA, made4 and RGGCA can help researchers achieve these goals. In order to further comprehensively mine the information in the data, meta-analysis using CNAmet, Rottper and other similar software is common [31]. Since a single genomic analysis cannot reflect transcriptional and translational changes, multi-omics analysis techniques are often used in studies to further explore the source of drug resistance through the comprehensive use of genomic, transcriptome, metabolome and proteome data. Both IPA and MetaCore provide multi-omics analysis capabilities. In addition, InCroMAP and 3Omics provide visualization functions to further facilitate understand multi-omics data [32].

12.4 Drug Repurposing for New Antimicrobial Treatments

Drug repurposing, also referred to as drug repositioning, reprofiling, or re-tasking, is a strategy employed to discover fresh applications for approved or investigational drugs that extend beyond the scope of their initial medical indications [33]. A well-known exemple of drug repurposing is sildenafil. Originally developed as an antihypertensive medication by Pfizer, sildenafil was repositioned to treat erectile dysfunction and has achieved substantial success in the global market [34]. When compared to the discovery and creation of new drugs, drug repurposing offers a reduced risk of failure, necessitates lessi time and financial investment, and hence presents an appealing route for developing antimicrobial drugs [33–35].

12.4.1 Omics-Based Approaches in Drug Repurposing and Repositioning

We will see three primary cases in the upcoming examples. First, where an already marketed drug approved for an unrelated effect was found to have antimicrobial activity and can be repurposed to treat AMR. In the second case of drug repositioning or rescue, a previously discovered drug that did not make it to market (for any financial or safety reason) is reconsidered as a second- or third-line treatment against MDR. In the third case, a compound with known broad therapeutic effects is re-assessed for its potential in influencing AMR or the potency of any other therapeutic. Note that a drug may be repurposed as an independent therapy, a tertiary drug, an excipient to be used in combination with another therapy, or even as part of a multi-drug regimen.

Existing drugs with potential for repurposing can be easily identified using omics techniques. This has been demonstrated in recent research on treating resistant *Staphylococcus aureus*, particularly MRSA. The clinical usage of semisynthetic antibiotics such as methicillin has swiftly led to the rise of methicillin-resistant *S. aureus* (MRSA), and MDR-MRSA [36]. In one progressive study, Shakilur Rahman [37] and colleagues aimed to screen possible therapeutic targets against MRSA (*S. aureus* USA300) using a *proteomics* technique. The staphylococcal FemA protein was subjected to structure-based virtual screening, and ten FDA-approved drugs against FemA were identified. These further displayed a comparable inhibition constant (Ki) through in silico screening. This is just one robust method showing how ten already approved drugs that could be of interest for antimicrobial repurposing have been quickly detected.

In another example, we see how *metabolomics* has revealed the potential for re-tasking drugs from a separate class, statins in particular, for use as adjuvants to traditional antibiotic compounds. *S. aureus* develops resistance to β-lactam antibiotics, including methicillin, through the expression of a low-affinity penicillin-binding protein (PBP2a) that operates conjointly with the general penicillin-binding protein PBP2. In one study, García et. al [38] used proteomics and lipid profiling to clarify the specific mechanisms by which functional membrane microdomains (FMMs) enable efficient oligomerization of PBP2a. Drawing on this discovery, they devised an approach to externally disrupt PBP2a oligomerization through the inhibition of FMM-constituent lipid biosynthesis using cholesterol-lowering drugs (statins). MSRA-infected mice treated with a synergistic combination of oxacillin and the statin zaragozic acid (ZA) demonstrated significantly lower bacterial loads compared to those treated solely with oxacillin, supporting the antimicrobial utility of statin compounds.

Mycobacterium tuberculosis is one of the most prevalent pathogens displaying MDR and XDR. AMR development over the years has already led to the standard four-drug regimen, but emerging strains display XDR. XDR-TB was redefined by the WHO in 2021 as TB resistant to the core four-drug regimen, all fluoroquinolones, and at least one of the group A second-line treatments (levofloxacin, moxifloxacin, bedaquiline, and linezolid) [39]. From antihypertensives to CNS drugs, many compounds have been researched and are in clinical trials for repurposing against TB. Bedaquiline, developed through drug repurposing, was approved by the WHO in 2013 for treating MDR-TB and had been the first anti-tuberculosis drug to be approved in over 40 years [40]. While originally discovered in 1983 and classified as an antimicrobial, it was not developed into a drug compound due to early research revealing safety concerns. However, with the growth of MDR-TB, the drug was reprofiled and is now a standard second-line drug, used only in cases of TB displaying MDR to the four drug, first-line therapy.

Mycobacterium as a class (including *M. tuberculosis*) displays extremely adaptive AMR acquisition and is a driving force behind the effort to repurpose drugs as second-line or tertiary treatments. *Mycobacterium abscessus* infections in particular are deemed the most challenging to treat due to an elevated level of resistance mechanisms (intrinsic and extrinsic) to diverse antimicrobials. Anjos and colleagues [41], employing an in silico genomics approach, compared the sequences of 4942 *M. abscessus subsp. abscessus* (*M. abscessus*) proteins with 5258 and 3473 therapeutic targets registered in the DrugBank and therapeutic target database, respectively. This comparison identified 446 drugs or drug candidates with targets homologous to *M. abscessus* proteins. Further screening and evaluation revealed miconazole (MCZ), a drug used for managing cutaneous fungal infections, and tamoxifen (TAM), a cancer drug, as candidates displaying bacteriostatic activity. Subsequent testing found both compounds decreased the bacterial load in the lungs of infected mice, indicating they could be of interest for repurposing to treat *M. abscessus* infection.

There are a number of compounds such as certain vitamins and nutrients that are known to be bioactive or have pharmaceutical properties. Many of these compounds are hard to research because they act through complex mechanisms and have many non-specific targets, thus their use is generally not FDA approved. Berberine for instance is a principal alkaloid harbored by many medicinal plants. This isoquinoline alkaloid showcases a broad range of pharmacological activities such as antimicrobial, antihypertensive, anti-inflammatory, antioxidant, anti-depressant, anti-cancer, anti-diarrheal, hepatoprotective, and anti-diabetic properties. Berberine forms a prime component of Coptis extract tablets, a traditional Chinese medicine employed for treating traveler's diarrhea. Omics techniques might help to categorize the complex mechanisms of such compounds, allowing them to be applied in a targeted manner, and further developed into practical therapeutics.

Karaosmanoglu and colleagues [42] used *transcriptomics* analysis to uncover effectors in DNA replication/repair and transcription mechanisms that were the main targets triggered by berberine. These targets are thought to regulate cell wall or cell membrane transportation and motility-related processes. Elucidating this berberine mechanism has justified more targeted studies in bacteria, yielding results showing a bacteriostatic effect against *Escherichia coli*. This research displays the extensive potential for omics-guided approaches in discovering the utility of numerous compounds that were previously considered unsuitable or too difficult to study.

Classical methods of drug repurposing need considerable manpower to evaluate the growth phenotype of bacteria and ascertain the target and mechanism. However, with the rapid development of in silico platforms for drug repurposing and the availability of omics data from public databases, it is possible to perform target screening in silico and obtain potential drugs via molecular docking analyses in a short duration. Wang et al. [43] devised auto in silico consensus inverse docking (ACID), a tool for drug repurposing that leverages a consensus inverse docking strategy. This tool enhances the success rate by approximately 10% compared to the existing leading methodologies. The in silico omics approach enables the screening of potential drug targets analogous to the therapeutic targets registered in public databases [41].

Even though drug repurposing proves to be a useful approach in curbing antimicrobial resistance (AMR), it presents some limitations. First, the capacity of FDA-approved drug libraries are limited. Preclinical drug libraries may also be hard to access due to lack of data sharing/public availability. Repurposing drugs still demands the verification of activity in vitro and in animal models, particularly for those drug candidates that have been screened through an in silico approach. Lastly, the employment of pathogen-directed therapies may affect the human microbiome, and these antibacterial compounds may incite drug resistance in other bacteria [31].

12.4.2 The Advantage of Repurposing: Financial and Regulatory Factors

While combating AMR and MDR infection is a large medical concern and a humanitarian crisis, the establishment of robust therapeutic developmental pipelines has been hindered by the financial burden underlying such endeavors. In the case of Bedaquilin's approval, we see a few regulatory initiatives that have come in place to ease financial roadblocks encountered by developers (pharmaceutical companies) to more quickly address prevalent MDR threats. First, as a repurposed drug fewer clinical trials were needed to gain approval as previous testing data could be used in some instances. For example, safety characterization may be supported by prior clinical data (such as trial data or post-marketing data). The evaluation of trial results was also fast-tracked, with policies such as standard waiting periods between applications and approvals being waived. This was part of an accelerated approval status given to the drug as it was determined to treat a *neglected disease*. Such fast-tracking is becoming more common through initiatives by regulatory agencies that have been formed to address AMR by introducing different criteria for drugs targeting specific infections as well as categorizing drugs for these infections as *orphan drugs*. Orphan product designation granted by the FDA now also provides financial incentives such as tax exemptions to pharmaceutical companies undertaking development [44]. Through these initiatives, pharmaceutical companies are encouraged to research and develop new antibiotics and therapeutics to circumvent AMR.

While many regulatory considerations aid in drug repurposing the process can also hit a roadblock when it comes to patented medications [33]. Regulations have been enacted in many countries distinguishing repurposed drugs as off-patent, allowing for their development as antimicrobial compounds circumventing violations [45]. Furthermore, as many repurposed drugs have different formulations or delivery mechanisms they may even be considered new therapeutic compounds (NTC) and are subsequently protected under intellectual property (IP) laws giving the company that developed the application exclusive rights for a certain number of years—a further financial incentive [46]. In America and other nations, government and regulatory agencies are encouraging data sharing so that drugs may be repurposed for novel antimicrobial applications. For instance, pharmaceutical companies might be incentivized to share proprietary data with a research institution. Despite these endeavors navigating the field of intellectual property and patent regulations remains a significant challenge for researchers and companies attempting to repurpose already marketed compounds.

12.5 Omics and Systems Biology in Discovery of Novel Drug Combinations

Drug combination, just as its name implies, means treating diseases with multiple drugs rather than a single drug and has already been widely applied in treating various infectious diseases. A typical example is the treatment of tuberculosis which warrants a four-drug regimen including isoniazid, rifampin, ethambutol, and pyrazinamide [47].

The basic principle of drug combinations can be interpreted straightforwardly: for two drugs, if the probability of spontaneous resistance to the first drug is given by *A*, and the probability of spontaneous resistance to the second drug is given by *B*, then the probability of spontaneous resistance to the two-drug combination is *A* multiplied by *B* where every incidence of spontaneous resistance is (considered) an independent event [48]. According to this model, theoretically, the more drugs that are combined, the less likely they are to be impacted by resistance. However, note interactions between drugs may have different effects and the number of drugs combined will be subject to various constraints, which make drug combinations in practice far more complicated than the model.

In order to introduce drug combinations more clearly, we divide them into three categories based on the pathway and target of action: independent pathways, varying targets within the same pathway, and the same target but by different mechanisms.

12.5.1 Drugs Targeting Independent Pathways

For a particular pathogen, there are many metabolic pathways producing essential molecules that are each independently necessary for bacterial survival. We continue with the example of isoniazid, rifampin, ethambutol, and pyrazinamide used for traditional tuberculosis treatment given in the previous section. It can be seen that these four drugs act on at least four different pathways. Isoniazid is an inhibitor of the inhA gene, which encodes for a NADH-specific enoyl-acyl carrier protein reductase fatty acid synthase [49]. Rifampicin is an inhibitor of DNA-dependent RNA polymerase, the central enzyme of bacterial transcription [50]. Ethambutol exhibits its mode of action by inhibiting the biosynthesis of arabinogalactan, which is involved in cell wall biosynthesis [51]. Pyrazinamide interrupts the biosynthesis of coenzyme A by binding to an aspartate decarboxylase [52]. In addition, two newer drugs, bedaquiline and delamanid, can also be used in combination. The former acts on ATP synthase involved in energy metabolism and the latter inhibits mycolic acid synthesis [53, 54]. By using different drugs to independently target multiple essential pathways, the pathogen cannot adapt fast enough to develop a workaround for each targeted pathway.

It is worth mentioning that the range of different pathways is not limited to the pathogen itself but can also include the metabolic pathways of the host. For example, Ritonavir, an anti-viral medication, was initially used as a protease inhibitor but was later found to inhibit the host drug metabolism pathway and prolong the half-life of the drug in the host, thereby improving treatment regimens [55].

12.5.2 Multiple Drugs Targeting a Single Pathway

If we look at a metabolic pathway as a sequence of biochemical reactions, there are multiple regulatory elements that can independently disrupt the pathway. An early example taking advantage of this is the combined use of sulfamethoxazole and trimethoprim (TMP-SMX) in the 1960s, which both act on folate metabolic pathways. Sulfamethoxazole is a sulfonamide antibiotic, and its target is dihydropteroate synthase, an enzyme in the biosynthetic pathway for the cofactor folic acid. Meanwhile, trimethoprim is an inhibitor of dihydrofolate reductase [56].

As discussed in previous chapters, bacteria can display alterations in metabolic activity that lead to drug resistance by producing enzyme variation. For instance, a particular AMR bacteria may contain a mutated enzyme with increased activity, leading to faster efflux or breakdown of a drug, causing resistance. The combination of antibiotics and corresponding inhibitors of resistant enzymes has thus become a focus for developing drug combinations. Among the most successful examples of such treatment is the use of the β-lactam antibiotic amoxicillin in combination with the β-lactamase inhibitor clavulanic acid, a formulation known as augmentin [56].

12.5.3 Multiple Drugs Targeting a Single Locus via Different Mechanisms of Action

Two compounds may both act on a single enzyme but alter the enzyme in different ways, or at different sites. If the effectiveness of the combination is greater than the additive effects of each drug, they are said to display a synergistic relationship. An example of this is the drug marketed as Synercid, a formulation consisting of Quinupristin and Dalfoprisdn. These two ingredients bind adjacent sites near the peptidyl transferase center of the 50S subunit and their combination displays a 10–100 fold increase in potency compared to when either molecule is used as a single agent [57].

12.5.3.1 Current Drug Combination Therapies

This section presents in tabular form some of the representative drug combinations discovered in recent years for the treatment of some serious infectious diseases (Table 12.1).

Table 12.1 Some examples of drug combinations have been approved or put in clinical trial against infectious diseases

Drug combination	Disease	Mechanism	Approved	References
Sulbactam, durlobactam	Drug-resistant pneumonia	Durlobactam blocks beta-lactamase enzymes and restores sulbactam's bactericidal activity	Yes	[58]
Stribild (elvitegravir, cobicistat, emtricitabine, tenofovir disoproxil fumarate)	HIV infection	Emtricitabine and tenofovir disoproxil fumarate are two kinds of nucleoside reverse transcriptase inhibitors, elvitegravir is a new integrase inhibitor and cobicistat inhibits cytochrome P-450 isoenzyme 3A activity	Yes	[59]
Vosevi (sofosbuvir, velpatasvir, and voxilaprevir)	Hepatitis C virus	The first treatment approved for patients who do not respond to the direct-acting antiviral drug sofosbuvir or other HCV drugs that inhibit nonstructural protein 5A	Yes	[60]
Triple artemisinin-based combination	Plasmodium falciparum malaria	The extreme rarity of parasites acquiring mutations conferring resistance against both partner drugs over a short period of time	Clinic trial	[61]
Bedaquiline and delamanid	Drug-resistant tuberculosis	Bedaquiline acts on ATP synthase involved in energy metabolism and delamanid inhibits mycolic acid synthesis	Clinic trial	[54]

12.6 Omics in Novel Therapeutic Discovery

As discussed previously a number of therapies that depart from traditional pharmaceutical small molecule drug (SMD) therapeutics are being researched for potential in fighting AMR. Existing antimicrobials, in fact, only encompass a small range of physiochemical properties affecting a miniscule number of targets considering the diverse structure and vast target spectrum of some alternative approaches. The novel ability to make moderations to many therapeutics using Omics approaches allows many therapies to be reconsidered as candidates in fighting AMR. Particularly, even in cases where the number of useful compounds is relatively low, novel computational technology allows these directions to be explored. Furthermore, a deeper understanding of expression pathways and metabolism allows for functional modifications that may alter the therapeutic delivery, and action or potency.

Given how fast some bacteria can develop AMR, a number of experts suggest that even with drug repurposing and development of combination therapies, we will not achieve a singular effective treatment method [62]. Rather, an integrated strategy that includes novel antibiotics in combination with vaccines, gene-based therapeutics, monoclonal antibodies, and microbiota interventions among other strategies is required to combat AMR effectively. In this section, we will consider a number of alternative therapeutics, some of which are already being applied and others representing interesting new potential avenues for inquiry.

12.6.1 Unidentified Natural Antibiotics: SMDs and AMPs

While the synthesis of compounds based on actinomycetes has yielded many important antimicrobials, we hit a proverbial wall in development as many synthetic compounds are unable to permeate the bacterial cell. There are still a large number of organisms producing natural (small molecule/small biologic) antimicrobials that can penetrate the cellular envelope. There may indeed be many of these compounds yet undiscovered as less than 1% of cells in the natural environment have been successfully cultured in a medium. With the advent of omics technology, it is possible to reconstruct the metabolic pathways of organisms based on genomic data. This will be of great help in exploring the growth conditions of microorganisms that cannot currently be cultured, assessing the biosynthetic potential of these bacteria, and facilitating the discovery of antimicrobial producers []. Furthermore, even when cultures are possible many organisms such as soil actinomycetes contain important clusters of biosynthetic genes that cannot be expressed in vitro. Therefore, combining genomic, transcriptomic, and proteomic techniques to find and express these silent operons may be yet another avenue in the discovery of antimicrobials.

Another important potential source of antimicrobial compounds is a vast pool of antimicrobial peptides (AMPs). In the past, these molecules were harder to analyze and had less favorable pharmacologic interactions than SMDs. Of some 3000 AMPs identified (sec. Antimicrobial Peptide Database (APD)) only 7 have been clinically approved including Colistin (1959), Gramicidin (1955), Vancomycin (1958), Daptomycin (2003), Telavancin (2009 and 2013), Dalbavancin (2014), and Oritavancin (2014) [63]. Due to significant toxicity, most of these are only applied topically such as gramicidin, an active ingredient in Neosporin (a common non-prescription ointment in America); while vancomycin is given IV mainly for serious MRSA or MRSE internal infections.

Unlike the small molecule drugs we have previously discussed that target specific proteins, AMPs act through a more general mechanism that typically disrupts the cell membrane of bacteria but may also act on host cells. (sc. While AMPs can extend beyond 100 peptides and display a variety of structures, identification of these molecules has been limited. The approved AMPs are quite similar being natural products of Gram-positive soil bacteria, containing fewer than 50 peptides, having low molecular weights, and being cyclical in structure.) However, new computational technology makes discovery and optimization of AMPs much more feasible.

Using AI programs, compounds can be quickly screened and those with desirable properties can be identified with relative success. Using subsequent technology, optimization of stable formulas such as those utilizing lipid nanoparticle (LNP) carriers can be quickly performed. Each AMP might further act as a scaffold for the design of modified synthetic compounds. Lastly, the resulting predicted useful formulations can be tested *in silico* to further screen candidates. Cyclic peptide drug conjugates (CDCs) are formulations that bind cyclic AMPs to other antibiotics like SMDs, which facilitate targeted delivery and may lead to improved action for AMPs resulting in many viable drug candidates [64].

The vast abilities of computational power and AI have also made screening huge amounts of culture samples for the production of compounds with antibiotic effects a renewed route of discovery, similar to the process of therapeutic discovery in the early phenotypic era [65]. This method led to the discovery of the potentially novel antimicrobial Teixobactin in 2015, another AMP molecule [66]. Similar genome mining techniques strive to identify biosynthetic gene clusters encoding compounds homologous to known antibiotic scaffolds that may represent new antibiotic compounds by applying various bioinformatics techniques to screen libraries of genomes and microbe samples [67].

12.6.2 Gene Therapies: ASO/siRNA and CRISPR-Cas9 Laden Bacteriophage

By analyzing the gene expression patterns of drug-resistant strains, researchers can pinpoint potential drug targets that are either overexpressed or essential for survival. This information can be harnessed to engineer novel therapeutics aimed at inhibiting specific resistance mechanisms, such as antisense oligonucleotides (ASO) and small interfering RNA (siRNA). These molecules can be designed to target and silence the expression of resistance genes, effectively neutralizing the mechanisms responsible for drug resistance. Specifically, these act by binding a complementary bacterial mRNA, preventing translation to any essential protein. This is the first gene therapy we have discussed and represents a distinctive development strategy compared to traditional antimicrobials [68].

Research into ASO use for antibacterial application has been limited due to the inability of these molecules to penetrate the bacterial cell wall, and their instability and rapid degradation by ubiquitous RNases. Recent research on transportation methods identifies the conjugation of ASOs to cell-penetrating peptides as a strategy to increase uptake and an avenue for future inquiry [69]. Although there are many hurdles involved in developing a possible ASO drug, achieving such a breakthrough could mean a magic bullet, theoretically able to silence any gene. Perspective on the number of therapeutic avenues this would unveil is demonstrated in a study by Garcia et al [70] that identifies over 600 fitness genes with 293 essential genes in uropathogenic *E. coli* that could all function as ASO targets.

The bacterial immune system is composed of restriction enzymes, toxin-antitoxin systems, and the clustered regularly interspaced short palindromic repeats (CRISPR)-associated protein (Cas) system. The CRISPR-Cas9 complex targets and modifies DNA through a sequence-specific DNA probe paired with a modifying enzyme capable of DNA cleavage. By hijacking this system, the site-specific cleavage action can be leveraged for antimicrobial effects via two strategies: the inactivation of chromosomal gene pathways or the elimination of plasmids encoding resistance to antibiotics (I.E. a barrier to MGE and HGT) [71]. The first example is demonstrated in an experiment by Kim et al. [72] that used the CRISPR-Cas9 system to effectively target a conserved sequence within extended-spectrum β-lactamases, thereby reinstating the vulnerability of *E. coli* to β-lactam antibiotics. In application, a significant challenge in utilizing CRISPR/Cas9 is its delivery to the proper cells [73].

Bacteriophages are lytic/lysogenic phages that inject their genetic material into host bacteria, ultimately leading to bacterial cell lysis. These viral organisms are thus being studied for possible antimicrobial applications. Although the direct therapeutic application of phage therapy remains under scrutiny, engineered phages demonstrate the potential to knock down a variety of targets by delivering specially designed Cas9 sequences to form CRISPR-Cas9 systems within target bacteria. Engineered phages containing the cas9 gene and an associated guide RNA sequence can inject these into bacteria to form the cleavage machinery [74]. Non-replicating phages can thus be used as vectors to target and disrupt antibiotic resistance genes, or might even be programmed to eliminate bacteria by targeting essential genes [75]. The second method of antimicrobial action mentioned above was displayed in a study employing phage delivery to introduce synthetic Cas9 genes into *E. coli* cells. The CRISPR-Cas9 system subsequently targeted plasmids containing the mcr1 gene in the bacterial samples, leading to plasmid loss and re-establishing sensitivity to polymyxins. While these strategies show immense promise, the real-world clinical value is yet to be determined.

12.6.3 Immunomodulation: Vaccines, mAB, Antibacterial Antibodies, and DAC

Vaccines are effective in the case of drug resistance as resistance to vaccines is nearly impossible to develop. This is because, as vaccines are prophylactic in character, they prevent pathogens from reaching population sizes where selective adaptation can have significant effects, and furthermore act through a concerted immune response instead of targeting one aspect of bacterial physiology [76]. The WHO has undertaken a campaign to increase vaccine development efforts against MDR organisms as the problem continues to accelerate. Vaccinomics is the application of omics approaches (pharmacogenomics and bioinformatics) in vaccine development, and a powerful tool for combating AMR. Particularly, proteomics plays a key role in identifying important surface antigens that might be vaccine targets while genomics helps quickly identify which of these proteins might be conserved across a species.

The most effective vaccine should have broad coverage of a bacterial species, and so should stimulate immunity against a particular antigen that is ubiquitous throughout all strains. In other words, the antigen should be a structure that is conserved across strains/serotypes and stably maintained within natural bacterial populations. In the pursuit of a broadly protective vaccine, Maione et al. [77] conducted a comprehensive analysis of the genomes of eight *S. agalactiae* strains, cloning and testing 312 surface proteins. Notably, the combination of four proteins (GBS322, GBS104, GBS67, and GBS80) proved highly protective against multiple strains, offering promise as a potential universal vaccine against *S. agalactiae*. Significantly, only one of these proteins belonged to the core genome, challenging the intuitive concept of relying solely on conserved antigens for broad protection.

Reverse vaccinology is a structural-based method that identifies likely antigenic sequences, such as those coding surface proteins, from a whole microbial genome [78]. The encoded antigens are then synthesized and incorporated into vaccine candidates that are tested in in vivo infection models. This technique has been described for producing an experimentally successful vaccine against resistant *E. coli*, particularly *EHEC* [79], and *P. aeruginosa* [80]. A number of other vaccine strategies such as the use of an mRNA-LNP vaccine against the capsular antigen of *Y. Pestis* have shown promise through initial research [81]. Another strategy, the use of recombinant protein vaccines, has led to more effective variations of the BCG vaccine against *M. tuberculosis* [82, 83].

Metabolomics emerges as another discipline key to understanding the action of vaccines against bacterial infections. As seen in the case of *M. tuberculosis*, the BCG vaccine acts not only through immune mechanisms but also by priming the host metabolic response to infection. *M. tuberculosis* shields itself from the immune system in part by hijacking the host's regular cellular activity. The host metabolome plays a part in stimulating host immunity as well as affecting *M. tuberculosis'* ability to self-regulate and create its own microenvironment—establishing latent infection. Metabolomics as a vaccinomics tool provides a broader picture of vaccine-induced biochemical changes that can inform the selection and design of vaccine candidates [83].

Another class of immunomodulatory therapeutics comprises the variety of monoclonal antibodies (mABs). These target specific epitopes on targets of interest, including those on bacteria, and can inhibit the target or cause complement-mediated lysis and opsonization/phagocytosis [6]. Three antibacterial mAB drugs have been approved to date including raxibacumab, obiltoxaximab, and bezlotoxumab. Raxibacumab and obiltoxaximab function to neutralize the *Bacillus anthracis* protective antigen while bezlotoxumab neutralizes toxin B from *Clostridium difficile*. An upside of mABs is that they may be able to stimulate an immune response against infection in immunocompromised patients who do not develop immunity on vaccine exposure. Despite some success, research in this area is limited as these compounds can only target secreted virulence factors to combat pathogenicity or surface proteins that help to mobilize a host immune response to aid in pathogen clearance [84]. More extensive research is being applied to study drug antibody conjugates with a combined activity allowing the antibody to act almost like an adjuvant or excipient by aiding in targeted delivery [85].

12.6.4 Microbiome Modulation

Therapeutic strategies that exploit the human microbiota introduce a plethora of bacteria into the body. These bacteria collectively express a vast array of genes and generate innumerable metabolic products. Such interventions are designed to augment the natural microbial defense against pathogenic colonization via various mechanisms. The microbiota plays a pivotal role in the host's overall health and its immunological responses.

The use of antibiotics, while essential for combating infections, can inadvertently destroy beneficial commensal bacterial populations. This decimation can pave the way for pathogenic microbes to establish themselves within the host, potentially leading to opportunistic infections. Therapeutics that aim to restore the integrity of the gut microbiota have surfaced to address conditions such as recurrent Clostridioides difficile-associated diarrhea.

For instance, there have been reports where orally administered bacterial spore concoctions, derived from the fecal matter of a healthy individual, effectively prevent subsequent *C. difficile* infections [86]. Similarly, the introduction of non-toxicogenic strains of *C. difficile* serves a competitive advantage over their toxicogenic counterparts. Such interventions have been linked to a marked decrease in the recurrence of *C. difficile* infections and offer a potential avenue for microbiota restoration [87].

12.7 Concluding Remarks and Perspective: The Role of Omics and Multi-omics

> Verba ita sunt intelligenda ut res magis valeat quam pereat – words are to be understood such that the subject matter may be more effective than wasted – ancient legal teaching

In a previous section, we introduced strategies for drug combination but in the actual development process, finding the right drug combination and the corresponding optimal ratio has been difficult. Considering the wide variety of drugs that have been discovered, possible combinations are astronomical. Therefore, traditional drug combination screening usually stays within the scope of existing antibiotics. However, because the dosage and proportion of each drug in the combination also affects its efficacy, the amount of work required to screen a combination using traditional experimental methods is not feasible.

Modern omics techniques, including genomics, proteomics, metabolomics, etc., provide promising approaches to solve this intractable problem. By comparing the omics data of different drugs, we can predict in advance which drug combinations are more likely to work. Compared with the traditional direct screening of all combinations one by one, this will greatly reduce the workload and improve the research and development efficiency of new drug combinations.

Compared with traditional biology, which studies a single pathway or molecule independently, the advantage of such systems biology approaches is that they can incorporate related pathways or molecules into the scope of the study. The holistic view produces results that more closely model actual biology because the pathways in organisms are related through networks rather than linearly. In addition, through the construction of mathematical models and the introduction of computer simulation technology, the screening efficiency and accuracy of drug combinations have been further improved.

There are many reasons for the development of resistance, including target variation, effusion, dormancy, and so on [88]. Because omics can provide information at all levels of expression, such as genes, transcription, translation, metabolism, etc., researchers can find the causes of drug resistance through this information and develop new drugs. This could greatly speed up the development of new drugs and drug combinations, compared to the previous development approach that resembled random trials.

These strategies exist not only in theory but have been widely used in research. By combining the knockout profile of *E. coli* with its metabolic profile for a variety of drugs that would otherwise have no antibacterial activity, many new drug interactions were predicted. This greatly expands the study of drug combinations and provides a way to study compounds whose mechanism of action is not to inhibit growth [89]. Not limited to bacteria, omics techniques have also played an important role in the development of drugs for viral diseases such as influenza, as well as non-communicable diseases such as cancer [90, 91]. Moreover, for diseases with heterogeneous clinical outcomes and disease severity, such as COVID-19 and cancer, the comparison of different individual omics data can better reveal the causes of diseases and enable personalized and precise treatment [92, 93].

There are limitations to the computational technologies underlying many omics techniques. Due to the high complexity of metabolic pathways in living organisms, computers sometimes do not fully simulate in vivo phenomena (relying on dimensional reduction). Moreover, the establishment of various criteria in mathematical models is very subjective, which leads to many differences in screening results obtained through different models. Therefore, these methods should not be completely separated from traditional experiments. Experimental verification after obtaining screening results is a good strategy to balance workload and credibility.

In any case, these technologies are promising for the future. With the accumulation of omics data and the reduction of costs, drug development will no longer be limited to the development of one-size-fits-all treatments but will focus more on personalized medication and treatment for each patient, which has very significant implications for patients. With the help of these technologies, drug development can enter an unprecedented new stage.

References

1. Lewis K. Recover the lost art of drug discovery. Nature. 2012;485:439–40. https://doi.org/10.1038/485439a.
2. Lewis K. The science of antibiotic discovery. Cell. 2020;181:29–45. https://doi.org/10.1016/j.cell.2020.02.056.
3. Taylor SD, Palmer M. The action mechanism of daptomycin. Bioorg Med Chem. 2016;24:6253–68. https://doi.org/10.1016/j.bmc.2016.05.052.
4. Committee on the Review of Omics-Based Tests for Predicting Patient Outcomes in Clinical Trials, Board on Health Care Services, Board on Health Sciences Policy & Institute of Medicine. Evolution of translational omics: lessons learned and the path forward (Micheel CM, Nass SJ, Omenn GS, editors.). Washington, DC: National Academies Press (US); 2012.
5. Rutledge PJ, Challis GL. Discovery of microbial natural products by activation of silent biosynthetic gene clusters. Nat Rev Microbiol. 2015;13:509–23. https://doi.org/10.1038/nrmicro3496.
6. Rosenn EH, et al. Monoclonal war: the antibody arsenal and targets for expanded application. Immuno. 2023;3:346–57.
7. Wong F, et al. Discovery of a structural class of antibiotics with explainable deep learning. Nature. 2024;626:177–85. https://doi.org/10.1038/s41586-023-06887-8.
8. Suay-Garcia B, Falcó A, Bueso-Bordils JI, Anton-Fos GM, Pérez-Gracia MT, Alemán-López PA. Tree-based QSAR model for drug repurposing in the discovery of new antibacterial compounds against Escherichia coli. Pharmaceuticals (Basel). 2020;13:431. https://doi.org/10.3390/ph13120431.
9. Danko D, et al. A global metagenomic map of urban microbiomes and antimicrobial resistance. Cell. 2021;184:3376–3393.e3317. https://doi.org/10.1016/j.cell.2021.05.002.
10. Foudraine DE, et al. Exploring antimicrobial resistance to beta-lactams, aminoglycosides and fluoroquinolones in E. coli and K. pneumoniae using proteogenomics. Sci Rep. 2021;11:12472. https://doi.org/10.1038/s41598-021-91905-w.
11. Theuretzbacher U, et al. Critical analysis of antibacterial agents in clinical development. Nat Rev Microbiol. 2020;18:286–98. https://doi.org/10.1038/s41579-020-0340-0.
12. Modjarrad K, Moorthy VS, Millett P, Gsell PS, Roth C, Kieny MP. Developing global norms for sharing data and results during public health emergencies. PLoS Med. 2016;13:e1001935. https://doi.org/10.1371/journal.pmed.1001935.
13. Groschel MI, Walker TM, van der Werf TS, Lange C, Niemann S, Merker M. Pathogen-based precision medicine for drug-resistant tuberculosis. PLoS Pathog. 2018;14:e1007297. https://doi.org/10.1371/journal.ppat.1007297.
14. Meehan CJ, et al. Whole genome sequencing of Mycobacterium tuberculosis: current standards and open issues. Nat Rev Microbiol. 2019;17:533–45. https://doi.org/10.1038/s41579-019-0214-5.
15. Pucci MJ. Use of genomics to select antibacterial targets. Biochem Pharmacol. 2006;71:1066–72. https://doi.org/10.1016/j.bcp.2005.12.004.
16. Clements J, Ayscough A, Keavey K, East S. Peptide deformylase inhibitors, potential for a new class of broad spectrum antibacterials. Curr Med Chem Anti-Infect Agents. 2002;1:239–49. https://doi.org/10.2174/1568012023354866.
17. Schmid MB. Seeing is believing: the impact of structural genomics on antimicrobial drug discovery. Nat Rev Microbiol. 2004;2:739–46. https://doi.org/10.1038/nrmicro978.
18. Rozwandowicz M, et al. Plasmids carrying antimicrobial resistance genes in Enterobacteriaceae. J Antimicrob Chemother. 2018;73:1121–37. https://doi.org/10.1093/jac/dkx488.
19. Suzuki S, Horinouchi T, Furusawa C. Prediction of antibiotic resistance by gene expression profiles. Nat Commun. 2014;5:5792. https://doi.org/10.1038/ncomms6792.
20. Zhang S, et al. Diclofenac resensitizes methicillin-resistant Staphylococcus aureus to beta-lactams and prevents implant infections. Adv Sci (Weinh). 2021;8:2100681. https://doi.org/10.1002/advs.202100681.

21. Agostini M, Traldi P, Hamdan M. Mass spectrometry-based proteomics: analyses related to drug-resistance and disease biomarkers. Medicina (Kaunas). 2023;59:1722. https://doi.org/10.3390/medicina59101722.
22. Arora G, Bothra A, Prosser G, Arora K, Sajid A. Role of post-translational modifications in the acquisition of drug resistance in Mycobacterium tuberculosis. FEBS J. 2021;288:3375–93. https://doi.org/10.1111/febs.15582.
23. Wilhelm M, et al. Mass-spectrometry-based draft of the human proteome. Nature. 2014;509:582–7. https://doi.org/10.1038/nature13319.
24. Herold M, et al. Integration of time-series meta-omics data reveals how microbial ecosystems respond to disturbance. Nat Commun. 2020;11:5281. https://doi.org/10.1038/s41467-020-19006-2.
25. Kok M, Maton L, van der Peet M, Hankemeier T, van Hasselt JGC. Unraveling antimicrobial resistance using metabolomics. Drug Discov Today. 2022;27:1774–83. https://doi.org/10.1016/j.drudis.2022.03.015.
26. Chaiyachat P, et al. Metabolomic analysis of Mycobacterium tuberculosis reveals metabolic profiles for identification of drug-resistant tuberculosis. Sci Rep. 2023;13:8655. https://doi.org/10.1038/s41598-023-35882-2.
27. Wang L, Yang Q, Peng S, Liu X. The combination of the glycolysis inhibitor 2-DG and sorafenib can be effective against sorafenib-tolerant persister cancer cells. Onco Targets Ther. 2019;12:5359–73. https://doi.org/10.2147/OTT.S212465.
28. Borah K, Xu Y, McFadden J. Dissecting host-pathogen interactions in TB using systems-based omic approaches. Front Immunol. 2021;12:762315. https://doi.org/10.3389/fimmu.2021.762315.
29. Aminov R. Metabolomics in antimicrobial drug discovery. Expert Opin Drug Discov. 2022;17:1047–59. https://doi.org/10.1080/17460441.2022.2113774.
30. Rahman M, Schellhorn HE. Metabolomics of infectious diseases in the era of personalized medicine. Front Mol Biosci. 2023;10:1120376. https://doi.org/10.3389/fmolb.2023.1120376.
31. Kannan L, et al. Public data and open source tools for multi-assay genomic investigation of disease. Brief Bioinform. 2016;17:603–15. https://doi.org/10.1093/bib/bbv080.
32. Cambiaghi A, Ferrario M, Masseroli M. Analysis of metabolomic data: tools, current strategies and future challenges for omics data integration. Brief Bioinform. 2017;18:498–510. https://doi.org/10.1093/bib/bbw031.
33. Pushpakom S, et al. Drug repurposing: progress, challenges and recommendations. Nat Rev Drug Discov. 2019;18:41–58. https://doi.org/10.1038/nrd.2018.168.
34. Langtry HD, Markham A. Sildenafil: a review of its use in erectile dysfunction. Drugs. 1999;57:967–89. https://doi.org/10.2165/00003495-199957060-00015.
35. Breckenridge A, Jacob R. Overcoming the legal and regulatory barriers to drug repurposing. Nat Rev Drug Discov. 2019;18:1–2. https://doi.org/10.1038/nrd.2018.92.
36. Lakhundi S, Zhang K. Methicillin-resistant Staphylococcus aureus: molecular characterization, evolution, and epidemiology. Clin Microbiol Rev. 2018;31:e00020-18. https://doi.org/10.1128/CMR.00020-18.
37. Rahman S, Das AK. Integrated multi-omics, virtual screening and molecular docking analysis of methicillin-resistant Staphylococcus aureus USA300 for the identification of potential therapeutic targets: an in-silico approach. Int J Pept Res Ther. 2021;27:2735–55. https://doi.org/10.1007/s10989-021-10287-9.
38. Garcia-Fernandez E, et al. Membrane microdomain disassembly inhibits MRSA antibiotic resistance. Cell. 2017;171:1354–1367.e1320. https://doi.org/10.1016/j.cell.2017.10.012.
39. Viney K, et al. New definitions of pre-extensively and extensively drug-resistant tuberculosis: update from the World Health Organization. Eur Respir J. 2021;57:2100361. https://doi.org/10.1183/13993003.00361-2021.
40. Sharma K, Ahmed F, Sharma T, Grover A, Agarwal M, Grover S. Potential repurposed drug candidates for tuberculosis treatment: progress and update of drugs identified in over a decade. ACS Omega. 2023;8:17362–80. https://doi.org/10.1021/acsomega.2c05511.

41. Anjos L, Costa VAF, Neves BJ, Junqueira-Kipnis AP, Kipnis A. Repurposing miconazole and tamoxifen for the treatment of Mycobacterium abscessus complex infections through in silico chemogenomics approach. World J Microbiol Biotechnol. 2023;39:273. https://doi.org/10.1007/s11274-023-03718-w.
42. Karaosmanoglu K, Sayar NA, Kurnaz IA, Akbulut BS. Assessment of berberine as a multi-target antimicrobial: a multi-omics study for drug discovery and repositioning. OMICS. 2014;18:42–53. https://doi.org/10.1089/omi.2013.0100.
43. Wang F, et al. ACID: a free tool for drug repurposing using consensus inverse docking strategy. J Cheminform. 2019;11:73. https://doi.org/10.1186/s13321-019-0394-z.
44. Miller KL, Kraft S, Ipe A, Fermaglich L. Drugs and biologics receiving FDA orphan drug designation: an analysis of the most frequently designated products and their repositioning strategies. Expert Opin Orphan Drugs. 2022;9:265–72. https://doi.org/10.1080/21678707.2021.2047021.
45. Murteira S, Millier A, Ghezaiel Z, Lamure M. Drug reformulations and repositioning in the pharmaceutical industry and their impact on market access: regulatory implications. J Mark Access Health Policy. 2014;2. https://doi.org/10.3402/jmahp.v2.22813.
46. Begley CG, et al. Drug repurposing: misconceptions, challenges, and opportunities for academic researchers. Sci Transl Med. 2021;13:eabd5524. https://doi.org/10.1126/scitranslmed.abd5524.
47. Cano-Muniz S, Anthony R, Niemann S, Alffenaar JC. New approaches and therapeutic options for Mycobacterium tuberculosis in a dormant state. Clin Microbiol Rev. 2018;31:e00060-17. https://doi.org/10.1128/CMR.00060-17.
48. Cottarel G, Wierzbowski J. Combination drugs, an emerging option for antibacterial therapy. Trends Biotechnol. 2007;25:547–55. https://doi.org/10.1016/j.tibtech.2007.09.004.
49. Vilcheze C, et al. Transfer of a point mutation in Mycobacterium tuberculosis inhA resolves the target of isoniazid. Nat Med. 2006;12:1027–9. https://doi.org/10.1038/nm1466.
50. Kirsch SH, Haeckl FPJ, Muller R. Beyond the approved: target sites and inhibitors of bacterial RNA polymerase from bacteria and fungi. Nat Prod Rep. 2022;39:1226–63. https://doi.org/10.1039/d1np00067e.
51. Zhang L, et al. Cryo-EM snapshots of mycobacterial arabinosyltransferase complex EmbB(2)-AcpM(2). Protein Cell. 2020;11:505–17. https://doi.org/10.1007/s13238-020-00726-6.
52. Gopal P, et al. Pyrazinamide triggers degradation of its target aspartate decarboxylase. Nat Commun. 2020;11:1661. https://doi.org/10.1038/s41467-020-15516-1.
53. Bahuguna A, Rawat DS. An overview of new antitubercular drugs, drug candidates, and their targets. Med Res Rev. 2020;40:263–92. https://doi.org/10.1002/med.21602.
54. Mohr E, Ferlazzo G, Hewison C, De Azevedo V, Isaakidis P. Bedaquiline and delamanid in combination for treatment of drug-resistant tuberculosis. Lancet Infect Dis. 2019;19:470.
55. Bierman WF, van Agtmael MA, Nijhuis M, Danner SA, Boucher CA. HIV monotherapy with ritonavir-boosted protease inhibitors: a systematic review. AIDS. 2009;23:279–91.
56. Wormser GP, Keusch GT, Heel RC. Co-trimoxazole (trimethoprim-sulfamethoxazole): an updated review of its antibacterial activity and clinical efficacy. Drugs. 1982;24:459–518. https://doi.org/10.2165/00003495-198224060-00002.
57. Cocito C. Antibiotics of the virginiamycin family, inhibitors which contain synergistic components. Microbiol Rev. 1979;43:145–92. https://doi.org/10.1128/mr.43.2.145-192.1979.
58. Mullard A. FDA approves new antibiotic combination for drug-resistant pneumonia. Nat Rev Drug Discov. 2023;22:527. https://doi.org/10.1038/d41573-023-00096-8.
59. Traynor K. FDA approves four-drug anti-HIV combination tablet. Am J Health Syst Pharm. 2012;69:1616. https://doi.org/10.2146/news120068.
60. Voelker R. Combination drug for HCV infection. JAMA. 2017;318:790. https://doi.org/10.1001/jama.2017.10930.
61. Nguyen TD, et al. Preventing antimalarial drug resistance with triple artemisinin-based combination therapies. Nat Commun. 2023;14:4568. https://doi.org/10.1038/s41467-023-39914-3.

62. Micoli F, Bagnoli F, Rappuoli R, Serruto D. The role of vaccines in combatting antimicrobial resistance. Nat Rev Microbiol. 2021;19:287–302. https://doi.org/10.1038/s41579-020-00506-3.
63. Chen CH, Lu TK. Development and challenges of antimicrobial peptides for therapeutic applications. Antibiotics (Basel). 2020;9:24.
64. Fu C, Yu L, Miao Y, Liu X, Yu Z, Wei M. Peptide–drug conjugates (PDCs): a novel trend of research and development on targeted therapy, hype or hope? Acta Pharm Sin B. 2023;13:498–516. https://doi.org/10.1016/j.apsb.2022.07.020.
65. Stokes JM, et al. A deep learning approach to antibiotic discovery. Cell. 2020;180:688–702.e613. https://doi.org/10.1016/j.cell.2020.01.021.
66. Shukla R, et al. Teixobactin kills bacteria by a two-pronged attack on the cell envelope. Nature. 2022;608:390–6. https://doi.org/10.1038/s41586-022-05019-y.
67. Durand GA, Raoult D, Dubourg G. Antibiotic discovery: history, methods and perspectives. Int J Antimicrob Agents. 2019;53:371–82. https://doi.org/10.1016/j.ijantimicag.2018.11.010.
68. Ngamcherdtrakul W, Yantasee W. siRNA therapeutics for breast cancer: recent efforts in targeting metastasis, drug resistance, and immune evasion. Transl Res. 2019;214:105–20. https://doi.org/10.1016/j.trsl.2019.08.005.
69. Hadjicharalambous A, Bournakas N, Newman H, Skynner MJ, Beswick P. Antimicrobial and cell-penetrating peptides: understanding penetration for the design of novel conjugate antibiotics. Antibiotics (Basel). 2022;11:1636. https://doi.org/10.3390/antibiotics11111636.
70. García V, et al. Genome-wide analysis of fitness-factors in uropathogenic Escherichia coli during growth in laboratory media and during urinary tract infections. Microb Genom. 2021;7:000719. https://doi.org/10.1099/mgen.0.000719.
71. Mancuso G, De Gaetano S, Midiri A, Zummo S, Biondo C. The challenge of overcoming antibiotic resistance in carbapenem-resistant gram-negative bacteria: "attack on titan". Microorganisms. 2023;11:1912. https://doi.org/10.3390/microorganisms11081912.
72. Kim JS, et al. CRISPR/Cas9-mediated re-sensitization of antibiotic-resistant Escherichia coli harboring extended-spectrum β-lactamases. J Microbiol Biotechnol. 2016;26:394–401. https://doi.org/10.4014/jmb.1508.08080.
73. Redman M, King A, Watson C, King D. What is CRISPR/Cas9? Arch Dis Child Educ Pract Ed. 2016;101:213–5. https://doi.org/10.1136/archdischild-2016-310459.
74. Mayorga-Ramos A, Zúñiga-Miranda J, Carrera-Pacheco SE, Barba-Ostria C, Guamán LP. CRISPR-Cas-based antimicrobials: design, challenges, and bacterial mechanisms of resistance. ACS Infect Dis. 2023;9:1283–302. https://doi.org/10.1021/acsinfecdis.2c00649.
75. Tao S, Chen H, Li N, Liang W. The application of the CRISPR-Cas system in antibiotic resistance. Infect Drug Resist. 2022;15:4155–68. https://doi.org/10.2147/idr.S370869.
76. Kennedy DA, Read AF. Why the evolution of vaccine resistance is less of a concern than the evolution of drug resistance. Proc Natl Acad Sci USA. 2018;115:12878–86. https://doi.org/10.1073/pnas.1717159115.
77. Maione D, et al. Identification of a universal Group B streptococcus vaccine by multiple genome screen. Science. 2005;309:148–50. https://doi.org/10.1126/science.1109869.
78. Kanampalliwar AM. Reverse vaccinology and its applications. Methods Mol Biol. 2020;2131:1–16. https://doi.org/10.1007/978-1-0716-0389-5_1.
79. Sanchez-Villamil JI, Tapia D, Torres AG. Optimization of multivalent gold nanoparticle vaccines eliciting humoral and cellular immunity in an in vivo model of enterohemorrhagic Escherichia coli O157:H7 colonization. mSphere. 2022;7:e0093421. https://doi.org/10.1128/msphere.00934-21.
80. Khalid K, Poh CL. The promising potential of reverse vaccinology-based next-generation vaccine development over conventional vaccines against antibiotic-resistant bacteria. Vaccines (Basel). 2023;11:1264. https://doi.org/10.3390/vaccines11071264.
81. Kon E, et al. A single-dose F1-based mRNA-LNP vaccine provides protection against the lethal plague bacterium. Sci Adv. 2023;9:eadg1036. https://doi.org/10.1126/sciadv.adg1036.

82. Srivastava S, Dey S, Mukhopadhyay S. Vaccines against tuberculosis: where are we now? Vaccines (Basel). 2023;11:1013. https://doi.org/10.3390/vaccines11051013.
83. Rawat BS, Kumar D, Soni V, Rosenn EH. Therapeutic potentials of immunometabolomic modulations induced by tuberculosis vaccination. Vaccines (Basel). 2022;10:2127.
84. McConnell MJ. Where are we with monoclonal antibodies for multidrug-resistant infections? Drug Discov Today. 2019;24:1132–8. https://doi.org/10.1016/j.drudis.2019.03.002.
85. MacNair CR, Rutherford ST, Tan M-W. Alternative therapeutic strategies to treat antibiotic-resistant pathogens. Nat Rev Microbiol. 2023;22:262–75. https://doi.org/10.1038/s41579-023-00993-0.
86. Khanna S, et al. A novel microbiome therapeutic increases gut microbial diversity and prevents recurrent Clostridium difficile infection. J Infect Dis. 2016;214:173–81. https://doi.org/10.1093/infdis/jiv766.
87. Wang R. Clostridioides difficile infection: microbe-microbe interactions and live biotherapeutics. Front Microbiol. 2023;14:1182612. https://doi.org/10.3389/fmicb.2023.1182612.
88. Hughes D, Andersson DI. Evolutionary consequences of drug resistance: shared principles across diverse targets and organisms. Nat Rev Genet. 2015;16:459–71. https://doi.org/10.1038/nrg3922.
89. Campos AI, Zampieri M. Metabolomics-driven exploration of the chemical drug space to predict combination antimicrobial therapies. Mol Cell. 2019;74:1291–1303.e1296. https://doi.org/10.1016/j.molcel.2019.04.001.
90. Huang L, et al. Driver network as a biomarker: systematic integration and network modeling of multi-omics data to derive driver signaling pathways for drug combination prediction. Bioinformatics. 2019;35:3709–17. https://doi.org/10.1093/bioinformatics/btz109.
91. Xin Y, Chen S, Tang K, Wu Y, Guo Y. Identification of nifurtimox and chrysin as anti-influenza virus agents by clinical transcriptome signature reversion. Int J Mol Sci. 2022;23:2372. https://doi.org/10.3390/ijms23042372.
92. Rosenblum D, Peer D. Omics-based nanomedicine: the future of personalized oncology. Cancer Lett. 2014;352:126–36. https://doi.org/10.1016/j.canlet.2013.07.029.
93. Ambikan AT, et al. Multi-omics personalized network analyses highlight progressive disruption of central metabolism associated with COVID-19 severity. Cell Syst. 2022;13:665–681.e664. https://doi.org/10.1016/j.cels.2022.06.006.

Chapter 13
Future Prospective of Omics-System Biology to Control AMR: Recommendations and Directions

Nicholas Bartelo, Saurabh Mishra, Prabhat Ranjan Singh, Murugesh Padmanarayana, and Vijay Soni

Abbreviations

AMR	Antimicrobial resistance
AST	Antibiotic susceptibility test
DNA	Deoxyribonucleic acid
E. coli	Escherichia coli
M. smegmatis	Mycobacterium smegmatis
MIC	Minimum inhibitory concentration
RNA	Ribonucleic acid
RNA-seq	RNA-sequencing
rRNA	Ribosomal RNA
STRING	Search tool for the retrieval of interacting genes/proteins
TCA	Tricarboxylic acid cycle

N. Bartelo
Department of Physiology and Biophysics, Weill Cornell Medicine, Institute for Computational Biomedicine, Englander Institute for Precision Medicine, New York, NY, USA

S. Mishra · P. R. Singh
Department of Microbiology and Immunology, Weill Cornell Medicine, New York, NY, USA

M. Padmanarayana
Department of Biochemistry, Stanford University, Beckman Center for Molecular and Genetic Medicine, Stanford, PA, USA

V. Soni (✉)
Division of Infectious Diseases, Weill Department of Medicine, Weill Cornell Medicine, New York, NY, USA
e-mail: vis2032@med.cornell.edu

V. Soni, A. S. Akhade (eds.), *Antimicrobial Resistance: Factors to Findings*,
https://doi.org/10.1007/978-3-031-65986-7_13

13.1 Introduction

Antimicrobial resistance (AMR) poses a significant global health risk, as microorganisms undergo adaptive changes to withstand the effects of antibiotics and other antimicrobial substances, thereby undermining the efficacy of standard medical treatments. A recent study published in The Lancet (2022) associates bacterial AMR with a substantial toll of 4.95 million global deaths in 2019, of which 1.27 million were directly linked to bacterial AMR [1]. The conventional methods employed to counteract AMR encounter several challenges, notably the excessive use of antimicrobials in healthcare, agriculture, and aquaculture, leading to an expedited development of resistance (Fig. 13.1) [2].

The pharmaceutical industry encounters difficulties in developing new antibiotics due to elevated research and development costs, limited market appeal, and the swift emergence of resistance, resulting in a scarcity of novel antimicrobial agents. Inadequate infection prevention and control measures within healthcare settings, coupled with a deficiency in surveillance systems and the ease of global travel, contribute to the widespread dissemination of resistant strains. Environmental pollution stemming from pharmaceutical production and agricultural runoff amplifies the selection of resistance genes.

The adoption of a one-size-fits-all strategy, particularly the widespread use of broad-spectrum antibiotics, exacerbates the problem by encouraging unnecessary and indiscriminate utilization of potent drugs when milder alternatives may suffice [3]. Effectively addressing these multifaceted challenges demands a comprehensive

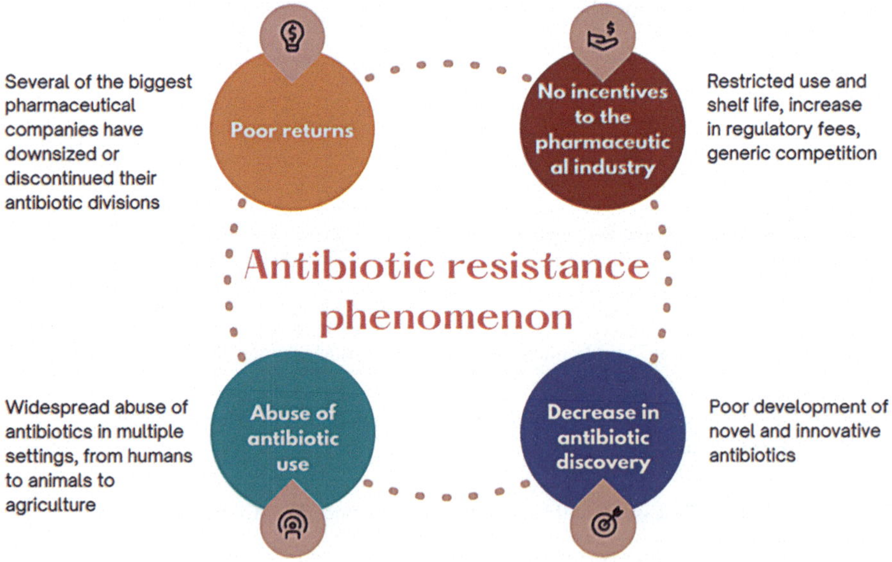

Fig. 13.1 Conceptual illustration of the multivariate issues involved in the phenomenon of antibiotic resistance. (Figure and caption are adopted from Ref. [2])

approach encompassing the development of novel antibiotics, enhanced infection prevention measures, judicious antimicrobial use, and collaborative efforts on a global scale, as advocated by O'Neill in 2016. This holistic strategy is imperative for successfully navigating the intricate landscape of AMR and safeguarding the efficacy of antimicrobial agents for future generations.

Omics-systems biology is well-known for understanding all cellular processes by combining genomics, transcriptomics, proteomics, and metabolomics. Several groundbreaking examples illustrate the application of omics-systems biology, collectively emphasizing its multidisciplinary significance. Suthers and Maranas (2017) showcase the pivotal role of systems biology in understanding microbial metabolism through the reconstruction and analysis of metabolic networks, essential for comprehending complex biological processes [4]. Wang et al. emphasize the revolutionary impact of RNA-seq in transforming transcriptomics, providing comprehensive insights into gene expression patterns [5]. Pioneering works by Kitano [6], Ideker et al. [7], Hood and Perlmutter [8], and Fiehn [9] underscore the historical development and potential of systems biology in deciphering complex biological systems.

The comprehensive exploration of AMR involves cutting-edge approaches across various disciplines. Utilizing omics atlas, researchers employ genomics, proteomics, and metabolomics to unravel the molecular intricacies of AMR. Deep learning methodologies are then applied in interactomics analysis, facilitating the understanding of complex molecular interactions for the development of innovative therapies and diagnostic tools. Single-cell omics emerges as a key player, emphasizing the importance of studying AMR heterogeneity, providing insights into individual cell behavior, uncovering specific therapeutic targets, and enabling real-time monitoring for adaptive intervention.

Recent advances in omics-system biology for AMR have showcased unprecedented achievements in unraveling the molecular basis of resistance, identifying novel drug targets, and predicting the emergence of resistant strains. This chapter provides an overview of key topics related to the application of omics-system biology in combating AMR, including the integration of diverse omics data, the identification of biomarkers for resistance, and the exploration of host-pathogen interactions. By synthesizing information from various omics levels, researchers can unravel the intricate dynamics of AMR and design more effective therapeutic interventions.

Comparative omics, addressing challenges in drug development, employs a systematic pipeline integrating multiple omics data, utilizes machine learning for data analysis, and employs network analysis to identify crucial drug development targets (Fig. 13.2) [10]. However, challenges such as the complexity of biological systems, cost, interpretation hurdles, and ethical considerations must be acknowledged. To overcome these challenges, recommendations include integrating omics-system biology with other disciplines, establishing standardized protocols and databases, promoting collaborative research efforts, and emphasizing education and training for professionals engaged in advanced omics-systems biology approaches. A systems biology approach can predict AMR through microbial evolution experiments,

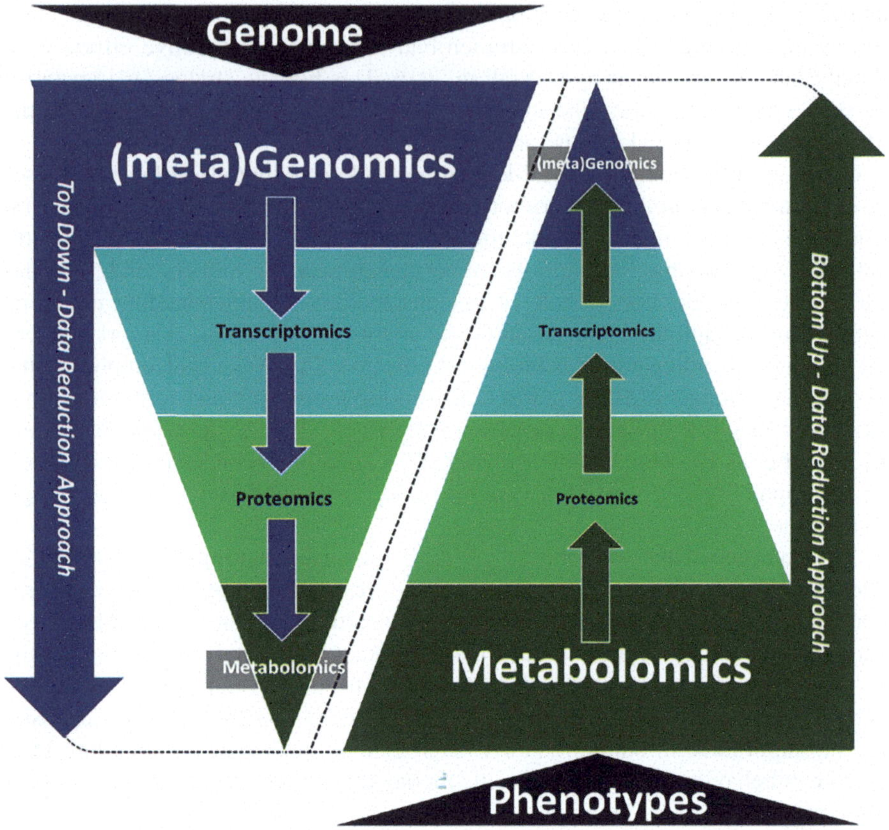

Fig. 13.2 Top-down and bottom-up data reduction integration approaches used in system biology. (Figure and caption are adopted from Ref. [10])

and genomic surveillance helps in the battle against AMR [11, 12]. These initiatives aim to pave the way for a more effective and collaborative approach to tackle AMR.

13.2 Application of Multi-Omics Methods to Investigate Antimicrobial Resistance

The advancement of high-throughput technology used to measure biological molecules from cells and tissues to determine their function in biological processes and disease has allowed for an enhanced ability to probe the mechanisms by which a phenotype of interest develops. This method of investigating different small molecules is termed "omics" and involves the study of DNA (genomics/epigenomics/metagenomics), RNA (transcriptomics), proteins (proteomics), metabolites (metabolomics), lipids (lipidomics), and others. The central dogma of biology states there is a one-way flow of genetic information, from genes to RNA to proteins. By studying these small molecules in a biological system, our understanding of how disease affects patients, how drugs interact with a host, and even how the environment is affected by human actions has increased substantially. As the availability and reliability of omics-measuring technology has increased, the examination of multiple modalities of omics data has become more common, termed "multi-omics." With the measurement of more than one type of small molecule, researchers can begin to understand the connections from DNA to molecules directly related to phenotype. Each modality of omics data has a different process by which the data are measured and analyzed, and a workflow of these differences is shown (Fig. 13.3) [13].

To understand the biological impact of antimicrobial resistance on bacteria, multi-omics approaches have been utilized. During the early years of omics technology development, careful planning of large studies led to discoveries using measurements of these small molecules. In 2013, the Tailored Treatment Project, a prospective clinical study measuring genomic, transcriptomic, proteomic, and clinical data, was launched with goals of increasing antimicrobial therapy, reducing adverse events, and limiting the spread of AMR by developing new insights into

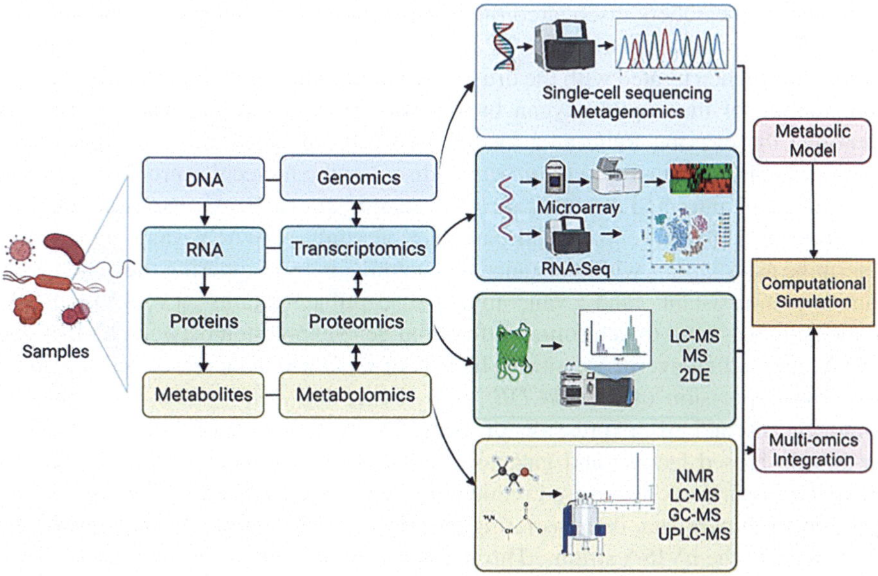

Fig. 13.3 Schematic representation of omics workflow for genomics, transcriptomics, proteomics, and metabolomics approaches. Multi-omics data integration can be used for refining and reconciling modeling predictions to construct computational simulations. RNA-seq: RNA-sequencing; LC-MS: liquid chromatography-mass spectrometry; MS: mass spectrometry; 2DE: two-dimensional electrophoresis; NMR: nuclear magnetic resonance; GC-MS: gas chromatography coupled to mass spectrometry; UPLC-MS: ultra-high-performance liquid chromatography coupled to mass spectrometry. (Figure and caption are adopted from Ref. [13])

host-pathogen interactions and biomarker discovery for patients with lower respiratory tract infections (LRTI) and/or sepsis [14]. After 4 years, the data collection ended in 2017. Since this time, 23 publications from this work and one developed software have been cited on the project page. The achievements stemming from this project include the creation of new tools for rRNA gene sequencing [15], discovery of proteomic biomarkers to discriminate bacteria types [16] and antibiotic resistance profiles [17], and validation of short-read and long-read genome sequencing [18]. The Tailored Treatment Project exemplifies how large studies not only provide answers to questions they originally were proposed to solve, but also generate new hypotheses which lead to a greater understanding of the general topic of interest. In addition, it is possible for these types of endeavors to result in the construction of new tools to best probe the data. These technological advances can often be applied to numerous biological systems, creating discoveries in many domains.

As the measurement of omics data has decreased in both time and expense, smaller collaborations have been able to advance our knowledge of antibiotic response in resistant and susceptible bacteria strains. Jo et al. conducted whole-genome sequencing, proteomics, and targeted metabolomics on methicillin-resistant *Staphylococcus aureus* (MRSA) and methicillin-susceptible *Staphylococcus aureus* (MSSA) [19]. Genetic identicality was verified by using whole-genome sequencing, while biomarker identification was determined by proteomic and metabolomic analyses. These researchers discovered the upregulation of peptidoglycan biosynthesis in oxacillin-treated resistant bacteria, suggesting the mechanism by which the bacteria survive when treated with the drug is by maintaining cell wall integrity through upregulation of the peptidoglycan biosynthetic pathway. Additional experiments identified the enzyme RelQ as a mediator of β-lactam resistance. A multi-omics study of methicillin-resistant *S. aureus* to discover the molecular profile of clinical heterogeneous intermediate resistance to vancomycin (hVISA) was conducted by Castro et al. [20]. A genome-wide association study (GWAS) was conducted to determine associations with resistance to vancomycin by comparing the genomes of vancomycin-resistant and vancomycin-susceptible strains (VSSA). RNA-sequencing was used to determine differential gene expression between hVISA and VSSA, concluding overexpression in hVISA of *sdrD* which increases cellular adhesion, overexpression of *capABCDE*, *esaA*, *essA*, and *esaD* which are related to pathogenesis, and overexpression of *ssaA*. Downregulation of *lacABCDEFG* in hVISA decreased lactose and galactose metabolism. By incorporating metabolomics, the researchers discovered reduction of aromatic acids and amino acids synthesized from pyruvate, demonstrating an alteration of the glycolysis pathway and TCA cycle in the hVISA strains. Through the combination of different omics measurements, the hVISA phenotype was determined to have changes in bacteria cell wall synthesis, increased biofilm formation, expression of virulence factors, and reduction in glycolysis via the TCA cycle. This summary of how the molecular interactions change due to vancomycin resistance demonstrates the great capacity with which multi-omics studies can culminate information from different steps of the central dogma of biology, creating a unified picture of how changes in genetics lead to phenotype. Another important outcome of this approach is the ability to find a multitude of biomarkers of resistance and therapeutic targets that could be perturbed to destroy resistant bacteria.

The use of in silico models has complemented biological experiments by providing an expectation for how the biology will evolve. This allows for the reduction of possibilities for testing in laboratory trials, reducing both cost and time spent. Zhu et al. created a genome-scale metabolic model (GSMM) to probe the metabolic changes resulting from polymyxin treatment, a last-resort class of antibiotics, on *Pseudomonas aeruginosa* [21]. The model consists of 3022 metabolites, 4265 reactions, and 1458 genes. Simulations with lipid A modifications associated with polymyxin resistance suggested changes in the outer membrane with only a small contrast in metabolism and *P. aeruginosa* growth. Future experiments should focus on biological validation to increase the confidence in the model predictions. Another in silico model, produced by Hanafin et al., incorporated transcriptomics and metabolomics profiles of polymyxin B and chloramphenicol monotherapy- and combination therapy-treated *Klebsiella pneumoniae* isolates [22]. The mechanism-based model was created with the goal of predicting the pharmacodynamics, i.e., the bacterial proliferation, after the introduction of one or more drugs. The model uses a partial least squares regression analysis to discover which genes and/or metabolites are predictive of the pharmacodynamics of drug-treated bacteria. Using the predictive model, the authors concluded a decrease in uridine diphosphate glucose utilization, which was observed in previous experiments, using combination therapy of polymyxin B and chloramphenicol, a process ultimately leading to bacteria proliferation. Overall, L-Ara4N modification of lipid A was determined to be the driver for the synergistic effect demonstrated through combination therapy. While this model made predictions in accordance with what was observed from previous experiments, validation of the model architecture must be performed in additional biological systems. However, this study reveals the power of a multi-omics approach to predict the outcome of bacterial proliferation when treated with different drugs. In addition, pathway analysis of differential genes and metabolites can lead to the development of new combination therapies. If the researchers determine a combination therapy elicits a certain response which makes the resistant bacteria susceptible to killing by another known drug, a new combination can be made to efficiently kill the bacteria. Through this method, it may be possible to synthesize targeted therapeutics for specific patients based on how the bacteria infecting them respond to different drug combinations. However, to do this, there must be a descriptive database made available for clinicians to successfully and quickly determine the optimal drug combinations for the tested bacteria.

The spread of antimicrobial resistance is not limited to humans, and multi-omics approaches have revealed its impact on the environment. Current farming techniques have created environments in which disease is spread easily and rapidly between animals. In livestock, methicillin-resistant *S. aureus* causes infection in those who consume infected animals. Through a multi-omics method utilizing whole-genome sequencing, RNA-sequencing, and gel-based comparative proteomics, samples from humans and a wild boar determined similarities and differences in differentially expressed genes and proteins [23]. Transcriptomics detected the most highly expressed gene in the wild boar strain to encode RNAIII, which was consistent with one of the two human strains of focus. A large difference between human and wild boar strains was that proteins involving virulence were much more abundant in wild boar. This implies there could be an increased chance for the bacteria to spread to humans. Another livestock-related pathway in which antimicrobial resistance is proliferating is through deposits of antibiotic resistance genes in manure. A meta-omics study by Zhang et al. conducted metagenomics and metaproteomics to compare three management strategies of manure treatment: aerobic composting, mechanical drying, and precipitation [24]. Comparison of remaining antibiotic resistance genes after treatments revealed only 192 and 572 total antibiotic resistance genes in winter and summer after aerobic composting, respectively, whereas mechanical drying and precipitation had over 10,000 antibiotic resistance genes remaining after treatment. Focusing on the aerobic composting manure, the aadA1 gene was the most dominant resistance gene. Metaproteomics identified known proteins in composting samples that are responsible for maintaining cell survival when antibiotics are introduced to bacteria. Pathway analysis of 601 proteins for composting samples observed the greatest number in the metabolic pathway, the microbial metabolism in diverse environments, the biosynthesis of secondary metabolites, and the biosynthesis of antibiotics. The authors conclude the last pathway portrays the selection pressure of the antibiotic resistance genes. As the metabolome was greatly perturbed, shown from the results of the pathway analysis, the incorporation of metabolomics to this study may reveal important biomarkers for the mechanism by which aerobic composting results in much fewer antibiotic resistance genes and how other treatments do not filter many of these genes.

While antibiotics are added to feed to prevent the growth of infection, some medicated feed is released into the surrounding environment and contacts nearby bacteria, which then develop resistance [25]. Additionally, manure can also be transported to different regions via rainwater. The extent to which antimicrobial resistance populates European freshwater lakes was determined by Spänig et al. [26]. Samples from 274 lakes were subjected to 16S rRNA amplicon sequencing to identify taxonomy followed by metagenomics analysis. These methods were used to characterize the resistome of each lake by counting the number of gene fragments traditionally coding for drug resistance. Genes encoding for antimicrobial resistance were commonly found for medication classes tetracyclines, cephalosporins, and quinolones. Although sulfonamides were present, fragments mapped to genes encoding sulfonamide resistance were in lower abundance than the other drugs. This study is the first to map the location of antibiotic resistance genes across a large area. The use of maps to visualize the distributions of resistant genes clearly demonstrates the most problematic medications and regions which can be targeted in the future for the removal of pathogens. In addition, the method by which these lakes were infected with antibiotic resistance genes can be investigated further, and better practices can be developed, as well as legislation passed, to mitigate these effects. These studies illuminate the increase in antimicrobial resistance in different environmental setting due to human actions. Even though multi-omics research on our food and water supply aids in the determination of biomarkers driving phenotypic change which will ultimately effect humans, there have not been many published works focused on antimicrobial resistance in animals or the environment. As this issue rapidly increases, projects involving multi-omics will multiply to develop a better understanding of how the bacterial strains develop and spread, leading to the creation of new antibiotics and policies to diminish potential harm to humans.

Due to the achievements resulting from analyzing omics, the creation of an omics atlas for many different biological systems has been designed to determine how information is passed throughout the given system [27–30]. The outcome is an experimentally established insight into the systematic changes for other researchers to compare their results with and build from to culminate a greater understanding of the actions leading to the phenotype. Imwattana et al. created a genomic atlas for multiple clades of *Clostridioides difficile* from different continents [31]. Prevalence of antimicrobial resistance genes for important antimicrobial classes conferred mainly by accessory antimicrobial resistance genes was reported. Australia had a much lower prevalence than the other continents studied: North America, Asia, and Europe. These researchers further identified which clades composed the majority of resistant strains, demonstrating a clear specificity of clades to antimicrobial classes. While other research focuses have been able to create encompassing multi-omics atlases, antimicrobial resistance has not yet achieved the same overarching success. A review by Chernov et al. discusses genomics, transcriptomics, proteomics, and metabolomics of antimicrobial resistance [32]. Although not an omics atlas, this publication provides a reference for new studies to use in order to create a framework for the development of an omics atlas focused on antimicrobial resistance. The authors include a section that describes integrating omics to develop new antimicrobials, in which they state the most important obstacle to drug discovery is the difficulty of specifying bacterial cell targets. Without confidence that the biomarkers are coming from bacteria, targeted therapeutics are unable to be synthesized. As multi-omics investigations of antimicrobial resistance in different species are continually conducted, it may be possible to pool the knowledge learned from these projects to manufacture a multi-omics atlas that will be used to compare future experiments to, as well as act as a tool for researchers to access in order to find novel biomarkers of resistance.

13.3 The Potential and Benefits of Single-Cell Omics to Study AMR Heterogeneity

The development of single-cell omics approaches for studying different aspects of bacterial physiology has been remarkable [33]. As compared to bulk omics studies, which are unable to distinguish bacterial heterogeneity, single-cell omics approaches can identify small but clinically relevant persister populations. BacDrop, a droplet-based technology for bacterial single-cell RNA-seq, has been recently shown to elucidate heterogeneous responses of *Klebsiella pneumoniae* clinical isolates to antibiotic stress and can be applied to thousands to millions of cells from both Gram-negative and Gram-positive species [34]. Single-cell transcriptomics using an improved poly(A)-independent sequencing protocol captured growth-dependent gene expression patterns in individual *Salmonella* and *Pseudomonas* bacteria across all RNA classes and genomic regions [35]. Such technological advancements have the potential to address transcriptional heterogeneity in persisters that resist antibiotic treatment.

13.3.1 Advances in Single-Cell Omics Technology

In the past decade, there have been a lot of improvements in classical methods, leading to the development of single-cell techniques that can be applied to study phenotypic heterogeneity and bacterial persisters [36]. These belong to different experimental approaches like fluorescent probes, microscopy, mass spectrometry, Raman spectroscopy, and next-generation sequencing (NGS) [36, 37]. In this section, we will provide a brief overview of these methods and their applications:

13.3.1.1 Fluorescent-Based Methods

Fluorescence-based studies have been extensively used for studying heterogeneity at the single-cell level using a wide variety of techniques, such as flow cytometry or microscopy, using transcriptional fluorescent protein (FP) fusion constructs [38, 39]. Förster resonance energy transfer (FRET), which is a protein-FP fusion technique, has been applied to analyze phenotypic heterogeneity in chemotaxis signaling in *E. coli* [40]. Small-molecule fluorescent probes such as Hoechst 33342 demonstrated at the single-cell level that a higher genome copy number correlates with increased persistence to levofloxacin [41]. Fluorescent-activated cell sorting (FACS) has emerged as a powerful technique because of its ability to resolve and isolate single cells for further downstream analysis [42, 43]. Imaging flow cytometry (IFC), which combines flow cytometry with digital microscopy, has been applied for the characterization and quantification of bacterial persistence and resistance [44–46]. Fluorescent microscopy, in conjunction with live cell imaging, has also been extensively used for examining morphological and physiological heterogeneity at the single-cell level [47–55].

13.3.1.2 Mass Spectrometry

Mass spectrometry is a powerful technique used for determining the cellular composition via metabolomic and proteomic analysis. Specialized variations like nanoscale secondary-ion mass spectrometry (NanoSIMS) in combination with flow cell sorting have been developed that use isotopically labeled nutrient sources, antibiotics, or other substrates to analyze various single-cell populations [56] and can also be applied for studying AMR heterogeneity. Like the principle of Imaging flow cytometry, high-resolution mass spectrometry can also be combined with imaging techniques to study the spatial distribution of analytes within single cells. For example, nanoscale imaging (using cluster ToFSIMS) has been used to chemically map small molecules and antibiotics (tetracycline and ampicillin) within the *E. coli* cells [57].

13.3.1.3 Raman Spectroscopy

Raman spectroscopy could be used for single-cell identification based on the distinct spectral profile of each species [58, 59]. Application of "single-cell Raman spectroscopy" (SCRS) which employs a homochromous laser as a light source has been used for the investigation of persisters in *E. coli* upon ampicillin treatment [60]. Its nonintrusive nature allows the identification of the biochemical characteristics of a single cell and the profiling of antibiotic resistance in diverse bacterial species [61–65].

13.3.1.4 Next-Generation Sequencing

Advancements in the field of next-generation sequencing probably have most prominently revolutionized the study of single cells [66–68]. In the last few years, different versions of this technology have been applied to study the transcriptomes of many bacterial species [34, 35, 69–78]. Single-cell sequencing has also revealed heterogeneity in AMR [34, 79–82]. Single-gene targeted nanopore sequencing can reveal cell-to-cell copy number heterogeneity, which can drive resistance during antibiotic exposure [83]. Single-cell genomics can also reveal the presence of uncultured bacteria in the intestinal microbiome and the presence of antibiotic resistance modules [84]. Single-cell RNA-sequencing methods like scDual-Seq simultaneously capture both host and pathogen transcriptomes [85].

13.3.2 Applications of Single-Cell Omics in Identifying New Targets for Therapies

Due to recent advances in bacterial single-cell omics studies, researchers have just begun to apply these strategies to identify new targets for therapy. One such example is the application of a single-cell genomic sequencing (SiC-seq) platform to analyze the distributions of antibiotic-resistance genes, virulence factors, and phage sequences in microbial communities from an environmental sample [86]. FACS sorting combined with NGS has revealed novel mechanisms responsible for redox heterogeneity and drug tolerance in *Mycobacterium tuberculosis* [43, 87]. A single-cell analysis of *E. coli* was used to develop a method to examine the antimicrobial and bactericidal activities of antimicrobial peptides [88]. Phenome-genome-combined strategy called single-cell identification, viability and vitality tests, and source tracking (SCIVVS) integrates Raman micro-spectroscopy and single-cell sorting, lysis, and DNA amplification in a Raman-activated gravity-driven cell encapsulation (RAGE) chip which allows direct coupling of SCRS-based, noninvasive metabolic phenome profiling to downstream single-cell genome sequencing [89].

13.3.3 Real-Time Monitoring of AMR Evolution

The application of different omics technologies to track antimicrobial resistance is well documented [90–97]. Although the application of these technologies at the single-cell level is still under development, significant advancements have been made in using them for monitoring AMR evolution. For example. a novel single-cell nanoscale technique, atomic force microscopy-infrared spectroscopy (AFM-IR), combined with chemometric analysis, was used to study changes in the chemical composition of clinical isolates of vancomycin and daptomycin-resistant strains of *S. aureus* [98]. Antimicrobial susceptibility testing systems based on single-cell imaging have been used for the determination of urinary tract infections (UTIs) caused by resistant bacteria [99, 100]. It can also be used to detect bacterial pathogens in urine, blood cultures, and whole blood and to analyze polymicrobial samples [101, 102]. A microfluidic agarose channel (MAC) system determines MICs by single-bacterial time-lapse imaging under different antibiotic culture conditions [103]. Droplet-based high-throughput cultivation from stool samples revealed 21 populations that evaded detection in plate-based assessments of antibiotic resistance [104]. The Raman-assisted antibiotic susceptibility test (FRAST) was used to analyze four Gram-negative and two Gram-positive bacteria strains in response to 38 antibiotics, and the results showed 88.0% agreement between the FRAST and the conventional AST assay [105]. Single-cell Raman with isotope labeling was used to profile active antibiotic-resistant bacteria (ARB) inhabiting different soils [106]. Application of sensitive NGS technologies for antimicrobial susceptibility testing at the single-cell level can be used to improve culture-based methods [107, 108].

13.3.4 Challenges and Limitations of Single-Cell Omics in AMR Research

Developments in the field of single-cell omics in AMR research are limited by several factors, the most prominent being the smaller size compared to the eukaryotic cell, which requires the method of detection to be of higher sensitivity [109]. Other factors that pose challenges are difficulty in permeabilization and lysis, bacterial cultures being prone to contamination, a low signal-to-noise ratio, higher costs, etc. [110]. One additional disadvantage associated with the study of AMR research is the pathogenic nature of the microorganisms being studied and their low abundance in various biological samples, which requires the presence of sophisticated biological facilities.

13.3.5 Combining Single-Cell Omics with Other Approaches for a More Comprehensive Understanding of AMR

Advancement in the field of engineering at the microscale has given rise to technologies like microfluidics, which have made it easier to study single-bacterial cells [111–113]. Gel-based microdroplets have also been developed for capturing single cells [114–116]. Cells in these entrapped structures can then be processed for targeted amplification of genes and barcode the individual genomes [86, 117]. For the amplification of the femtogram-range genome of individual microbial cells, multiple displacement amplification (MDA) methods are most commonly used, which improves genome recovery [118–120]. Other methods used include multiple annealing and looping-based amplification cycles (MALBAC) and the PicoPLEX single-cell whole-genome amplification kit (NEB-WGA) [121]. Droplet microfluidics in combination with fluorogenic peptide nucleic acid (PNA) probes have been applied as a diagnostic tool for combating antimicrobial resistance [122]. Single-cell-level microfluidic visualization has also been applied to *M. smegmatis* subjected to drug treatment regimens during continuous culture [123].

13.4 Comparative Omics and Phylogeny Show Specificity of Bacterium Types

Most of the omics applications we have discussed have been involved with finding differential gene expression, protein expression, or abundances of molecules between two phenotypes. An additional application of omics data is through comparative genomics, which uses genomic information to determine the evolutionary conservation between two organisms. Genome sequences are aligned between organisms to determine the extent to which the genomes overlap, which is then used to predict genome and molecular evolution. Norsigian et al. compared the genomes of 22 *K. pneumoniae* strains, discovering a core genome of 3965 shared open reading frames across all strains, as well as 2805 unique genes attributable to only one strain [124]. The resistome of strains was investigated, encompassing 122 antimicrobial resistance genes or mutations determined, 12 of which were common to all strains and 35 of which were unique to a single strain. Focused analysis of five strains is shown below (Fig. 13.4) [124]. The phylogenetic tree shown in panel (a)

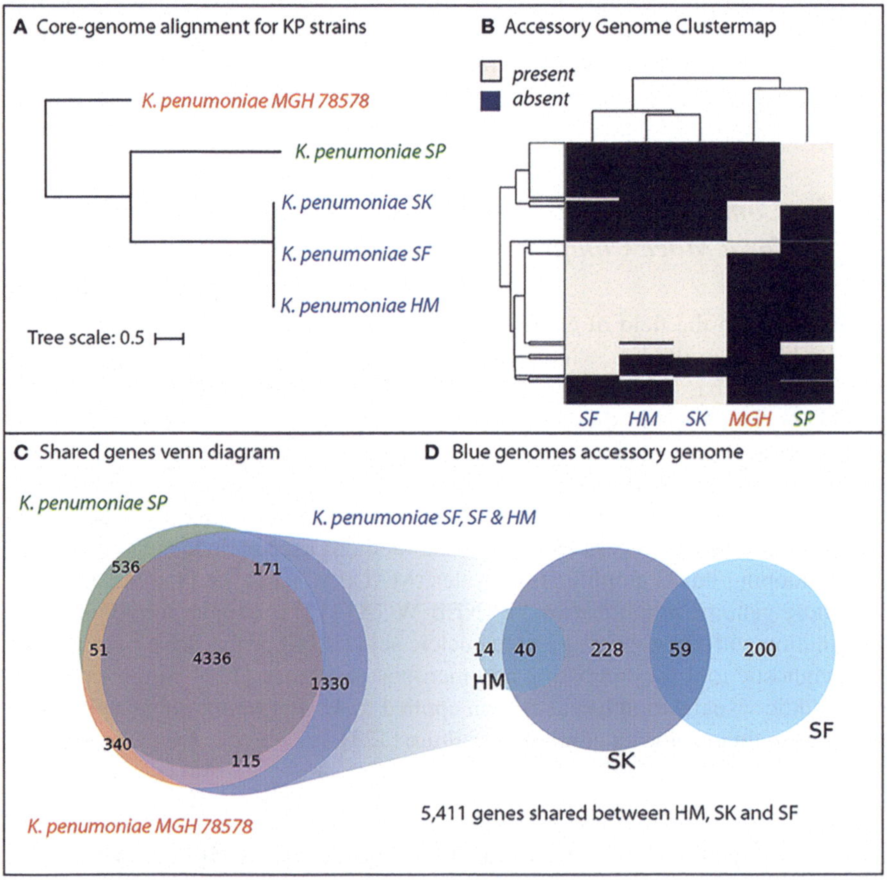

Fig. 13.4 Genomic analyses of five *K. pneumoniae* isolates including four isolated in Cairo, Egypt (SP, SK, SF, HM). (**a**) Core-genome phylogenetic tree demonstrates that three of the *K. pneumoniae* isolates (SK, SF, HM) are most similar to each other, with SP and the model strain MGH78578 more distantly related. (**b**) Hierarchical clustering of the accessory genome of the five strains demonstrates that the three closely related KP strains also share the most genes. (**c**) The core genome of all five strains is composed of 4366 genes shared by all five strains. (**d**) The three strains, SD, SK, and HM possess an additional 1330 unique genes not shared by SP or MGH78579. (Figure and caption are adopted from Ref. [124])

describes the similarity between the strains, with each branch of the tree representing a greater difference between two strains. The clustermap in panel (b) portrays the genomic structure of different accessory genes across the strains, revealing clear differences in genetic makeup determined by a clustering algorithm. The Venn diagrams in panels (c) and (d) visualize the overlap of the three groups' output by the phylogenetic tree, illustrating a quantitative representation of the genes shared and unique to the strains. These three types of plots are common to comparative genomic analyses and are therefore important to comprehend to gain an understanding of how to interpret the results of this method.

Importantly, this analysis suggests there could be unique methods of resistance development from genes not shared by strains and has identified the unique genes in each strain for potential future targeting. In addition, this work has developed a common gene profile that could potentially be used to target multiple strains if the resistance phenotypes in the future arise from shared genes.

A comparative genomics profiling of Elizabethkingia was conducted by Burnard et al. for 128 strains to determine closest ancestors [125]. The creation of a phylogeny using 127,236 SNPs allowed for the discovery of which strains were more similar evolutionarily. The authors state that susceptibility to antimicrobial therapy is clearly strain dependent, suggesting that it is possible that therapeutics created for one strain could be reliable for related strains with resistance, while prescribing patients the therapeutic with disparate strains from the same species could result in little to no response. For the best-targeted therapy for a patient, it may be necessary to identify the strain of bacteria and compare its position in a phylogenetic tree with other strains to choose the optimal drug to kill the resistant bacteria. To show the effectiveness of comparative genomics in the clinic, Ye et al. published a first-of-a-kind case study in which the researchers isolated two multidrug-resistant *Citrobacter freundii* isolates from the same patient with bloodstream and urinary tract infections [126]. Comparative genomics analyses revealed a high overlap between the genomes of the two isolates, with only five single nucleotide polymorphisms measured to be unmatched. This study demonstrates the ability of comparative genomics to be used as a tool for determining how similar bacterial infections are within a patient with more than one infection. If the isolates were discovered to have a large genetic variability between them, the administration of multiple drugs may have been necessary for the precise treatment of each infection. If there are multiple infections that each need a specific drug to kill the resistant bacteria efficiently, important considerations regarding toxicity and targeting of bacteria must be discussed to ensure patient safety. Identification of strains in patients with multiple infections can help to advance the precision medicine field.

Interactomics is the study of functional interactions of proteins at the molecular level within a cell or between multiple cells. Extensive networks are created to determine potential exchanges in the biological system, often with edges between nodes of the network weighted with data that directly influence the phenotype of the organism, such as gene expression. One of the most widely used databases harboring protein-protein interactions is the STRING database [127]. Numerous functions are served by STRING including prediction of protein-protein interactions, enrichment analysis, and clustering of the association network. By developing interactome networks for drug-resistant bacteria, researchers have determined novel biomarkers of resistance development. Raman et al. engineered a genome-scale protein-protein interaction network for *M. tuberculosis* H37Rv using the STRING database [128]. The edges are weighted by gene expression and centrality, i.e., how integral they are to connect a path of nodes. The results determined paths to son of sevenless (SOS) proteins were the optimal scorers, followed by paths to cytochromes, while horizontal gene transfer and pumps ranked lower. Examples of a path for each of these four resistance classes are visualized below (Fig. 13.5) [128]. Through inspection of the interactome, the authors noticed upregulated pathways for resistant bacteria and concluded that targeting *Rv2737c* or *Rv1547* or *Rv0823c* together with *Rv0824c* could be a strategy to counter emergence of drug resistance for *M. tuberculosis* H37Rv. The reason these would be good candidates for targeting is because they are nodes (in the upper left-hand quadrant of Fig. 13.5 [128]) which

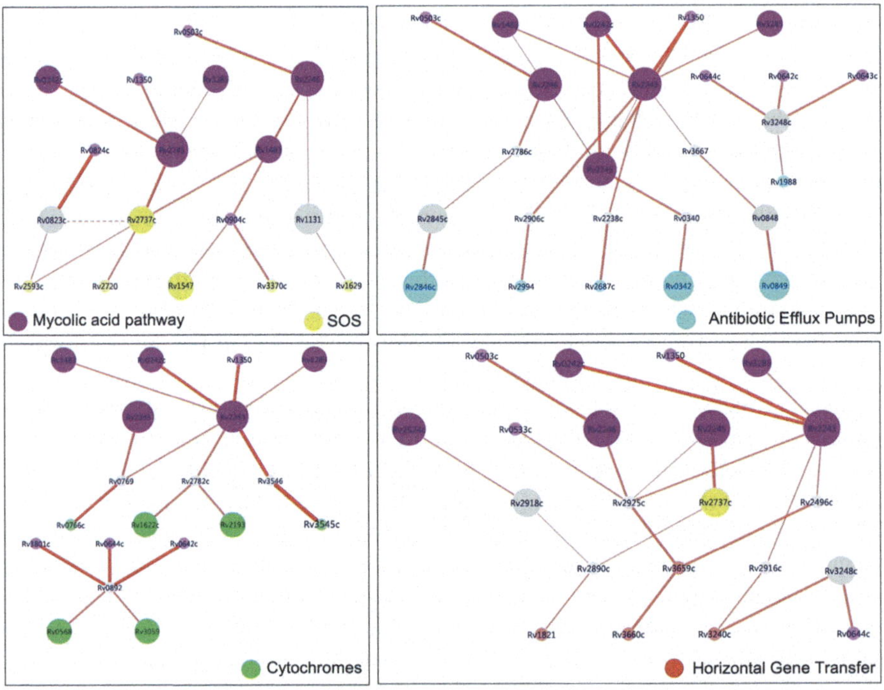

Fig. 13.5 Top-scoring paths from MAP to each of the four resistance classes. Nodes are labeled by their Rv IDs, as obtained from TubercuList. Nodes correspond to the individual proteins in the network while the edges indicate interactions between them. Each class of nodes is colored differently as indicated. Gray nodes indicate those that do not belong to any of the marked classes. The nodes are sized in proportion to the number of MAP drugs that induce its upregulation. The thickness of an edge is proportional to the number of times a shortest path is traversed through that edge. The dotted edge is not a high-scoring path but is of significance, as discussed in the text. (Figure and caption are adopted from Ref. [128])

either provide integral connections to dysregulated proteins or themselves are dysregulated proteins.

Hwang et al. created an interactome called PseudomonasNet containing 203,118 edges between 5456 genes with nine total networks combined [129]. Pseudomonas aeruginosa gene pairs with shared annotations from the gene ontology biological process database were used to benchmark co-functional gene pairs. Genes differentially expressed after ceftazidime treatment were queried in the network and results showed predicted genes related to ceftazidime resistance. Four predicted genes were verified to associate with ceftazidime resistance using a disk diffusion assay, demonstrating the network's ability to discover biomarkers of resistant *P. aeruginosa*. In addition, a network of antibiotic-resistant genes against multiple drugs using PseudomonasNet was constructed to probe antibiotic resistance. The authors concluded a significant percentage of genes can switch their interaction bias toward the same direction of mutational effect between different drug conditions, thus demonstrating a specificity of genes to various antibiotics. Lee et al. developed KlebNet, a co-functional network created using 13 different networks, which focused on *K. pneumoniae* strain HS11286 and was composed of 5316 coding genes with 160,450 total edges [130]. The network was able to prioritize 15 candidate genes predicted to be involved with carbapenem resistance using the *K. pneumoniae* strain with acquired carbapenem resistance. Taking 15 mutated genes with significant expression changes in the local subnetworks of KlebNet within the carbapenem-resistant strain, four genes were successfully validated in vivo to have increased virulence. These studies demonstrate the ability of the interactome of resistant bacteria to be utilized to discover new biomarkers that aide in the development of resistance. These models also allow researchers to predict genes that are the most likely candidates to be mutated based on current resistance genes, therefore providing a method by which therapeutics may be developed in advance of the resistant strain's spread. In addition, by developing networks for different bacteria species, it may be possible that there is overlap between predicted targets involved with the evolution of resistance, creating drug targets that can be used for multiple bacteria species.

13.5 Therapeutic Discovery Pipeline and Database Curation

While there are many applications of omics data, one of the most common goals is to create therapies. The era of precision medicine, decreasing cost and time taken to measure patient data, and increased use of statistical approaches that can handle large amounts of data such as machine learning and deep learning has created an exciting environment in healthcare in which many researchers are focused on developing drugs that work optimally for a patient based on his/her specific omics profiles. A precision medicine style of prescribing treatments to patients in the field of antimicrobial resistance is logical due to the harm inflicted on a patient if the incorrect antibiotic is taken, resulting from the bacterial resistance to the drug which will not stop the infection. Nicolás et al. published a chapter discussing the integration of omics to aide in the development of therapies for pathogenic bacteria, with an emphasis on drug resistance [131]. The protocol of the traditional route to drug discovery, involving the high-throughput screening against fragment libraries, is suggested to be enhanced by the in silico analyses added to the traditional route which involves the filtering of drug candidates to those which have a genome-derived response to an antagonist. The traditional and in silico-powered therapeutic pipelines are shown below (Fig. 13.6).

Different methods of discovering lead candidates for therapies are discussed including graph-based approaches to model static representations of metabolism, genome-scale network reconstruction to organize the full metabolic repertoire of the sample, flux balance analysis to optimize the metabolic formulation of a metabolite, transcriptional regulatory modeling to understand the flow of genetic information through transcription and translation, and interactions used to develop an overall idea of a genome-scale model. Each of these methods provides targets based on a different objective. Integration of the potential small molecules determined to negatively impact the pathogen for multiple methods may lead to the reduction in time and cost needed to develop novel therapies. The authors provide applications of combining multiple layers of omics data to create genome-scale models which include genetic knowledge, molecular information including structural, functional, and cellular processes, metabolic context, and expression data to prioritize relevant proteins, to demonstrate how to find lead targets for new antimicrobials.

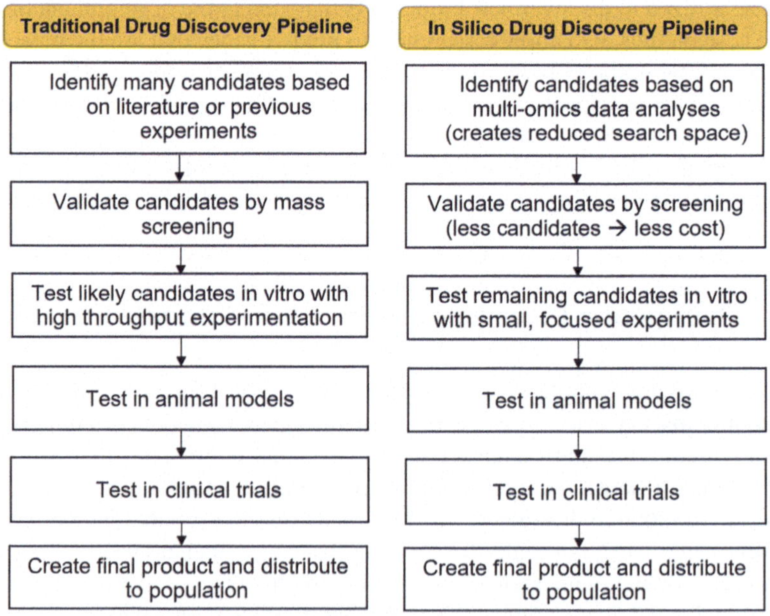

Fig. 13.6 Difference between traditional drug discovery and in silico drug discovery pipelines demonstrating multi-omics reduces the initial targets needed to be screened

Omics technologies can aide in the construction of new types of therapies to counteract the spread of antimicrobial resistance. Martin-Galiano and McConnell wrote a review in which the authors suggest to develop monoclonal antibodies as a new cure for bacterial infections due to their lack of susceptibility to current bacterial resistance mechanisms, facilitation of immune-mediated clearance of bacterial pathogens, ability to interact with only their target thereby not affecting other bacteria in the surrounding area, previously studied safety and efficiency, and passive immunization [132]. The development of and thorough analysis of large omics datasets focused on the study of monoclonal antibody epitopes could provide an organized list of the epitopes grouped according to attributes that can be used to identify optimal candidates for different therapeutic applications. In addition, once a database is curated with features describing the monoclonal antibody epitopes, it may be advantageous to use combinations of the epitopes. The reasons for this include the possible increase in the rate of success of neutralization of a process resulting from the reduction of the network tolerance of the resistant bacteria, decreased probability of immune evasion due to the use of several concerted epitope switches, and stage-dependent designs of the therapy that could be used to block the progression of the bacteria based on its current virulence.

13.5.1 Prediction of Drug Targets Using Protein-Protein Interaction Networks

Protein-protein interactions have been utilized to determine drug targets for resistant bacteria by revealing the connections between the host and pathogen with the goal of creating therapies to inhibit the proteins responsible for the crosstalk. Uddin et al. constructed a network of host-pathogen protein-protein interactions between methicillin-resistant *S. aureus* and humans for protein homologs using PSI-BLAST [133]. Predictions from the network concluded that histone deacetylases with gene identification number 49483975 were the protein with the greatest number of interactions, making it the best drug target. This protein's structure was built using the SWISS-MODEL server, and analyses using a Ramachandran plot identified the regions in which binding was favorable, while docking simulation using AutoDock suggested the most likely molecules for binding determined by favorable energetic reactions. This study provides an example of the process by which a novel therapeutic target can be predicted using protein-protein interactions. Further experiments probing the biological system in the laboratory to validate the in silico conclusions should be performed.

To predict drug targets of *P. aeruginosa*, Zhang et al. used a random forest machine learning algorithm to build a protein-protein interaction network by integrating genomic features of proteins and genes, encompassing 54,107 interactions among 4181 proteins [134]. The interactome network was validated using 3343 proteins and 19,416 interactions known to exist. A total of 28 proteins were determined to be good candidates for drug targets if they possessed a relatively well-elucidated function, lack of close homologs in humans to reduce toxicity, and easy accessibility by drugs such as those in the cytoplasmic membrane, periplasm, or outer membrane. Constructing a *P. aeruginosa*-human interactome revealed the presence of three proteins previously identified as strong candidates for drug targets within the study, providing extra priority for the testing of their efficacy in the biological model. By finding proteins that are continually predicted to be important to the interactomes, a greater confidence for the use of specific proteins can be argued for. However, in silico research must always be carefully interpreted, as sometimes further experimentation finds no direct relationship between the actual biology and that expected from the predictive model.

Multiple protein-protein interaction models for *M. tuberculosis* H37Rv were created by Mei, constructed based on different datasets, namely STRING, two-hybrid alone, copurification and two-hybrid together, and copurification alone, which were compared based on the area under the receiver operating characteristic of L2-regularized logistic regression for a classification model between *M. tuberculosis* H37Rv protein pairs and random protein pairs [135]. The final interactome was visualized to determine potential drug targets, defined by their number of connections, which when inhibited could disconnect essential pathways and stop communication systems within the bacteria. Good protein candidates for drug targets included Rv2264c, Rv1023, Rv2857c, Rv2737c, and Rv1344, each having greater than 400 edges to other proteins. Mei also showed known resistance genes had large network degrees. This could be worrisome because as more proteins that are central to pathways become resistant, the ability to identify targets likely to have a large impact on the resistant bacteria may become more difficult. As more data are collected for different types of resistant bacteria, protein-protein interaction networks may have enhanced predictive power, and there may emerge a common pathway of resistance among multiple bacteria species which could lead to the creation of a multipurpose therapeutic.

13.5.2 Drug Combinations Can Be Developed to Kill Resistant Bacteria

An exciting approach to counter antibiotic resistance is the development of new therapies involving drug combinations from those that are currently used. The goal of this work is to determine two therapeutics that, when combined, act synergistically to kill the bacteria. While it may be known how the bacteria will react to each drug alone, it is often difficult to correctly predict how the system will respond to a combination of drugs. Maifiah et al. investigated combination therapy treating *Acinetobacter baumannii* with colistin and doripenem [136]. After testing each therapy separately and then combined at times of 15 min, 1 h, and 4 h, the combined therapy exhibited the greatest number of significantly changed metabolites at the first two time points, with 31 metabolites not altered by either of the individual treatments suggesting a synergistic response. The doripenem alone was characterized by the most significantly different changed metabolites at the last time point, suggesting a time-dependent effect. Analysis of which lipids were significantly perturbed was conducted. For colistin at 15 min and 1 h, mostly membrane lipids were significantly different dominated by glycerophospholipids and fatty acids. For doripenem at 4 h, results revealed substantial perturbation of cellular lipids dominated by fatty acids. For combination therapy, glycerophospholipids were significantly changed at all time points, while only few fatty acids were affected, suggesting less of an overlap with doripenem perturbations than with colistin perturbations. Significant reduction in the levels of three pentose phosphate pathway metabolites, D-ribose 5-phosphate, D-sedoheptulose 7-phosphate, and D-erythrose 4-phosphate, key precursors for biosynthesis of nucleotides, lipopolysaccharides (LPS), and aromatic amino acids, was observed across all time points for the combination therapy. However, these metabolites were only depleted in colistin monotherapy at 15 min and 1 h and only depleted in doripenem monotherapy at 4 h. This suggests that the effects of both colistin and doripenem therapies alone were also shown by the combination therapy. This investigation of synergistic effects reveals that while aspects of the individual treatments are recapitulated by the combination therapy, there are additional perturbations resulting from administering both drugs at the same time, illustrated by the increased number of significant metabolites in the first two time points.

13.6 Challenges and Limitations of Omics-System Biology Approaches to AMR Control

While there are numerous applications through which the incorporation of omics data has been shown to improve our understanding of the biological mechanisms resulting in antimicrobial resistance, there are many challenges that need to be overcome to implement omics data on a large scale. A constant issue in the biomedical field is the cost incurred to conduct research and to care for patients. In 2015, Ventola reported that the attempt to synthesize new antibiotics was abandoned by 15 of the largest 18 pharmaceutical companies due to economic and regulatory obstacles, and academic research funding for antibiotics was reduced due to economic crisis [3]. While the price for genomic sequencing has decreased as technological advances were made, it is still expensive to measure and analyze large numbers of samples for other omics data [137]. As technological advancement and increases in competition between companies lead to the decrease in cost of equipment, there may be more researchers who will have enough funding to incorporate multi-omics into their labs. Price reduction of laboratory instruments may also lead to the development of core facilities, where researchers can bring their samples to be measured, and the core can assist with data analysis, allowing for access to omics measurements otherwise unattainable. An additional challenge to the large-scale deployment of new omics technology in the clinic is the difficulty to estimate the cost-effectiveness due to uncertainties in the data used for current approximations of cost [138]. As this issue has been discussed mostly for genomics due to this type of omics being the most developed, it is important to learn from these trends and use the knowledge to correctly analyze the cost-effectiveness of new omics technologies when they begin to be tested in the clinic at a higher rate.

Another challenge of omics-based projects is that the choice of method to process the data at each step of the analysis, including extraction methods and reagents, sequencing technology used, processing of data, and statistical analysis tools used, may lead to variable results [138]. For example, the platform used for whole-genome sequencing can impact the phylogenies inferred from the experiment [139]. The lack of standardization among omics methods leads to difficulty in assessing the quality of the results, thereby slowing any implementation that could be helpful to patients. To overcome this issue, it is necessary for antimicrobial resistance experts in both academic research and the clinic to come together to create a set of guidelines for methods in the laboratory in academia and discuss how to make these feasible in a clinical setting.

Two of the main goals of AMR research are to diagnose resistant bacteria and determine the correct therapeutic to eradicate a specific strain of bacteria in a patient. These processes are time-sensitive, as the spread of the pathogen can be lethal. Therefore, the incorporation of omics in the clinic to solve these problems must be developed such that the results are quickly given. This poses a difficult challenge to the technological innovators creating diagnostic omics tools, as newly integrated tools have an additional constraint that they will likely need to give a recommendation faster than current protocols. To construct a useful omics application in the clinic, only a small number of biomarkers must be measured to formulate the guidance for the healthcare professional. One reason for this is because there must be a clear interpretation of the patient's data which will be used to make the decision and which the clinician will likely need to discuss with the patient to help with the patient's understanding of why they are receiving a specific prescription. Additionally, newly proposed omics devices for the clinic must have a minimal number of steps taken to achieve a recommendation. By measuring only a few biomarkers, it is more likely that there can be a faster analysis and simpler explanation for how the final result was synthesized. These limitations may provide large constraints on the deployment of omics data in a clinical setting because it requires the discovery of biological changes to be observed with large effect sizes. However, with the measurement of more data and the development of standardized, powerful computational methods, large-scale omics analyses have the ability to identify the most likely candidates to classify a bacteria sample and target to make tailored therapeutic recommendations.

A harsh reality of AMR is that it may never cease to be an issue due to the high metabolic diversity and evolutionarily plastic characteristics of microorganisms [32]. As new antibiotics are developed, approved, and deployed to the public, it is likely that there will be a resistance developed at some point in time. The processes through which resistance are developed are mainly driven by genetic mutations and horizontal transfer of antimicrobial resistance genes, the modeling of which may optimally come from genomics and metabolomics experiments. However, while the development of resistance may be inevitable in many cases, omics techniques may aide in the determination of the most likely mechanisms by which the bacteria may become resistant to an antibiotic. The use of multi-omics analyses to discover the molecular changes in response to the introduction of an antibiotic could facilitate the detection of the resistance signal. If this signal is common to the onset of resistance in many bacteria species, it could become a beneficial practice to test clinical isolates for the signs of resistance development, thus leading to preemptive determination of potential resistance outbreaks.

With the increasing interest in and applications of omics, an important consideration is the consent of subjects, data management, and data privacy. These topics are critical as proper execution leads not only to the practice of good scientific ethics, but also to an increase of trust in the scientific community from the population. Due to easy access to misinformation and biased interpretations of data, increased emphasis must be put on the proper and ethical conduct of researchers to gain a favorable opinion from patients who may otherwise be misled toward not wanting to participate in a trial that would provide great benefit to their health. Soares et al. wrote a review in which they describe a structured framework to follow in order to create omics research partnerships with Indigenous Australians [140]. While this publication is focused on the Australian population, the points outlined in the table below should be considered where applicable in all research projects in the biomedical field, especially those involving human subjects (Table 13.1) [140].

13.7 Future Outlook and Recommendations

Despite challenges to incorporating omics data into clinical care, the future outlook of antimicrobial resistance treatment should be looked upon with optimism. One of the main obstacles that recent technological advances may soon overcome is the determination of the exact genetic location of antimicrobial resistance genes, which is challenging when using short-read sequencing, given their long, repetitive DNA sequences [138]. As long-read sequencing platforms continue to decrease in cost and increase in availability, the genome structure of bacteria will be more efficiently investigated. Currently, short-read sequencing is generally more accurate and higher throughput than long-read sequencing. Therefore, one strategy to target antimicrobial resistance genes may be through the combination of both short-read and long-read techniques, by which the researcher can determine identity, gene content, and relatedness and resolve the structure of the bacteria genome. The establishment of a better resolved picture of the genome may provide the basis for more efficient targeted therapeutics, as well as an enhanced understanding of the evolution and transmission of antimicrobial resistance. Another promising modality of omics data is spatial omics. Spatial techniques allow researchers to visualize the differences in

Table 13.1 Framework of emerging issues for co-designing culturally appropriate omics research protocols with Indigenous Australians

Governance	1. Model	Identify the preferred model of governance and input within each community
	2. Roles	Outline the roles of the Indigenous governance body
	3. Issues	Define mechanisms for addressing unanticipated issues related to the project as they arise
Consent	1. Representatives	Identify community representatives who can speak on behalf of collectives and provide community consent
	2. Community consent	Identify the processes through which community consent should be obtained
	3. Dynamic consent	Develop mechanisms for a dynamic consent model (providing opportunities at different stages of the research for the participants to withdraw their consent, "renew" their consent, or grant consent for further use of their data) according to community preferences and expectations
	4. Information	Outline the information that should be provided for participants before initiating the collection of blood samples (identify the questions regarding the handling of the blood samples that participants would want answered)
	5. Delivery	Identify communities' preferences regarding how information should be delivered, presented, and recorded during the consent process
	6. Disposal and repatriation	Identify whether preferences related to repatriation and disposal of blood samples should be recorded as part of the individual consent process or as a community agreement
	7. Concerns, risks, and benefits	Explore the communities' worries and concerns regarding the potential uses of biological samples, potential harms, and perceived benefits of the research
	8. Individual results	Develop mechanisms for recording participants' preferences regarding obtaining individual tests results
	9. Rights of nonparticipants	Develop mechanisms for protecting the rights of family and community members who did not provide consent but share genetic information with participants regarding privacy and access to information

(continued)

Table 13.1 (continued)

Respectful handling of biological samples	1. Ownership	Explore the communities' understandings regarding the ownership of biological materials
	2. Meaning	Explore the communities' understandings regarding the cultural and spiritual meanings of biological samples
	3. Collection of samples	Explore the communities' preferences regarding how samples should be collected, by which professionals, and in which settings
	4. Handling and storage	Develop protocols for safe and respectful handling and storage of biological samples throughout all stages of the research
	5. Secondary uses	Identify in which circumstances, if any, it is acceptable to use samples for secondary purposes or further research
	6. Withdraw	Develop mechanisms for withdrawing samples at request of participants or communities
	7. Return of samples	Develop culturally sensitive protocols for the repatriation of samples if requested by communities or participants, with expected timeframes
	8. Disposal	Develop mechanisms for continuous community advice regarding the appropriate disposal of blood samples
	10. Participants who passed	Develop protocols for consulting family members regarding the use, repatriation, or disposal of blood samples from participants who passed away
Data management	1. Custody	Define how the data should be stored, who will be the data custodian(s), and who might have access to the data
	2. Access	Define whether access to data and/or samples might be considered in the future and which protocols should be followed to grant access
	3. Community	Outline how the data should be made available and accessible for communities and participants, and at which level (individual, community, or whole population)
Communication	1. Communication strategy	Build a communication strategy that outlines the preferred methods of reporting the study findings to communities, the frequency and expected timeframes
	2. Community input	Define mechanisms for incorporating community input on study reports
	3. Knowledge translation	Develop a knowledge translation plan that considers the community expectations regarding the translation of research findings to practice, policy, and public dissemination

Table and caption are adopted from Ref. [140]

the molecular makeup of samples with varying phenotypes. Through the process of segmentation, defining a specific region of the sample, it is also possible to compare parts of the sample which are of the greatest interest. This may allow for the identification of areas driving antimicrobial resistance, of which statistical testing may provide the molecular mechanism that is changed in pathogenic samples and its corresponding location may be targeted by therapeutics.

As bacteria have a multitude of mechanisms by which they can develop resistance to a drug, the focus on one type of therapy may be problematic. Therefore, to combat the spread of resistance, it is important to work in parallel to synthesize new drugs that act on their targets in unique ways. One method proposed to stop the evolution of resistance is to create treatments that do not kill microbes. Examples of such techniques include disarming pathogens so that they cannot cause disease when present, removing nutrients from the bacteria to starve them, changing the response of the host to pathogens, and using probiotics to protect microbiota [141]. Benefits of these approaches include that there are likely less undesired side effects inflicted on the patient and that the cost may be low if an abundant molecule can be determined to successfully interfere with infectious processes. One of the difficulties is that it may take a lot of time to discover a molecule that can remove or stop the effects of the bacteria. Advancements in other therapies such as drug combinations and monoclonal antibodies may also aide in the solution of eradicating pathogens from patients. The incorporation of large-scale omics studies may lead to the identification of current drugs that produce a large synergistic effect when combined, effecting a bacteria's metabolic function more so than when each therapy is given separately. In addition, machine learning and deep learning can be used to create predictive models that output potential synergistic drug combinations, reducing the time and cost spent to test all possible therapeutic synergies. However, the risk of using drug combinations is that there could be a high toxicity resulting from the synergistic effect which may also harm the patient. Therefore, a high attention to adverse effects must be stressed when testing this type of treatment. The future of monoclonal antibody incorporation into therapies may be in the manufacture of cocktails which are composed of multiple monoclonal antibodies [132]. With current limited use in antimicrobial resistance, successes have been found in other disease models, and their ability to be highly directed makes them important candidates for use as therapeutics, as the measurement and analysis of new omics data identify targets in different bacteria species.

As omics technologies continue to be developed and analysis pipelines are refined, other biological tools have also advanced. The integration of omics results with other resources, such as 3D modeling of structures, prediction of protein folding using AlphaFold, and ligand-binding simulations will allow for the greatest understanding of the biological mechanisms. Omics and docking approaches can be used to identify and test in silico likely candidates for protein ligands to be used as novel therapeutics [142]. Through the creation of a semi-automated pipeline, with automation occurring at every step possible, starting from sample preparation and measurement to omics analysis to biomarker and drug target prediction to structural predictions and simulations using software that does not use omics and to testing and validation in animal models, a path is provided to streamline the development of new diagnostics and therapeutics.

Antimicrobial resistance has been studied since the identification of penicillinase in 1940 [143]. A publication in 2014 by Spellberg and Gilbert, 74 years after the first antibiotic resistance mechanism was detected, provided numerous tasks for what needed to be considered to overcome the global issue [141]. The necessity to collect real-time data on antibiotic resistance rates and antibiotic use was forefront. Predictive models of the phenotypic behavior of a hazard are only as good as the input data. Currently, there is still uncertainty in the degree of correlation between genetic profile and phenotype [139]. As data collection and curation rises, the more powerful our predictive models will become, as their discriminative ability, sensitivity, and specificity will increase, resulting in more accurate risk prediction, ultimately enhancing patient care and potentially lowering cost due to better prognosis. To tackle the most difficult and time-pressured issues, such as extinguishing outbreaks, there must be more collaborations among healthcare institutions to identify early signatures of resistance, clearly communicate the extent to which the bacteria could spread, and discuss and share the results from tests on any samples that were isolated. By providing a communication platform among nearby hospitals and clinics, it may be possible to greatly slow the spread of a new strain of resistant bacteria, allowing for the other medical centers to prepare for any outbreak, as well as work toward creating therapeutics to kill the bacteria. To increase new breakthroughs, a helpful idea may be to start de-identifying and making data public for more researchers to use. While there must be regulations in place to ensure the safety of patients, this process could lead to the creation of new methods to work with the large amounts of public data, as well as insights into the biological mechanisms leading to the resistance phenotype by culminating the data from multiple experiments.

Educational outreach needs to be expanded to stop the growth of antimicrobial resistance, as there is an abuse of antibiotics. One reason for overuse is that in some places around the globe, antibiotics are cheap and easy to obtain, and the population believes they will help with any pathogen such as viruses, which is not true. Therefore, the implementation of educational programs in these areas to aide in the understanding of when to use antibiotics, and the dangers associated with the incorrect usage, will stop the development of antimicrobial resistance in patients who possibly never had a bacterial pathogen. Agricultural practices also overuse antibiotics, accounting for 80% of overall use in the United States [141]. In agriculture, environmental contamination continues to drive resistance, shown to contribute to resistance in humans. Public officials need to intervene to impose alternate strategies to combat disease in livestock, such as enforcing that there be limits on the number of animals allowed to be in a space to ensure less contact with sick animals, and the removal of sick animals from others.

13.7.1 Wearables Will Enhance Treatment and Diagnostics

With an increasing focus on real-time diagnosis and precision therapeutics, wearables have been developed to help achieve these goals. Roth et al. manufactured a device to kill bacteria by using ozone therapy, providing convenience to patients as they can receive treatment at home [144]. The method by which the device works is shown (Fig. 13.7) [144]. This innovative therapy may provide increased interest in the antimicrobial resistance space because it shows the ability for entrepreneurial opportunities to be pursued, possibly attracting the attention from entrepreneurs and investors.

Another wearable invented by Lin et al. functions to determine the correct dose of an antibiotic to avoid toxicity [145]. This technology can be used to enhance confidence that the prescription given will be likely to kill the pathogen, thereby removing the possibility of resistance. Applications of wearables result in the culmination of a large amount of data which can be used to tune predictive models to achieve better performance. The creation of omics-based wearables may allow for

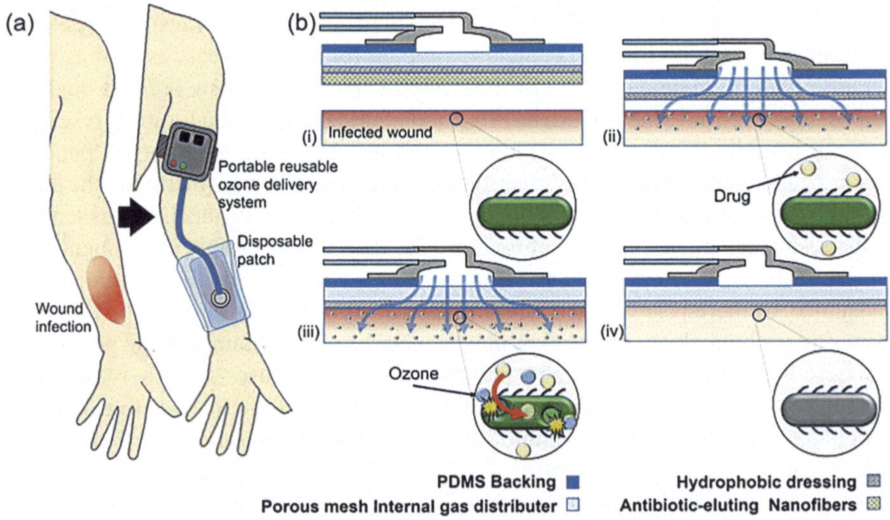

Fig. 13.7 Wearable adjunct ozone and topical antibiotic therapy system. (**a**) Ozone and antibiotic adjunct therapy can be used as an alternative treatment for skin and soft tissue infections that do not respond to traditional therapies. Ozone provides antimicrobial properties and enables antibiotic to enter cell and disrupt cell functions, such as protein production. (**b**) The system utilizes gaseous ozone and gas permeable and drug-eluting nanofiber mat for treating developing wounds in the following process: (i) All-in-one wound patch with drug-eluting nanofiber mesh and gas permeable membrane is applied to skin wound. (ii) NFs begin to dissolve and release the topical antibiotics. (iii) Ozone is applied to the system for full treatment duration as topical antibiotics are completely released from NFs. Ozone and antibiotics work together to eliminate infection. (iv) Once the wound has healed, the wound patch is removed from the area. Combination of ozone and antibiotic treatment can treat antibiotic resistance infections and prevent development of new infections, leading to faster healing times. (Figure and caption are adopted from Ref. [144])

the opportunity to test and refine which biomarkers work optimally to diagnose the presence of a resistance phenotype. As greater advancement in technology occurs and use of wearables becomes more common, the ability to monitor and treat antimicrobial resistance will advance rapidly.

13.8 Conclusion

This chapter has illustrated the bountiful applications of omics to understand the biological mechanisms leading to antimicrobial resistance. As the overuse of antibiotics continues to drive the development of new strains of resistant bacteria due to the inherent variation of microbes, it is imperative to investigate the molecular profiles of pathogenic bacteria to elucidate the processes responsible for the generation of resistance. With technological advancement in multiple omics disciplines leading to the decrease in cost and time needed to measure samples, establishment of new methods to data, creation of computational algorithms such as machine learning and deep learning designed to interpret large data inputs, and increased resolution of biological markers using single-cell techniques, new insights into antimicrobial resistance will be made. Novel therapeutics will be synthesized using the results of omics analyses, as well as efficient drug combinations, with the possibility that a multiuse treatment will be created for more than one resistant bacteria species. In the era of precision medicine, effective diagnostic tools may be manufactured that use omics measurements to quickly identify the pathogenic bacteria in a human sample, and a tailored treatment may be prescribed. Wearables have already been designed and are currently being tested to prescribe proper doses of antibiotics and provide treatments to patients at home. While the potential is great for omics to revolutionize the study of antimicrobial resistance, there are many challenges remaining such as current time and cost to measure biological samples, lack of standardization across academic and clinical pipelines, necessity of newly implemented omics tools to be not only more efficient than current methods but also faster, and incorporation of ethical and political considerations for experiments with human samples.

References

1. Antimicrobial Resistance Collaborators. Global burden of bacterial antimicrobial resistance in 2019: a systematic analysis. Lancet. 2022;399(10325):629–55.
2. Francine P. Systems biology: new insight into antibiotic resistance. Microorganisms. 2022;10(12):2362.
3. Ventola CL. The antibiotic resistance crisis: part 1: causes and threats. P T. 2015;40(4):277–83.
4. Kaya-Copur A, Schnorrer F. A guide to genome-wide in vivo RNAi applications in drosophila. Methods Mol Biol. 2016;1478:117–43.

5. Wang Z, Gerstein M, Snyder M. RNA-seq: a revolutionary tool for transcriptomics. Nat Rev Genet. 2009;10(1):57–63.
6. Kitano H. Systems biology: a brief overview. Science. 2002;295(5560):1662–4.
7. Monteverde M, et al. Pressure dependence of the superconducting transition temperature of magnesium diboride. Science. 2001;292(5514):75–7.
8. Butcher EC, Berg EL, Kunkel EJ. Systems biology in drug discovery. Nat Biotechnol. 2004;22(10):1253–9.
9. Fiehn O. Metabolomics--the link between genotypes and phenotypes. Plant Mol Biol. 2002;48(1–2):155–71.
10. Pinu FR, et al. Systems biology and multi-omics integration: viewpoints from the metabolomics research community. Metabolites. 2019;9(4):76.
11. Charlebois DA. Quantitative systems-based prediction of antimicrobial resistance evolution. NPJ Syst Biol Appl. 2023;9(1):40.
12. Jauneikaite E, et al. Genomics for antimicrobial resistance surveillance to support infection prevention and control in health-care facilities. Lancet Microbe. 2023;4(12):e1040–6.
13. Yow HY, et al. Optimizing antimicrobial therapy by integrating multi-omics with pharmacokinetic/pharmacodynamic models and precision dosing. Front Pharmacol. 2022;13:915355.
14. van Houten CB, et al. Observational multi-centre, prospective study to characterize novel pathogen-and host-related factors in hospitalized patients with lower respiratory tract infections and/or sepsis – the "TAILORED-Treatment" study. BMC Infect Dis. 2018;18(1):377.
15. Hiltemann SD, et al. Galaxy mothur Toolset (GmT): a user-friendly application for 16S rRNA gene sequencing analysis using mothur. Gigascience. 2019;8(2):giy166.
16. Karlsson R, et al. Proteotyping bacteria: characterization, differentiation and identification of pneumococcus and other species within the Mitis Group of the genus Streptococcus by tandem mass spectrometry proteomics. PLoS One. 2018;13(12):e0208804.
17. Boulund F, et al. Typing and characterization of bacteria using bottom-up tandem mass spectrometry proteomics. Mol Cell Proteomics. 2017;16(6):1052–63.
18. Salva-Serra F, et al. Complete genome sequences of Streptococcus pyogenes type strain reveal 100%-match between PacBio-solo and Illumina-Oxford Nanopore hybrid assemblies. Sci Rep. 2020;10(1):11656.
19. Jo SH, et al. Multi-omics based characterization of antibiotic response in clinical isogenic isolates of methicillin-susceptible/-resistant Staphylococcus aureus. RSC Adv. 2020;10(46):27864–73.
20. Castro BE, et al. Multiomics characterization of methicillin-resistant Staphylococcus aureus (MRSA) isolates with heterogeneous intermediate resistance to vancomycin (hVISA) in Latin America. J Antimicrob Chemother. 2022;78(1):122–32.
21. Zhu Y, et al. Genome-scale metabolic modeling of responses to polymyxins in Pseudomonas aeruginosa. Gigascience. 2018;7(4):giy021.
22. Hanafin PO, et al. Proof-of-concept for incorporating mechanistic insights from multi-omics analyses of polymyxin B in combination with chloramphenicol against Klebsiella pneumoniae. CPT Pharmacometrics Syst Pharmacol. 2023;12(3):387–400.
23. Ribeiro M, et al. Bioinformatics study of expression from genomes of epidemiologically related MRSA CC398 isolates from human and wild animal samples. J Proteome. 2022;268:104714.
24. Zhang L, et al. Aerobic composting as an effective cow manure management strategy for reducing the dissemination of antibiotic resistance genes: an integrated meta-omics study. J Hazard Mater. 2020;386:121895.
25. Seiler C, Berendonk TU. Heavy metal driven co-selection of antibiotic resistance in soil and water bodies impacted by agriculture and aquaculture. Front Microbiol. 2012;3:399.
26. Spanig S, et al. A multi-omics study on quantifying antimicrobial resistance in European freshwater lakes. Environ Int. 2021;157:106821.
27. Xu Y, et al. An atlas of genetic scores to predict multi-omic traits. Nature. 2023;616(7955):123–31.

28. Sun Q, Li M, Wang X. The Cancer Omics Atlas: an integrative resource for cancer omics annotations. BMC Med Genet. 2018;11(1):63.
29. Liang Q, et al. A multi-omics atlas of the human retina at single-cell resolution. Cell Genom. 2023;3(6):100298.
30. Da Ros L, et al. Multi-omics atlas of combinatorial abiotic stress responses in wheat. Plant J. 2023;116:1118.
31. Imwattana K, et al. A species-wide genetic atlas of antimicrobial resistance in Clostridioides difficile. Microb Genom. 2021;7(11):000696.
32. Chernov VM, et al. Omics of antimicrobials and antimicrobial resistance. Expert Opin Drug Discov. 2019;14(5):455–68.
33. Woyke T, Doud DFR, Schulz F. The trajectory of microbial single-cell sequencing. Nat Methods. 2017;14(11):1045–54.
34. Ma P, et al. Bacterial droplet-based single-cell RNA-seq reveals antibiotic-associated heterogeneous cellular states. Cell. 2023;186(4):877–891.e14.
35. Imdahl F, et al. Single-cell RNA-sequencing reports growth-condition-specific global transcriptomes of individual bacteria. Nat Microbiol. 2020;5(10):1202–6.
36. Davis KM, Isberg RR. Defining heterogeneity within bacterial populations via single cell approaches. BioEssays. 2016;38(8):782–90.
37. Hare PJ, et al. Single-cell technologies to study phenotypic heterogeneity and bacterial persisters. Microorganisms. 2021;9(11):2277.
38. Ambriz-Avina V, Contreras-Garduno JA, Pedraza-Reyes M. Applications of flow cytometry to characterize bacterial physiological responses. Biomed Res Int. 2014;2014:461941.
39. Sanchez-Romero MA, Casadesus J. Contribution of phenotypic heterogeneity to adaptive antibiotic resistance. Proc Natl Acad Sci USA. 2014;111(1):355–60.
40. Keegstra JM, et al. Phenotypic diversity and temporal variability in a bacterial signaling network revealed by single-cell FRET. Elife. 2017;6:e27455.
41. Murawski AM, Brynildsen MP. Ploidy is an important determinant of fluoroquinolone persister survival. Curr Biol. 2021;31(10):2039–2050.e7.
42. Shan Y, et al. ATP-dependent persister formation in Escherichia coli. mBio. 2017;8(1):e02267-16.
43. Mishra R, et al. Targeting redox heterogeneity to counteract drug tolerance in replicating Mycobacterium tuberculosis. Sci Transl Med. 2019;11(518):eaaw6635.
44. Power AL, et al. The application of imaging flow cytometry for characterisation and quantification of bacterial phenotypes. Front Cell Infect Microbiol. 2021;11:716592.
45. Marutescu LG. Current and future flow cytometry applications contributing to antimicrobial resistance control. Microorganisms. 2023;11(5):1300.
46. Bongiorno D, et al. Detection of methicillin-resistant Staphylococcus aureus persistence in osteoblasts using imaging flow cytometry. Microbiology. 2020;9(5):e1017.
47. Wagley S, et al. Bacterial dormancy: a subpopulation of viable but non-culturable cells demonstrates better fitness for revival. PLoS Pathog. 2021;17(1):e1009194.
48. Goormaghtigh F, Van Melderen L. Single-cell imaging and characterization of Escherichia coli persister cells to ofloxacin in exponential cultures. Sci Adv. 2019;5(6):eaav9462.
49. Bos J, et al. Emergence of antibiotic resistance from multinucleated bacterial filaments. Proc Natl Acad Sci USA. 2015;112(1):178–83.
50. Santucci P, et al. Visualizing pyrazinamide action by live single-cell imaging of phagosome acidification and Mycobacterium tuberculosis pH homeostasis. mBio. 2022;13(2):e0011722.
51. Manina G, Dhar N, McKinney JD. Stress and host immunity amplify Mycobacterium tuberculosis phenotypic heterogeneity and induce nongrowing metabolically active forms. Cell Host Microbe. 2015;17(1):32–46.
52. Brandis G, Larsson J, Elf J. Antibiotic perseverance increases the risk of resistance development. Proc Natl Acad Sci USA. 2023;120(2):e2216216120.
53. Aldridge BB, et al. Asymmetry and aging of mycobacterial cells lead to variable growth and antibiotic susceptibility. Science. 2012;335(6064):100–4.

54. Rego EH, Audette RE, Rubin EJ. Deletion of a mycobacterial divisome factor collapses single-cell phenotypic heterogeneity. Nature. 2017;546(7656):153–7.
55. Volzing KG, Brynildsen MP. Stationary-phase persisters to ofloxacin sustain DNA damage and require repair systems only during recovery. mBio. 2015;6(5):e00731-15.
56. Zimmermann M, et al. Phenotypic heterogeneity in metabolic traits among single cells of a rare bacterial species in its natural environment quantified with a combination of flow cell sorting and nanoSIMS. Front Microbiol. 2015;6:243.
57. Tian H, et al. Subcellular chemical imaging of antibiotics in single bacteria using C(60)-secondary ion mass spectrometry. Anal Chem. 2017;89(9):5050–7.
58. Berry D, et al. Tracking heavy water (D2O) incorporation for identifying and sorting active microbial cells. Proc Natl Acad Sci USA. 2015;112(2):E194–203.
59. Huang WE, et al. Raman microscopic analysis of single microbial cells. Anal Chem. 2004;76(15):4452–8.
60. Wang C, et al. Single-cell Raman spectroscopy identifies Escherichia coli persisters and reveals their enhanced metabolic activities. Front Microbiol. 2022;13:936726.
61. Cui L, et al. Surface-enhanced Raman spectroscopy for identification of heavy metal arsenic(V)-mediated enhancing effect on antibiotic resistance. Anal Chem. 2016;88(6):3164–70.
62. Germond A, et al. Raman spectral signature reflects transcriptomic features of antibiotic resistance in Escherichia coli. Commun Biol. 2018;1:85.
63. Verma T, et al. Profiling antibiotic resistance in Escherichia coli strains displaying differential antibiotic susceptibilities using Raman spectroscopy. J Biophotonics. 2021;14(1):e202000231.
64. Lu W, et al. Identification of pathogens and detection of antibiotic susceptibility at single-cell resolution by Raman spectroscopy combined with machine learning. Front Microbiol. 2022;13:1076965.
65. Tao Y, et al. Metabolic-activity-based assessment of antimicrobial effects by D(2)O-labeled single-cell Raman microspectroscopy. Anal Chem. 2017;89(7):4108–15.
66. Llorens-Rico V, et al. Single-cell approaches in human microbiome research. Cell. 2022;185(15):2725–38.
67. Homberger C, Barquist L, Vogel J. Ushering in a new era of single-cell transcriptomics in bacteria. microLife. 2022;3:uqac020.
68. Rinke C, et al. Insights into the phylogeny and coding potential of microbial dark matter. Nature. 2013;499(7459):431–7.
69. Kuchina A, et al. Microbial single-cell RNA sequencing by split-pool barcoding. Science. 2021;371(6531):eaba5257.
70. Blattman SB, et al. Prokaryotic single-cell RNA sequencing by in situ combinatorial indexing. Nat Microbiol. 2020;5(10):1192–201.
71. McNulty R, et al. Probe-based bacterial single-cell RNA sequencing predicts toxin regulation. Nat Microbiol. 2023;8(5):934–45.
72. Homberger C, et al. Improved bacterial single-cell RNA-seq through automated MATQ-seq and Cas9-based removal of rRNA reads. mBio. 2023;14(2):e0355722.
73. Xu Z, et al. Droplet-based high-throughput single microbe RNA sequencing by smRandom-seq. Nat Commun. 2023;14(1):5130.
74. Liu Y, et al. Bacterial single cell whole transcriptome amplification in microfluidic platform shows putative gene expression heterogeneity. Anal Chem. 2019;91(13):8036–44.
75. Wang J, et al. RNA-seq based transcriptomic analysis of single bacterial cells. Integr Biol (Camb). 2015;7(11):1466–76.
76. Sheng K, et al. Effective detection of variation in single-cell transcriptomes using MATQ-seq. Nat Methods. 2017;14(3):267–70.
77. Dar D, et al. Spatial transcriptomics of planktonic and sessile bacterial populations at single-cell resolution. Science. 2021;373(6556):eabi4882.
78. Wang B, et al. Single-cell massively-parallel multiplexed microbial sequencing (M3-seq) identifies rare bacterial populations and profiles phage infection. Nat Microbiol. 2023;8(10):1846–62.

79. Zhou L, et al. Single gene targeted nanopore sequencing enables simultaneous identification and antimicrobial resistance detection of sexually transmitted infections. PLoS One. 2022;17(1):e0262242.
80. Henry TC, Brynildsen MP. Development of persister-FACSeq: a method to massively parallelize quantification of persister physiology and its heterogeneity. Sci Rep. 2016;6:25100.
81. Bawn M, et al. Single-cell genomics reveals population structures from in vitro evolutionary studies of Salmonella. Microb Genom. 2022;8(9):mgen000871.
82. Ueno H, et al. Revealing the metabolic activity of persisters in mycobacteria by single-cell D(2)O Raman imaging spectroscopy. Anal Chem. 2019;91(23):15171–8.
83. Girgis HS, et al. Single-molecule nanopore sequencing reveals extreme target copy number heterogeneity in arylomycin-resistant mutants. Proc Natl Acad Sci USA. 2021;118(1):e2021958118.
84. Chijiiwa R, et al. Single-cell genomics of uncultured bacteria reveals dietary fiber responders in the mouse gut microbiota. Microbiome. 2020;8(1):5.
85. Avital G, et al. scDual-seq: mapping the gene regulatory program of Salmonella infection by host and pathogen single-cell RNA-sequencing. Genome Biol. 2017;18(1):200.
86. Lan F, et al. Single-cell genome sequencing at ultra-high-throughput with microfluidic droplet barcoding. Nat Biotechnol. 2017;35(7):640–6.
87. Shee S, et al. Biosensor-integrated transposon mutagenesis reveals rv0158 as a coordinator of redox homeostasis in Mycobacterium tuberculosis. Elife. 2023;12:e80218.
88. Hossain F, Billah MM, Yamazaki M. Single-cell analysis of the antimicrobial and bactericidal activities of the antimicrobial peptide magainin 2. Microbiol Spectr. 2022;10(4):e0011422.
89. Xu J, Zhang J, Xu Y, Tang Y-W, Ma B, Wu Y. Single cell metabolic phenome and genome via the ramanome technology platform: precision medicine of infectious diseases at the ultimate precision? iLABMED. 2023;1:5–14.
90. Hendriksen RS, et al. Using genomics to track global antimicrobial resistance. Front Public Health. 2019;7:242.
91. Cuetero-Martinez Y, et al. Next-generation sequencing for surveillance of antimicrobial resistance and pathogenicity in municipal wastewater treatment plants. Curr Med Chem. 2022;30(1):5–29.
92. Luftinger L, et al. Predictive antibiotic susceptibility testing by next-generation sequencing for periprosthetic joint infections: potential and limitations. Biomedicines. 2021;9(8):910.
93. Otto M. Next-generation sequencing to monitor the spread of antimicrobial resistance. Genome Med. 2017;9(1):68.
94. Ramanathan B, et al. Next generation sequencing reveals the antibiotic resistant variants in the genome of Pseudomonas aeruginosa. PLoS One. 2017;12(8):e0182524.
95. Kullar R, et al. Next-generation sequencing supports targeted antibiotic treatment for culture negative orthopedic infections. Clin Infect Dis. 2023;76(2):359–64.
96. Chao L, et al. Application of next generation sequencing-based rapid detection platform for microbiological diagnosis and drug resistance prediction in acute lower respiratory infection. Ann Transl Med. 2020;8(24):1644.
97. Kaprou GD, et al. Rapid methods for antimicrobial resistance diagnostics. Antibiotics (Basel). 2021;10(2):209.
98. Kochan K, et al. Detection of antimicrobial resistance-related changes in biochemical composition of Staphylococcus aureus by means of atomic force microscopy-infrared spectroscopy. Anal Chem. 2019;91(24):15397–403.
99. Zhang F, et al. Direct antimicrobial susceptibility testing on clinical urine samples by optical tracking of single cell division events. Small. 2020;16(52):e2004148.
100. Baltekin O, et al. Antibiotic susceptibility testing in less than 30 min using direct single-cell imaging. Proc Natl Acad Sci USA. 2017;114(34):9170–5.
101. Li H, et al. Adaptable microfluidic system for single-cell pathogen classification and antimicrobial susceptibility testing. Proc Natl Acad Sci USA. 2019;116(21):10270–9.

102. Forsyth B, et al. A rapid single-cell antimicrobial susceptibility testing workflow for bloodstream infections. Biosensors (Basel). 2021;11(8):288.
103. Choi J, et al. Rapid antibiotic susceptibility testing by tracking single cell growth in a microfluidic agarose channel system. Lab Chip. 2013;13(2):280–7.
104. Watterson WJ, et al. Droplet-based high-throughput cultivation for accurate screening of antibiotic resistant gut microbes. Elife. 2020;9:e56998.
105. Yi X, et al. Development of a fast Raman-assisted antibiotic susceptibility test (FRAST) for the antibiotic resistance analysis of clinical urine and blood samples. Anal Chem. 2021;93(12):5098–106.
106. Li HZ, et al. Active antibiotic resistome in soils unraveled by single-cell isotope probing and targeted metagenomics. Proc Natl Acad Sci USA. 2022;119(40):e2201473119.
107. Yuan J, et al. Metagenomic NGS optimizes the use of antibiotics in appendicitis patients: bacterial culture is not suitable as the only guidance. Am J Transl Res. 2021;13(4):3010–21.
108. Hu B, et al. A comparison of blood pathogen detection among droplet digital PCR, metagenomic next-generation sequencing, and blood culture in critically ill patients with suspected bloodstream infections. Front Microbiol. 2021;12:641202.
109. Blainey PC. The future is now: single-cell genomics of bacteria and archaea. FEMS Microbiol Rev. 2013;37(3):407–27.
110. Kaster AK, Sobol MS. Microbial single-cell omics: the crux of the matter. Appl Microbiol Biotechnol. 2020;104(19):8209–20.
111. Marcy Y, et al. Dissecting biological "dark matter" with single-cell genetic analysis of rare and uncultivated TM7 microbes from the human mouth. Proc Natl Acad Sci USA. 2007;104(29):11889–94.
112. Leung K, et al. A programmable droplet-based microfluidic device applied to multiparameter analysis of single microbes and microbial communities. Proc Natl Acad Sci USA. 2012;109(20):7665–70.
113. Xu L, et al. Virtual microfluidics for digital quantification and single-cell sequencing. Nat Methods. 2016;13(9):759–62.
114. Zengler K, et al. Cultivating the uncultured. Proc Natl Acad Sci USA. 2002;99(24):15681–6.
115. Dichosa AE, et al. Capturing and cultivating single bacterial cells in gel microdroplets to obtain near-complete genomes. Nat Protoc. 2014;9(3):608–21.
116. Fitzsimons MS, et al. Nearly finished genomes produced using gel microdroplet culturing reveal substantial intraspecies genomic diversity within the human microbiome. Genome Res. 2013;23(5):878–88.
117. Spencer SJ, et al. Massively parallel sequencing of single cells by epicPCR links functional genes with phylogenetic markers. ISME J. 2016;10(2):427–36.
118. Dean FB, et al. Comprehensive human genome amplification using multiple displacement amplification. Proc Natl Acad Sci USA. 2002;99(8):5261–6.
119. Stepanauskas R, et al. Improved genome recovery and integrated cell-size analyses of individual uncultured microbial cells and viral particles. Nat Commun. 2017;8(1):84.
120. Povilaitis T, et al. In vitro evolution of phi29 DNA polymerase using isothermal compartmentalized self replication technique. Protein Eng Des Sel. 2016;29(12):617–28.
121. de Bourcy CF, et al. A quantitative comparison of single-cell whole genome amplification methods. PLoS One. 2014;9(8):e105585.
122. Hsieh K, et al. Combating antimicrobial resistance via single-cell diagnostic technologies powered by droplet microfluidics. Acc Chem Res. 2022;55(2):123–33.
123. Wang H, et al. Development of single-cell-level microfluidic technology for long-term growth visualization of living cultures of Mycobacterium smegmatis. Microsyst Nanoeng. 2021;7:37.
124. Norsigian CJ, et al. Comparative genome-scale metabolic modeling of metallo-beta-lactamase-producing multidrug-resistant Klebsiella pneumoniae clinical isolates. Front Cell Infect Microbiol. 2019;9:161.

125. Burnard D, et al. Comparative genomics and antimicrobial resistance profiling of Elizabethkingia isolates reveal nosocomial transmission and in vitro susceptibility to fluoroquinolones, tetracyclines, and trimethoprim-sulfamethoxazole. J Clin Microbiol. 2020;58(9):e00730-20.
126. Ye J, et al. Complete-genome sequencing and comparative genomic characterization of bla(NDM-5) carrying Citrobacter freundii isolates from a patient with multiple infections. BMC Genomics. 2023;24(1):506.
127. Szklarczyk D, et al. STRING v11: protein-protein association networks with increased coverage, supporting functional discovery in genome-wide experimental datasets. Nucleic Acids Res. 2019;47(D1):D607–13.
128. Raman K, Chandra N. Mycobacterium tuberculosis interactome analysis unravels potential pathways to drug resistance. BMC Microbiol. 2008;8:234.
129. Hwang S, et al. Network-assisted investigation of virulence and antibiotic-resistance systems in Pseudomonas aeruginosa. Sci Rep. 2016;6:26223.
130. Lee M, et al. Network integrative genomic and transcriptomic analysis of carbapenem-resistant Klebsiella pneumoniae strains identifies genes for antibiotic resistance and virulence. mSystems. 2019;4(4):e00202-19.
131. Nicolás MF, et al. Integrating omics data to prioritize target genes in pathogenic bacteria. In: da Silva FAB, et al., editors. Networks in systems biology: applications for disease modeling. Cham: Springer International Publishing; 2020. p. 217–76.
132. Martin-Galiano AJ, McConnell MJ. Using omics technologies and systems biology to identify epitope targets for the development of monoclonal antibodies against antibiotic-resistant bacteria. Front Immunol. 2019;10:2841.
133. Uddin R, et al. Identification of Histone Deacetylase (HDAC) as a drug target against MRSA via interolog method of protein-protein interaction prediction. Eur J Pharm Sci. 2017;106:198–211.
134. Zhang M, et al. Prediction and analysis of the protein interactome in Pseudomonas aeruginosa to enable network-based drug target selection. PLoS One. 2012;7(7):e41202.
135. Mei S. In silico enhancing M. Tuberculosis protein interaction networks in STRING to predict drug-resistance pathways and pharmacological risks. J Proteome Res. 2018;17(5):1749–60.
136. Maifiah MH, et al. Untargeted metabolomics analysis reveals key pathways responsible for the synergistic killing of colistin and doripenem combination against Acinetobacter baumannii. Sci Rep. 2017;7:45527.
137. Mohammadi-Shemirani P, Sood T, Pare G. From ‘omics to multi-omics technologies: the discovery of novel causal mediators. Curr Atheroscler Rep. 2023;25(2):55–65.
138. Waddington C, et al. Exploiting genomics to mitigate the public health impact of antimicrobial resistance. Genome Med. 2022;14(1):15.
139. Collineau L, et al. Integrating whole-genome sequencing data into quantitative risk assessment of foodborne antimicrobial resistance: a review of opportunities and challenges. Front Microbiol. 2019;10:1107.
140. Soares GH, et al. From biocolonialism to emancipation: considerations on ethical and culturally respectful omics research with indigenous Australians. Med Health Care Philos. 2023;26(3):487–96.
141. Spellberg B, Gilbert DN. The future of antibiotics and resistance: a tribute to a career of leadership by John Bartlett. Clin Infect Dis. 2014;59(Suppl 2):S71–5.
142. Krell T, Matilla MA. Antimicrobial resistance: progress and challenges in antibiotic discovery and anti-infective therapy. Microb Biotechnol. 2022;15(1):70–8.
143. Davies J, Davies D. Origins and evolution of antibiotic resistance. Microbiol Mol Biol Rev. 2010;74(3):417–33.
144. Roth A, et al. Wearable adjunct ozone and antibiotic therapy system for treatment of Gram-negative dermal bacterial infection. Sci Rep. 2022;12(1):13927.
145. Lin S, et al. Wearable microneedle-based electrochemical aptamer biosensing for precision dosing of drugs with narrow therapeutic windows. Sci Adv. 2022;8(38):eabq4539.

Index

V. Soni, A. S. Akhade (eds.), *Antimicrobial Resistance: Factors to Findings*, https://doi.org/10.1007/978-3-031-65986-7

M

N

O

P

SPRINGER NATURE

GPSR Compliance

The European Union's (EU) General Product Safety Regulation (GPSR) is a set of rules that requires consumer products to be safe and our obligations to ensure this.

If you have any concerns about our products, you can contact us on ProductSafety@springernature.com

In case Publisher is established outside the EU, the EU authorized representative is:

Springer Nature Customer Service Center GmbH
Europaplatz 3
69115 Heidelberg, Germany

The manufacturer's authorised representative in the EU is Springer Nature Customer Service Centre GmbH, Europaplatz 3, 69115 Heidelberg, Germany. If you have any concerns regarding our products, please contact ProductSafety@springernature.com

Printed and bound by CPI Group (UK) Ltd, Croydon, CR0 4YY
01/10/2026
02225859-0017